CONTROL STRUCTURES

GENERAL FORM	EXAMPLE	REFERENCE (Page)
CALL subroutine-name(var1,var2,...)	CALL SWITCH(A,B)	306
DO label var=initial,final,step body of loop label CONTINUE	DO 10 I= 0, 2, 1 PRINT *,I 10 CONTINUE	161
DO WHIILE(conditional) body of loop ENDDO	DO WHILE(X.GE.Y) X=X-0.1 ENDDO	167
GOTO statement-label	GOTO 25	142
IF (condition) statement	IF (DENO.EQ.0) STOP	152

IF (condition) THEN
execute if true
[ELSE
execute if false]
ENDIF
	IF (X.GT.Y) THEN
	Z=X
	ELSE
	Z=Y
	ENDIF
		143

RETURN	RETURN	299
variable-name = expression	PI=3.1415927 AREA=PI*R**2	73
DATA variable-list/data-list/	DATA X,Y,Z/1,2,3/ DATA ((A(I,J),J=1,3),I=1,3)/9*0/	246
PARAMETER(variable-name=exp,...)	PARAMETER(PI=3.1415927)	231 , 347

INTRODUCTION TO COMPUTING FOR ENGINEERS

INTRODUCTION TO COMPUTING FOR ENGINEERS

William E. Mayo, Ph.D.
Martin Cwiakala, Ph.D.

Rutgers University

McGRAW-HILL, INC.
New York St. Louis San Francisco Auckland Bogotá Caracas
Hamburg Lisbon London Madrid Mexico Milan Montreal New Delhi
Paris San Juan São Paulo Singapore Sydney Tokyo Toronto

INTRODUCTION TO COMPUTING FOR ENGINEERS

2 3 4 5 6 7 8 9 0 DOH DOH 9 0 9 8 7 6 5 4 3 2 1

ISBN 0-07-041139-5

This book was set in Garamond Book ITC by Beacon Graphics Corporation.
The editors were B. J. Clark, Lyn Beamesderfer, and Eleanor Castellano;
the designer was Charles A. Carson; the cover illustrator was Circa 86, Inc.
R. R. Donnelley & Sons Company was printer and binder.

Library of Congress Cataloging-in-Publication Data

Mayo, William E.
Introduction to computing for engineers / William E. Mayo, Martin Cwiakala.
p. cm.
Includes bibliographical references and index.
ISBN 0-07-041139-5
1. Engineering—Data processing. I. Cwiakala, Martin. II. Title.
TA345.M388 1991
620'.00285—dc20 90-49601

ABOUT THE AUTHORS

William E. Mayo is an Associate Professor of Materials Science at Rutgers University. He received his Ph.D. degree from Rutgers and served as a post-doctoral fellow at Bell Laboratories before joining the faculty at Rutgers. As an educator, Professor Mayo has trained over 8,000 engineering students in the FORTRAN language and its applications. Professor Mayo is also actively involved in developing computer models of strengthening methods of metal alloys. His most recent work involved a supercomputer model of the effects of grinding of metal and ceramic surfaces.

Martin Cwiakala is an Adjunct Professor in Mechanical Engineering at Rutgers University, as well as a Project Engineer at GEMCO, in Middlesex, New Jersey. He received his Ph.D. in Mechanical Engineering from Rutgers University. Dr. Cwiakala is the author of a number of papers in the field of Robotics and Computer Aided Design. His most current work involves developing extensions to existing CAD packages to generate main assembly drawing and P & I diagrams by utilizing a graphic interface to select desired features.

CONTENTS

PREFACE

We have designed this textbook to meet the needs of today's first-year students. During the past five years great changes have occurred among engineering students as they have become increasingly proficient in computer programming. This results from microcomputers that are introduced at the high school level. As a result of the wider availability of these machines, students intending to pursue engineering and science careers enter college with more computer competency than similar students of just five years ago. Therefore, this text has shifted from the predominately all FORTRAN text to one including a significant portion of introductory material concerning numerical methods and engineering applications. In addition to introductory numerical methods, the development of good programming practices are stressed. Our objectives in this text are:

1. To acquaint the student with the fundamentals of an operating system and text editor. By stressing functions to be performed, the student is armed with the necessary questions to tackle any system.
2. To present the basic organization and the objectives of a program. "Top-down" design is presented, and a novel method for fleshing out a problem is offered. This establishes good problem-solving techniques.
3. To teach students FORTRAN 77.
4. To provide the student with the necessary debugging tools so that they can correct their own programs. Two chapters are devoted to methods used to locate and correct errors. Commonly encountered errors are also discussed.
5. To present methods for good programming documentation and validation. By students' documenting programs in an efficient and useful manner, the program written in the FORTRAN class will be one that they can use throughout their academic and professional career.
6. To provide the student with a foundation of numerical methods.

Distinguishing Features

There are a number of features distinguishing this textbook from other FORTRAN texts. These features include:

Engineering and Science Applications Over 300 problems have been compiled from a wide range of disciplines. Topics include statics, kinematics, fluid dynamics, statistics, equilibrium (chemical, thermal, and mechanical), electrical networks, filtering, and many others.

Learning Tools The students are provided with a number of learning tools. Most important is the use of tables that we term *information maps.* As students learn various commands, they are requested to keep an updated table of what they have learned. In this fashion, the student is provided with a learning aid for use in future chapters (and for use after they leave the course).

Problem-Solving Design Process and Top-Down Algorithm Refinement A novel method for fleshing out a problem is presented along with top-down design. The method presented asks the student to look for key words in a problem's specification. This leads to a chain of questions based on the identified key words. Answering those questions results in one level of refinement in the top-down scheme. By documenting this process, the student is able to inform the instructor of where a problem is encountered. This becomes a diagnostic tool for the student as well as for the instructor. The fleshing-out process continues by generating questions from key words in the previous answers. The process ends when the answers can be implemented using FORTRAN code directly.

Debugging Techniques Two full chapters are devoted to the art of debugging. Students are provided with methods to locate and correct errors. They are taught about the importance of compiler list-files. Syntax errors, run-time errors, and logical errors are discussed. A number of faulty program examples are given, and the method to locate the errors is discussed. Code is developed to help students trace errors in nested subroutines. Our recommendations to the students on debugging are summarized in the form of *debugging proverbs* that provide important tools for debugging faulty code. More importantly, we stress the design of bug-free code and offer ways to help accomplish this.

Good Programming Practice Program documentation, structure, and validation are stressed. Why programs are documented and what information is relevant for the process are discussed. The process of validation is stressed, especially with the numerical methods section. Various methods of validation are discussed. Using these methods, students are able to assess their own work critically and identify weaknesses in their algorithms.

Introduction to Numerical Methods Topics include data reduction, curve fitting, root solving, systems of linear equations, and numerical integration and differentiation. The students are provided with code to perform these various operations. The objective is for the student to be able to apply these methods to solve practical problems.

Organization

This text is organized into three sections: Introduction, FORTRAN, and Numerical Methods. In the Introductory section Chap. 1 provides a history of high-level languages and an introduction to the operating system and text editor. Chapter 2 presents the organization and general process for developing programs. It is here that students are introduced to the fleshing-out process. Chapter 3 presents the common threads of high-level programming languages. The purpose of this chapter is to shed light on many of the common operations present in high-level languages. Since most engineering students will have had programming experience before using this book, this chapter provides an overview of the common threads of programming. This provides a reference point for their previous experience. It also allows them to see the "big picture," so that when they learn their next language after FORTRAN, students will know what to look for.

The FORTRAN section provides the necessary background to construct programs. The topics covered in Chaps. 4 through 11 include the standard topics such as data types, assignment statements, and control structures. Two chapters of debugging techniques are also presented. An important chapter here is the discussion on the advanced features of the new FORTRAN standard. It provides an overview of the important changes that will occur over the next few years.

The last section, Numerical Methods, covers data analysis, curve fitting, root solving, systems of linear equations, and numerical integration and differentiation. The topics have been arranged so that the most difficult material appears at the end of the section. This allows the more difficult topics to be delayed until the freshman has covered these topics in his or her other courses. The book is designed, however, with the freshman engineering student in mind. So, the mathematics level is kept to an introductory level.

Supplements

Fortran 90 Workbook A textbook supplement is available from the publisher to help students with the FORTRAN syntax.

Solutions Manual and Instructor's Guide A solutions manual is available on request from the publisher. This manual provides complete answers to the exercises at the back of each chapter. The instructor's guide can be used for creating overhead transparencies for large lectures.

Test Bank A test bank is also available from the authors. This contains typical questions for quizzes and course examinations.

Acknowledgments

The authors would like to thank Peggy Cwiakala for her professional assistance throughout the development of this text. We are indebted to Dr. Chris Smith who provided many of the problems in the text and also provided critical feedback on the manuscript. We would also like to thank the following people for their helpful comments, suggestions, and assistance: Dr. Chi Fung Lo, Gwan-Chong Joo, Shyam Rao, and Michele Bellinger. Finally, we wish to thank the many students over the last ten years who provided us with the inspiration to write this book.

McGraw-Hill and the authors would like to thank the following reviewers for their careful critiques of early drafts of the manuscript: Dorothy Attaway, Boston University; Jerry Bayless, University of Missouri at Rolla; George N. Facas, Trenton State University; Jack Foster, Broome Community College; Jim Jacobs, Norfolk State University; Melton Miller, University of Massachusetts; Robert Probasco, University of Idaho; John Sheffield, University of Missouri at Rolla; and Joseph F. Shelley, Trenton State College.

William E. Mayo
Martin Cwiakala

INTRODUCTION TO COMPUTING FOR ENGINEERS

CHAPTER 1

INTRODUCTION

1.1 A Brief History of High-Level Languages

Until the mid 1950s programming a computer was very complex. Only those willing to master the intricacies of machine or assembler language could use a computer. Casual users had no choice but to leave programming to the experts. Sadly, the user was the one who best understood the problem and could best devise a solution. Assigning the task to a professional often resulted in less than ideal solutions. As a result, there was an urgent need to develop a method that would permit greater access to computers.

To show the early obstacles in programming, consider Table 1.1. We show here a simple program written in three different computer languages (machine, assembler and FORTRAN) to divide the number 13 by 6 and print the results. Note the complexity of the programs written in the low-level machine and assembler languages. The program written in the high-level language, on the other hand, is brief and easy to follow. Also, the FORTRAN program is *portable,* which means that the program will run unchanged on another machine. This would not be possible with the lower-level languages since the instructions that make up the commands vary for each machine. As a result, low-level languages are strongly machine-dependent, which greatly limits their use for everyday programming.

To be accurate, we must point out that there are times when low-level languages are desirable. Examples might include programs demanding high-speed or access to low-level computer functions. Usually, though, programmers prefer the high-level language since its use results in great savings in programming costs. For example, it took almost 15 minutes to construct the two low-level language examples in Table 1.1, but only a few seconds to write the equivalent FORTRAN instructions.

TABLE 1.1

Comparison of Machine, Assembler, and FORTRAN Language Commands to Divide Two Numbers on an IBM PC Computer

Machine	Assembler	FORTRAN
1011100000001101	MOV AX,13	J=13/6
1011001100000110	MOV BL,6	PRINT *, J
1111011111101001	DIV L	
1011010000000000	MOV AH,0	
1011101100000000	MOV BX,0	
1100110100001010	INT 10H	

There will be times when the benefits of assembler outweigh the convenience of high-level languages. Therefore, many of you will undoubtedly go on to learn one or more assembler languages. In fact, most engineers will learn two to three different languages during their college career. Thus, it is usual, that engineers learn an easy language like FORTRAN as an important base before proceeding to some more advanced languages.

FORTRAN appeared in 1957 and was the first of the high-level languages. This version of FORTRAN ran on the IBM 704 computer and programmers quickly hailed it as a great advance. It featured English-like commands that permitted novice users to write a complex program in only a small fraction of the time previously required. Although FORTRAN contained many math functions that were useful for scientific calculations, the original version would be considered crude by today's standards. It lacked the ability to use subroutines or a structured approach to programming. Today programmers demand these features as essential tools to write efficient and powerful code. Over the last 30 years many changes to the language have overcome these early problems and many newer programming concepts are now part of FORTRAN. As a result, it has matured to become a preferred computer language for solving engineering problems.

Shortly after FORTRAN appeared many other first-generation languages were introduced. Two of these that gained the widest favor and have endured to this day are BASIC and COBOL. COBOL is the business equivalent to FORTRAN that the business world uses widely. It has good input/output abilities and is excellent for handling files. In spite of this, COBOL has few engineering uses.

The developers of BASIC, on the other hand, had no specific application in mind when they wrote their language. Their chief aim was to teach a first high-level language in a climate that gives rapid feedback. Although not originally built into the language, today's BASIC is *an interpreted* language rather than a *compiled* language. Interpreters convert each line into machine instructions as the programmer types each line. Compiled languages (like FORTRAN) by contrast convert the program into machine instructions only after reading the entire program. Thus, an interpreted language detects errors as they occur. As a result, the instant feedback reinforces student understanding.

There is a penalty for the benefit of an interpreted language in the tendency to develop poor programming skills. Since the interpreter reports errors one line at a time, it invites a quick fix to the problem, which may conceal larger problems in logic stretching over many lines. Part of this problem lies in the lack of structure in the BASIC language. A *structured* language, among other things, reduces the use of unconditional transfers (GOTO statements). Programmers who abuse the **GOTO** create code whose logic is hard to follow. We often call this *Spaghetti Code* to dramatize the disorganized appearance of the program. You can see this in the example programs of Table 1.2, which compare structured code (in FORTRAN) and unstructured code (in BASIC) to do the same thing. The programs compare two numbers to see if they are equal. Note that both programs require five program lines, but the structured code is much easier to follow. For longer programs, the benefits of structured programming greatly multiply.

Programmers who need to write a simple program often use BASIC. This is especially true when only a single programmer is writing the program. More complex problems that may be written by a team of programmers usually rely on compiled languages such as FORTRAN, Pascal, C, or MODULA-2.

The 1970s saw the development of a second wave of computer languages. Pascal and C are the most famous of these. Pascal is a widely used first language for microcomputer courses taught in high school. It is a powerful language that has many constructs of a structured language. A good programmer developing a large-scale program will find this language useful since its design forces good programming style. Nonengineering applications often use this language because of its ability to handle nonnumerical data. But it is important to note that Pascal is limited for intensive computation. This limitation may be addressed in future revisions.

The C language is one of the most powerful high-level languages available today. This power comes from the blend of functions commonly found in high-level languages with functions found only in low-level languages. To use C effectively, the programmer should understand the underlying principles of how a computer operates. As a result, the C language is beyond the skill level of most users.

The engineering student is likely to use many different computer languages. Besides those already mentioned, these may include Ada, APL, FORTH, LISP, and MODULA-2. Table 1.3 briefly reviews the salient points of each.

Table 1.2

Comparison of Structured Program in FORTRAN to Unstructured Program in BASIC

BASIC	FORTRAN
`10 IF A=B THEN GOTO 40`	`IF(A.EQ.B) THEN`
`20 PRINT "NOT EQUAL"`	`PRINT *,' EQUAL'`
`30 GOTO 50`	`ELSE`
`40 PRINT "EQUAL"`	`PRINT *,' NOT EQUAL'`
`50 END`	`ENDIF`

TABLE 1.3

Partial List of Programming Languages of Interest to Engineers

Language	Description
FORTRAN	A compiled language widely used for engineering applications. Contains a full range of structured commands and uses of subroutines which promote good style. Periodic revisions update the language to include new advances.
BASIC	An interpreted language primarily used for its simplicity. The lack of structured commands tends to produce the problematic spaghetti code. The structure of a subroutine using the GOSUB statement prevents development of long programs. Proprietary dialects, such as QuickBASIC eliminate some of these problems.
Pascal	A compiled language which utilizes structured constructs. Promotes good programming practice. Very good for handling nonnumeric data, but somewhat limited for computational work.
C	Very powerful compiled language which combines system level and high-level commands. Requires an advanced level of programming experience and can be somewhat difficult to debug. Debugging problems are most severe for lengthy programs.
Ada	Very powerful, sophisticated language designed by the Department of Defense as the *official* language for technical applications. Combines functions from most of the other languages that preceded it. Especially useful for large programs written by teams of programmers. Requires advanced programming skills.
APL	Special purpose language specially adapted to manipulating large data arrays. However, simple operations are cumbersome.
FORTH	Similar to C in many respects. This language gives good low-level control of a computer. Used for controlling robots, motors, and computer peripherals.
LISP	A special purpose language used for artificial intelligence. Has the unique ability to modify language elements in addition to data.
MODULA-2	A descendant of Pascal, this language allows the programmer to combine different modules to create a complete program. Each of the modules resides in a previously created library.

1.2 Which Language to Choose?

Choosing a language is almost like choosing a religion. No other topic will generate an argument as quickly as a discussion about which programming language is best. The question should not be so much "Which is better?", but rather, "Which one can do the job?". When choosing a language you should consider three things:

Prior experience
Unique problem needs
Features of the language

Many programmers make the mistake of choosing a pet language for *all* applications without weighing their specific problem. Each language has its strong points, but they may not be appropriate for every problem. For example, FORTRAN is useful for solving number-crunching problems often found in engineering simulations. But FORTRAN would be a poor choice for controlling hardware, such as a robot arm, or for generating sophisticated graphics. Each problem should be judged on its own merits, which then forms the selection basis for the "best" language.

In an ideal world the language choice would be based on the preceding rule. In other words, only the benefits of each language would be weighed. Clearly, serious thought should be given to the programmer's prior training. For example, if the preferred languages were Pascal, FORTRAN, and APL for a specific problem, the programmer would likely select the most familiar one. Usually, the extra time to learn the new language would far outweigh the savings gained from the special features of the best language. Since the time to write and debug a program is usually the most costly element of a project, the choice of the familiar language would be justified. There are many situations, of course, where the reverse is true.

There is no perfect programming language. Rather, students should learn a few languages that are suitable for the general range of problems they are likely to encounter. A good system would be to master a simple language like FORTRAN or Pascal for everyday problems and an advanced language like C or assembler for special occasions. Many engineering curricula now follow this approach by teaching FORTRAN in the freshman year, followed by a more advanced language in the junior year.

1.3 Computers and Engineering

Engineers and scientists were among the first large-scale computer users dating back to the earliest days of the invention of computers. But, even for this select group, access to computing facilities was limited until only recently. Part of this increased usage is due to the arrival of cheap microcomputers and data acquisition systems. It is now rare to see any engineering laboratory without several microcomputers. Over the last ten years the uses for computers have greatly expanded. In the early days engineers used computers mostly for intensive numerical calculations (number-crunching). Today a single desktop microcomputer can collect data from an experiment, compare that data to a proposed theory, graph the results, and help in the production of the report.

Engineers use computers in many ways, but the software that they use usually falls into two broad categories. The first contains *utilities*, such as word processing and graphics programs, that are low-cost packages and require no programming. The second category includes programs devoted specifically to engineering applications. Often these are programs written for a unique problem, for example, a program written to control an experiment. In many cases these special applications require that the engineers write their own programs. Sometimes it is possible to modify an existing program, but some programming is unavoidable.

Programs in the second group further divide into five subdivisions. At times the distinctions between the areas might blur, but engineers generally agree that there are five major uses of computers in their profession today:

Modeling and simulations
Data acquisition
Data analysis
Process control
Design

Examples of each will be given in the following sections. These examples depict only a few problems that engineers work on. The field is broad and no survey such as this one can attempt to cover the entire field, so we use these examples only to describe typical problems.

MODELING

Modeling uses a mathematical basis to describe a physical process. A simple example is the model describing the motion of a ball in a gravitational field. The math is simple and can be stated in a few equations. Generally, modeling requires a good understanding of the underlying scientific laws. The engineer can then use these models to develop a framework for solving the problem at hand. Often the engineer uses the model to ask "what if..." questions. In the falling ball example the modeler might ask how the flight path would change in the presence of a crosswind or if the ball loses mass. More meaningful problems might then use these simple descriptions to describe more complex phenomena. For example, the simple dropping ball model might be used to understand what occurs when a satellite enters the atmosphere.

Figure 1.1 shows the results of a molecular dynamics model of the structure of vitreous silica. Molecular dynamics is a complex mathematical process that considers the nature of chemical bonds between different elements to predict the atomic structure. The figure shows that the large spheres are oxygen atoms and the small spheres are silicon atoms with an average Si-O bond length of less than 2 angstroms (Å). Notice that the silicon atoms bind to 4 oxygens while each oxygen atom binds to 2 silicon atoms. What is of interest here, though, is that a ring structure is apparent. The central part of the figure has a ring that may have important implications in understanding the properties of the material.

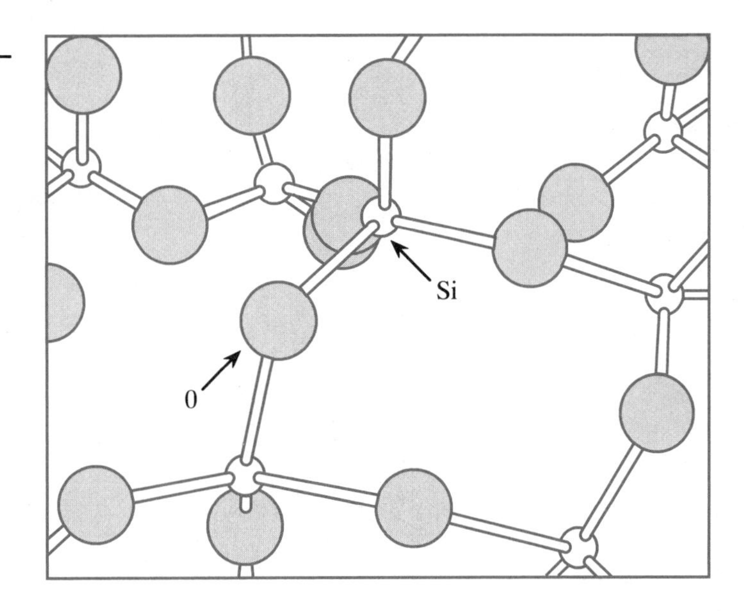

Figure 1.1

Molecular Dynamics Simulation of Structure of Vitreous Silica, SiO_2 (Courtesy of S. Garofalini, Rutgers University)

One advantage of molecular dynamics study is that it is easy to change conditions slightly to see what happens to the structure. For example, we might allow the chemistry to be slightly off the perfect 2 oxygen to 1 silicon ratio. This might produce interesting changes in the structure. Also, we might want to introduce "impurities" such as boron or calcium to see what their effect is. These simulations are useful tools in the study of many classes of materials.

SIMULATIONS

Simulations are similar to models in the sense that both rely on mathematics to describe how something works. But whereas models refer to a specific process, simulations refer to large systems. A rocket launching would be a good example. A launching simulation could be run in which the simple gravitational model described previously is only a small part. A large number of such models would then be the basis for the complete simulation. For example, there might be a model for the fuel consumption, a model for the guidance control, and so on. The system simulation combines each of these independent models. Frequently, students use the terms of *modeling* and *simulation* interchangeably. Keep in mind that modeling refers to the part, whereas simulations refer to the whole.

Figure 1.2 illustrates the results of a computer simulation. This simulation examines the stresses in a mechanism designed to hold a metal sample during machining. The purpose of the simulation is to calculate the deflection in the fixture and to measure the local stresses when placed in the machine. The simulation uses a model that divides the fixture into small elements. These elements are easier to handle mathematically than the whole fixture.

FIGURE 1.2

Computer Simulation of a Dynamometer Used for Measuring and Controlling Grinding Parameters

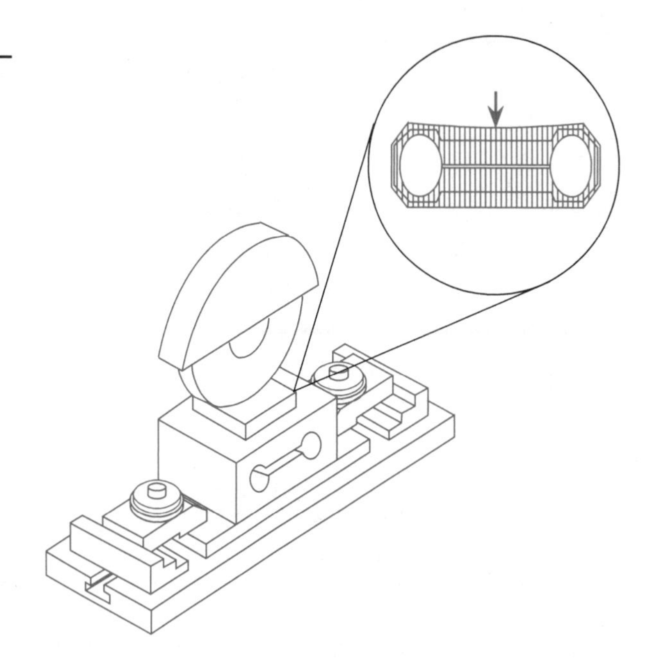

The first step is to find the desired properties for one element according to the appropriate physics. Once the program finishes one element, it then moves from element to element until it calculates the properties of the whole fixture. The simulation then uses this model to examine the maximum load that can be carried before the fixture exceeds its design stress.

Note in this example that the simulation contains many different models. First, there is the model that calculates the stresses in the fixture. However, to calculate the stress, this model must know the applied load. This comes from a second model of the machine design and its operating conditions. Finally, a third model will calculate the maximum stress that the fixture can support. This will vary from material to material, so that a separate model may be necessary for each material. Only when the programmer combines all three models can we call the whole program a simulation. Each model can exist by itself and provide a single answer. But only when the programmer combines all three, can we understand how the complex system behaves.

DATA ACQUISITION

Computer control is an ideal tool for control of many experiments run in an engineering lab. The repetitious data collection, the required data collection speed, and hostile (sometimes dangerous) conditions make computer-controlled data acquisition an ideal tool. Usually, these experiments can be controlled with a single low-cost microcomputer. This area now represents one of the largest uses for computers within engineering.

The term *data acquisition* applies to more than just idle data collection. It also applies to experimental control. The computer can control external equipment through a variety of means, such as interfaces and digital-to-analog (D/A) boards. Both types of devices are boards placed into the computer itself that enable it to communicate with control equipment able to decode its instructions. Often the computer and the external device can "talk" to each other to organize their activities. In this way the computer can manipulate a variety of equipment, which in turn can control an experiment.

The data created during an experiment can be sent to the computer by interfaces similar to those just mentioned, or by use of analog-to-digital (A/D) boards. A common experimental arrangement uses a mix of data collection and equipment control. The computer may be used to set the initial conditions, begin the data collection, check to see if there are any safety hazards, and, if necessary, end the experiment.

Figure 1.3 shows an example of a computer-controlled experiment to study the brain response of insects to visual stimuli. The goal of the research is to understand how the optic nerve within the eye transmits visual signals to the brain. Details of the stimulus sent to the insect such as the intensity, the duration, and the color of the light are under the control of the computer. In addition, the computer controls the light pulse timing. So, it knows when to look for brain response. This is not a simple timing problem since the stimulus and response occur only milliseconds apart.

Computer-aided data acquisition and experimental control is an important element in engineering. Engineers custom design many of these tests, as in the preceding example. Increasingly, manufacturers are designing evermore test equipment with microcomputers built into it. For example, there are now computer-controlled microscopes. Undoubtedly, automated testing will continuously expand in scope in the near future, and the well-trained engineer of tomorrow must be trained in this aspect of engineering.

DATA ANALYSIS

Once the computer collects data from an experiment, the engineer may wish to run an analysis on the results. Experimental data always has small errors that are due to equipment errors or to statistical fluctuations. Thus, the first step in an analysis is to prepare the data by removing, if possible, these errors.

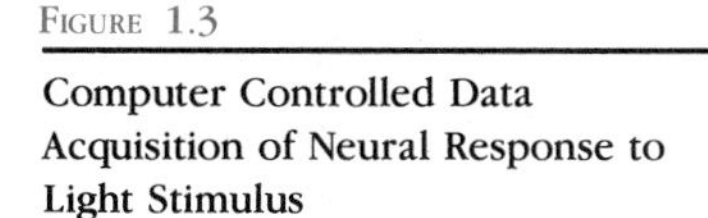
Figure 1.3

Computer Controlled Data Acquisition of Neural Response to Light Stimulus

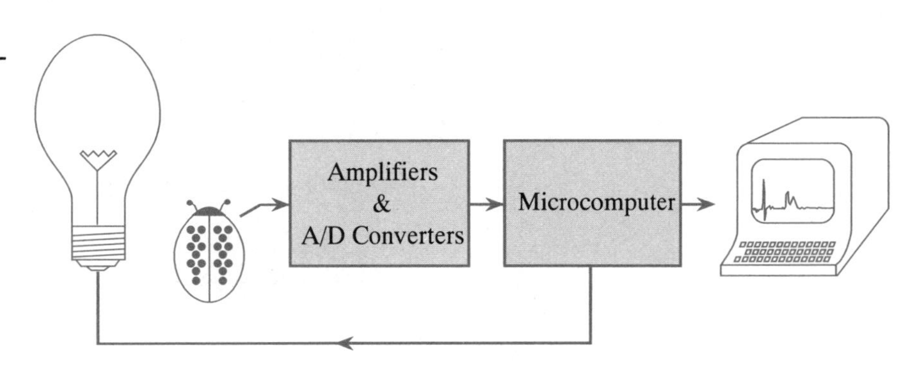

An example of a typical source of experimental error might be a voltmeter that always reads 0.5 volts (V) too high. Here error removal is easy. Each voltage reading would simply have 0.5 V subtracted from its value. Most equipment has an error source of this type, but this does not present a serious problem since the investigator can correct the data easily.

The second source of error is more difficult to remove since the errors occur randomly. One source is due to counting errors. As an example, consider how you would weigh a small amount of a chemical for an experiment. You would place the material carefully on a scale for weighing and then take a reading. If you make a second measurement of the weight, it is usually slightly different from the first time. Similarly, if you make a third measurement, it also will vary slightly. The usual procedure in such cases is to average all the measurements and use the average value for the true value. This kind of error is impossible to eliminate, but its contribution can be reduced by careful test methods and better equipment. For example, when weighing chemicals, it helps to keep air currents from flowing near the scale. These currents perturb the weighing pan and add to the noise. Also, use of a better scale would help to reduce the error.

The first type of error can be eliminated by careful calibration of the equipment. The second type of error, though, presents a more complex problem. Mathematicians have developed methods to remove the noise, an example of which is shown in Fig. 1.4. An experimenter can remove the noise (or at least minimize it) by a process known as ***digital filtering***. Note that the filter has done an excellent job in removing the noise. The filter is a mathematical tool that can remove high-frequency components in a collection of data. Since noise falls into this category, the filter, if properly designed, can minimize the noise. Precise details of a digital filter are beyond the scope of this book. But there is a simpler method, known as ***smoothing***, which we will present in a later chapter. Smoothing is a much simpler mathematical process involving averaging of adjacent data points. This helps to smooth out the random fluctuations due to the noise.

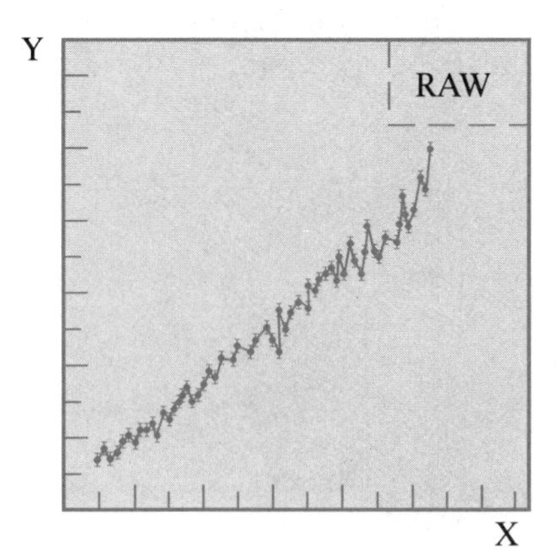

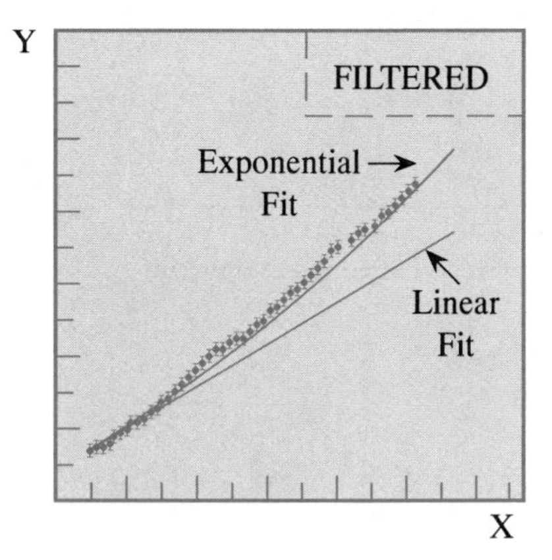

FIGURE 1.4

Removal of Random Noise by Use of a Digital Filter Followed by Curve Fitting

Once the computer prepares the data by reducing the noise, analysis can proceed. This may consist of trying to fit the data to a model under study by curve fitting. Great care must be taken in these types of analyses since data can be fit to many different types of curves simultaneously. The task then reduces to finding the *best* fit. Thus, in Fig. 1.4 the data is fit to two different types of curve—a straight line and an exponential. Over the first part of the curve both fitting functions describe the relationship well. However, for the higher values, the exponential curve fits the data better. Thus, we would accept the exponential curve as the correct model. This figure should stress the need to survey a variety of fitting functions. We will discuss this in greater detail later.

PROCESS CONTROL

In the industrial world engineers spend much time designing computer-aided manufacturing processes (CAM). A plant manufacturing a product may have many different stations at which different components are assembled. Sometimes these stations number over a thousand and are interconnected by an intricate network of computers. At each station computers control equipment such as a robot or a positioning device. In addition, computers organize the flow between the individual stations to improve the total plant efficiency. Often there may be a central computer that oversees all functions of the plant and acts as a supervisor. Meanwhile the localized microcomputers concentrate on the specific tasks assigned to it. Such plants are characterized by a high degree of independence in that one part of a plant may still operate while another part shuts down.

A simple example of a small-scale computer-aided manufacturing process is a device to fill containers on an assembly line such as that illustrated in Fig. 1.5. The computer records the volume of the material deposited into the

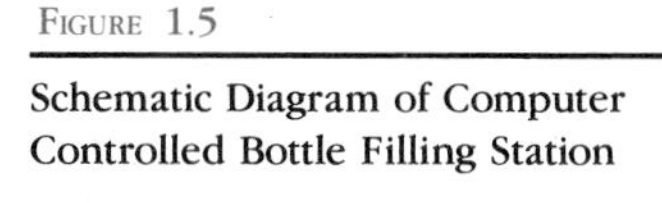

FIGURE 1.5

Schematic Diagram of Computer Controlled Bottle Filling Station

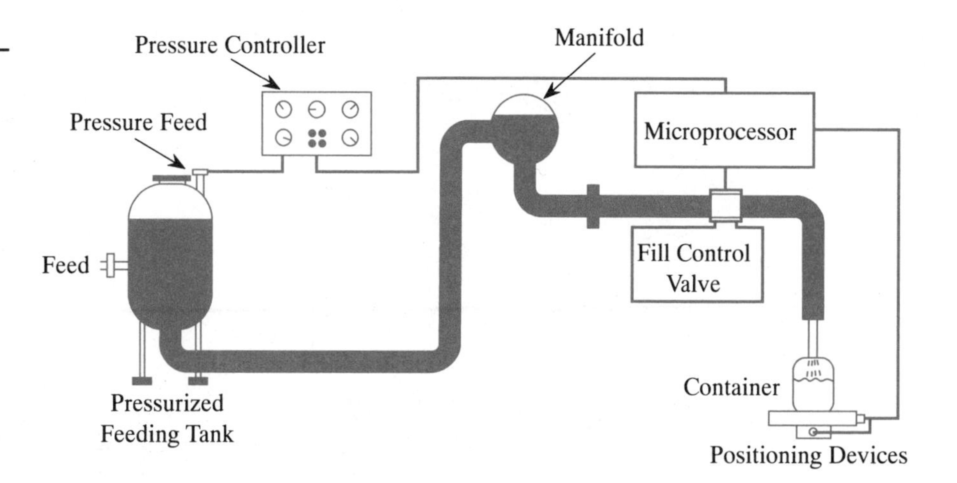

container accurately by measuring the flow time through a special fill control valve. This valve is carefully designed to maintain constant pressure. One way to control it is with a microprocessor-based controller. Every few milliseconds the computer measures the pressure. If there is any deviation from the expected value, the computer takes corrective action. This results in a tighter control over the pressure than any mechanical device can possibly achieve.

Maintaining the pressure in the valve is not a difficult task for the computer to do. Therefore, the microprocessor also has time to do other things. These might include such things as monitoring and adjusting the pressure in the main feed tank and controlling a positioning device for moving the bottle into position. Also, the small microcomputer probably will communicate every few seconds with a central computer regarding its status. If there are difficulties at the filling station that the local computer cannot handle, it may ask the central computer to suspend delivery of material to the feed tank.

DESIGN

An important application of computers is in the area of computer-aided design (CAD). Here, the engineer can design various products or systems with the aid of the computer and associated graphics facilities. The computer relieves the engineer of the need to worry about details and allows the engineer to focus on the overall design. The computer supplies all the detailed information such as dimensions. In such systems it is easy to make changes since the computer will recalculate all dimensions of a part when the designer makes changes.

Figure 1.6 shows an example of CAD usage. Here, the engineer can use the CAD system to create a frame model of a small model airplane engine. This model is a common form of a CAD figure since it shows the position of the major structural components. Also shown in the figure is the solid model, where the surface drawn onto the structure hides the frame. This figure is closer to the final appearance of the part or system.

FIGURE 1.6

Use of CAD System for Design of Model Airplane Engine (Courtesy of Small Business Computers, Inc.)

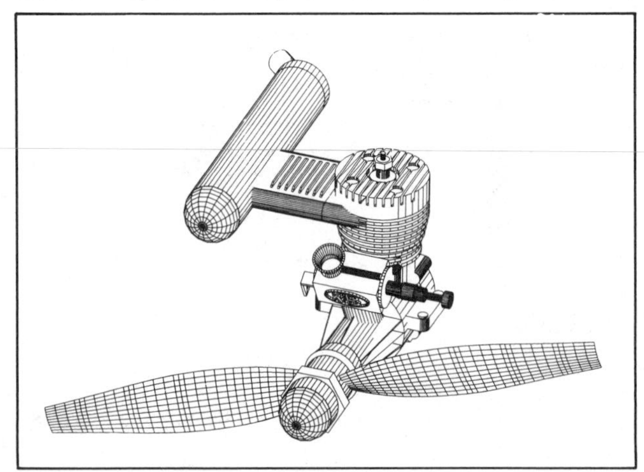

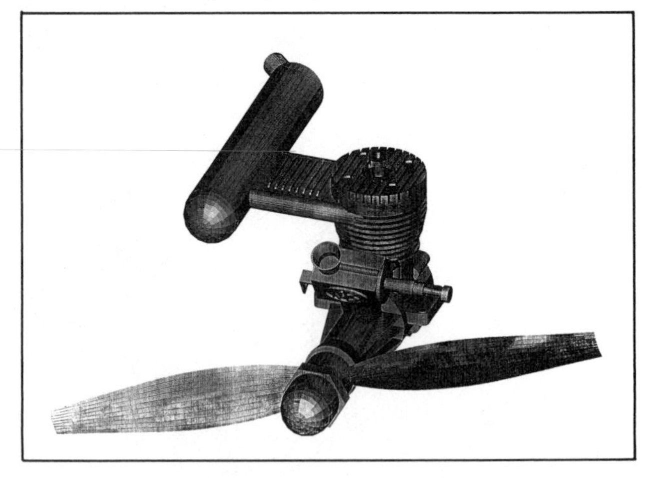

1.4 The Computing Environment

We often use a diagram such as the one shown in Fig. 1.7 to show the different types of *software* used on a computer. The system software is a collection of computer programs whose function is to control the *hardware* that actually carries out the program instructions. Direct control of the hardware with machine language is tedious as described in the previous sections. So, software developers have introduced several types of software to permit the programmer to use the hardware more effectively.

The different types of software are schematically arranged in a shell around the hardware as shown in the figure. The inner shells have specific functions that the user cannot alter. In contrast, the outer shells provide the user a simplified environment to create solutions for the specific problem. A programmer constructs the solution in the outer shells, then passes it to the inner shells for translation and execution.

The innermost shell is the *operating system*, OS, which supervises the most basic computer functions. It is a program usually supplied by the computer manufacturer that passes commands from the user to the machine hardware. It performs such functions as program loading, file maintenance, security, hardware control, error checking, and so forth. A portion of the operating system always resides in the computer's memory. Other functions are loaded as needed.

Many different operating systems are now in use that the engineering student is likely to encounter. The smallest computers, microcomputers (micros), often use either the MS-DOS or PC-DOS operating systems, although many others exist. Increasingly, micros are using the more sophisticated UNIX, WINDOWS, or OS/2 operating systems. The latter systems permit *multitasking*, which allows more than one program (or user) to access the system simultaneously. Many other micros use an operating system based on *icons*.

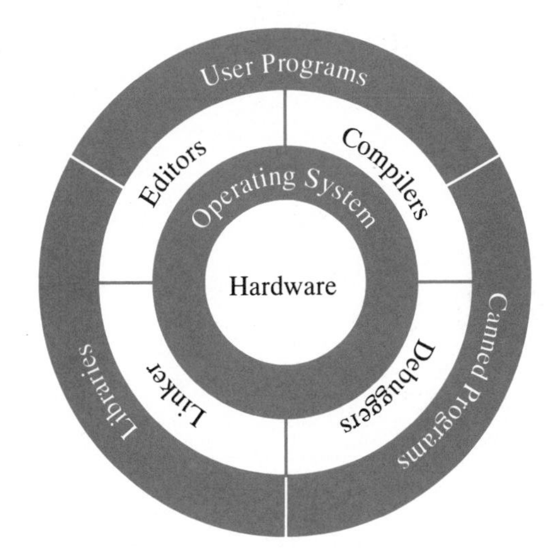

FIGURE 1.7

Schematic Description of the Computing Environment

Icons are small figures drawn on the screen that represent a function. The user then activates the function by moving the cursor to the desired icon and selecting it. Such *user-friendly* systems are easy to use and have become popular. To date icon-based operating systems are largely limited to the microcomputer environment, although they are beginning to appear on more powerful computers. Larger computers, such as minicomputers, use more powerful operating systems such as VMS or UNIX. The largest computers (mainframes and supercomputers) use still different operating systems such as OS/MVS and UNIX.

Surrounding the operating system is a series of utilities consisting of editors, compilers, linkers, and debuggers, among others. Programmers use these utilities to create or modify a program, prepare the program for running on the computer, or help to locate program errors. We describe each of these major steps briefly in the following sections.

THE EDITOR

The process of creating a program, Fig. 1.8, begins with the *editor*. The editor is essentially a wordprocessor you use to create or modify a program called the *source code*. You will write the source code in FORTRAN, although you could use the same editor to write a BASIC, assembler, or C program also. The editor allows you to enter the program, store it for future reference, and modify it later as needed.

The editor lacks the sophistication of a full-blown modern wordprocessor, but it is much easier to use. The goal is not to produce fancy graphics or an elegant layout but to enter text with a minimum of effort. With such simple goals, we suggest that an editor needs only ten basic instructions to function:

Insert a letter	Move cursor down one line
Delete a letter	View the document
Move cursor left one space	Retrieving/saving a file

FIGURE 1.8

Process for Creating a Program

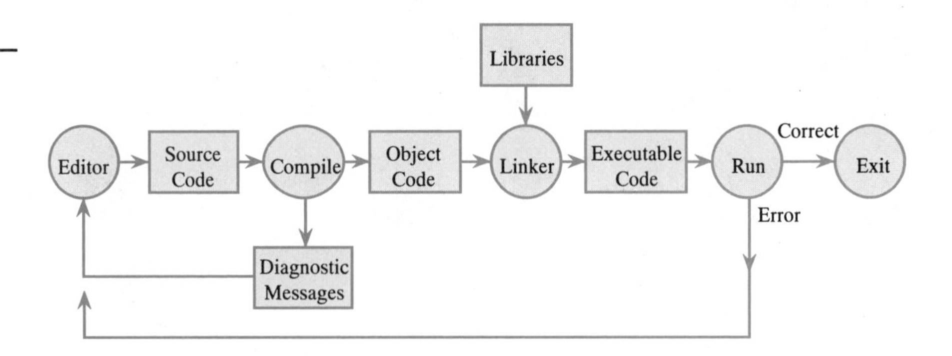

Move cursor right one space	Oops! Command to undue an error
Move cursor up one line	Exit the editor

You can create more complicated functions by combinations of these basic ten. For example, to DELETE A LINE, you use the DELETE A LETTER command as often as needed. Of course, your editor will have a single command to DELETE A LINE, but this only increases the number of commands that you must learn. Every editor is different. Therefore, we cannot give you a summary of the common commands. You must do this yourself. To help you, the table in Exercise 1.6 at the end of this chapter will guide you through the common editor commands. Also shown for comparison in this table are the commands for a common editor, EMACS.

When you create a program with the editor, it will be stored in a *file*. Each file will be given a name, usually consisting of two or more parts:

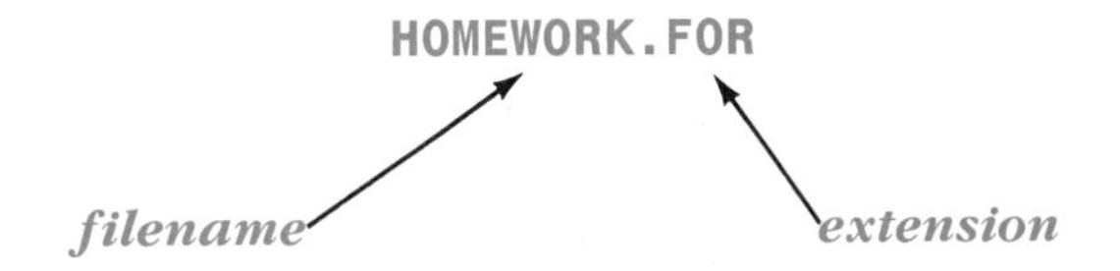

This file has two parts, the *filename* and the *extension*. The filename is the principal means to identify a file. Thus, in the preceding example HOMEWORK is the filename and is sufficient for most purposes to identify it. The extension provides additional descriptive information about a file. The most common extensions that you are likely to see are:

.FOR	A FORTRAN file
.BAS	A BASIC file
.OBJ	An OBJECT file written in machine language and created by the FORTRAN compiler
.EXE	An EXECUTABLE file written in machine language and created by the LINKER
.DAT	A data file containing only data

There are many rules for naming files, but they vary widely from system to system. Therefore, in Exercise 1.4 we ask you to find out the file-naming conventions of your system.

THE COMPILER

The second step in creating a running program is the *compiler* that translates the source code into a machine language version called the *object module*. Thus, the object module's primary job is to take the English-like program and convert it into the 0 and 1's that the computer understands. Beside this primary task, the compiler also helps to detect simple *syntax* errors. Syntax errors are mistakes in the grammar of the programming language, such as typographical errors or missing parentheses among others.

If the compiler detects any errors, you must return to the editor to correct them before you attempt to compile the program again. Thus, there may be several edit-compile cycles (Fig. 1.8) before you can go to the third step. The compiler, unfortunately, cannot detect more complicated errors such as *logic* errors, which are faults in the program logic. These are usually more subtle and difficult to detect than the simple syntax errors and require extra effort to uncover. There is also a third type of error, the *run-time* error, which can be detected only when the machine executes the program. Later chapters will focus on methods for detecting and eliminating these types of errors.

If the compiler detects no errors, it creates a machine language version of your program. The compiler is nothing more than a translator whose primary job is to free you from the tedium of writing a program in machine language. When it completes its job, the compiler will give this newly created file the extension, .OBJ, with the filename. For example, if the original name of your file was X.FOR, the compiler creates the file, X.OBJ. This OBJECT file is not yet ready for execution on the computer, however; there is still another step.

THE LINKER

After you eliminate all the errors reported by the compiler diagnostic messages, the translated program (*object module*) passes to the linker. This is another program that prepares the program for execution. It links together the object code and any additional libraries that may be needed. These libraries are collections of previously written programs that perform common operations. Examples might include libraries for statistical analysis, mathematical analysis, or graphic output. The use of libraries greatly reduces the amount of work the programmer needs to do since solutions to many of these common problems already exist.

When you complete the linking, the final version of the program is assembled into the *executable code*. It is this code that you submit to the hardware for execution rather than the original FORTRAN program. Once the linker produces the executable code, you may run it as often as desired without the need to repeat the compiling and linking steps. This is in contrast to interpreted BASIC where the interpreter translates the source code into executable code *every* time you run the program.

1.5 A Typical Terminal Session

To summarize the important topics presented in this chapter, we present a typical terminal session to create, correct, and execute a program. We can only give you a general idea of the process since every computer facility uses a different operating system, different editor, different printer commands, and so forth. Therefore, you will need to find corresponding commands for your system.

The session will consist of the following major steps:

LOGIN	A process to get onto the computer. This has two components, the account name and the password. If you are using a microcomputer, you may not need to login.
EDIT	Used to create the FORTRAN file FUBAR.FOR
COMPILE	Used to convert the file FUBAR.FOR to machine language. The product is the file FUBAR.OBJ if there are no errors.
REEDIT	If the compiler reports error messages, you must return to the editor to revise the program.
LINK	Once you eliminate all the errors, link the file FUBAR.OBJ to produce the executable file, FUBAR.EXE.
RUN	To run the program, the computer will use the executable file created by the linker.
CLEANUP	After you finish, eliminate unnecessary files.
LOGOFF	When you complete your work, you must close your account.

In the dialogue the computer responses are in *italics*, whereas our input is in **bold**. The commands here may vary for your system, so focus on the general sequence of events, not on specific details.

The first two lines show the login procedure. Once the computer accepts the password, it will allow you to proceed. The first step is to create a new program, FUBAR.FOR, with the editor, EMACS, in this example. We do this with the command $EMACS followed by the program. When we finish, we save the file with <CTRL>X <CTRL>S, where <CTRL> indicates that you press the CONTROL KEY simultaneously with the X and S keys. Finally, to leave the editor, we use the <CTRL>X <CTRL>C command. Note that the computer uses the symbol, $, as a *prompt*. A prompt is a symbol that indicates that the computer is ready to accept commands. Some computers use other symbols such as %, or _ as prompts.

After we finish entering the program, we save it and then try to compile. But the compiler detects several errors that it prints on the screen for your viewing. Note that the compiler reports that it could not produce the .OBJ file.

```
Login            USERNAME: MAYO
                 PASSWORD:

System Response  WELCOME TO THE VAX CLUSTER

Edit             $EMACS FUBAR.FOR          <CR>
                 EMACS EDITOR FUBAR.FOR, VERSION 1

Program             SUM=0.0
                    X=1.0
                    SUM=SUM+(1./X***3)*S
                    S=-S
                    X=X+2
                    IF(X.LE.61)GOTO 6
                    PI=(SUM*32)**(1./3.)
                    WRITE(5,*) PI
                    STOP

Save file        <CTRL>X <CTRL>S
                 <CTRL>X <CTRL>C
                 FILE SAVED IN FUBAR.FOR, VERSION 1

Compile          $FORT FUBAR.FOR                             <CR>

Error Message       %FORT-F-OPENOTPER-operation not permissible on
                          data types [SUM+1.X***3]
                    %FORT-F-MISSEND-Missing END statement, END is
                          assumed in module FUBAR$MAIN at line 9
                    %FORT-F-UNDSTALAB-Undefined statement label [3]
                          in module FUBAR$MAIN at line 3
                    %FORT-F-UNDSTALAB-Undefined statement label [6]
                          in module DUMMY$MAIN at line 6
                    COMPILATION ABORTED-TOTAL OF 4
                    DIAGNOSTICS PRODUCED
```

If syntax errors are present in the program, the computer will print out appropriate error messages similar to those shown above. In addition, the computer stops the creation of the .OBJ file. You cannot proceed to the linking process until you remove the errors and the computer successfully creates the .OBJ file. Therefore, you must return to the editor, correct the errors and then recompile the program.

The above program contains four errors in the FORTRAN syntax. Details of how to correct these errors are not important at this point. It is essential, though, that you recognize the need to return to the editor. Here is what that dialogue will look like:

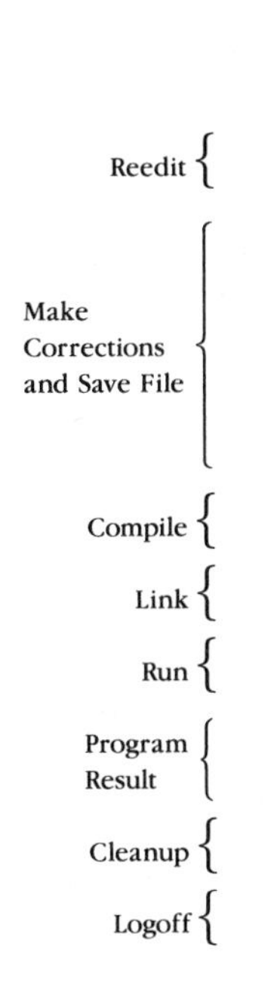

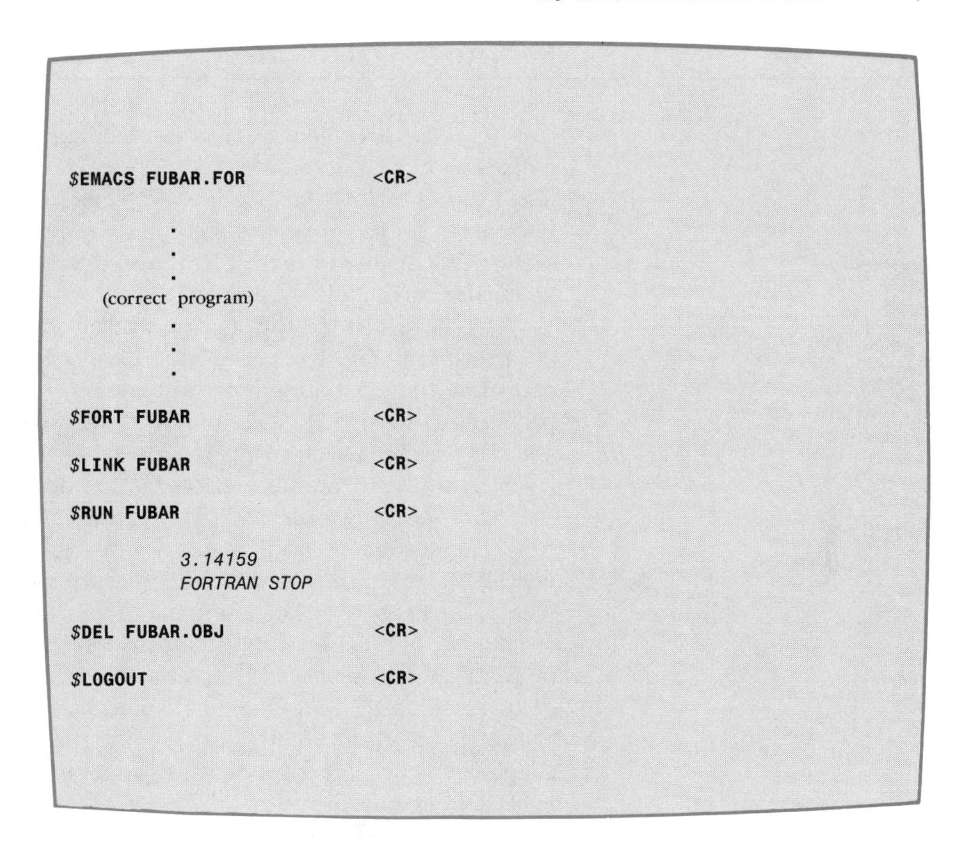

We do not show the actual commands to make changes in the editor since editors vary so much. Note that after editing the file, you must save it before leaving the editor. After correcting all the errors, you can complete the compilation. What surprises many students is that the system prints nothing on the screen to indicate success. In fact, the only time compilers print something on the screen is if it detects errors. Its only response is the prompt, $. The next steps that you execute are the link and run steps, after which the machine prints the computational result of the program.

Before you leave the computer, you should get rid of files that are no longer necessary. The command for this is usually DEL*filename*. In the example we deleted the .OBJ file since we do not need it anymore. But we leave the .FOR and .EXE files in case we need them again. Finally, the last step is to close the account with the LOGOUT command.

Again, every computer system has somewhat different ways of doing similar things. Check with your system operators to find out any differences. You will find, though, that the preceding exercise contains most of the ideas that you will need to create and execute programs.

1.6 Purpose of This Text

This text has three goals: (1) to teach a method of program development that helps the novice and intermediate programmer to write programs; (2) to teach the FORTRAN language to implement the programs; and (3) to introduce many useful numerical method concepts. You will need these methods to solve many engineering problems and they form the basis for much of the engineering curriculum.

We have divided the text into three sections, roughly corresponding to the preceding objectives. The first, Chaps. 1 through 3, presents information regarding the computing environment and provides procedures to develop a program. For many reasons, this section is the most important since developing a logical approach to problem solving is difficult for many students. Accordingly, we expend much effort here to develop good problem-solving skills before proceeding with the FORTRAN language.

The second section Chaps. 4 through 10 focuses on the syntax of the FORTRAN language. The various constructs of the language will be introduced with usage examples. The emphasis here will be on the language itself, and no attempt will be made at that time to present engineering methods. That will be reserved for the last section, Chaps. 11 through 17. As suggested in the previous sections, FORTRAN is a dynamic language and changes periodically. These changes will greatly enhance the power of FORTRAN, and therefore a chapter will be presented that introduces the advanced features of the new FORTRAN standard.

The third and final section of this text will examine numerical methods and their application to engineering problems. This section will introduce important tools that are common to all fields of engineering and show you how to carry them out. The important point of this section is to focus on the numerical method itself and not on the FORTRAN program used to implement it. Since you will have mastered the language in the second section, the emphasis on the language will be greatly reduced. The point of section three is to learn how to do something with your newly learned computer skills.

Further Reading

1. S. L. Edgar, *Advanced Problem Solving with FORTRAN 77,* 1st ed., Science Research Associates, Chicago, 1989.
2. H. Ledgard and M. Marcotty, *The Programming Language Landscape,* 1st ed., Science Research Associates, Chicago, 1981.
3. M. Metcalf and J. Reid, *FORTRAN 8x Explained,* 1st ed., Clarendon Press–Oxford, Oxford, England, 1987.
4. M. Stiegler and B. Hansen, *Programming Languages: Featuring the IBM PC and Compatibles,* 1st ed., Baen Books, New York, 1984.

EXERCISES

Information maps are tables set up in a particular way that enables you to organize information for easy retrieval. They store a matrix of answers to specific questions. In these exercises you should complete the information map for the various computer functions.

1.1 Information Map of Operating Systems Shown here is a partial information map for operating systems. We give many common functions and the actual commands in MS-DOS and VAX/VMS. MS-DOS is common on microcomputers, especially IBM's and their clones; whereas, the VAX/VMS is the operating system on the most common minicomputer. Study this table and then determine what the appropriate commands should be on your system:

	SUMMARY OF OPERATING SYSTEM COMMANDS		
	MS-DOS	VAX/VMS	Your System
Login?	none prompt: >	account name password prompt: $	
EDIT A FILE?	**EDLIN**	**EMACS *filename***	
LIST FILES?	**DIR/*option***	**DIR/*option***	
DELETE FILE?	**DEL *filename***	**DEL *filename***	
COMPILE?	**FORT *filename***	**FORT *filename***	
LINK?	**LINK *filename***	**LINK *filename***	
RUN?	***filename***	**RUN *filename***	
PRINT A FILE?	**PRINT *filename***	**PRINT/*que filename***	
COPY A FILE?	**COPY *orig dup***	**COPY *orig dup***	
RENAME FILE?	**RENAME *old new***	**RENAME *old new***	
LOGOUT?	**none**	**LOGOUT**	

1.2 Information Map for Operating System Options Many operating system commands have options associated with them. In this exercise you are to complete the appropriate command options for your specific operating system:

	SUMMARY OF OS OPTIONS		
	MS-DOS	VAX/VMS	Your System
DELETE? /DBL CHK WILDCARD	**DEL *file*** **DEL/PROMPT *file*** **DEL **.ext*** or **DEL *file.****	**DEL *file*** **DEL/CONFIRM *file*** **DEL **.ext*** or **DEL *file.****	
DIRECTORY? /PAGE /SIZE /WIDE WILDCARD	**DIR** **DIR/P** ***(default)*** **DIR/W** **DIR **.ext*** or **DIR *file.****	**DIR** **DIR/PAGE** **DIR/SIZE** **DIR/BRIEF** **DIR **.ext*** or **DIR *file.****	
COMPILE? /LIST	**FORT *file*** ***(none)***	**FORT *file*** **FORT/LIST *file***	

1.3 Supplementary Operating System Commands For the following information map you are to find the instructions for your system to perform the following functions:

	SUMMARY OF SUPPLEMENTARY OS COMMANDS		
	MS-DOS	VAX/VMS	Your System
SEND MAIL?		**MAIL**	
CREATE A SUBDIRECTORY?	**MKDIR *name***	**CREATE/DIR *name***	
TRANSFER TO SUBDIRECTORY?	**CD *name***	**SET DEFAULT *name***	
COMBINE TWO FILES?	**COPY** ***file1+file2:file3***	**COPY/CONCAT** ***file1,file2***	

1.4 File Specification In the text the convention for naming files was presented as follows:

filename.extension

in which the *filename* was limited to eight letters or numbers and the *extension* was limited to three letters. On many systems you will have more flexibility in naming files since many other characters may be allowed. For example, the symbol, _, may be allowed in a filename such as:

```
FU_BAR.FOR
```

These symbols are useful because they act as spaces that make it easier for you to read the filename. On many systems *version numbers* will also be added to filenames:

filename.extension;version

The machine will automatically keep track of new versions as they are created and also retain older versions.

Determine if either of these options for naming files is available for your system. Also, check to see if other options not mentioned here are used.

1.5 Controlling File Printing Obtaining a printout of your files and output is a daily process in programming. The commands that are used vary considerably from system to system. On your system find the instructions to do the following:

	SUMMARY OF PRINTING INSTRUCTIONS		
	MS-DOS	VAX/VMS	Your System
TYPE A FILE ON SCREEN?	**TYPE *file***	**TYPE *file***	
FREEZE SCREEN?	<**CTRL**> **S**	<**CTRL**> **S**	
RELEASE SCREEN?	<**CTRL**> **Q**	<**CTRL**> **Q**	
ABORT COMMAND?	<**CTRL**> **C**	<**CTRL**> **C**	
HARD COPY OF FILE?	**PRINT *file***	**(system-dependent)**	

1.6 Text Editor Prepare an information map for the editor that you will use. Here for comparison are the commands for a popular editor, EMACS:

	EDITOR COMMAND	
	EMACS	Your System
Insert Letter	SPACE	
Insert Line	<CR>	
Delete A Letter	<CTRL> D	
Delete A Line	<CTRL> K	
Back One Letter	<CTRL> B	
Back One Word	<ESC> B	
Start of Line	<CTRL> A	
Forward One Letter	<CTRL> F	
Forward One Word	<ESC> F	
End of Line	<CTRL> E	
Up One Line	<CTRL> P	
Start of File	<ESC> <	
Down One Line	<CTRL> N	
End of File	<ESC> >	
View Document	<CTRL> L	
Save File	<CTRL> X <CTRL>S	
Oops!	<CTRL> Y	
Exit Editor	<CTRL> X <CTRL>C	
Help!	<CTRL> H	

CHAPTER

2

PROGRAM DESIGN

2.1 Introduction

Writing a computer program to solve a problem is a complex multistep process consisting of at least the four major elements shown in Fig. 2.1. Note that only one portion of this process (coding) involves the use of a specific programming language. The other three steps relate to the development of a problem-solving approach. Students often incorrectly assume that once they master a programming language, they can solve almost any problem. This could not be further from the truth. In fact, most experienced programmers spend up to 90 percent of their time working on the logic of their programs, not on the coding. Therefore, it is crucial that you understand this process and develop good problem-solving skills before attempting to write any programs. In many respects this chapter is the most important one in the text since this is where you will learn a general approach to problem solving.

As trivial as it may sound, the first step in program development is to *define the problem*. In this phase of the process there may be many different factors you need to consider. Sometimes this process is simple; other times it is a major task. Among the factors you will usually consider are:

Goals of the program
Necessary input data
Mathematical tools required
Expected results
Method of presenting results
Special cases

All programs, no matter how simple or complex, perform three functions—collect input data, process the data, and return results. Collecting input

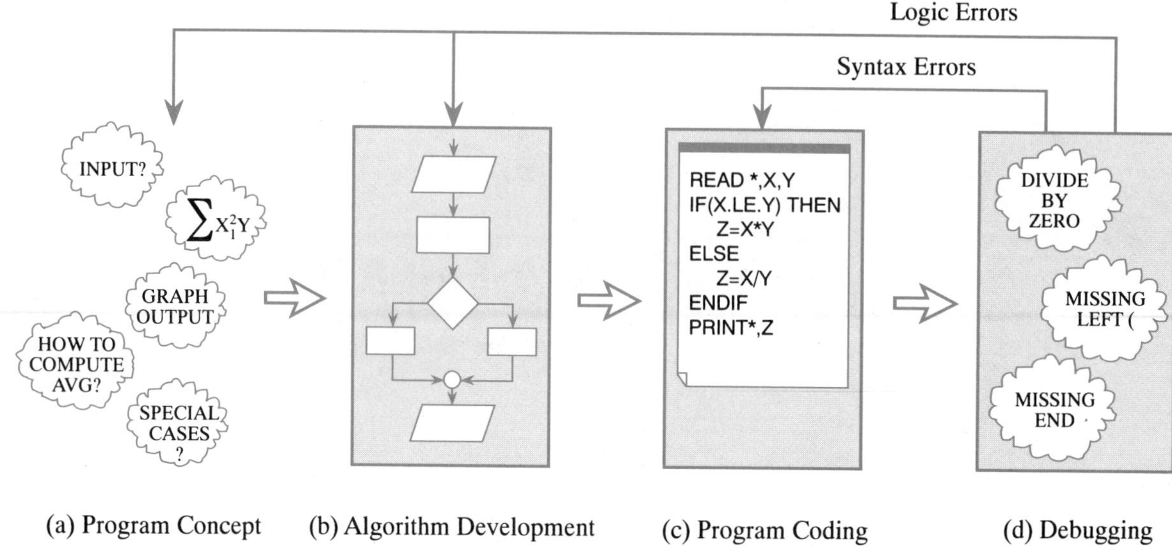

FIGURE 2.1

Stages of Program Development

data is necessary for the program to be useful. If a program only had one set of data directly programmed into it, the program would only provide one solution. Once we know the solution, there would be little need for that program afterward. Therefore, a need to enter multiple sets of data exists. These might come from within the system itself, for example, from a random number generator. More likely, the data will come from a piece of laboratory equipment or from the user who types in the input on the terminal keyboard or perhaps from a previously created data file (see Fig. 2.2).

Processing is the procedure of manipulating the input data to produce some output. This stage can be complex or simple. For example, the use of finite element techniques to solve differential equations will consist of hundreds of pages of code. At the other extreme, some simple programs that you will write in your first exercises will contain only a few lines.

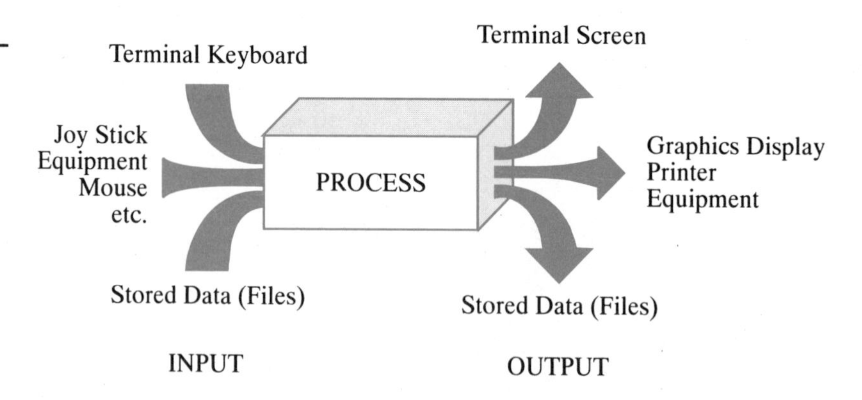

FIGURE 2.2

Information Processing by a Program

The final step, reporting results, is a process as general as collecting data. Returning results can take on many forms. It ranges from printing numbers and text on the user's terminal screen, plotting results in a graphical format, or making a robot arm move in a desired path. You should consider any desired action arising from your data processing to be output.

A video game is one example to illustrate the general forms of input and output. The program collects input data from various input devices such as joysticks, buttons, and control pedals. The output is in the form of graphics and sound effects. An example of unusual output would be a microprocessor-controlled traffic light system where switches buried in the road generate the input. By using the road switches to count the cars that pass through an intersection over a fixed period of time, a microprocessor controls the timing of the lights. Here the output is how long a light should be left on before changing colors.

This first stage of developing a program is a very crucial step. If you take care to define the problem carefully, you will find that the remainder of the program development process will be greatly simplified. Conversely, if you fail to define the problem adequately you will usually have great difficulties in later phases of the program development.

The second major step in the program development process is the *generation of the algorithm*. An algorithm is a map to a problem solution. This outline shows the precise order in which the program will execute individual functions to arrive at the solution. As we will see very shortly, all problem solutions, no matter how complex, can be reduced to combinations of only three basic *control functions*. These are:

Sequential	→ (DO THIS) → (DO THIS NEXT) → (etc....) →
Branching	→ (WHICH WAY?) ─┬→ (DO THIS) → (DO THIS) → └→ (DO THIS) → (DO THIS) →
Looping	→ (DO THIS) → (FINISHED?) *(yes)* → (DO THIS) → ↑ ← *(no)* ← ┘ (return to DO THIS)

How we use these basic building blocks to arrive at a problem solution will be discussed in great depth in Sec. 2.2. It should be emphasized here that the algorithm is *language-independent*. This means that it does not depend on the details of the language. For example, a FORTRAN programmer can use an algorithm developed by a Pascal programmer although the two programmers do not speak each other's language. Thus, the process of developing algorithms can be taught even before we begin a discussion of FORTRAN.

The third step in the program development process is the *coding* or conversion of the algorithm into the desired programming language. The algorithm allows the programmer to visualize the path to a solution, but it is too imprecise to run intact on a computer. Accordingly, we need to convert the algorithm into a more structured device, *the source code.* The structure of the program will follow very specific rules required by the specific language. Every programming language has its own set of rules, called *syntax*, which are very rigid and somewhat artificial. Nonetheless, you will find these rules easy to master, after which the coding process will be almost automatic. The syntax of the FORTRAN language will be the subject of the second section of this book, Chaps. 4 through 10.

The final step in the programming process is *debugging*. Every program contains *bugs* that can range from simple mistakes in the language usage (syntax errors) up to complex flaws in the algorithm (logic errors). Removing syntax errors from your programs is very simple since the compiler will give you diagnostic messages highlighting the problem. Logic errors, on the other hand, are much more difficult to remove since the computer will give you no clues. After all, the computer is simply executing the instructions that you give it. The computer cannot know what the answer is, and so you should not expect it to help you isolate the problem. Consequently, removal of logic errors is an *art* and you must develop techniques to discover and correct these problems. Problems of this sort usually can be traced to faults in the problem definition as described earlier. To minimize such problems, it is therefore imperative that you develop your algorithms with great care.

This chapter will focus on the first two steps in the program development process, defining the problem and writing the algorithm. A discussion of how to implement the algorithm in a program and how to debug it will be delayed until Chap. 4.

2.2 Basic Programming Tools

You will find that only three basic building blocks are needed to develop a solution to a problem. These building blocks are independent of the language and can be used to develop a conceptual plan to tackle a problem. As we described earlier, these blocks are shown in Fig. 2.3.

Sequential commands are those that perform a task and then continue to execute the next command in the program. Branching allows for the outcome of a test, in which the only possible answers are true or false, to control which instructions to execute next. Finally, the loop is the process of repeating a certain block of instructions until a certain condition is satisfied. We show two forms of the loops in this figure. The first executes repeatedly until a condition becomes true. The second loop executes a predetermined number of times.

Figure 2.3 shows the corresponding *flowchart symbols* for each basic function. The flowchart symbols are standardized figures used to show specific programming functions. By combining these symbols, you can gener-

FIGURE 2.3

Basic Building Blocks of Programming

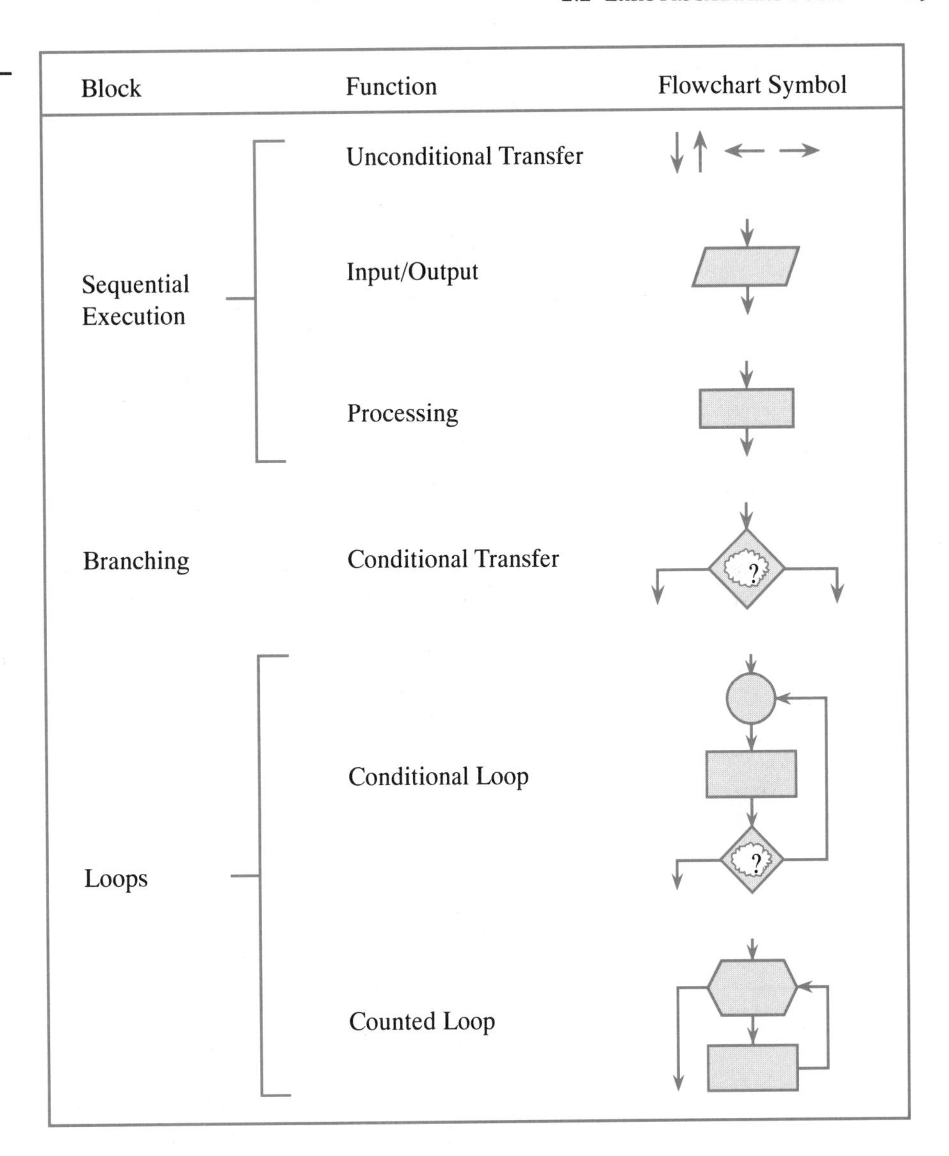

ate a map of the algorithm, which makes it easier to visualize the algorithm. Inside each symbol you can write a notation regarding the specific action to be taken. We present here only a few of the more common symbols. Additional symbols will be presented as the need arises.

To illustrate the use of these basic building blocks, let's examine a small portion of the set of instructions to assemble a model car engine. These instructions, with the appropriate flowchart symbols might look like those shown in Fig. 2.4.

In this example the computer executes the instructions one after another. This is called *sequential execution*. Note that there is no possibility of

FIGURE 2.4

Algorithm and Flowchart Showing Use of Simple Sequential Execution

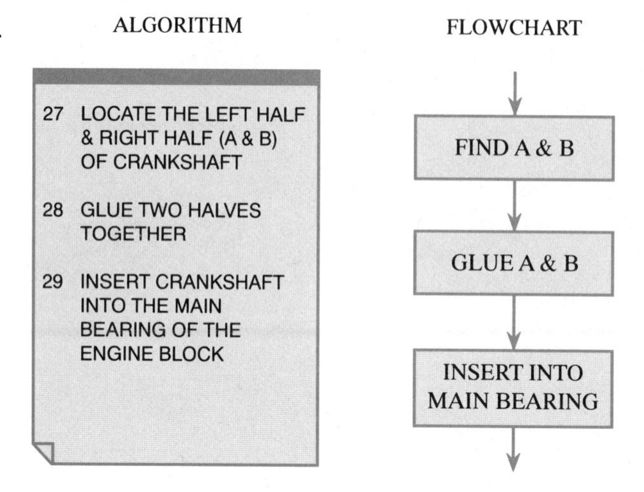

skipping over one instruction with sequential execution. Many times, though, there is a need to allow branching. For example, if the model engine for our model car had an optional turbo engine, then a decision would need to be made. We can do this with the simple set of instructions that allow a branch such as that illustrated in Fig. 2.5.

Note that in this figure the question that controls the subsequent branching must be set up so the only possible answers are true or false. All branching operations must be set up in this way, which is sometimes an awkward construct since it might be more natural to devise a test with more than two outcomes. This can be implemented with the previous branching structure by *nesting*. Nesting is the process by which we place one conditional transfer inside another conditional transfer. We will discuss this in more detail shortly. There is also a newer construct, called the *Case Select* structure that was specifically designed to handle complex branching operations where more than two outcomes are possible. We will discuss this in some detail in Chap. 11. We want to emphasize that the simple two alternative branch shown in Fig. 2.5 is the most fundamental form and that all branching problems can be solved with it. Some of these other types of branching operations are simply special cases that may occasionally simplify your code.

The final basic building block is the *Loop*, which allows repetition of several programming steps. Of course, loops must have a branching operation built into it to control how many times the loop executes. Sometimes we want the loop to execute a predetermined number of times. On other occasions we want the loop to execute indefinitely until some particular condition is satisfied. An example of the latter case is a loop that is to be repeated until there is no more data left for analysis. To illustrate how a loop operates, let's return to our simple example of an auto assembly line. Assume that while we are assembling the engine that we need to put together a piston-connecting rod assembly for each cylinder. Since the engine has many cylinders, usually between four and eight, the same assembly instructions will be executed many times for each engine, as illustrated in Fig. 2.6.

Figure 2.5

Algorithm and Flowchart Demonstrating Branching Operation

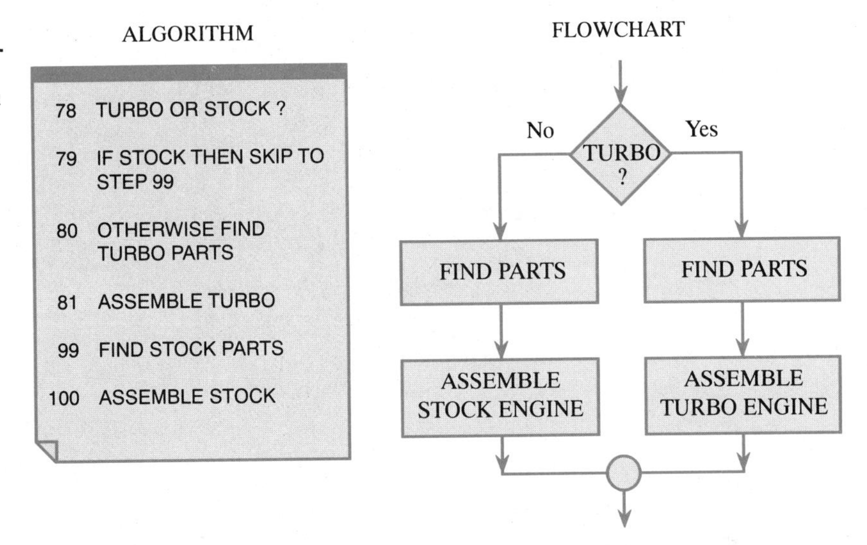

Many details in this algorithm need to be addressed before you can convert it into a program. For example, the problem of how to figure out if we are finished has not been spelled out. This is an easy problem to solve since we know exactly how many cylinders the car will have. Thus, we know exactly how many times the loop should execute. But this first version of the algorithm and the corresponding flowchart is sufficient to allow us to see if this algorithm will work. Of course, we can go back and *flesh out* our algorithm. We will discuss this problem in greater detail in the following section.

There are many different forms of the three basic blocks just presented. For example, the loop structure can have two general forms. The first will loop indefinitely, whereas the second will loop a predetermined number of times. Similarly, the branching function also can take on many different forms. These include the nested structure previously mentioned along with special cases of the standard branching operation.

Despite the large number of options associated with the three basic building blocks, you should attempt to keep your algorithms simple for now. As you mature in your algorithm-developing skills, you can add the advanced features to streamline your solutions and improve the efficiency of your code. As you become more proficient in programming, you will select automatically the most efficient implementation of the basic building block. For now, you should be content with the simplest forms.

2.3 Top-Down Approach to Algorithm Development

All the commands required to develop an algorithm can be constructed from the basic building blocks described previously. Usually, the first version of an algorithm is too vague to implement directly on the computer. But this first draft is often closest to the way that humans think. Therefore, the first draft is

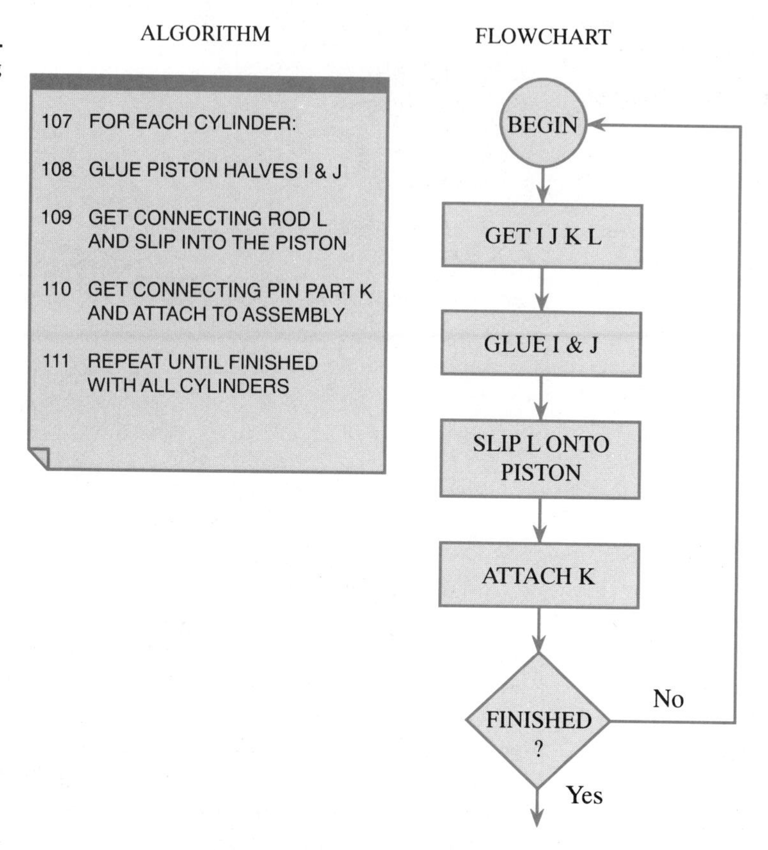

FIGURE 2.6

Algorithm and Flowchart Illustrating Use of a Loop

extremely useful to help develop some general details of the problem solution. Once we lay down these general ideas, we can begin to refine the ideas with increasing attention to details. Thus, in the example in Fig. 2.6 we were not concerned about how to determine when the loop finishes. In a second version of the algorithm we would be concerned about how this information could be obtained. The important point to recognize is that a refinement process is a necessary and desirable step in developing good programming skills. Programmers refer to this approach to algorithm development as the *top-down* approach.

Top-down design involves taking a large problem and breaking it up into a series of simpler, smaller problems. We then repeat this process on the newly created smaller sections to reduce further the size of those components. By repeating this procedure, we convert a large problem into several smaller problems. The process ends when the programmer has reduced the complexity of the problem down to a level for which previously developed solutions exist.

Figure 2.7 illustrates the use of top-down design to make a Christmas list. The general problem is who gets Christmas presents and what should they

Figure 2.7

Using Top-Down Design Process

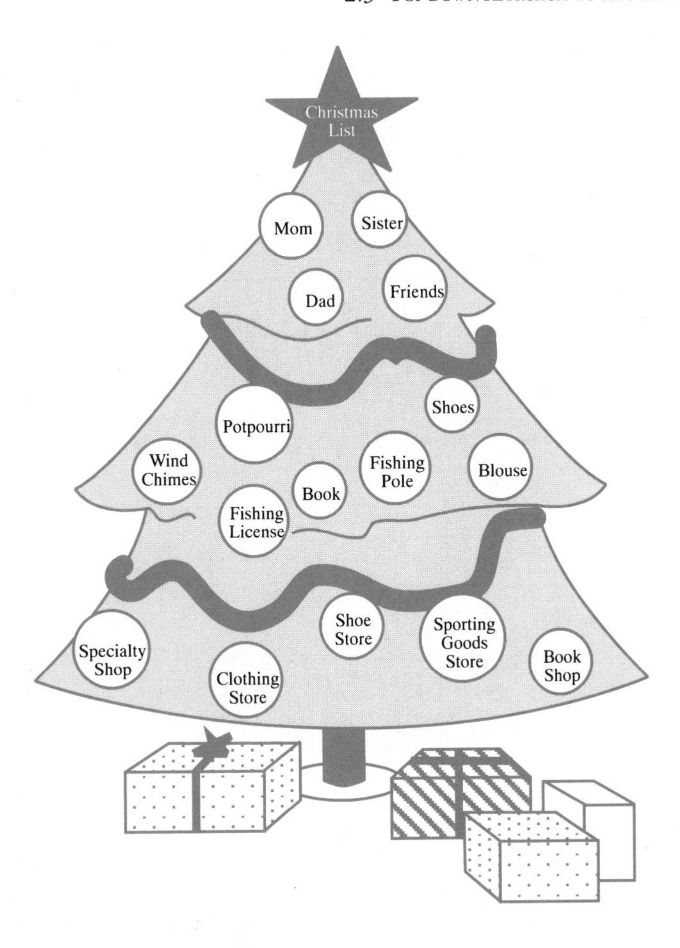

be? The process starts by determining the people who should get presents. This represents the first draft of the algorithm. We generate the second draft by asking how much money can we afford and how to divide it. Next, we propose ideas for presents, and finally suggest places to purchase them. Notice that in this problem we have made the task much easier by focusing on one small problem at a time. Thus, when we focused on a present for Dad, we felt that a fishing pole and a license would be nice gifts. A logical extension would be to include fishing tackle, but this would put us over our budget. So, we limit ourselves to the first two items.

Although the example seems like an obvious approach, it is useful to go through the exercise nonetheless. After all, everyone has generated a similar shopping list at some time, but it is unlikely that you have thought about it in as much detail as this example does. Figure 2.8 illustrates this process in a more general form. The multistage refinement process results in a tree structure, with the ending branches representing the class of subproblems you know how to solve. The boxes with a "*" represent solutions to problems that we previously solved.

FIGURE 2.8

General Form of the Top-Down Process

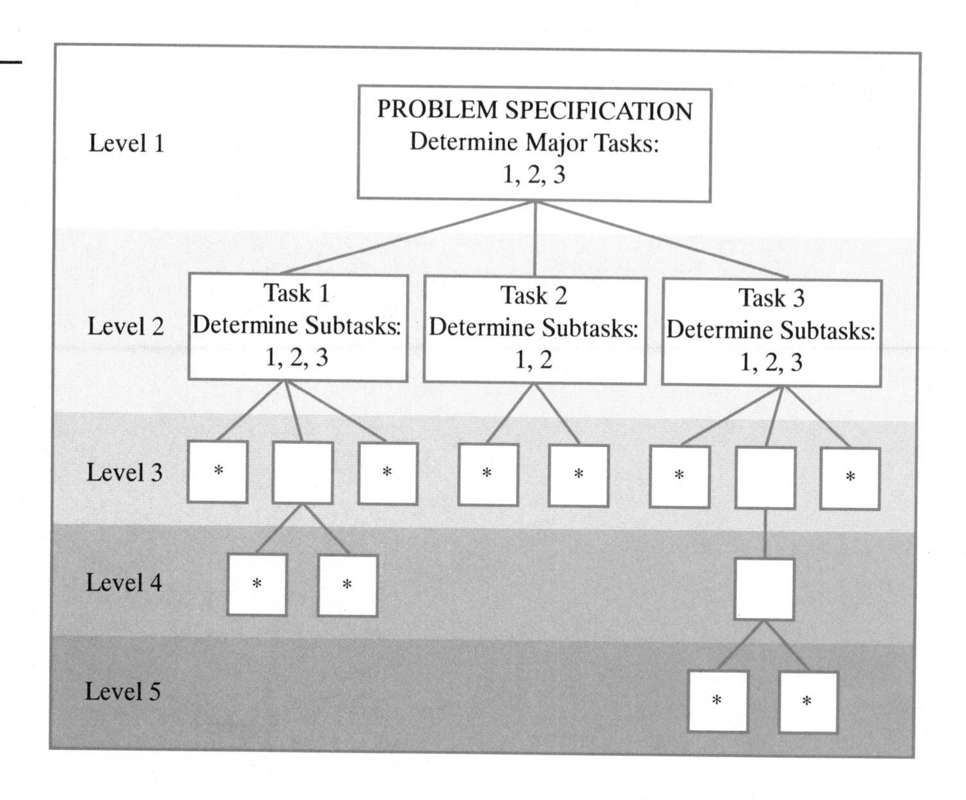

To use top-down design effectively, the programmer must have a library of previously solved smaller problems and have the solutions readily available. Not only must the programmer have a library of solutions, but those solutions must be accessible also. Having a method to organize solutions of previously solved problems is important when using top-down design. Thus, the top-down process will be very different for beginning and advanced programmers. The beginning programmer will need to carry out the algorithm refinement process further since previous solutions are not available. The advanced programmer, on the other hand, may need to refine the algorithm only a few levels deep because many solutions to the subproblems already exist.

2.4 Implementing the Top-Down Design

Breaking up a problem into a series of smaller problems requires some skill. We present a method here that will help you in this process. The method that we have chosen involves a question-and-answer scheme that we have adopted from educational techniques used in developing English composition.

FLESHING OUT A PROBLEM

The process of *fleshing out a problem* involves generating questions from key words within a problem. Ask the right questions, and the problem begins to solve itself. Therefore, recognizing key words is an important step in this process. From those key words come the questions for the refinement process. The answers to these questions then become the basis for further refinements.

One method of looking for key words is to use all the **nouns** and **verbs** in the problem statement. Processing will be suggested in the problem statement as verbs or action words. Nouns will suggest what is to be processed. The process that we use in this book to develop an algorithm is to take the problem statement and identify all the verbs and nouns. This will form the basis for the first version of the algorithm. You then construct questions based on these items such as:

What do we mean by...?
How do I calculate...?
When are we finished with...?

The answers to these questions will then provide a second batch of verbs and nouns that form the basis for the second version of the algorithm. To show how these guidelines can be used to generate an algorithm, consider the simple problem shown in the following example:

Example #1

PROBLEM STATEMENT: An all-news radio station announces an average temperature during the weather portion of its program. This average temperature is calculated from measurements made at three locations in the city and the local suburbs. Write a program to calculate the average temperature every hour.

ALGORITHM DEVELOPMENT:

Step 1a: Review the problem statement. Highlight the nouns in **bold** and the verbs in *italics:*

Develop an algorithm to *calculate* the **average temperature** every **hour**.

Generate questions based on key words:

Q1: What is an average temperature?
Q2: How do I calculate an average temperature?
Q3: When is the average temperature needed?

Step 1b: Answer the questions:

A1: The average temperature is obtained by taking all temperature values, adding them, and dividing by the number of values.

A2: To calculate an average temperature, get all three temperature readings and perform steps outlined in A1.

A3: Every hour.

Step 2a: No need to refine the answers since these are elementary functions.

ALGORITHM:

1. Check the time. If it is on the hour:

Get the three temperature readings.
Add the three temperatures and divide by 3.
Report the average temperature.

2. Go back to step 1.

FLOWCHART:

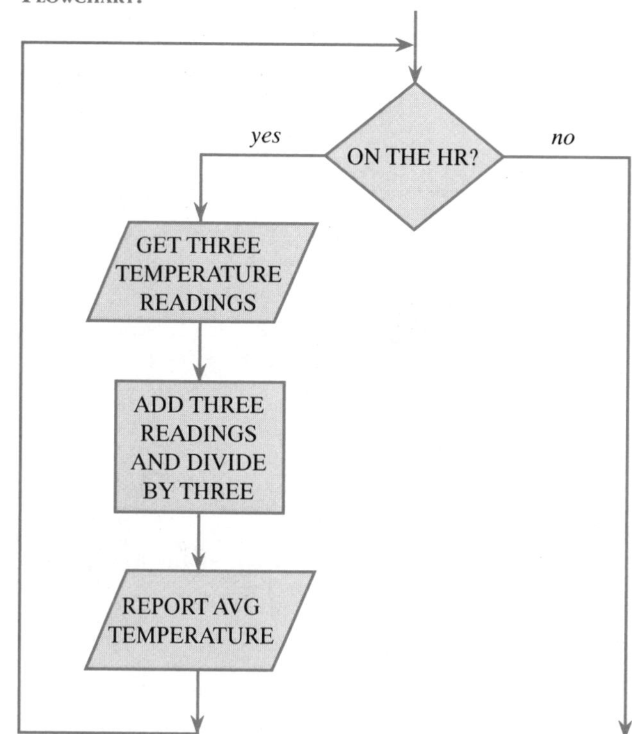

Example 1 contains all three basic building blocks. Step 1 of the algorithm contains a conditional statement that tests for the time. When the condition is satisfied (on the hour?), the program executes a series of sequential instructions. Finally, step 2 of the algorithm is a loop that instructs the computer to begin over. Note that the loop is infinite since there is no way for the program to stop. A good exercise for you at this point is to devise a way to modify the algorithm to run between the hours of 6 A.M. and 11 P.M. only.

LANGUAGE STRUCTURES

Understanding the requirements of a program is only one bit of information necessary to develop a program. In addition, you need to know about the specific information peculiar to the language you will use and some commonly used procedures. We present three types of these in this section: sequential commands, control statements, and language structures. Sequential commands carry out the fundamental operations of the programming language. For example, the command to print results to the terminal screen (the PRINT command) would be a sequential command. But these types of commands are only a small part of what we need to write a program. The ability to make decisions, and repeat parts of a program, is of great importance.

Control statements represent groups of commands used to manipulate sequential commands. The most important control structures include the conditional branching and looping instructions. For example, if we wanted to repeat a set of sequential commands 100 times, we can do this with a loop structure. This loop structure (a DO loop in FORTRAN) is an example of a control structure. The other major control structure is the branching instruction, which we implement in FORTRAN with the IF-THEN-ELSE statements.

Language structures, or LSs, are combinations of commands that the programmer considers to be important. These may be your own simple programs to do simple tasks or complex groups of code generated by other programmers to perform sophisticated calculations. The important point about LSs is that they are basic building blocks for problem solving. We will go to great lengths to construct many LSs that will make many of your programming tasks simpler. Whenever you have a complex problem to solve, you probably will find most of the basic building blocks are among the LSs presented here. All you need to do is to assemble these in the proper order to solve problems that might otherwise seem formidable.

TOOLS NEEDED TO ORGANIZE

By breaking down the various commands and structures into the classes stated, we establish a foundation for developing programs in FORTRAN 77. This foundation will also allow you to learn other computer languages more easily. As you work through problems, you will be gaining knowledge about programming techniques and methods. This is useful not only when constructing FORTRAN programs, but also when learning any new language.

Figures 2.9 and 2.10 show examples of commonly used sequential and control statements in the form of *information maps*. Information maps are a useful device to help you understand how to use the different programming commands. They contain all the information about a particular command or

FIGURE 2.9

Example of an Information Map for Sequential Commands

	TERMINAL INPUT	TERMINAL OUTPUT	ASSIGNMENT
What is the purpose of the command?	To accept input from the terminal and assign the value to a variable	To print messages and variable values to the terminal screen	To calculate an expression and assign the result to a variable
INPUT, OUTPUT, or PROCESS?	INPUT	OUTPUT	PROCESS
What input or arguments are needed?	List of variables to receive the data	List of variables and/or messages to be printed	Expression to be evaluated and a target variable
What is the general form of this command?	**READ *, *var1,...var2***	**PRINT *, *mes1, var1,...***	***target = expression***
What would an example look like?	**READ *, X, Y, Z**	**PRINT *, 'X IS', X**	**L = 2.0 * 3.1416 * R**
What are the machine-dependent attributes?	None	None	None
What do the flowchart symbols look like?	X, Y	X, Y	target = exp

structure that you will need. If you set them up properly, your maps should contain the following:

The syntax rules for the command
Examples of their use
The flowchart symbols
Special features such as things to watch out for

As you progress through the book, the information maps you generate will grow by adding new commands and structures as you learn them. Effectively, you are distilling the information from 250 pages of text into a compact form and creating a reference library of what you have learned. It is important, however, that you are the one to construct these information maps. Thus, we will not present information introduced in this text as an information map.

Figure 2.10

Example of an Information Map for Control Statements

	IF-THEN-ELSE	COUNTED LOOP
What is the purpose of the control statement?	Based on the results of a Boolean equation, select the appropriate list of instructions to execute.	Repeat a process while adjusting a variable from some starting value to a final value using a prescribed step size.
Branch, Loop, or combo?	Branching	Looping
What input or arguments are needed?	Boolean expression and corresponding procedures to be executed when true or false	Variable to be adjusted, lower and upper limits of variable value, step size and label that marks end of the loop.

What is the general form of this command?	**IF (*test*) THEN**
	procedure if TRUE
	ELSE
	procedure if FALSE
	ENDIF
		DO *label* var = start, stop, step
		procedure to be repeated
		***label* CONTINUE**

What would an example look like?	**IF(TEMP.GT.32)THEN**
	PRINT*, 'TEMP IS ABOVE 32'
	ELSE
	PRINT*, 'TEMP IS BELOW 32'
	ENDIF
		DO 10 J=1,10,1
		PRINT *, 'J=', J
		10 CONTINUE

What do the flowchart symbols look like?	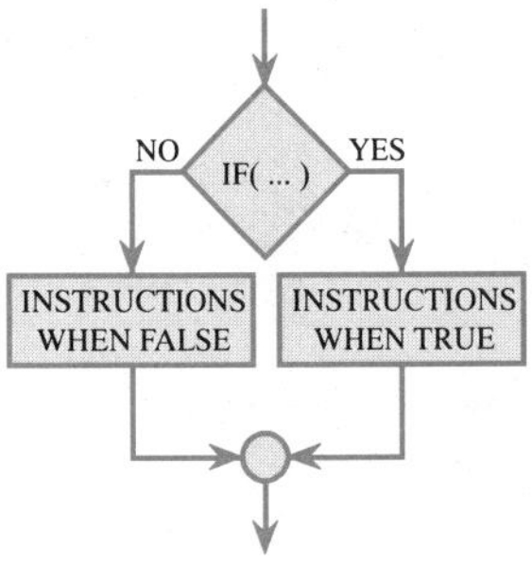 NO IF(...) YES INSTRUCTIONS WHEN FALSE INSTRUCTIONS WHEN TRUE	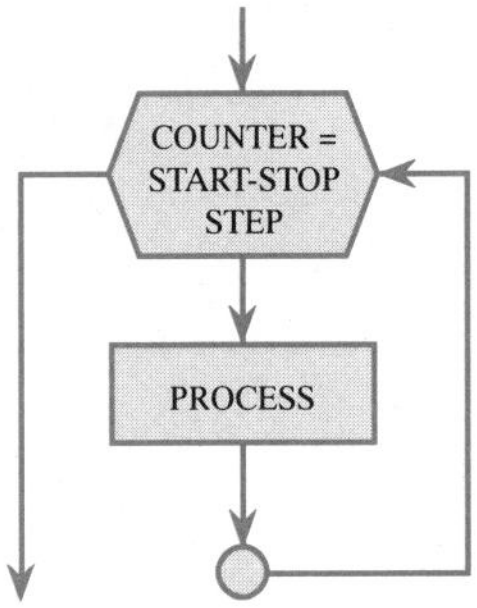 COUNTER = START-STOP STEP PROCESS

Rather, we suggest that you take the information and organize it in the indicated form.

BACKGROUND INFORMATION

The organization of the information map shown in Fig. 2.9 illustrates the topics that will be covered for future commands. As we present additional sequential and control statements, you will be extending this information map. This effectively means adding additional columns (on additional pages) to what we have already started for you. This is an exercise that must be done by you if it is to be effective.

The first question on the sequential statement information map asks what is its purpose? Here, we make a brief descriptive comment that for some commands might be a simple restatement of its title. However, for others, a more complete description may be required.

The second question asks about the function of the command. As mentioned earlier, all programs input, process, or output information. Therefore, we specify which of these three categories a command belongs to.

Question three asks for the arguments needed by the command. Arguments are data that the command requires to fulfill its function. Most commands will have arguments. Those that do not will always provide a single action or function. For example, the END statement always goes at the end of a program. Its purpose is merely to indicate the end of the program, and it has no arguments. The PRINT statement, however, requires a list of variables to output.

The purpose of illustrating the general form of the sequential command is to indicate the command and its arguments. It also helps to give a specific example of its use. Thus, we include questions four and five.

The next question asks if the statement has any machine-dependent features. For most of the commands this answer will be "none." After all, FORTRAN is a portable language. Thus, if your program runs on a mainframe, it should run also on a personal computer. There are times, though, when some small changes may be necessary. By knowing about these machine dependencies, we can write a program that makes the process of switching machines as painless as possible.

The last question on the information map asks for the flowchart symbol. Programmers usually use flowcharts after completing the algorithm as a final aid in writing the program. Because the flowchart represents common commands (which are common to many programming languages), a programmer who uses flowcharts can construct a program in whatever language is available.

SEQUENTIAL VS. CONTROL INFORMATION MAPS

We use control statements to control the execution of other commands. By comparing the information maps for sequential commands and control statements, you will see only one difference. The difference is that the function of the control statement is not to input, process, or output information, but rather, to branch, loop, or provide some combination of these. Otherwise the organization of the two maps is similar.

INFORMATION MAPS AND COMMANDS

The sequential commands and control statements presented so far are sufficient to write simple input, output, calculations, looping, and branching instructions (not bad for two pages of notes!). The sequential commands covered include terminal input, terminal output, and assignment statements. The control statements presented are incremental loops and IF-THEN-ELSE structures.

The purpose of the TERMINAL INPUT command is to provide a means to get information into the program. By using a READ statement, you can assign values to variables during execution of the program. FORTRAN variables can be thought of as providing the same function as algebraic variables.

The TERMINAL OUTPUT command provides a means to get information out of the program. Use of the PRINT statement is one method of displaying answers on the terminal screen.

We use the assignment statement, the third type of sequential statement presented, to evaluate equations. An expression can be thought of as a formula you want to have evaluated. The target variable receives the answer once the computer calculates it. The following illustration shows this graphically:

TARGET ⟵ RESULT OF CALCULATION OF AN EXPRESSION

an example of which might be:

$$X = Y + Z$$

The program executes the right-hand side of the statement first and then makes the assignment. The computer retrieves the values for Y and Z, adds them, and stores the result in the variable X. Thus, X is the target of the assignment statement.

The IF-THEN-ELSE structure is the first control structure presented. Its purpose is to check a Boolean equation and execute an appropriate list of instructions that depend on the outcome of the test. Boolean equations are expressions that can have only a true or false answer. You have probably used them in algebra. For example, a STEP FUNCTION is a function, where:

$$F(X) = \begin{cases} 0 & \text{for } X \leq 0 \\ 1 & \text{for } X > 0 \end{cases}$$

The two Boolean equations presented are:

$X \leq 0$ (Read as "Is X less than or equal to zero?")

and

$X > 0$ (Read as "Is X greater than zero?")

Thus, in the preceding example $F(X)$ is 0 when X is negative or zero, but $F(X)$ is 1 when X is positive. Note in these types of conditions that the only possible outcome of the test is true or false.

ORGANIZING A PROGRAM FROM A QUESTION/ANSWER SESSION

Using the background information provided by the two information maps, we can now develop a program using the fleshing-out process. It should be noted that the information provided in these information maps is somewhat incomplete for illustrative purposes. For example, the READ and PRINT statements have an additional argument that controls the appearance of the output. We will cover this process of formatting in depth later in the book. Fortunately, it has little to do with generating the logic for solving the problem.

To illustrate the technique of fleshing out a problem, we will solve a problem with a known solution. It is good practice whenever you are learning a new method to solve a problem for which you already know the answer. In this way you can see if you are using the method correctly. The previous example was a particularly easy one, which did not require any refinements. This second example is more complex and will require several refinement steps before you attempt to code it.

The second example that we give here shows the scale conversion for pressure of a gas. There are many ways to present the pressure. For example, you can quote pressure of a gas in *absolute pressure, gauge pressure,* or *torr of vacuum*. The absolute pressure is the actual pressure defined by the force of the gas molecules per unit area of an object or container. But since we live in an environment that always exerts a pressure of 14.7 pounds per square inch (psi) of absolute pressure, we tend to refer gas pressure to this atmospheric value. Therefore, the thing that interests us most is the pressure *above* atmospheric. We call this the *gauge* pressure, which is simply the absolute pressure minus the atmospheric pressure. For example, the pressure commonly used to fill a radial tire is 36 psi gauge or 50.7 psi absolute. We sometimes abbreviate these as psia and psig, respectively. The third type of pressure measurement occurs when the pressure inside a container is lower than the pressure outside. We measure the degree of vacuum inside the container with the unit, torr of vacuum. In this unit of measurement 760 torr of vacuum represents full vacuum or 0 psia. Also, zero torr of vacuum represents atmospheric pressure or 0 psig. The following equations put these relationships into a mathematical form:

$$\text{psig} = \text{psia} - 14.7$$

$$\text{torr} = \frac{760}{14.7} \quad (14.7 - \text{psia})$$

When we set up the conversions, we must keep in mind that the psig unit of measure is valid only at or above atmospheric pressure. In a similar way, the vacuum system is valid only at pressures below atmospheric. In the algorithm and program to follow we will read in a pressure in psia units, decide which scale to use (psig or torr), and then carry out the conversion:

Example #2

PROBLEM STATEMENT: Read in a pressure in psia units and calculate the gauge pressure for values greater than atmospheric or torr of vacuum for pressures below atmospheric.

Algorithm Development:

Step 1a: Review the problem statement. Indicate all nouns in **bold**, and verbs in *italics:*

Read in a **pressure in psia units** and *calculate* the **gauge pressure** for **values greater than atmospheric** or **torr of vacuum** for **pressures below atmospheric**.

Formulate questions based on the key words:

Q1: What is absolute pressure?
Q2: What is gauge pressure?
Q3: What is torr of vacuum?
Q4: How do I read in a value for absolute pressure?
Q5: How do I calculate gauge pressure and torr of vacuum from absolute pressure?

Step 1b: Construct answers to the above questions. Rather than repeat the questions, we use corresponding numbers to indicate to which question the answer belongs:

A1: Absolute pressure is the value given to the program.
A2: Gauge pressure is pressure above atmospheric.
A3: Torr of vacuum is pressure below atmospheric.
***A4:** Use the READ sequential statement to assign the value of psia.
A5: Gauge pressure in psig = psia − 14.7. If psig is negative, then a vacuum exists and Torr = 760/14.7(14.7 − psia).

In this example steps 1a and 1b represent the first iteration (or version) through the refinement process. Questions 1, 2, and 3 come from the nouns in the problem statement, whereas Questions 4 and 5 come from the verbs. Most of the time verb-based questions will be the ones that we focus on to develop the program since they define how to process the data. The nouns suggest what to process. Note that answer 4 has a "*" in front of it. This is what we use to show that the answer is something from an information map. So, this question needs no further refinement. However, notice that the other answers need some additional thought.

If you review the answers, you will see that we can break answer #5 into smaller tasks. We use the same process as that which we used in the first refinement. This time, though, the problem statement will be from answer #5. We refine this answer by focusing on nouns and verbs in the new statement until all the answers can be found in the information maps:

Example #2 (cont'd)

Algorithm Development:

Step 2a: Review previous answers to see if refinement is needed. Answer A4 will be reviewed by outlining nouns and verbs.

A5: **Gauge Pressure in psig** = **psi** − **14.7**. If **psig** *is* negative, then a **vacuum** *exists* and **TORR** = **760/14.7** · **(14.7** − **psia)**.

Formulate question based on the key words:

Q6: How do I calculate PSIG?
Q7: How do I determine if PSIG is negative?
Q8: How do I calculate torr of vacuum?

Step 2b: Construct answers to the new questions:

***A6:** Use an assignment statement to calculate PSIG.
***A7:** Use an IF-THEN-ELSE control statement to see if PSIG is negative.
***A8:** Use an assignment statement to calculate torr of vacuum

Notice that after this refinement all the answers have a "*" next to them, showing that the programming command to implement them can be found in the information maps. If there were any answers that did not have an asterisk next to them, we would have to go back and do a third refinement. Figure 2.11 illustrates the steps in the refinement process. The key to this process is to start out with a general idea of how to solve the process and then progress to greater and greater detail.

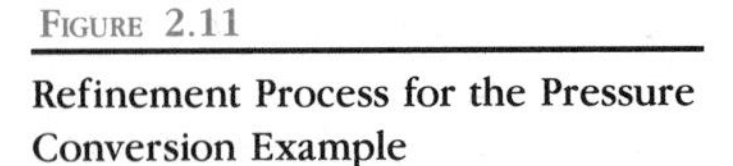

Figure 2.11

Refinement Process for the Pressure Conversion Example

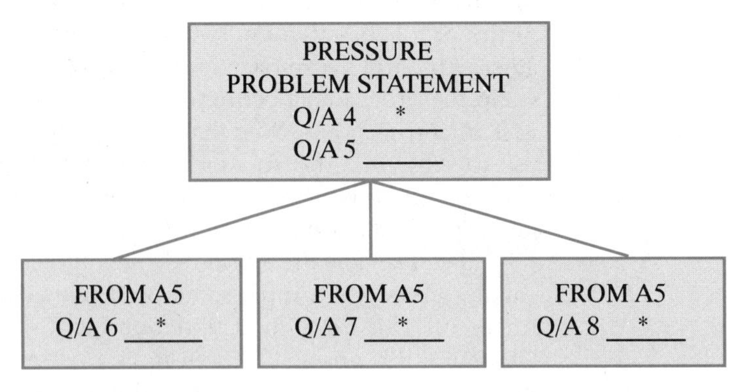

Only after you have completed the refinement process are you ready to construct the flowchart and program itself. But before creating the flowchart, you should check on the three fundamental functions of a program:

How do we get data into the program?
How will the program process the data?
How will we get the results out?

Our procedure has already identified A4 as the answer to the first question. Similarly, A6, A7, and A8 answer the second question. But note that we forgot to specify how to output the data. The easiest thing to do for this type of problem is to print the data to the terminal screen:

Example #2 (cont'd)

FLOWCHART:

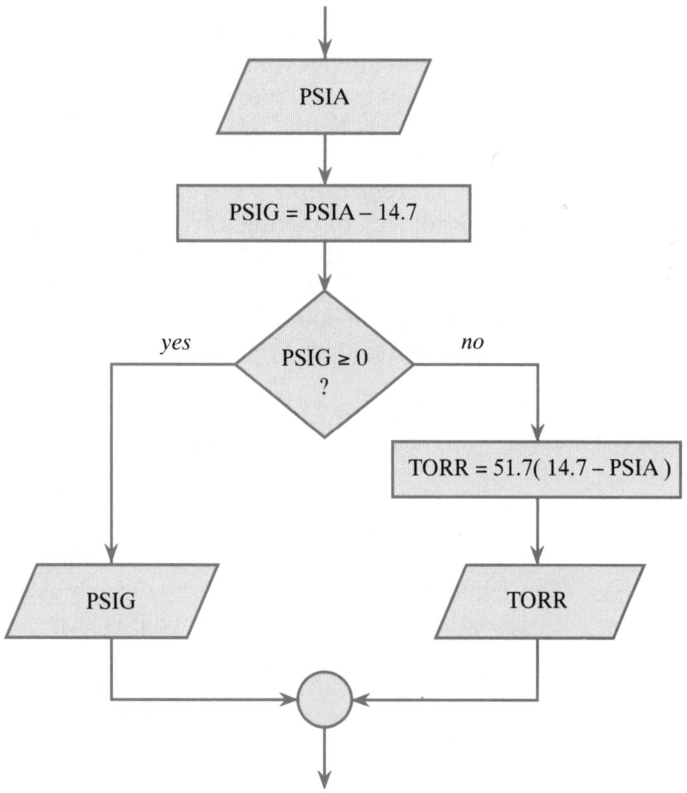

PSIA

PSIG = PSIA – 14.7

PSIG ≥ 0 ?

yes

no

TORR = 51.7(14.7 – PSIA)

PSIG

TORR

PROGRAM:

```
C    THIS PROGRAM READS IN A PRESSURE IN ABSOLUTE UNITS
C    AND CONVERTS IT INTO GAUGE PRESSURE OR TORR
C    OF VACUUM DEPENDING UPON WHETHER THE PRESSURE
C    IS ABOVE OR BELOW ATMOSPHERIC PRESSURE.
C    ________________________________________________
     READ*, PSIA
     PSIG=PSIA-14.7
     IF(PSIG.GT.0.0) THEN
            PRINT*,' PSIG= ', PSIG
     ELSE
            TORR=51.7*(14.7-PSIA)
            PRINT*, 'VACUUM= ', TORR
     ENDIF
     STOP
     END
```

An experienced programmer, as you will soon be, could generate the preceding code from the flowchart in only a few minutes. But generating the flowchart would take substantially more time. An important point regarding this process is that you draw the flowchart with a pencil and paper away from a terminal. *You should attempt to begin coding only after you have finished the flowchart and are convinced of its correctness.* Students often make the mistake of attempting to type in a program first and then try to figure out the logic errors later. This process is doomed to failure (or at least great inefficiency) when programs begin to approach a few hundred lines in length. In the long run you will find programming much easier if you develop the habit of refining algorithms before you write a single line of code.

HINTS ON FLESHING OUT

The refinement process stops when answers to the generated questions become sequential commands, control statements, or language structures. The tree diagram in Fig. 2.12 illustrates this point. To keep track of the tree structure, you can use several techniques. Some students prefer to write the question-and-answer process directly as a tree structure. Used computer paper serves well for this task since it is large and readily available. Another method is to write each question/answer on separate pieces of paper or cards. In this fashion, the cards can be laid out to form the tree structure. Ending branches (questions answered using the building blocks presented) can be marked as such, showing the end of the refinement process for that branch. To aid in organizing the cards, you can show the level of refinement with the source question number, illustrated in Fig. 2.12.

FIGURE 2.12

Illustration of the Q/A Process Using Index Cards

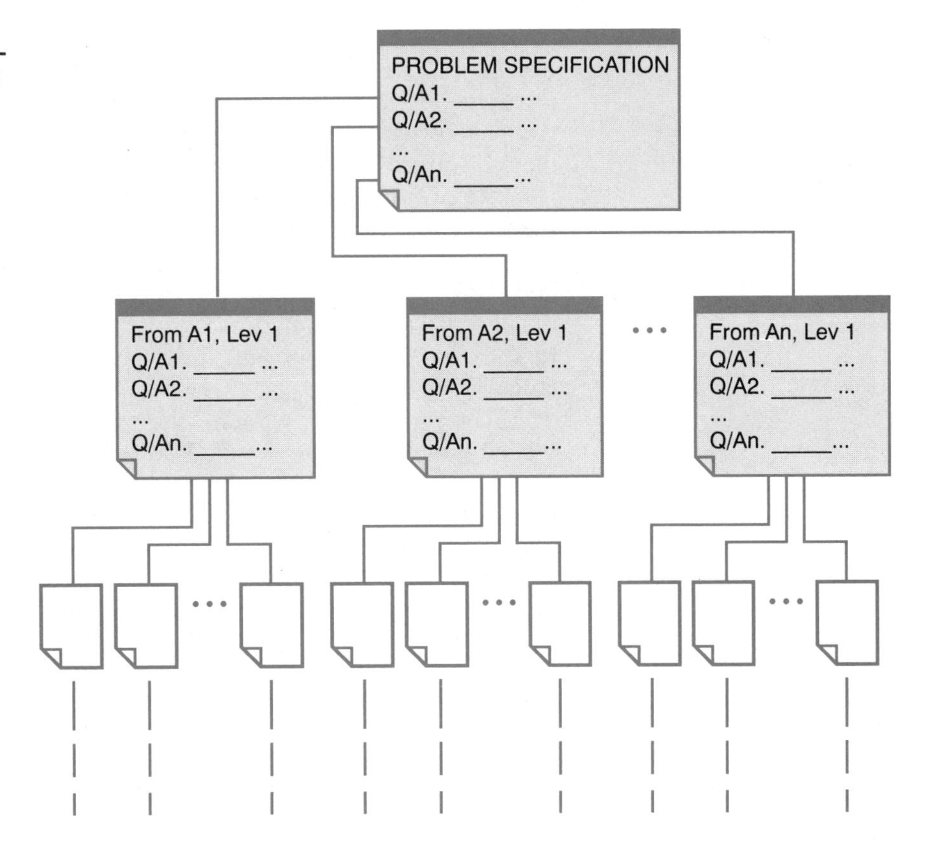

2.5 Summary

In this chapter we presented the idea of top-down design with some tools to help you implement it. Top-down design involves taking a large problem and breaking it up into a series of smaller, simpler problems. You repeat this process until you can solve all the smaller problems individually. We also presented the requirements of a program to collect data, process it, and return results. To help you refine your ideas into a creditable program, we introduced a question-and-answer scheme for fleshing out a problem. Using the technique requires a library of solutions or knowledge base.

It is common to organize the computer language commands into one of three groups. By using those classifications, you could record information learned as information maps, thus creating the library of solutions. We presented three classifications: sequential commands, control structures, and language structures. The sequential commands are the simple commands found in the computer language that do not cause any transfer of control. Those commands, such as branching or looping, are more powerful and control the execution of sequential commands. The third group, the language structures, are solutions that the programmer has previously solved. These are usually complete programs for solving small tasks. By combining these basic building blocks, you can assemble solutions to complex problems.

We presented a convenient means for you to store the information about the sequential commands and the control structures as an information map. This is a table of answers generated from a list of topics with a list of general questions. By using information maps, you can assemble a structure for organizing a library of solutions. Besides assembling information that you need to solve a current problem, it also creates a foundation for learning other languages.

The process that we have shown you may seem like a lot of work, and indeed it is. We intend this method to be most helpful to those who have no idea where to begin to develop a program. If you follow this process in all its detail, you should always be able to develop the algorithm. But if you feel that you have enough experience, you may want to skip the outlining stage. There are times, though, when even the best programmer will get stuck. You then have the outlining process to fall back on.

Exercises

For the following problems, identify the sequential, conditional, and looping operations. Also, construct a flowchart to illustrate the processes involved.

2.1 COMPUTING AN AVERAGE FOR A LARGE LIST OF NUMBERS The process of computing the average for a list of numbers involves counting the number of numbers entered, adding them together, and then dividing the total by the numbers of items in the list. For a large list of numbers, though, this process is somewhat difficult to do. Therefore, the following algorithm is more commonly used:

a. Add the first two numbers together and get an answer.
b. Take the previous answer and add the next number to it.
c. Repeat step b until no more numbers are present.
d. Divide the last value of the total by the number of numbers to get the average.

Construct the flowchart for the process described here.

2.2 STARTING A CAR Develop an algorithm to start a car. The algorithm should start with you sitting in the driver's seat with keys in hand. Be sure to indicate what procedure is necessary in the event that the car fails to start. Develop the algorithm to the extent that the car has started or that you need to call for additional help.

2.3 REGISTRATION FOR CLASS Develop an algorithm for the process of registering for class. Also, in many schools your term costs would be based on the number of credits taken. Therefore, include in your algorithm a method for computing the costs once registration is completed.

2.4 COMPUTING THE FACTORIAL Computers do not have the factorial function, $N!$, built into it. This is in stark contrast to a $5 hand calculator which does. Therefore, the factorial function must be done *in software*; that is, a program must exist which will calculate the factorial in order for it to be evaluated. In this problem you will construct an algorithm to compute $N!$. Recall the definition of the factorial:

$$N! = (N)(N - 1)(N - 2)(N - 3)\cdots(3)(2)(1)$$

Thus, for 7!, the expression to be evaluated is:

$$7! = (7)(6)(5)(4)(3)(2)(1) = 5040$$

Construct an algorithm for $N!$ by fleshing out the preceding steps. Your algorithm should take into consideration:

a. N must be a whole number. No fractional numbers allowed.
b. Under normal conditions $N!$ equals the product of N times each number less than N.
c. Special cases exist. When $N = 0$, $N! = 1$. Also, when N is negative, the factorial is given by $-(-N)!$.

2.5 APPROXIMATING THE SINE FUNCTION A very useful approximation for the sine function of the angle x measured in radians is given by the following series:

$$\sin(x) = \frac{x^1}{1!} - \frac{x^3}{3!} + \frac{x^5}{5!} - \frac{x^7}{7!} + \cdots$$

As an example, if $x = 0.7$, then each term in the series is:

$$\begin{aligned}\sin(0.7) &= \frac{0.7^1}{1!} - \frac{0.7^3}{3!} + \frac{0.7^5}{5!} - \frac{0.7^7}{7!} \\ &= 0.7 - 0.05717 + 0.001401 - 0.00001634 \\ &= 0.6442\end{aligned}$$

Note for the example given for $x = 0.7$ that the answer is correct to four decimal places after only three terms in the series. Write an algorithm to compute the approximation to the sine functions. Your algorithm should take into consideration the following:

a. A method that will calculate each term in the series and add it to the total until the total changes by less than 0.01%. Use this criterion to decide how many terms to use in the approximation. Thus, in the preceding example each term makes progressively smaller contributions:

Term	Value Series	Total Change	
1	+0.70000000	0.70000000	—
2	−0.05716667	0.64283333	−8.17%
3	+0.00140058	0.64423391	+0.22%
4	−0.00001634	0.64421757	−0.0025%

b. Note that the series includes the factorial, $n!$. Use the loop developed in the previous problem to calculate the factorial. (Note that this is an example of a LANGUAGE STRUCTURE where you can use the solution to an earlier problem).
c. The alternating series (+−+−) is best handled with the mathematical function $(-1)^{n-1}$ where n = nth term.

2.6 CHANGE FROM A DOLLAR Construct an algorithm that will determine the minimum number of coins returned as change from a dollar. For example, if a purchase were $0.43, the minimum number of coins is 4 (half dollar, nickel, and 2 pennies).

2.7 COEFFICIENT OF RESTITUTION When you drop a ball from a height, it will not bounce back to the original release point, but rather, to an intermediate point. The ratio (return height/original height), is related to the (coefficient of restitution, COR)^2. Each successive bounce will return to the same percentage of its original height. Thus, for a ball released from an initial height of 1 meter (m), with a coefficient of restitution of 0.836, the height of each bounce will be:

Bounce	Return Height (mm)
1	700
2	490
3	343
4	240
5	168
6	118
7	82
.	.
.	.
.	.

Construct an algorithm that will calculate the return height of the ball for each bounce, and determine how many bounces it takes for the ball to come to rest. Note that the ball never really comes to rest in this simple model. However, we may say that the ball is *effectively* at rest when the return height is less than a critical value, say, 0.001 mm. Your algorithm should allow for different heights, different CORs and different critical height values.

2.8 REFINED COEFFICIENT OF RESTITUTION The simple description of a bouncing ball in the previous problem is not very realistic since the coefficient of restitution is not constant. Actually, it will decrease after each bounce according to the approximation:

$$(COR\ CURRENT\ VALUE) = (COR\ OF\ PREVIOUS\ BOUNCE)^2$$

Thus, for the same 1-m drop height and initial COR of 0.836 of Prob. 2.7 the ball will bounce as follows:

Bounce	COR	Return Height (mm)
1	0.836	700
2	0.700	343
3	0.490	82
4	0.240	5
.	.	.
.	.	.
.	.	.

Note that the bounces damp out much more rapidly than the situation of the previous problem. Revise your algorithm of Prob. 2.7 to include the variable coefficient of restitution.

2.9 GRADE ASSIGNMENTS An algorithm often used for assigning course grades is as follows:

$$COMPOSITE\ SCORE = 0.45*(FINAL) + 0.35*(MIDTERM) + 0.20*(HOMEWORK)$$

where

FINAL = grade on the final exam,

MIDTERM = grade on the midterm exam, and

HOMEWORK = average grade on the ten best homework assignments.

Usually, there are 12 to 14 homework assignments, so that the lowest scores are dropped before determining the homework grade. Once the composite grade is determined, each composite score is *normalized* by dividing by the highest score in the course. In this way the normalized scores will be between 0 and 100%. From these normalized scores the final letter grades are determined according to the following schedule:

Normalized Composite Score	Letter Grade
90–100	A
80– 90	B+
70– 80	B
60– 70	C+
50– 60	C
40– 50	D
0– 40	F

In addition to these guidelines, students are given a second chance by comparing their midterm and final exam scores. If the student has done 25% better on the final exam as compared to the midterm exam, the grade is increased by an extra half grade (B to B+, for example). This encourages students not to be discouraged by a poor performance in the first exam. Construct an algorithm to implement this grading strategy. Use the following steps:

a. For each student:
   1. Enter final exam, midterm exam, and homework scores.
   2. Throw out lowest homework assignments.
   3. Calculate composite scores.
b. Find largest composite score and normalize all scores:
   1. Scan all composite scores and find the highest.
   2. Divide all composite scores by the highest.
c. Assign letter grades:
   1. Use above scale for grading.
   2. Consider special cases and change grade if necessary.

2.10 THREE-POINT SMOOTHING A simple way to remove random noise from data is to use a three-point smoothing procedure. The idea is very simple. All you need to do is to replace the value of a data point by the average of the original point plus its two nearest neighbor. Thus:

$$NEW\ VALUE = (OLD\ VALUE + RIGHT\ NEIGHBOR \\ + LEFT\ NEIGHBOR)/3$$

To see how this works, examine the following data before and after smoothing:

Old Data Point	New Data Point
11	—
19	15
15	20
27	28
42	35
36	44
53	46
50	48
40	43
39	35
27	29
20	—

Graph both sets of data and you will see that the second set is indeed "smoother." Construct an algorithm which will implement the three-point smoothing scheme. Note that there is at least one special case in the preceding example and that the first data point has only one "nearest" neighbor. So, you must develop some ideas on how to handle this special case.

2.11 RANGE OF A ROCKET A rocket ship excluding its fuel weighs 205,000 kilograms (kg), and it carries 150,000 kg of fuel. It consumes fuel at a constant rate of 7.2 kg/s, which produces a propulsive force of 2,000,000 Newtons (N). Note that as the fuel is burned, the total mass of the system (rocket + fuel) decreases. Consequently, the rate of acceleration increases. At the earliest stages, when the rocket is full, acceleration is sluggish. But at later times, when a portion of the fuel has been consumed, the acceleration has increased.

Because of the variable acceleration, it is somewhat difficult to calculate the range of the rocket with a simple closed form formula. One alternate way of determining the range is to calculate it numerically as outlined in the following procedure. Use the hints given here to construct an algorithm to calculate the expected range of our rocket. If the rocket is unable to reach the desired destination, a warning message should be printed:

a. Assume that for a time increment of 1.0 s of burn time the total mass of the system is constant.
b. Acceleration = (force from combustion/total mass of system).
c. Change in velocity = (acceleration *x* time period).
d. Total velocity = (previous velocity + change in velocity).
e. Change in distance = (total velocity *x* time period).
f. Total distance = (previous distance + change in distance).

CHAPTER 3

COMMON THREADS OF HIGH-LEVEL LANGUAGES

3.1 Overview

Before beginning our presentation of FORTRAN, we first want to present an overview of the common threads of high-level programming languages. Since most of you have previously studied programming (most likely BASIC or Pascal), you already know some of the fundamentals. Therefore, when we present a new FORTRAN topic, you will have something with which to compare it. If you do this often enough, you will find that there are many similarities among the different languages. We hope that a general discussion of this sort will identify the key ideas to look for, making it much easier for you to learn a second or third programming language.

Our second objective in this chapter is to review the general structure of a program. We do this to establish a framework for all that follows. The following eight chapters merely flesh out the basic elements discussed here by providing details and examples.

3.2 Elements of Programming Languages

In Chaps. 4 through 11 we will explore the basic elements common to most high-level programming languages. Few languages offer every feature shown in Fig. 3.1 and some languages are better than others in treating individual elements. But you will find that you can break down most languages such as FORTRAN, BASIC, and Pascal into these fundamental building blocks.

The analogy portrayed in the figure is an apt one. Textbooks, including this one, tend to focus on the individual elements, and students can lose sight

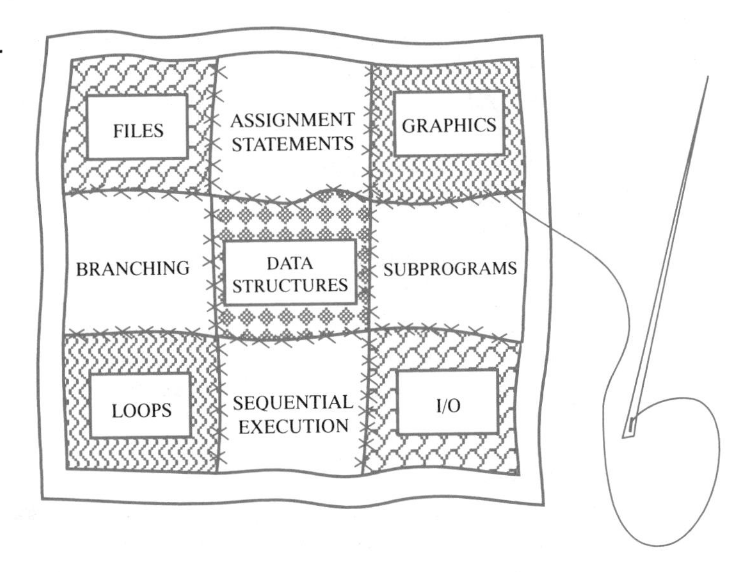

FIGURE 3.1

Common Threads of High-Level Languages

of the objective. Only when you combine all the elements in exactly the right manner, will you have a program. One advantage to organizing the elements of high-level languages into the various categories is that it makes learning a new language easier. In effect, Fig. 3.1 is in a form of an information map, similar to those that we introduced in Chap. 1. Whenever you need to learn a new language, you should prepare a list of the basic elements given in the figure and the associated subcategories. Next to this, list the appropriate commands in the language that you already know. As you obtain the equivalent information in the new language, add it to your map. We saw a simple example of this in Table 1.2, where we compared a BASIC segment with the equivalent FORTRAN segment. Also, we saw a more complete example in the exercises in Chap. 1, where we summarized the commands for the text editor, EMACS. Here is an example of how a small section of the information map might look for comparing equivalent BASIC and FORTRAN commands:

	SUMMARY OF LANGUAGE COMMANDS	
	BASIC	**FORTRAN**
COUNTED LOOP	**FOR** *counter=start* **TO** *stop* . . . **NEXT** *counter*	**DO** *label counter=start,stop* . . . *label* **CONTINUE**
CONDITIONAL LOOP	**WHILE** *condition is .true.* . . . **WEND**	**DOWHILE** *(condition is true)* . . . **ENDDO**

We have shown here two different types of loops—the counted loop and the conditional loop. The counted loop is available in all versions of BASIC and FORTRAN and executes a predetermined number of times. The conditional loop is a useful device, where the loop executes repeatedly until it satisfies a certain condition. Not all versions of BASIC and FORTRAN may support this useful device, however, and you must find other means to do the same thing.

The preceding loops do the same thing in either BASIC or FORTRAN. The only difference is the syntax. Therefore, once you have mastered one language, it is an easy matter with the aid of the map to learn the comparable structure in the other language. Using an information map approach forces you to think about categories of functions without focusing too heavily on the syntax. After completing the information map for all the other functions you will need, you will find the map to be an indispensable document until you become completely comfortable in the new language.

As you fill out your information map, you will find instances where your compiler does not support the desired function, but that other compilers do. In addition, you may find that the new language cannot do things that the old language could. Conversely, the new language may do things that the old language could not. Usually, however, you will find many similarities among the high-level languages. There are exceptions, of course, such as PROLOG and MACSYMA that follow few of the conventions discussed here. But you will find that most of the common languages you are likely to use in your career have much in common.

Each element of Fig. 3.1 may have several subcategories, but we will try to limit our present discussion to general ideas. The following paragraphs outline each idea and lay the groundwork for subsequent chapters.

DATA TYPES

Data are the most fundamental building blocks of a program. After all, the primary objective of all programs is to process data. For example, a program may base a person's pay on several pieces of data—the number of hours worked, tax bracket, number of dependents, and so forth. Based on this simple example, therefore, you might think that the idea of data is a simple one—and indeed it is. All that you need to know to calculate a person's pay is how to store and manipulate numbers on a computer. There are other forms of data, though, which are just as important as numbers. For example, we need a way to store a person's name. For this we need *character* data. Character data are *nonnumeric* data made up of things such as letters of the alphabet, special characters such as ←, $, #, @, !, and so forth. Thus, data types consist of many things besides the usual numbers that you may think of. In addition to numerical and character data, there are many other data types at which we need to look. Here's a short list of the simple types that we will examine in this book:

Constants and variables:	INTEGER
	REAL
	CHARACTER
	DOUBLE PRECISION
	COMPLEX
	LOGICAL
Simple data structures:	DATA STATEMENTS
	PARAMETER STATEMENTS
	ARRAYS

Further, there are several advanced data structures that we will not discuss in this book. If you want to learn more about these sophisticated structures, consult one of the references in the recommended reading list. The most widely used of these advanced data types are:

Advanced data structures:	STACKS
	QUEUES
	DIGRAPHS
	TREES
	LINKED LISTS

Notice in the preceding list that there are many different types of constants and variables. Recall that constants are numbers that do not change, whereas variables can take on any value during execution of a program. The first two types, *INTEGER* and *REAL*, are the ones that you already know as whole numbers and fractional numbers, respectively. Each is stored differently inside the computer and has special rules for storage and manipulation. As an example, if you divide two *INTEGER* numbers, the result also must be an *INTEGER*. This can result in very strange results sometimes. For example, division of the *INTEGERs* 3 by 2 gives an answer of 1, not 1.5. The true answer, 1.5, is a fractional number that the computer truncates to the whole number 1.

The third data type, *CHARACTER*, is occasionally used to store nonnumeric data. There is little manipulation to be done with *CHARACTER* data other than input or output. Occasionally, we may need to combine two *CHARACTERs*, search through a list of *CHARACTERs* or compare *CHARACTERs*. For example, we may need to scan a list of names and put them into alphabetical order:

Before Ordering	After Ordering
SMITH	ADAMS
JONES	JONES
ADAMS	KENNEDY
KENNEDY	SMITH
SMITHSON	SMITHSON

Putting the preceding list in alphabetical order requires that we have the means to compare two CHARACTERs to see which one is "less than" the other. For example, *ADAMS* is less than *JONES*. Similarly, *SMITH* is less than *SMITHSON*. The computer accomplishes this by referring to a *COLLATING SEQUENCE*, which is a convention that establishes the order for CHARACTERs. When we use only letters of the alphabet, there is little problem. But CHARACTERs also can be such symbols as # or &. So, how do we compare % to Q? The COLLATING SEQUENCE known as the ASCII code will tell us that % is 37th in the list and Q is 81st. Thus, % is less than Q. There are several other codes such as BCD and EBCDIC, the use of which can have an important impact upon the ordering. For example, if we ask, "Is *q* less than *Q*?" the answer is no in the ASCII sequence (113 vs. 81). Yet, the answer is yes in the EBCDIC sequence (44 vs. 71). If you refer to the Appendix, you will find listed the most common of these sequences.

DOUBLE PRECISION data types are *REAL* numbers, except that the computer stores them with a greater degree of accuracy. REAL numbers are usually stored with a finite number of significant digits that varies from machine to machine. In general, though, computers store REALs with seven to eight digits of accuracy. DOUBLE PRECISION numbers can generally store twice as many digits because they use twice as much memory. When we use DOUBLE PRECISION numbers, the computer will treat them differently from REALs. There are different mathematical procedures for DOUBLE PRECISION numbers and sometimes even different hardware is needed to ensure greater accuracy.

Engineers occasionally use COMPLEX data types in their calculations. These are numbers that contain both a real and an imaginary part. Thus, a COMPLEX number has two numbers associated with it—(3.2, 2.5); for example. The first number represents the REAL part and the second number is the IMAGINARY part ($2.5i$, where $i^2 = -1$).

LOGICAL data has only two possible values, *true* or *false*. Programmers often use them as flags or switches to change the course of a computation. For example, if a particular LOGICAL variable is *true*, then the program may proceed with the calculation. But, if the variable is *false*, the program may be told to print out an error message and stop the computation.

By far, the most important of the simple data structures is the array, which mathematicians refer to as a subscripted variable. It is a shorthand way to represent and manipulate large amounts of data. Arrays are organized in terms of their dimension. A one-dimensional array represents a list, a two-dimensional array represents a table, and so on for higher dimensions. Without arrays it is very difficult to manipulate anything more than just a few variables.

Try to think of an array as a single variable with many components. Mathematics describes these as a subscripted variable, which for a one-dimensional array looks like:

$$X_i$$

This represents a list of items. If we refer to $i = 3$, for example, we are referring to the value stored in the third location in a list, or $X_3 = 0.00$ for the list in Fig. 3.2. Similarly, if $i = 6$, we want the sixth item or $X_6 = 40.71$.

A two-dimensional array uses a similar arrangement, shown in Fig. 3.3 as a table of data. Of course, with a table of data, we need two subscripts to locate the desired item. The first index is used to locate the row and the other index locates the column. Thus, $X_{5,3}$ refers to the data in the fifth row and the third column of the table, or $X_{5,3} = 2.1$.

Most languages allow even higher-dimensional arrays. It is not uncommon for high-level languages to allow up to seven or more dimensions, although these are not often used. Continuing this line of thought, a three-dimensional array represents pages of tables; a four-dimensional array represents books of pages of tables of data, and so on.

The remaining two simple data structures, PARAMETER and DATA statements, are devices that allow you to enter values for constants and variables, respectively. This is sometimes useful when you use the same constant in many places thoughout the program. If you wanted to change the value of this constant, then you would need to change it everywhere in the program if you did not use these devices. However, if you replace the constant in the program with the variable, then you would need to change the value only once. To see this, compare the following two FORTRAN programs:

FIGURE 3.2

Illustration of One-Dimensional Array

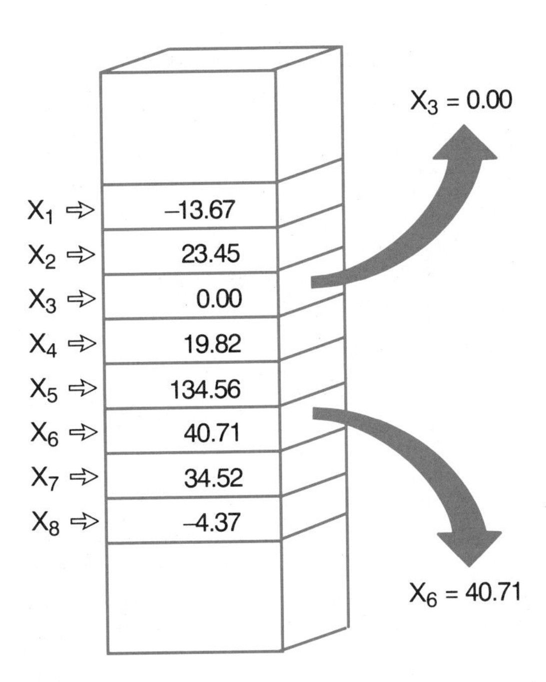

FIGURE 3.3

Illustration of a Two-Dimensional Array

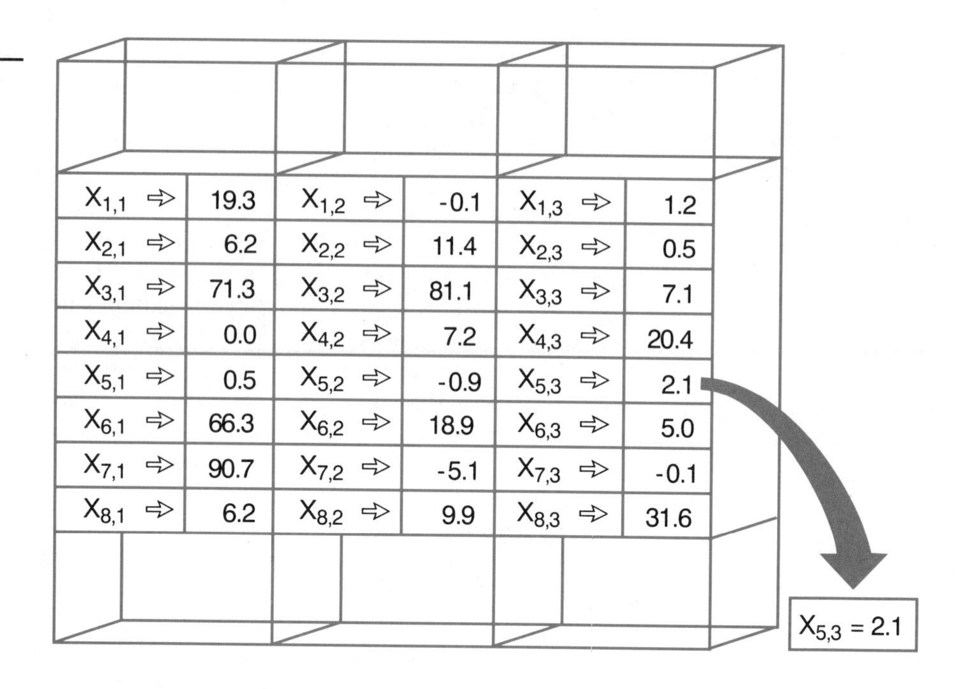


```
VOLUME = 1.333*3.14159*R*R*R
AREA = 3.14159*R*R
CIRCUM = 3.14159*R*2.0
```

```
PARAMETER (PI=3.14159)
VOLUME = 1.333*PI*R*R*R
AREA = PI*R*R
CIRCUM = PI*R*2.0
```

In these programs we calculate the volume of a sphere with a radius R, the area of the cross section and the circumference. Notice in the first program that we type in the value of π every time we need it, which is in every formula. In the second example we add one extra line at the beginning of the program, which defines the variable PI and gives it a value. After that point we use the variable PI instead of 3.14159. By doing this, we only need to make one change if we want to improve accuracy by changing π to 3.141592654 instead of to 3.14159. In the first program we would need to make three changes.

INPUT AND OUTPUT AND FILES

An essential element of high-level languages is the ease of passing data into and out of a program. Without such capabilities a program is essentially useless. One of the principal concerns of input and output (I/O) is the ability to control the appearance of the output, or formatting. For example, you may want to print data with a specific number of decimal places such as $1.98 instead of $1.979994. The two-digit number is the customary way of printing monetary amounts, while the second number is how the computer actually stores it. Notice also that the formatting rounds the number.

Besides the control of the appearance of individual data items, it is highly desirable to control the appearance of groups of data. Consider, for example, the following data printed without any attention to the format:

```
0.5000000E + 01   0.1240000E + 02   0.0000000E + 02   0.1000000E + 02
0.1273610E + 02   0.2710484E - 03   0.1500000E + 02   0.1305540E + 02
0.2507047E - 03   0.2000000E + 02   0.1334277E + 02   0.2201124E - 03
0.2500000E + 02   0.1358703E + 02   0.1830661E - 03
```

Now, we format the same data into a neat table:

TEMP (°C)	LENGTH OF BAR (CM)	EXPANSION (%)
5.0	12.400	0.000
10.0	12.736	2.710
15.0	13.055	2.507
20.0	13.343	2.201
25.0	13.587	1.831

Besides improving the appearance of the output, formatting makes it easier for you to scan the results of the calculation. You will find this an invaluable tool for checking the proper operation of your programs. For example, if all the data items in the third column were zero, you would immediately see that something was wrong. If you printed the same data with the first format, you would likely miss the error.

A second important element of I/O is the ability to direct output to a device other than the CRT screen. In particular, it is highly desirable to have access to data files from within a program. A data file is like any of those discussed in Chap. 1, except that its only content is data to be used by a program. Alternatively, the data file can be used to store the output of a program. Among the functions that a language should have in order to use data files are the abilities to:

Create a new file
Open an old file
Read from an old file
Write to a new data file
Append data to an old file
Skip around in a file
Close a file
Erase a file

Don't confuse these functions with the commands issued from within the operating system (OS). Our concern here is the ability to access files from *within* your application program. Accordingly, the commands probably will be different from those in the OS.

Most of these functions should be self-explanatory except the ability to skip around within the file. Sometimes the file contains thousands of data items, but we may wish to read only a very few of them. Here, it is essential that we be able to jump directly to the item. This saves us the need of reading all the thousands of items to get to the important one. We call this type of file a *DIRECT ACCESS FILE*. It is commonly used in business applications, but less frequently in scientific applications.

The other type of file is the *SEQUENTIAL FILE*. As its name implies, data in these files must be read in order without any skipping around. These files are easier to use and used more heavily in scientific applications than in business applications.

ASSIGNMENT STATEMENTS

The *ASSIGNMENT STATEMENT* is the primary means by which programs transfer data to variables in a program. These have the general form:

$$VARIABLE \longleftarrow EXPRESSION$$

The expression on the right-hand side may be as complex as a lengthy mathematical equation or as simple as a constant value. We use the symbol $\leftarrow$ to emphasize the flow of the data. Once the program evaluates the expression, its value is *assigned* to the variable on the left-hand side. This means that the numerical value of the expression will be stored in the memory location whose name corresponds to the variable.

High-level languages do not use the $\leftarrow$ symbol. Instead, they use = or := to suggest assignment. Sometimes the = sign creates confusion since students think of this as indicating an algebraic equation to be solved. As a result, the purpose of the statement may not be clear. To avoid confusion, you should read the = sign in a statement as "assigned to."

SEQUENTIAL EXECUTION

Computer programs execute statements sequentially. This means that the computer executes them in the order listed unless a branching instruction appears as we discuss later. Thus, the order in which we list statements is important. For example, these two program segments will give very different answers because of the ordering of the statements (note that * indicates multiplication):

a.	b.
$X \longleftarrow 2.3$	$Y \longleftarrow X * X$
$Y \longleftarrow X * X$	$X \longleftarrow 2.3$
$X \longleftarrow Y$	$X \longleftarrow Y$

In (a) the final value of X is 5.29, but in example (b) the final value of X is 0.0. Do you see why? The reason is that in the first line of (b), X is undefined since there was no previous assignment. Therefore, most computers will automatically assign a value of zero. Thus, the first line produces $Y = 0.0$. The second line now assigns a value to X, but it is too late.

BRANCHING

Branching is the second category of logic control that is fundamental to programming. It allows us to carry out a test at a critical point in our program. Depending on the results, control will then transfer or branch to a specific location within the program.

There are two general types of branches—the *UNCONDITIONAL* and *CONDITIONAL TRANSFER*. The *UNCONDITIONAL TRANSFER*, usually implemented with a form of **GOTO**, requires no testing. When the computer encounters this instruction, control will immediately transfer to the indicated spot. This is the easiest of the control structures to use. It also is the easiest to abuse. Overuse of the unconditional transfer results in convoluted logic within a program and results in code that is difficult to debug.

The second branching mechanism, the *CONDITIONAL TRANSFER*, (Fig. 3.4) is a powerful tool that has two general implementations. The first is the simple inverted Y structure shown in the figure. In this structure the computer performs a test in which the only possible outcomes are *true* or *false*. If the answer is *true*, then control goes in one direction. But if the answer is *false*, then control goes the other way. Most languages implement this *block* structure with the **IF-THEN-ELSE** structure of the form:

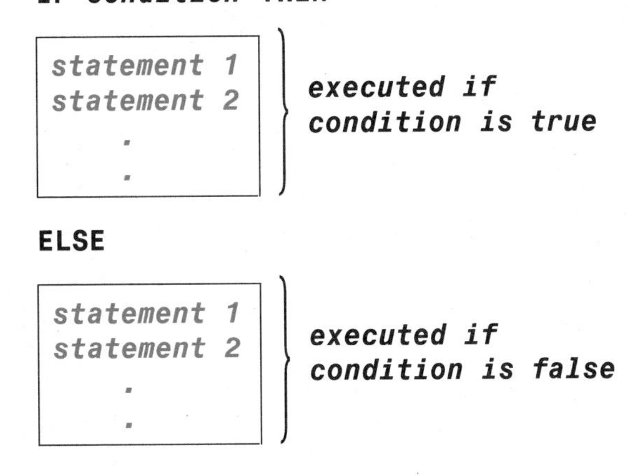

While this structure is very useful, it is somewhat limited since there are only two alternatives available. In some languages a more general branching

Figure 3.4

The Conditional Transfer

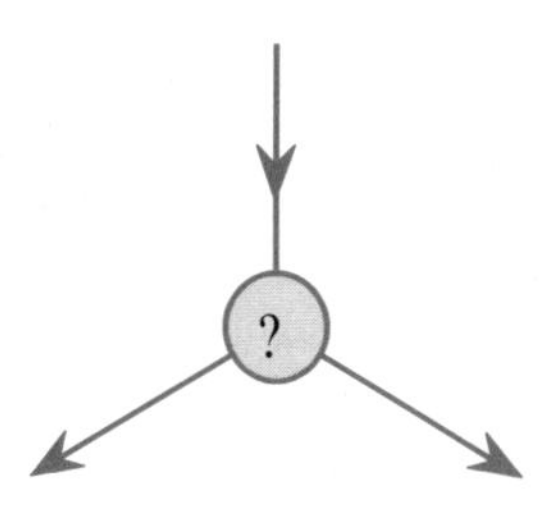

operation is available, called the SELECT-CASE structure. (See Fig. 3.5.) This structure allows you to set up as many alternatives as you like, but still only a single condition or question appears. In this structure the answer from the test can have a numerical, character, or logical value that matches the label for the block of instructions. Think of this as a switch where the switch can have many different positions, and once you set the switch, control will go in that direction. Languages that support this, do it with the form:

```
SELECT CASE (expression)
     CASE(selector 1)
┌─────────────┐
│instruction 1│
│instruction 2│
│      .      │
│      .      │
│      .      │
└─────────────┘
     CASE (selector 2)
┌─────────────┐
│instruction 1│
│instruction 2│
│      .      │
│      .      │
│      .      │
└─────────────┘
          .
          .
          .
END SELECT
```

Figure 3.5

The Select Case Structure

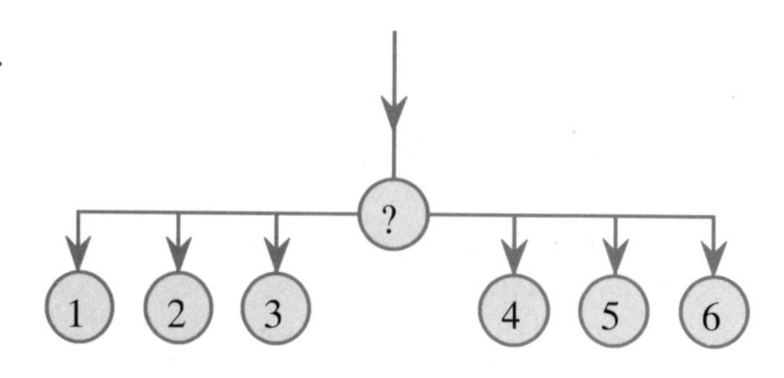

The idea of the SELECT CASE is very simple. The *expression* yields a value such as 1, 2, 3, or 4. These then direct the output to the corresponding blocks of instructions listed as specific cases. For example, if the expression evaluates to the number 6, then the instructions listed under CASE (6) will be executed. Note that this does not mean to transfer to the sixth CASE, but to transfer to the one with a label of 6.

Standard FORTRAN 77 compilers usually do not have the SELECT-CASE option, but different commercial versions may offer it as an extension. In addition, the new FORTRAN standard will include this structure as part of the language. Therefore, we introduce it at this time. Many other languages, such as Pascal, C, and Ada support this important function, but other languages like standard BASIC do not.

LOOPS

LOOPS are the third major type of control structure that we need to look at. There are two different types of loops, the *COUNTED LOOP* and the *CONDITIONAL LOOP*. As the names imply, a COUNTED LOOP repeats a predetermined number of times, whereas the CONDITIONAL LOOP repeats until a condition is satisfied. The schematic in Fig. 3.6 illustrates this.

The differences between these two structures are important. In the *COUNTED LOOP*, a counter keeps track of the number of times that the loop executes. Once the counter reaches the set number, 4, in Fig. 3.6 the loop terminates and the process will stop. The important point is that the limit for the number of loop executions is set before the loop begins and cannot be changed while the loop is running. The *CONDITIONAL LOOP*, on the otherhand, lacks a predetermined stopping point. Rather, each time it goes through

FIGURE 3.6

Loop Structures

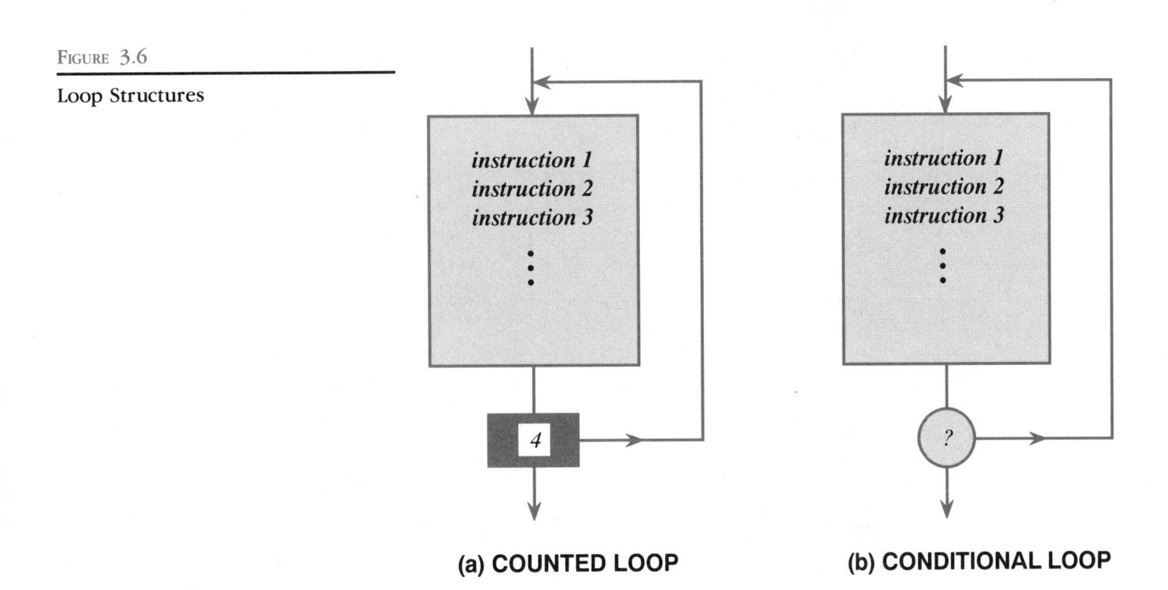

the loop the program performs a test, and if this test is satisfied, only then will the loop stop. Also, the thing to be tested can change (and, in fact, *must change*) while the loop executes. An example of this kind of test might be "Is today still Tuesday? If it is, then go back and continue doing homework."

Actually, there are two different forms of the CONDITIONAL LOOP. The first one asks the question at the beginning of the loop, whereas the second asks the question at the end (see Fig. 3.7).

We term these two loop structures the *REPEAT-UNTIL* and the *DO-WHILE* loops. The REPEAT-UNTIL loop has the condition evaluated at the end of the loop, and thus will always execute at least once. The second loop structure, the DO-WHILE loop, has the condition evaluated at the beginning. Thus, it is possible for the loop never to execute. In both cases, however, it is possible for the loop to execute indefinitely, so be careful when using either of them. Here's an example to show how you can get into trouble:

```
REPEAT
        CHECK ASSIGNMENT BOOK
        DO HOMEWORK ASSIGNMENTS
        DO LAUNDRY
UNTIL TIME > MIDNIGHT
```

If you executed these instructions, you would be caught in an infinite loop of homework assignments and laundry because there is no instruction to check the time. The proper structure should be:

FIGURE 3.7

Two Forms of the Conditional Loop

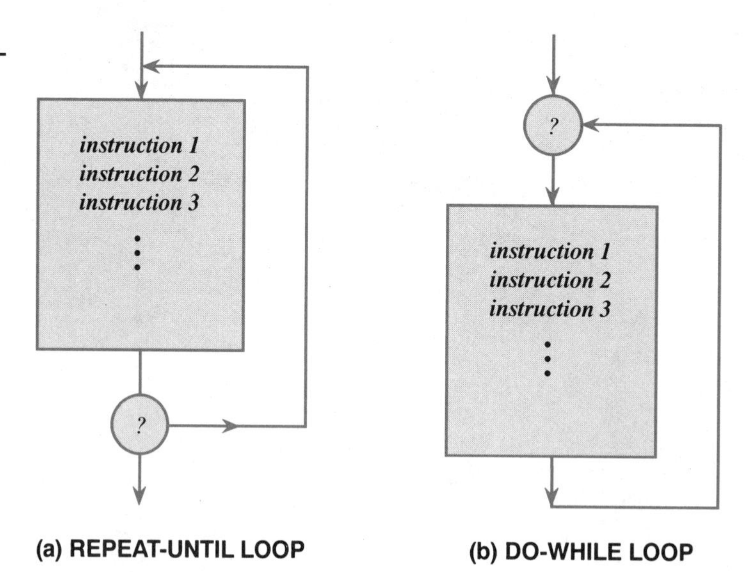

```
REPEAT
        CHECK ASSIGNMENT BOOK
        DO HOMEWORK ASSIGNMENTS
        DO LAUNDRY
        CHECK TIME
UNTIL TIME  >  MIDNIGHT
```

Notice that the test involves the variable TIME that changes while the loop executes. This is a key to using conditional loops. The control variable must have a mechanism for changing. Otherwise you will be trapped in an infinite loop. In the example just given we were fortunate that TIME is changing automatically. Thus, all you need to do is to check it occasionally so the test condition (TIME > MIDNIGHT ?) can eventually be satisfied.

Standard FORTRAN 77 includes only the counted loop and does not allow either of the conditional loops. Commercial vendors, however, sometimes have added one or both of the conditional loops to their compilers. So, look for these extensions in your compiler.

SUBPROGRAMS

An important idea in all programming languages is the ability to break a single large problem down into several smaller ones. By doing so, what started out as an intractable problem becomes a solvable one. Breaking the problem down into several smaller components allows you to focus on a more manageable part. This will reduce the defects in your program logic and improve your programming speed.

Every program consists of many identifiable subtasks. All that we are doing with subprograms is isolating them for closer scrutiny as Fig. 3.8 suggests.

Each task in the original program is broken out and written as an essentially self-contained program. Once these subprograms are written correctly, they can be linked together by a *MAIN* program. The primary purpose of the MAIN is to control the flow of data to and from the individual subprograms and to control the sequencing. Thus, the MAIN acts as a supervisor over the subroutines. To do this, we must have a means of transferring data back and forth and a means of calling the appropriate subprogram. In FORTRAN, for example, we give each subroutine a name, and the MAIN executes it by using the CALL statement followed by a list of variables:

```
CALL SORT (A, B, C, D)
```

The subroutine SORT is a separate program to sort a group of numbers, and the variables A, B, C, and D are the variables to be sorted. After the subroutine finishes with the sorting, the data returns to the MAIN, which then assumes control.

Subprograms are extremely useful for several reasons. First, they help you to break a large program into several smaller ones as discussed previ-

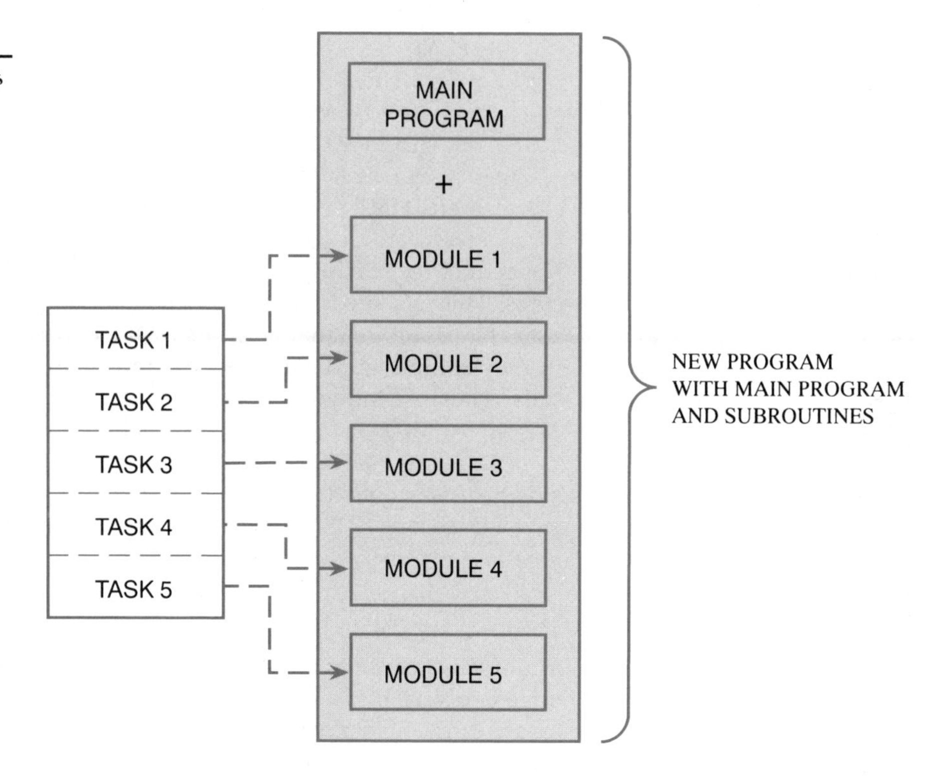

FIGURE 3.8

Schematic Drawing of Subprograms

ously. Secondly, subprograms can be used repeatedly with different sets of data without having to rewrite any code. Consider the following two CALLs:

```
CALL SORT (A, B, C, D)
CALL SORT (U, V, W, X)
```

which use the subroutine SORT twice, but each time with different sets of data. Both FORTRAN and Pascal are very good in transferring data between subroutines. BASIC, on the other hand, is very weak. Consequently, those of you who have a background in BASIC will find this aspect of FORTRAN quite different from what you have previously learned.

GRAPHICS

Graphics support varies widely between different languages. FORTRAN and Pascal, for example, do not support graphics at all, whereas BASIC has extensive graphics routines. Unfortunately, graphics commands vary widely from machine to machine and are strongly hardware-dependent. Even different commercial versions of the same BASIC language will have markedly different graphics commands. Thus, there are few universal guidelines on graphic functions. So, it does not make much sense to discuss this topic at this point. At various points in the text we may present ideas on graphics, but there will be no organized discussion.

Although FORTRAN itself does not support graphics, commercial software packages are available that can be used as subroutines in a FORTRAN program. Thus, for example, if you want to draw a circle, you make a call to the appropriate subroutine in the graphics package. But, again, these packages are very system-dependent, so if you need them you must consult the documentation that came with the package.

3.3 Summary

When learning a new computer language, it is always helpful to try to find elements in common with the language that you already know. Among the things to look for are:

- Simple data types—Constants and variables
- Simple data structures—Arrays
- Complex data structures
- Input and Output
- File access
- Assignment statements
- Control structures
  - Sequential execution
  - Branching—Conditional and Unconditional
  - Loops—Counted and Conditional
- Subprograms
- Graphics

We strongly suggest that you create an information map similar to the one shown earlier in this chapter. Include in it all the preceding elements from the language that you already know. Then, as we introduce you to the equivalent structure in FORTRAN, add it to your map. This function serves two purposes. (1) It will refresh your memory about your previous experience. (2) It will force you to find the parallel structure between FORTRAN and your current language.

Further Reading

1. S. L. Edgar, *Advanced Problem Solving with FORTRAN 77,* 1st ed., Science Research Associates, Chicago, 1989.
2. M. Metcalf and J. Reid, *FORTRAN 8x Explained,* 1st ed., Clarendon Press–Oxford, Oxford, England, 1987.
3. C. F. Taylor, Jr., *Master Handbook of Microcomputer Languages,* 2nd ed., TAB Books, Blue Ridge Summit, PA, 1988.

CHAPTER

4

DATA TYPES AND ASSIGNMENT STATEMENTS

4.1 Organization of a FORTRAN Program

Lines within a FORTRAN program have a very well-defined organization, as we show in Fig. 4.1. A standard line of FORTRAN has 80 columns, which coincides with the width of most computer terminals. We reserve some of these columns for the special functions summarized in the table:

Column	Purpose
1	A C or an * placed in column 1 shows that what follows on that line is a *comment* and is not a part of the program instruction. As a result, the compiler ignores comments.
1–5	We reserve these columns for *statement labels*. Statement labels are integer numbers between 1 and 99999 that are used to mark a line for reference. For example, the FORTRAN statement, GOTO 1020, would transfer control to the line with statement label 1020.
6	Typing any character in this column, other than zero or a blank, indicates that this line is a *continuation* of the previous line. You will often use this when long mathematical expressions cannot fit into the allotted space in an 80-column line.
7–72	Reserved for the FORTRAN statements. The line can begin or end within this region since spaces are ignored.
73–80	Ignored by the compiler during translation into machine language. May be used for comments, but usually left blank.

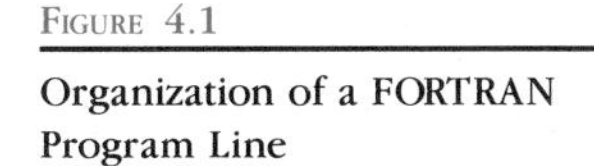

Figure 4.1

Organization of a FORTRAN Program Line

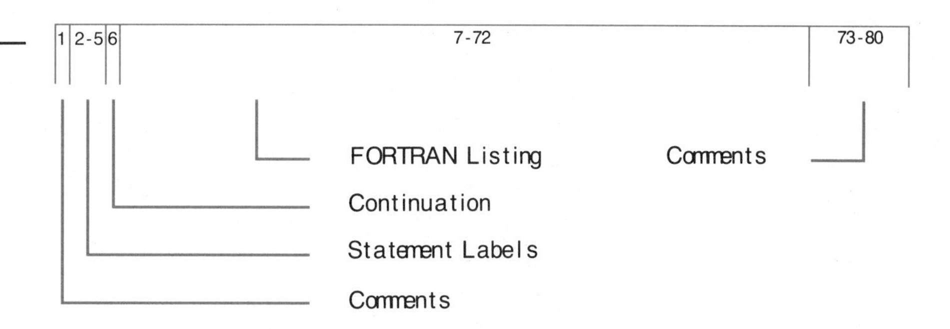

The following example in Fig. 4.2 shows a simple FORTRAN program that illustrates all the preceding rules. We demonstrate the use of comments in two different ways. The first four lines show the most common way, which is to place a C in column 1 and use the entire line for the comment. A short paragraph of comments is frequently placed at the beginning of each program to describe the function of the program, meaning of the variables, and so forth. Note also that comment statements can be placed within the body of the program, as illustrated in this example. Comment statements can be of any length and are not compiled. So, we encourage you to use them liberally throughout your programs to improve your documentation.

Subsequent lines of the program illustrate another way of inserting comments by using columns 73 to 80. Obviously, only short comments can be placed in this small space. Usually, programmers leave these columns blank,

Figure 4.2

Example Program Showing Line Organization

```
1 2-5 6                            7-72                                 73-80

C
C  LINE 003 READS IN THE VALUES A,B,C,X & Y AND COMPUTES THE VALUE
C  FOR Z ACCORDING TO A COMPLEX FORMULA IN LINES 5 & 6. IT THEN PRINTS
C  OUT A VALUE FOR Z ON LINE 007.
C
      REAL A,B,C,X,Y,Z                                           LINE 001
   10 PRINT*, 'ENTER VALUES FOR A,B,C,X,Y'                       LINE 002
      READ*, A,B,C,X,Y                                           LINE 003
      IF(A.EQ.0) GOTO 10                                         LINE 004
C
C  COMMENTS CAN GO ANYWHERE IN THE PROGRAM. LINE 002 ABOVE PRINTS A
C  PROMPT ON THE SCREEN TELLING YOU TO ENTER DATA. IT THEN READS IN
C  VALUES. IF A EQUALS 0 (LINE 004), THE EQUATION WILL NOT WORK, SO YOU
C  MUST RETURN TO LINE 002 TO REENTER THE VALUES. NOTE THAT THE
C  STATEMENT GOTO 10 INDICATES THE LINE WITH THE STATEMENT LABEL 10,
C  NOT THE 10TH LINE OF THE PROGRAM.
C
      Z=A*(B–C)*X+2.0*(X–Y)*(A–B)–37.2*(A–X)*(Y–B)*A*B           LINE 005
     1         *6.4/A+2.9**(A–C)                                 LINE 006
      PRINT*, 'Z=', Z                                            LINE 007
      STOP                                                       LINE 008
      END                                                        LINE 009
```

but some choose to enter line numbers as the following example shows. Note that FORTRAN does not require line numbers (unlike BASIC). We use them only for easy identification.

The program itself begins on the sixth line with the *declaration statement* REAL A, B, C, X, Y, Z. We will discuss the purpose of this statement shortly. Note that this line and all other program lines begin in column 7. In most editors the <TAB> key will automatically move the cursor over to the proper column. Note that line 002 has a statement label in front of it. This label must appear in columns 1 to 5. As we will see in a moment, another line in the program will refer to this label. Later in the program the instruction GOTO 10 appears. This instructs the computer to transfer control back to the line with the label number 10. Note that this does not mean (as it does in BASIC) to transfer to line number 10.

Finally, line 006 of the example contains a continuation mark in column 6. Thus, the equation defining the variable Z stretches over two lines since it is too long to fit on a single line. Also, you may use as many as 19 continuation lines. But be careful in placing continuation marks since the compiler will interpret them as statement labels if placed in the wrong column!

Most programs including the example in Fig. 4.2 are organized into the following main sections:

Declaration statements
Data input
Data manipulation
Data output
Termination

The beginning of your program must contain all the declaration statements, if any. These statements tell the compiler how to treat each variable named. Thus, the declaration statement, REAL A, B, C, X, Y, Z, lists all those variables that are to be treated as *single precision real numbers*. Typically, this means that each number will be stored with seven significant digits, although this will vary from machine to machine. If we had used instead DOUBLE PRECISION A, B, C, X, Y, Z, each variable would be stored with approximately 14 significant digits. There are many other types of declaration statements not yet discussed which will appear in this section of every program. Note, however, that *all* declaration-type statements must appear before any other statements other than comment statements.

The program uses lines 002 to 004 to input the desired data. The first line is a *prompt* since its only function is to print a message on the CRT screen reminding the user what data to type in. Thus, the line PRINT*, 'ENTER VALUES FOR A, B, C, X, Y' would cause the following instruction to appear on the screen:

ENTER VALUES FOR A, B, C, X, Y

to which you would type in the appropriate values such as:

1.23, 2.45, 3.46, 1.02, 1.99 <CR>

This line would result in assigning a number to the corresponding variable as illustrated here:

$A \leftarrow 1.23$

$B \leftarrow 2.45$

$C \leftarrow 3.46$

$X \leftarrow 1.02$

$Y \leftarrow 1.99$

The final part of the input section of the program is the statement:

```
IF(A.EQ.0) GOTO 10
```

which you read as "If A equals zero, go to statement label 10." This statement checks to see if you accidentally entered zero for the value of A. Note that if you did, a forbidden mathematical operation (division by zero) would occur in line 006. Thus, to head off this problem, we check the value of A first. If you did enter A incorrectly, the program returns to the beginning of the input section and asks for the data again. Thus, a well-designed input section should consist of three parts—prompt, input, and verification.

Once you successfully enter the input data, the program will go to the data manipulation section, lines 005 to 006. Here, the program calculates the value of Z according to the formula given on these two lines. In this example this section of the program is only two lines long. In extreme cases this section of the program might be several thousand lines long with complex loops and branches.

The program uses the output section, line 007, to print out a message with the numerical value of Z. Thus, for the values of A, B, C, X, and Y given earlier, the printout on the CRT screen would look like:

$Z = -55.15797$

The print statement (PRINT*, 'Z=', Z) prints out the message inside the quote marks intact, followed by the numerical value of this variable.

Finally, the program terminates when it comes to the two statements:

```
STOP
END
```

These two statements may appear to have the same function, but the program uses them quite differently. The *END* statement *must* appear at the end of every program and is used by the compiler to mark the point at which the compilation for a program segment terminates. The computer uses the *STOP* statement, on the other hand, at execution time to stop a computation. Unlike the *END* statement, one or more *STOP* statements may be used anywhere in a program. Thus, for example, you may want to stop execution if a particular set of conditions arises. Other times, you will not need the *STOP* statement, and so it is optional. Note, however, that you must always have an END statement.

4.2 Constants

Constants are numbers within a program whose value do not change. In theory, constants can take on any value, but because of the limited storage capacity of a computer's memory, there are practical limits. In addition, different types of constants correspond to the different data types. For example, there are data types for storing whole or fractional numbers, but there are also data types for storing *character strings* or complex numbers with real and imaginary parts. We give the rules for each with several examples in the following sections.

INTEGER CONSTANTS

INTEGER constants are positive or negative whole numbers written without either decimal points or commas. They may range between approximately -2^{31} and $+2^{31} - 1$, although this may vary from system to system. The CRAY supercomputer, for example, has a range of -2^{63} to $+2^{64} - 1$. In either case the range of INTEGER constants is generally large enough for most applications and you will not need to worry about the limits of integers. The following table lists some examples of INTEGER constants showing legal and illegal use:

Valid	Invalid	Comments
+174		Plus sign optional
−23		Negative sign required
111 111 111		Spaces ignored; useful for entering large numbers
	111,111,111	No commas
	+174.0	No decimal points
	−7 1/2	No fractions

REAL CONSTANTS

REAL constants are those numbers that we think of as fractional numbers, which may be either positive or negative and always have a decimal point. Also, we tend to write REAL numbers with commas in our everyday writing, but the FORTRAN language prohibits commas. Examples of proper and improper REAL constants are shown in the table:

Valid	Invalid	Comments
+13.7		Positive sign optional
−21.4		Negative sign okay
0.0034		Small numbers okay
123 456.0		Spaces ignored
	$ 1.23	Only numbers permitted
	0	Even 0 needs decimal point
	123,456.0	No commas
	π	Actual value must be entered

You can also write *REAL* constants in *scientific notation*. Recall that this is a convenient method for noting either very large or very small numbers. It relies on the use of a *mantissa* and an *exponent* that is a power of 10:

$$<mantissa> \times 10^{<exponent>}$$

For example, you can write the number 0.0034567 in scientific notation as:

$$0.34567 \times 10^{-2}$$

Note that the mantissa is a number less than 1 and the power of 10 adjusts accordingly. This method is especially useful when the number to be represented is extremely large or small. Without scientific notation, Avogadro's number (0.6023×10^{24}) would be written as:

602300000000000000000000.

To use scientific notation in FORTRAN, we find that the method is similar except that the letter E substitutes for the power of 10. Thus, the general form is:

<mantissa> E <exponent>

which for Avogadro's number results in:

0.6023E + 24

Here are some examples of correct and incorrect use of the exponential notation:

Valid	Invalid	Comments
0.625E − 23		Negative exponents okay
−0.321E 42		Negative mantissa okay; omit + sign on exponent optional
−0.0321E 41		Equivalent to previous example
0.0E 0		That's right, zero!
	0.241E − 12.5	Exponent must be whole number
	123E 23	Mantissa must have a decimal point

The method of storing a number on a computer limits the accuracy and magnitude of REAL constants. This limits the accuracy of REAL constants to approximately seven significant digits and the magnitude to the range of 10^{-39} to 10^{+38}. However, both limitations are very machine-dependent, so check with your system operator. The CRAY supercomputer, where the range of REAL constants is 10^{-2466} to 10^{+2466}, is an extreme example of this variability.

DOUBLE PRECISION CONSTANTS

The way that computers store fractional numbers places severe restrictions on the accuracy of numbers. In most systems the limit of accuracy is seven significant digits, although the maximum number that can be stored is much larger than this. To illustrate how this presents difficulties, consider the following problem of how two large numbers might be stored:

Entered on Keyboard	Stored Internally
12345678909876543.21	0.1234567E 19
12345678900000000.00	0.1234567E 19

Although we went to the trouble to type in all those extra significant digits in the first number, the computer stores both numbers internally as the same. In most situations an accuracy of seven digits is adequate. But in many situations this loss of accuracy is intolerable. For such situations *DOUBLE PRECISION* is required. As its name implies, this increases the amount of memory space allocated to store a number, thus increasing its accuracy. In most computers, the use of *DOUBLE PRECISION* results in an accuracy of 14 to 17 significant digits instead of the 7 to 8 with *REAL* constants. Note in the previous example that even *DOUBLE PRECISION* does not store the two numbers without loss of some least significant digits. This is not a serious problem since there are few applications that require greater precision.

DOUBLE PRECISION is very simple to use. It is similar to the exponential notation described earlier. Instead of using E for the exponent, DOUBLE PRECISION uses the letter D:

<mantissa> E <exponent> *single precision*

<mantissa> D <exponent> *double precision*

Here are some examples:

Valid	Invalid	Comments
0.0D 0		Okay for precise zero
0.23D − 178		DOUBLE PRECISION may give greater range, thus exponent of −178 is within range on many systems
	0.123456789E + 23	Not DOUBLE PRECISION, even with extra digits

There are two penalties for the privilege of using DOUBLE PRECISION. As you might expect, you will need additional memory. But, also, mathematics involving DOUBLE PRECISION requires two to ten times the computation time of an equivalent expression with REAL constants. Therefore, you should be careful to use DOUBLE PRECISION only when absolutely required.

CHARACTER CONSTANTS

There are many occasions when we need to work with nonnumerical data. An example would be a list of names or addresses. Accordingly, we will use a different type of constant, the *CHARACTER* constant. A CHARACTER constant is any set of the allowed symbols enclosed in single quote marks. The only allowed symbols in FORTRAN are:

Letters of the alphabet (upper- or lowercase)
Numbers 0 to 9
Special characters + − ( ) . , * / = ' $
Blank space

While you can generate other symbols on your computer terminal such as Ç ¥ or π, FORTRAN does not permit these as part of a CHARACTER constant. Here are some examples:

Valid	Invalid	Comments
'Helen'		Mixing upper-/lowercase okay
'PROBLEM1'		Mixing letters/numbers okay
'I"M OK'		If you want an apostrophe inside quote, must use " instead
'123456'		All numbers okay
	"Helen"	Single quote mark (apostrophe)
	Helen	Missing quote marks, even if this is obviously a character constant
	'I ♥ NY'	Illegal character (♥)

Be sure that you do not confuse CHARACTER constants such as '123' with their numerical counterpart, 123. Two INTEGER constants 123 and 456 can be added, for example, but their equivalent CHARACTER constants, '123' and '456', cannot.

COMPLEX CONSTANTS

Engineers often need to use *COMPLEX* numbers in their calculations. You may recall that COMPLEX numbers consist of two parts—REAL and IMAGINARY. Computers cannot work with imaginary numbers since this requires a square root of a negative number. Therefore, programmers have developed a convention where a COMPLEX constant contains two REAL constants such as:

(REAL1, REAL2)

The first number represents the real part, whereas the second number represents the imaginary part. Thus, the FORTRAN COMPLEX constant (1.23456, 9.8765) is equivalent to the standard notation with which you may be familiar:

$$(1.23456, 9.8765) = 1.23456 + 9.8765i$$

where, of course, $i^2 = (-1)$.

The rules for using COMPLEX constants are not standard since this is not a part of the FORTRAN standard adopted in 1977. Yet, most commercial versions of the FORTRAN language use a similar set of rules in defining COMPLEX. The following list summarizes these:

Valid	Invalid	Comments
(1.23, −3.45)		Either component may be negative
(+1.23, 0.0)		Positive sign is optional; equivalent to REAL 1.23
(1.23E − 2, 3.45)		Exponential format okay
	(1.23D − 128, 3.45)	Both numbers must be either REAL or DOUBLE PRECISION
	(12, 30)	REAL only, no INTEGERS

Note that the parentheses and comma are always required. Specific rules for using COMPLEX data types may vary widely from one version of FORTRAN to another. So, be sure to check the local rules on your system.

LOGICAL CONSTANTS

The final data type permitted by FORTRAN is the *LOGICAL* constant. There are only two of them, so the rules are simple. In programming, we must set up questions so that the only possible outcomes can be one of the two LOGICALS:

.TRUE.

.FALSE.

Each LOGICAL constant begins and ends with a period and the full word (true or false) is spelled out. Otherwise, there are no other rules. You may use these constants only with branching operations. An example would be:

ARE WE FINISHED WITH CONSTANTS YET?

If the answer to that question is .TRUE., we can move onto the next section. Actually, the answer will be .TRUE. after we give you a list of correct and incorrect usage:

Valid	Invalid	Comments
.True.		Mixing upper-/lowercase okay
	TRUE	Periods required
	.T.	Must spell out complete word

4.3 Variables

Variables are another way of introducing data into your programs besides using constants. When you studied algebra, you learned that variables could be used to represent a quantity to be manipulated by a mathematical formula. In programming, variables have this purpose also, but you use them to show a position in the memory where the computer stores a quantity. Thus, if the value stored in that particular memory cell changes, then the value of the associated variable also changes. By contrast, constants do not change their value. Consequently, variables offer much more flexibility, and therefore are the primary tool for performing calculations.

A common way to consider this process is to think of the computer's memory as a series of mailboxes (Fig. 4.3). Each mailbox will have a unique identification, and whenever we summon the information within that box, a numerical value will be returned. Thus, if we have a box named JONES, the value stored in the location is synonymous with the variable name. For exam-

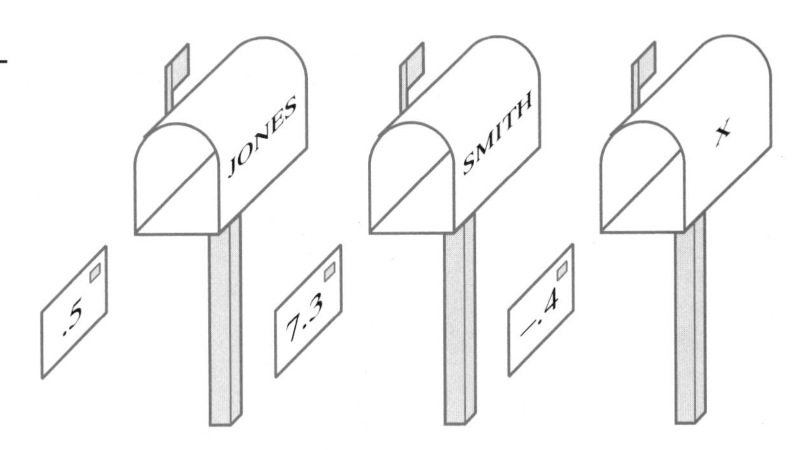

FIGURE 4.3

Schematic Representation of Variables

ple, if we store the value +5.0 in JONES, we could use either one in an equation such as:

```
Z = 13.0 + JONES
```

Here, the computer will go to the memory location called JONES, retrieve the numerical value stored there, and then add it as indicated. The result is, of course, Z = 18.0.

VARIABLE NAME CONVENTIONS

When you begin to name variables, try to construct names that show their function in the program. Avoid overuse of names like X, Y, or Z. This is a throwback to algebra where you may use only a few variables. In a lengthy program there may be hundreds of variables and it is easy to forget what each variable represents. You will be far better off to spend a little bit of additional time to create variable names that show their function. For example, in a program to calculate workers' hourly pay, good choices for variables would be HOURS, PAY, or TAX. The following list summarizes the rules for naming FORTRAN variables:

1 to 6 characters long
Only letters (*A* to *Z*) and numbers (0 to 9) allowed
First character must be a letter
Upper-/lowercase equivalent
Blank spaces ignored

Here are some examples:

Valid	Invalid	Comments
X		Okay, but not very illustrative
TAXDUE		Better since it describes function
TEMP1		Okay to mix letters/numbers
AMT DUE		Spaces ignored
Amt Due		Lowercase treated same as uppercase. Thus, this variable same as previous one.
	AMOUNTDUE	Too long—last three characters ignored
	AMOUNTOWED	Same as previous case. Characters after 6 are ignored. Thus, this variable is same as previous one
	$OWED	Illegal character ($)
	2BEES	First character must be a letter

Some high-level languages allow you to use many more than six characters in constructing a variable name. But in our opinion six is more than adequate to construct several hundred unique variable names. We cannot overemphasize the usefulness of using descriptive variable names. Use of descriptive names is an important guideline that we will introduce in a later chapter for debugging programs.

NEED FOR TYPING VARIABLES

In the previous discussion of the use of variables both the variable and the numerical constant must be of the same data type, also called attributes. That is, if the value stored is REAL, then the variable storing it also must be REAL. Similarly, if the value is a LOGICAL, the variable also must be LOGICAL. When we try to mix constants and variables with different attributes, trouble often occurs. So, you must take care that:

TYPE OF VARIABLE = *TYPE OF CONSTANT*

The reason the variable type must match the constant type is that each data type has different storage requirements. In effect, each has a different size mailbox. DOUBLE PRECISION requires the most amount of memory and LOGICALs require the least. If we did not match the type, then we may try to store a constant in a memory location not large enough to receive it.

In the previous section we showed that REAL, DOUBLE PRECISION, COMPLEX, CHARACTER, and LOGICAL constants each had a unique way to be identified. REALs, for example, used a decimal point, DOUBLE PRECISIONs used the D format, COMPLEXes used parentheses, CHARACTERs used apostrophes, and LOGICALs used double periods.

1.23456	(REAL since it uses decimal point)
0.12345E + 12	(REAL since it uses E format)
0.12345D + 12	(DOUBLE PRECISION since it uses D format)
'123456'	(CHARACTER since it uses ' ')
(1.2345,9.8765)	(COMPLEX since it uses (.., ..))
123456	(INTEGER since it is none of the above)

Thus, by looking at a number, you could tell immediately what type it is. No other identification was necessary. With variables, however, it is not so straightforward. There are two different methods available for us to inform the computer about the type of each variable. The first method uses *IMPLICIT TYPING*, which you use only for REAL and INTEGER variables. In IMPLICIT TYPING a variable will be treated as either REAL or INTEGER, based on the first letter of the variable name. You need to do nothing else. In the second method, *EXPLICIT TYPING*, the programmer tells the compiler directly at the beginning of the program how each variable is to be treated.

IMPLICIT TYPING

In the earliest versions of FORTRAN there were only two types of variables, REAL and INTEGER. All variables that began with the letters A to H or O to Z are REALs. All other variables, those that began with the letters I to N, were INTEGERs. Thus, it was always easy to identify the two variable types by looking at the first letter:

```
AMTDUE     (REAL)
ICOUNT     (INTEGER)
SUM        (REAL)
MONEY      (INTEGER)
```

IMPLICIT TYPING rules are still in effect, even after several revisions of the FORTRAN language. Thus, you need to do nothing else if these typing rules are adequate for your problem. There are times, though, when you may prefer that a variable, which is usually an INTEGER, should be a REAL. A good example comes from the preceding list where MONEY is an INTEGER according to IMPLICIT TYPING. If you now want this variable to become a REAL, you need to inform the compiler of your change. To do this, you must use the EXPLICIT TYPING instruction.

EXPLICIT TYPING

If you want a specific variable (or a group of variables) to be treated as a specific type, the general statement to be used is:

type variable1, variable2, variable3,...

where *type* can be any one of the six types discussed so far followed by a list of the variables concerned. These *type* statements are examples of *declaration statements* and as such must come first in the program. They must come before any *executable statement* so the compiler knows how much memory to reserve for each before it tries to manipulate any of them. Here are some examples:

```
REAL X , Y , Z         (Okay, but redundant. X, Y, and
                         Z are already REAL)
REAL I , J , K         (Okay; I, J, K usually INTEGERs,
                         but now REAL)
DOUBLE PRECISION I
LOGICAL ANSWER
INTEGER X , Y , Z
COMPLEX PHASE
```

Note in these examples that you can declare more than one variable in the same statement. The only potential pitfall in declaring variables is if you attempt to declare a variable twice in two different ways. For example, the following will create problems since the variable TAXES was defined twice, once as REAL and once as DOUBLE PRECISION:

```
REAL TIME, ICOUNT, PAY, TAXES
DOUBLE PRECISION TAXES, AMTDUE
```

Declaring CHARACTER variables requires that you supply more information than that in the preceding examples. When declaring a variable as CHARACTER, you must list the variable with the maximum number of characters for each. For example, assume that the variable NAME will have up to 32 characters and that the variable CITY will have up to 20 characters. Then the proper declaration statement is:

```
CHARACTER NAME*32, CITY*20
```

Each variable named in the statement must have its anticipated length expressed in the format, **(size)*. There is a shortcut that you will find useful if all variables in the list are to be the same size. Here, the *(size) is moved to the beginning:

```
CHARACTER*40 NAME, CITY, STATE, ZIP, PHONE
```

You must always use declaration statements for DOUBLE PRECISION, COMPLEX, CHARACTER, and LOGICAL variables. For REAL and INTEGER variables, though, declaration statements are optional if you are satisfied with the rules of IMPLICIT TYPING.

4.4 Assignment Statements

The assignment statement is the primary means of loading data into variables. As its name implies, we are telling the computer what to *assign* to a given variable. The most general form of the assignment statement is:

TARGET ⟵ VALUE FROM AN EXPRESSION

The way that we read this statement is "the target receives the numerical value obtained from an expression." The ← shows the flow of the value to the target, which here is simply a variable. The way that this is implemented in FORTRAN is:

VARIABLE = VALUE FROM AN EXPRESSION

In FORTRAN, the equals sign replaces the ← sign. This sometimes creates confusion because students think that this is an equation to be solved rather than a simple transfer of data. Some computer languages use symbols such as := or ← to avoid this problem symbolically, but in all languages the assignment statement has the same meaning.

The expression on the right-hand side (RHS) of the = or the ← signs can be one of several types, exampes of which are:

Constants	`(PAY = 5.12345)`
Variables	`(TAXES = CALC)`
Expression	`(PAY = GROSS - NET + 5.00)`
Functions	`(X = SQRT(Y))`

In the first example the program assigns the REAL constant 5.12345 to the variable PAY. The second example would first retrieve the numerical value stored in the memory location called CALC and then assign this value to the variable TAXES. The expression shown in the third example is a combination of variables or constants linked by mathematical *operators*. In the example shown the value for NET would be subtracted from the value for GROSS, after which the constant value of 5.00 would be added. Once completed, the program assigns the result to the variable PAY. In our final example expressions also may contain functions. These functions are comparable to the function buttons on a handheld calculator except that a computer does not have a physical way to access these functions. Instead, we access them through software instructions similar to the one shown. Usually, the desired function is already available, as in the example of a square root function. Yet, sometimes the desired function may not be available, and you must write it yourself. We will discuss these in more detail later. In the fourth example the program gets the numerical value of Y and takes the square root. The result is assigned to X.

Note in all the assignment statement examples given the expression is always on the right-hand side of = and the target is always on the left. You must not reverse the two. In algebra this would be acceptable, but in programming it is not. Also, note that the computer evaluates the expression first, no matter how trivial it may appear. Therefore, consider the classic assignment statement:

```
ICOUNT = ICOUNT + 1
```

In algebra this statement is meaningless, but in FORTRAN it has a very important function. First, examine the expression ICOUNT + 1. This tells us to go to the memory location ICOUNT, retrieve the value and then add the value of 1 to it. Finally, the result is returned to the memory location indicated, which happens to be ICOUNT. We call this process *incrementing* which is an important tool in programming.

Now, consider the statement:

```
X = -X
```

Again, in algebra this expression makes no sense. But in programming this statement has the effect of changing the sign of a number as illustrated in Fig. 4.4. First, the computer recalls the value for X, multiplies it by −1 and returns it to its memory location. You will see many examples of these two types of assignment statements, so be sure you understand them fully.

FIGURE 4.4

Schematic Representation of Assignment Statement

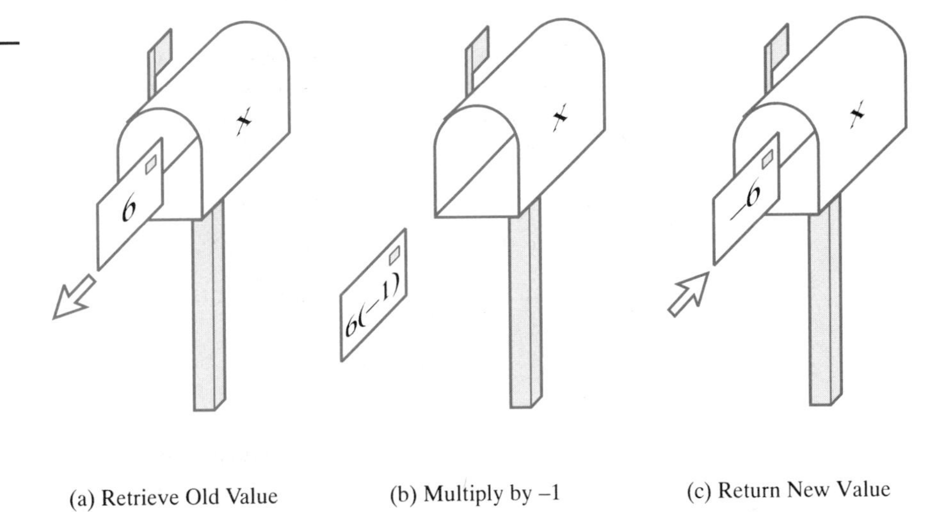

(a) Retrieve Old Value (b) Multiply by –1 (c) Return New Value

Here are some examples of correct and incorrect usage:

Valid	Invalid	Comments
X = 2.0*X		Okay to use same variable as target and part of expression
	I = 3.14159	Will work, but not as you think; I is an integer; the number is real. Value stored is 3, not 3.14159
	X + 1.0 = X	Expression on wrong side of = sign
	X = Y = Z = 2.5	Only one assignment per line

4.5 Arithmetic Operations

There are only five basic arithmetic operations that are possible with FORTRAN. These are addition, subtraction, multiplication, division, and exponentiation as summarized in the following table:

Priority	Algebraic Symbol	FORTRAN Symbol	Meaning
1	(............)	(............)	Parentheses
2	A^B	**	Exponentiation
3	×	*	Multiplication
	÷	/	Division
4	+	+	Addition
	–	–	Subtraction

In order to create mathematical expressions, we must use the symbols listed in the table on a single line of text entered into your program. Whereas algebra permitted the use of multiline expressions, FORTRAN requires you to express everything on a single line:

$$\text{Root} = \frac{A + 2B + C}{D} \qquad \text{(in algebra)}$$

`Root = (A + 2.0*B+C)/D` (in FORTRAN)

There are several key points that you should notice in this simple example:

- Implied operations are not allowed in FORTRAN. In algebra we know that $2B$ means $2.0 \times B$. In FORTRAN we must explicitly write $2.0 * B$.
- Everything is written on one line. In the algebraic expression the numerator is written above the denominator. In FORTRAN we separate them by a / and write it on the same line.

In both algebra and FORTRAN it is understood that you should perform the multiplication ($2B$) before any addition. Thus, in the expression $A + 2B + C$ the product $2B$ is determined first. Then the result is added to A and C. Therefore, all mathematical operations have a well-defined hierarchy as summarized in the preceding table. We give the highest priority to sets of parentheses. Everything inside a set of parentheses is evaluated first before any other operations are carried out. Consider the following example to see how the hierarchy of operations works:

9.2 − (2.0**3 − 14.0/7.0) + 14.0 * 0.1

Since parentheses have the highest priority in the hierarchy, we will evaluate the three operations inside the parentheses first: exponentiation, subtraction, and division. From the previous table notice that exponentiation will be done first, then division, and finally subtraction. Thus:

(2.0**3 − 14.0/7.0)

(8.0 − 2.0)

(6.0)

Now our expression becomes:

9.2 − 6.0 + 14.0 * 0.1

Again, we must decide which operation is to be done first. Of course, we do the multiplication first since it has a higher priority than the addition or the subtraction. Thus, the expression becomes:

9.2 − 6.0 + 1.4

which gives a final answer of 4.6.

An interesting situation occurs when multiplication and division occur as consecutive operations. Consider the following case:

64.0/4.0 * 8.0

If we do the multiplication first, the answer is 2.0. But if we do the division first, the answer is 128.0. Unfortunately, referring to the list of priorities does not help. It shows that both multiplication and division have the same priority. In situations like this the rule is that we evaluate the expression left to right. Thus, we do the division first and the correct answer is 128.0. The exception to this rule is two exponentiation operations in succession:

2 ** 3 ** 2

which should be evaluated as 2^9 and not 8^2. Examine the following expressions to see if you fully understand the rules for evaluating expressions:

Expression	Value
16.0 – 4.0 – 2.0	10.0
16.0 – (4.0 – 2.0)	14.0
16.0 + 4.0 * 2.0	24.0
16.0 / 4.0 / 2.0	2.0
16.0 ** 4.0 * 2.0	131072.0
16.0 ** (4.0 * 2.0)	4294967296.0

In all the previous examples we have been very careful to make all the constants and variables REAL. This is because there are special rules that govern INTEGER arithmetic. Also, when you try to mix REAL and INTEGER data types, some strange things can happen. We discuss these in the following two sections.

INTEGER ARITHMETIC

Recall that we defined INTEGERs as whole numbers. Fractional numbers such as 2-1/2 are REAL. Therefore, if we try to do arithmetic with integers, we must be careful since fractional results are dropped. This is especially important when division of INTEGER numbers takes place. The simplest examples are:

8/4 or 3/2

where we attempt to divide two whole numbers. In the first case the answer is 2 and the correct answer is reported. In the second example the actual result is 1.5. But this will be reported as 1 since INTEGERs cannot store fractional numbers. The rule that governs these situations is that if INTEGER division occurs, then any remainder will be dropped. Note that the computer makes no attempt to round up or down. The computer simply truncates any remainder. Thus, the result of 3/4 is 0! An interesting point is that if we had presented this as 3.0/4.0 (both REALS), the result would be 0.75. So, watch those decimal points! Because of this peculiar feature, avoid using INTEGERs in mathematical expressions. Reserve them for use as counters or other special effects that we will discuss shortly.

Try the following exercises to make sure you understand INTEGER arithmetic.

Expression	Value
6 + 2 / 4 * 5	6
6 * 2 * 4	48
6 * 2 - 4 * 5	−8
1 / 3 + 1 / 3 + 1 / 3	0

MIXED MODE ARITHMETIC

Since INTEGERs have special rules for arithmetic manipulation, you should keep INTEGERs separate from REALs. Sometimes, though, mixing these two data types occurs and strange effects will happen if you are not careful. Consider the situation when we multiply two REAL numbers together. The result is always REAL. Similarly, when you multiply two INTEGERs, the result is always an INTEGER. But what happens if you multiply an INTEGER and a REAL number? In this situation one number must be converted to the other type before the calculation can proceed. In FORTRAN the convention adopted is that the INTEGER will be converted to a REAL number. Once this is done, the multiplication will proceed.

The previous example is an example of the general problem of *mixed mode arithmetic.* You must be very careful to avoid mixed mode unless you fully understand how to use it. There are a few situations where mixed mode is desirable, but, generally, mixed mode should be avoided since unintentional errors can be generated quite easily.

Two types of mixed modes that we will consider are:

Mixed mode assignments
Mixed mode arithmetic

For a mixed mode assignment statement, the value to be stored may undergo a conversion before the computer performs the assignment. A typical example is an attempt to store a REAL number in an INTEGER variable:

```
J = 9.87654
```

Since an INTEGER variable can only store a whole number, the decimal part of the number 9.87654 will be *truncated.* The computer will retain only the whole part of the number. Thus, the variable J will contain the number 9. Note that the number is *truncated,* not *rounded.*

In the reverse case where the computer stores an INTEGER in a REAL variable the problem is not so severe since the computer will simply *pad* the number with zeros beyond the decimal point. Thus, in the example:

```
X = 3
```

the whole number 3 will be converted to the fractional number 3.000000 before it is assigned to the indicated variable. Note that this does not create the same problem as the first case.

The use of mixed mode arithmetic expressions also will create problems if you are not careful. Consider the following two expressions:

```
X = 3/2.0      (a)
X = 3/2        (b)
```

In the first example the expression is mixed mode since 3 is an INTEGER but 2.0 is REAL. As we stated earlier, the computer converts the INTEGER to the REAL number 3.000000 before the division. The result is 1.500000, which the computer assigns to the variable X. In the second example both 3 and 2 are INTEGERS. Thus, mixed mode is not involved (at least not yet). The INTEGER division yields the result (3/2 = 1). When the computer attempts assignment, however, it will find an attempt at mixed mode assignment in the form of X = 1. Thus, the rule of mixed mode is that the computer will convert INTEGERs to REALs, but only when needed to complete a calculation. To illustrate this further, consider the more complicated example:

```
J = 2.3 * (3/2) - (5 - 3.4) * 6.2
```

which will be evaluated as follows:

Starting expression:	2.3 * (3/2) - (5 - 3.4) * 6.2
(5 − 3.4) is mixed mode, convert 5 → 5.0	2.3 * (3/2) - (5.0 - 3.4) * 6.2
Neither (5.0 − 3.4) nor (3/2) is mixed mode, proceed	2.3 * (1) - (1.6) * 6.2
2.3*(1) is mixed mode, convert 1 → 1.0	2.3 * 1.0 - 1.6 * 6.2
No more mixed mode, proceed	2.3 - 9.92
No more mixed mode, proceed	-7.62

Even after completing the mixed mode calculations, there is still a mixed mode assignment statement:

```
J = -7.62
```

which, of course, results in the assignment J = −7. Trace through the mixed mode statements in the following table to make sure that you understand how to evaluate assignments and expressions:

Expression	Value
J = 1 / 2	0
X = 1 / 2.	0.5000000
Y = 2.0 ** (1/2)	1.0000000
Y = 2.0 ** (1/2.)	1.4142136
J = 5 * 3 / 4.0 * (−1)	−3
J = 7 / 3 / 2	1
V = 7 / 3./ 2	1.1666667
V = 7 / 3 / 2.	1.0000000

Note that slight changes in the positioning of a decimal point can radically change the value of a mathematical expression. So, be careful with those decimal points!

BUILT-IN FUNCTIONS

Anyone who owns a small hand calculator is already familiar with built-in functions. These are common mathematical functions such as x^2, e^x, $\sin(x)$, $\ln(x)$, x^y among others, which we use so frequently that it is desirable to have them available at a touch of a button. Sometimes we also call them *library* or *intrinsic* functions. Of course, on a computer, function buttons are not available as they are on a hand calculator. Instead, we implement them with *function calls* such as:

```
Y = SQRT(X)
```

This function call summons the previously stored instructions to compute the square root of the numerical value stored in X. Once the computer obtains the value, it assigns the value to the variable Y. There are many such functions which will vary from machine to machine. We have listed the most common ones in Table 4.1.

To use these functions, you must provide an *argument* for the function. For some, this argument is a single number or a variable such as SQRT(14.3).

TABLE 4.1

Most Common FORTRAN Intrinsic Functions

Name	Description	Argument	Result	Example
ABS( )	Takes absolute value	INTEGER	INTEGER	J = ABS(-51)=51
		REAL	REAL	X = ABS(-17.3)=17.3
		DOUBLE	DOUBLE	Z = ABS(-0.1D04)=0.1D04
ACOS( )	Calculates arccosine	REAL	REAL (radians)	X = ACOS(0.5) = 1.04712
		DOUBLE	DOUBLE (rad)	X = ACOS(0.5D0)=0.104712D01
ALOG( )	Natural logarithm	REAL	REAL	X = ALOG(2.71828)=1.000000
		DOUBLE	DOUBLE	X = ALOG(0.2718D01)=0.1D01
ALOG10( )	Logarithm base 10	REAL	REAL	X = ALOG10(10.0)=1.000000
		DOUBLE	DOUBLE	X = ALOG10(0.1D0)=-0.1D1
AMAX(,,)	Returns largest argument	INTEGER	INTEGER	I = AMAX(5,7,1,6,2)=7
		REAL	REAL	X = AMAX(0.2,-3.4,5.6)=5.6
		DOUBLE	DOUBLE	X = AMAX(0.1D0,-0.3D3)=0.1D0
AMIN(,,)	Returns smallest argument	INTEGER	INTEGER	I = AMIN(4,0,3,-4)=-4
		REAL	REAL	X = AMIN(0.2,-3.4,5.6)=-3.4
		DOUBLE	DOUBLE	X = AMIN(0.1D0,-0.3D3)=-0.3D3
ASIN( )	Arcsine	REAL	REAL (radians)	X = ASIN(0.5)=0.523560
		DOUBLE	DOUBLE (rad)	X = ASIN(0.5D0)=0.523560D0
ATAN( )	Arctangent	REAL	REAL (radians)	X = ATAN(1.0)=0.78534
		DOUBLE	DOUBLE (rad)	X = ATAN(1.0D0)=0.78534D0
COS( )	Cosine	REAL (radians)	REAL	X = COS(1.04712)=0.5
		DOUBLE (radians)	DOUBLE	X = COS(1.04712D0)=0.5D0
DBLE( )	Converts to DOUBLE	INTEGER	DOUBLE	X = DBLE(3)=0.3000D01
		REAL	DOUBLE	X = DBLE(3.0)=0.30000D01
EXP( )	Exponential, e^x	REAL	REAL	X = EXP(1.0)=2.718282
		DOUBLE	DOUBLE	X = EXP(1.0D0)=0.2718282D01
INT( )	Converts to INTEGER	REAL	INTEGER	J = INT(3.9999)=3
		DOUBLE	DOUBLE	J = INT(0.3999D01)=3
FLOAT( )	Converts to REAL	INTEGER	REAL	X = FLOAT(4)=4.000000
		DOUBLE	REAL	X = FLOAT(0.4D01)=4.000000
MOD(,)	Integer remainder from I/J	INTEGER	INTEGER	J = MOD(29,4)=1
NINT( )	Round to nearest integer	REAL	INTEGER	J = NINT(3.9999)=4
		DOUBLE	INTEGER	J = NINT(0.1678D01)=2
REAL( )	Convert to REAL	INTEGER	REAL	X = REAL(3)=3.000000
		DOUBLE	REAL	X = REAL(0.123D03)=123.0000
SIN( )	Sine function	REAL (radians)	REAL	X = SIN(0.5235602)=0.5
		DOUBLE (radians)	DOUBLE	X = SIN(0.5235602D0)=0.5D0
SQRT( )	Square root	REAL	REAL	X = SQRT(17.64)=4.2
		DOUBLE	DOUBLE	X = SQRT(0.1764D2)=0.42D1
TAN( )	Tangent function	REAL (radians)	REAL	X = TAN(0.78534)=1.000000
		DOUBLE	DOUBLE	X = TAN(0.78534D0)=1.000D0

For others, the argument may be two numbers or variables, such as in the function to return the remainder from integer division, MOD(7, 2). (In this example the result is 1 since $7/2 = 3$ with a remainder of 1). Finally, in a few other cases you supply a long list of variables as in the function that determines the largest number in the argument list, AMAX(2.2, 3.7, −1.6, 14.5). (The answer is 14.5).

The functions can be used in simple assignment statements such as:

```
Y = SQRT(X)
```

or even within another function. An example would be to take the absolute value of X, using the ABS(X) function, before attempting to take the square root:

```
Y = SQRT(ABS(X))
```

A large computer system may have up to a hundred of these functions available for your use. Small microcomputers, however, may be limited to as few as a dozen functions. So, be sure to check the manuals for your FORTRAN compiler. Here are a few examples of how to use functions:

Expression	Value
Z = SQRT(ABS(−17.64))	4.200000
W = ALOG10(100.0)**2	4.000000
Y = SIN(0.5235983)	0.500000
L = MOD(17,5)	2
X = FLOAT(9)	9.000000
J = INT(7.2)	7

Built-in functions are very easy to use. You must exercise some care, though, to match the argument type (REAL, INTEGER, and so forth) to the requirements of the function. It makes no sense to convert the number 4.2 to a REAL number using X=REAL(4.2), for example, since 4.2 is already REAL. Thus, it should be apparent that the only permissible argument of the REAL( ) function can be an INTEGER. So, be sure to check Table 4.1 for the permitted argument types when in doubt. Usually, the argument type matches the function type except for the mode conversions INT, REAL, and DBL. You also should note that all trigonometric functions (direct and inverse) use radians instead of degrees (1 radian = $180/\pi$ degrees). We illustrate how to do this in Table 4.1.

4.6 Basic Input and Output (I/O)

The easiest input and output (I/O) statements to use in FORTRAN are the so-called *list-directed* types. In these types of statements we are not concerned about the appearance of the data, but rather, we worry more about *what* is input or output. In the next chapter we will consider how to improve the appearance of the data. But for now we concentrate on simply getting information into and out of our programs.

Some of the example programs shown to you previously used the list-directed I/O statements. These are:

```
READ *, VARIABLE1 , VARIABLE2 , VARIABLE3 . . .
```

for input and

```
PRINT *, VARIABLE1 , VARIABLE2 , VARIABLE3 . . .
```

for output. The star, *, which appears in each of these statements, indicates that we are using a *free format*. Free format means that the computer will use a set of predetermined instructions to read or print data. For example, the machine will figure out how many decimal places to print, how many blank spaces to leave between each number, and so forth. Your only concern is to provide the proper instructions on what to read in or print out. In the next chapter we will replace the * with a series of instructions that will give us some control over the output appearance.

We use the *I/O list* in the READ * statement to indicate the variables to which we will assign the data. To illustrate how this works, consider the statement:

```
READ *, X, Y, Z
```

which will cause the first three numbers that you type in from the keyboard to be assigned to the corresponding variables. A typical response might look like:

14.3, −27.943, 0.0034567 ⟨CR⟩

which is equivalent to the four assignment statements:

```
X = 14.3
Y = -27.943
Z = 0.0034567
```

One advantage of the READ statement over the equivalent assignment statements is that you do not need to rewrite and recompile the program if you enter a different set of data. All you need to run the program with different data is merely enter the new data in response to the READ statement.

The print command, PRINT *, *output list,* indicates that the numerical values of each variable in the output list will be printed on the CRT screen. Thus, the statements:

```
X = 12.345
Y = 0.098765
Z = -3.9045
PRINT *, X, Y, Z
```

will produce the following output on the CRT screen:

```
12.34500    0.09876500    -3.904500
```

You have no control over the appearance of the output at this point since you have specified free format. Therefore, the computer will usually print the output of REALs with seven significant digits. Sometimes the computer will print the output in exponential format such as:

```
0.12345E+02    0.98765E-01    -0.39045E+01
```

which is somewhat awkward to read. In the next chapter we will examine ways to prevent such difficulties.

The PRINT statement also can contain character strings or arithmetic expressions in the output list. If you enclose a character string inside single quotation marks, the string will be printed intact. Also, it is permissible to write full mathematical expressions in the list. Here is an example that combines the two:

```
X = 2.4
PRINT *, 'X = ', X, 'X**2 = ', X * X
```

The character string, 'X = ', will print out exactly what is inside the quotes. The next segment of the list (, X ,) prints out the numerical value. The third segment, 'X**2', again prints out the message inside the quotes. Finally, the last segment (, X*X) prints out the product of X times X. The printed result looks like:

```
X = 2.400000    X ** 2 = 5.7600000
```

As we showed in one of our earliest examples, it is generally good practice to use PRINT statements liberally as prompts to remind you of what order to enter data. For example, the statements:

```
PRINT *, 'READY TO PROCEED. PLEASE ENTER X, Y, Z:'
READ *, X, Y, Z
```

will produce the screen prompt:

READY TO PROCEED. PLEASE ENTER X, Y, Z:

to which you will type in the appropriate values. Without prompt statements the computer will go instead to the READ statement and wait for you to enter data. No message will be printed to indicate that it is waiting.

4.7 Program Termination

Every FORTRAN program must end with the executable statement:

```
END
```

The purpose of this statement is to mark the physical end of the program and to show the compiler when it should stop the translation process. It is also used at execution time to stop the running of a program. If you forget the END statement in your program, an error message will be returned. A good compiler, however, will simply supply the required END statement and warn you about what it has done.

A similar but very different statement is the executable statement:

```
STOP
```

This statement, like the END statement, also terminates the execution of a program. But unlike the END statement, the STOP statement can go anywhere within a program. Also, you may have as many STOP statements as you need which will depend on the context. This is especially useful when you wish to stop the program if something goes wrong. For example, if you accidentally enter invalid data, you can terminate execution with the statement.

A few FORTRAN programmers still end their programs with both types of termination statements:

```
STOP
END
```

This is a throwback to the early days of the language when FORTRAN required both. More recent versions of FORTRAN have made these two statements at the end of the program redundant. Still, you may find some compilers that require both of them. Therefore, all example programs in this book will contain both. Try a simple program on your system to see if the STOP statement is required. If it is not, simply drop it. The double statements serve no purpose in this situation. To be sure, there are situations where the STOP statement is necessary. However, the end of a program is usually not one of them.

4.8 Algorithms

In Chap. 2 we talked about the process for developing algorithms. We presented a technique for fleshing out problem statements until all subproblems reduce to either sequential commands, control statements, or language structures. The success of this process, though, is dependent on the availability of a library of such tools. Programmers who are just beginning will have only a few tools in their libraries. So, they will need to flesh out their problems to a greater degree than more experienced programmers. As your programming experience expands, your library will keep pace. In this section and in corresponding sections in later chapters we will present several useful examples of basic programming functions that you should add to your library. Later you will begin to develop some of them on your own.

We present four examples of language structures in this section. These are:

Input of data
Output of data
Switching two numbers
Incrementing a variable

These structures consist of only a few FORTRAN lines, but our purpose here is not just to list the code. We want to give examples of the algorithm development process and examples of correct usage of the FORTRAN syntax. The first two examples for I/O will be used in almost every program, although in somewhat modified form. The second two examples will be used frequently as part of more complex processes. For example, switching two numbers is an integral part of the algorithm to put a list of numbers in numerical order. Similarly, the process for incrementing a counter is a basic component of the looping process.

In the first algorithm we read in three numbers and assign them to the variables *A*, *B*, and *C*. The full algorithm and program segment follow. The algorithm can also be represented with a flowchart. With simple input such as this, the variables to which data is to be assigned is written inside the symbol shown in Fig. 4.5. Note that this is only a program *segment,* that is, only a small portion of a total program.

FIGURE 4.5

Flowchart Symbol for Data Input and Assignment

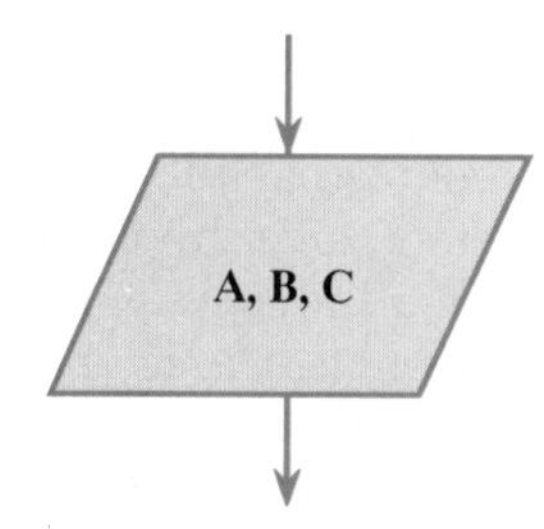

While this algorithm development may seem long for such a simple process, we are trying to focus on the process instead of the final result. It is important, therefore, that you follow the technique that we use:

TERMINAL INPUT LS

PROBLEM STATEMENT: Enter numerical values for variables A, B, and C.

ALGORITHM DEVELOPMENT:

Step 1a: Review the problem statement. Indicate all nouns in **bold** and verbs in *italics:*

Enter numerical **values** for **variables** A, B, and C.

Formulate questions based on nouns and verbs:

Q1: What do we mean by values?
Q2: What are variables?
Q3: How do we enter values?

Step 1b: Answer above questions:

A1: Values are constants to be assigned to appropriate variables.
A2: Variables are symbolic names representing memory locations.
A3: Values are typed in from the keyboard.

ALGORITHM:

1. Type in three numerical values.
2. Read in values and assign to variables A, B, and C.

SEGMENT:

```
READ *, A, B, C
```

TRACING: Assume that the three values 1.23, 4.56, and 7.89 are typed in on the CRT keyboard. The values will be assigned as follows:

$A \leftarrow 1.23$

$B \leftarrow 4.56$

$C \leftarrow 7.89$

A similar language primitive can be constructed for the process of printing data to a CRT screen. Both the preceding input example and the present example of output work with the terminal device only. In the next chapter we will see how to work with other I/O devices. The equivalent flowchart symbol for this simple output algorithm is shown in Fig. 4.6. Note that the symbol is the same as the input statement. You can tell the difference between the two in a larger flowchart by the context in which you use it. For example, the input

FIGURE 4.6

Flowchart Symbol for Output

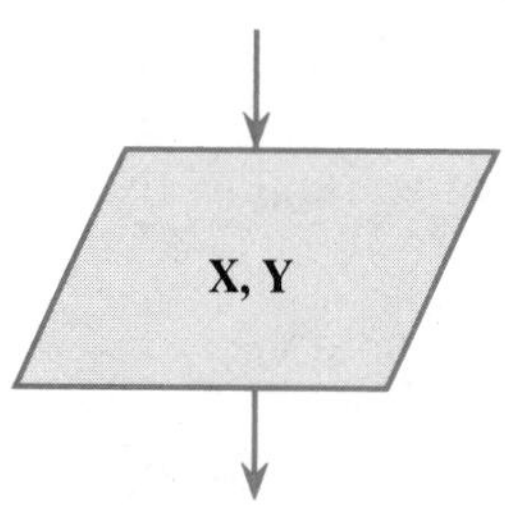

statements are usually at the beginning and the output statements at the end. If there is any doubt, you should show explicitly your intention either to input or output data:

TERMINAL OUTPUT LS

PROBLEM STATEMENT: Print out the numerical values of the two variables *X* and *Y*.

ALGORITHM DEVELOPMENT:

Step 1a: Review the problem statement. Indicate all nouns in **bold** and verbs in *italics*.

Print out the numerical **values** of the two **variables** *X* and *Y*.

Formulate questions based on nouns and verbs:

Q1: What do we mean by values?
Q2: What are variables?
Q3: How do we print out?

Step 1b: Construct answers to above questions:

A1: Values are constants stored by the variables.
A2: Variables are symbolic names representing memory locations.
A3: Use PRINT instruction to type values onto CRT screen.

ALGORITHM:

1. Print out desired values by listing variable names.

SEGMENT:

```
PRINT *, X, Y
```

TRACING: Assume that the values 9.87 and 6.54 are stored in memory locations called X and Y, respectively. Then the statement PRINT*, X, Y will result in the following to appear on the screen:

```
9.870000    6.540000
```

The algorithm for switching two numbers follows. You will use this process often in sorting problems when a list of numbers must be placed in either ascending or descending order. There are also other situations where this algorithm will come in handy.

Before we show you the algorithm for switching, there are a few things that you should observe. Some of you might be tempted to switch two numbers with two simple assignment statements:

```
A = B
B = A
```

To see why this will not work, let's trace through the lines, assuming A = 12.6 and B = −3.75. Note what happens to the two variables:

Program Line	A	B
START OF PROGRAM	12.6	−3.75
A = B	−3.75	−3.75
B = A	−3.75	−3.75

Although the previous two-line program intuitively looks correct, it does not achieve the desired goal. This is because a variable can only store one value at a time. When we make the assignment $A = B$, we erase the old value of A to make room for the new value. So, unless we store the old value of A somewhere else, it will be lost. One way of doing this is to store A in a *dummy variable*. You can think of the dummy variable as a scratch space where things may be stored temporarily, out of harm's way, until needed.

The flowchart representation of this simple algorithm can be drawn two different ways as illustrated in Fig. 4.7. The single assignment box with the three statements is preferred to the three individual boxes to save space:

LS FOR SWAPPING TWO NUMBERS

PROBLEM STATEMENT: Write a program segment to swap two numbers.

DESCRIPTION: The purpose of this problem is to swap the numerical value of two variables. Thus, if we begin with

$$A = 12.6 \qquad B = -3.75$$

after swapping, we should have

$$A = -3.75 \qquad B = 12.6$$

ALGORITHM DEVELOPMENT:

Step 1a: Review the problem statement. Indicate all nouns in **bold** and verbs in *italics:*

Write a program segment to *swap* two **numbers.**

Formulate questions based on nouns and verbs:

Q1: What two numbers?
Q2: How do we swap?

Step 1b: Construct answers to above questions:

A1: Numbers stored in variables *A* and *B*.
A2: Need to use a *DUMMY VARIABLE* to store temporarily one of the variables. Then the other one can be assigned to the old variable.

ALGORITHM:

1. Assign value of *A* to DUMMY
2. Assign value of *B* to *A*.
3. Assign value of DUMMY to *B*.

SEGMENT:

```
DUMMY = A
A = B
B = DUMMY
```

TRACING: Assume values for $A = 12.6$ and $B = -3.75$ as previously given. Value of each variable after each line is executed:

Program Line	A	B	Dummy
START OF PROGRAM	12.6	−3.75	0
DUMMY = A	12.6	−3.75	12.6
A = B	−3.75	−3.75	12.6
B = DUMMY	−3.75	12.6	12.6

FIGURE 4.7

Flowchart Symbol for Multiple Assignments

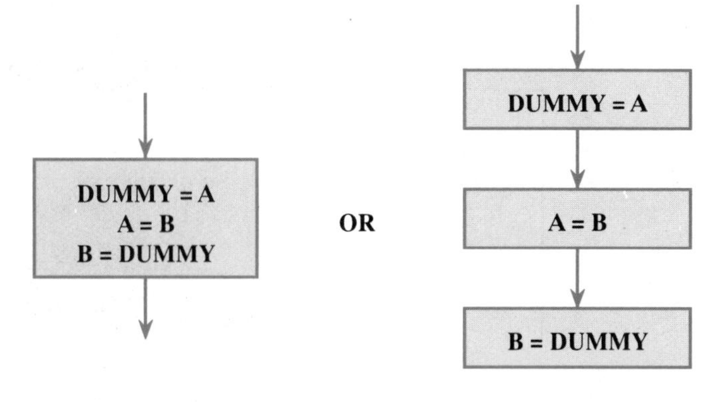

The final algorithm that we want to present is the INCREMENT A VARIABLE LS. As in the previous example, this is a language structure (LS) since we are using a combination of previous simple sequential commands and control statements to accomplish our task. The process of incrementing is an important idea in programming, which we will use primarily for counting. There are many situations where we wish to count the number of times that something has occurred—how many data points have we entered, for example. In general, the counter is not an integral part of any calculation. We use it primarily for controlling another process:

LS FOR INCREMENTING A VARIABLE

PROBLEM STATEMENT: Increment the variable ICOUNT by 1.

ALGORITHM DEVELOPMENT:

Step 1a: Review the problem statement. Indicate all nouns in **bold** and verbs in *italics:*

Increment the **variable** ICOUNT by 1.

Formulate questions based on nouns and verbs:

Q1: How do we increment?
Q2: Which variable do we increment?

Step 1b: Construct answers to above questions:

A1: We increment a variable by taking its old value and adding 1 to it. This new value is then stored back into the location from which it came.
A2: Value stored in ICOUNT is to be used. When final value is calculated, it will be stored in ICOUNT.

ALGORITHM:

1. Retrieve value stored in ICOUNT.
2. Add 1 to value.
3. Store result in ICOUNT.

SEGMENT:

```
ITEMP = ICOUNT
ICOUNT = ITEMP + 1
```

or

```
ICOUNT = ICOUNT + 1
```

Preferred method

TRACING: Assume ICOUNT has an initial value of 5. After each line of the program is executed, the variables will have the following values:

	Program Line	ICOUNT	ITEMP
	START OF PROGRAM	5	0
Method 1:	ITEMP = ICOUNT	5	5
	ICOUNT = ITEMP + 1	6	5
Method 2:	ICOUNT = ICOUNT + 1	(5+1)	

Note in this algorithm that we show two equivalent versions. We do this so that you can see how the preferred statement:

```
ICOUNT = ICOUNT + 1
```

executes. This kind of statement sometimes creates confusion as discussed earlier. Recall, the computer evaluates the right-hand side of the = sign first. This instructs the computer to recall the *old* value of ICOUNT, add 1 to it, and store it as the *new* value for ICOUNT. Make sure that you understand this process very well since it is an integral part of many other processes in programming.

4.9 Putting It All Together

Let's now apply some of the things discussed in this chapter to solve a simple engineering problem. Figure 4.8 shows a uniform beam of length, L, and weight, W_2, which is hinged at one end. From the other end a weight W_1 is supported. There is a tie rope connected to the beam at point X, and this rope makes an angle, θ, with the beam. If the system is at equilibrium, we can calculate the horizontal, H, and vertical, V, components of the forces at the hinge. Also, we can calculate the tension in the rope. Once we know these values,

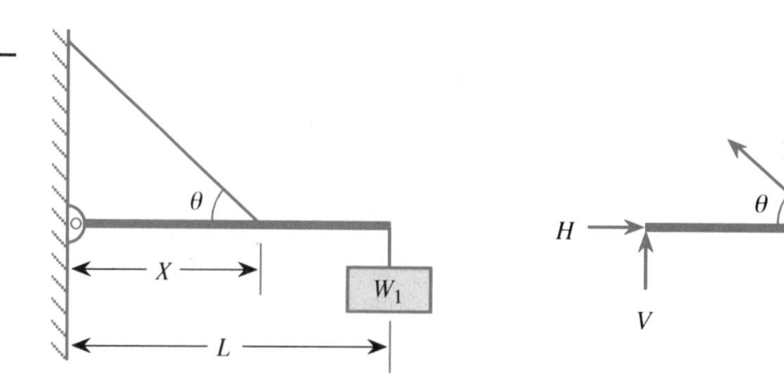

FIGURE 4.8

Force Diagram of End Loaded, Hinged Beam

the components can be designed to support the load safely. Figure 4.8 shows a schematic of the beam and the force diagram.

The force balance and torque balance equations give the following sets of equations for this problem:

$$T(X \sin(\theta)) - W_1 L - W_2(L/2) = 0$$
$$H - T\cos(\theta) = 0$$
$$V + T\sin(\theta) - W_1 - W_2 = 0$$

The three quantities that we are trying to solve for are the rope tension, T, the horizontal force, H, and the vertical force, V, at the hinge. Rearranging these equations, we easily find these quantities:

$$T = \frac{W_1 L + W_2(L/2)}{X \sin(\theta)}$$
$$H = T\cos(\theta)$$
$$V = W_1 + W_2 - T\sin(\theta)$$

The quantities, X, L, W_1, W_2, and θ are all constants and are given. Note from these equations that in order to calculate V and H, you must first calculate the rope tension, T. This will establish the order of calculation in the program. Therefore, the procedure for calculating the forces is:

Read in the constants: X, L, W_1, W_2, and θ.
Calculate T according to the preceding formula.
Calculate V and H, using the value of T.
Print out the results.

EXAMPLE #3

PROBLEM STATEMENT: Read in the values for a beam length, L, its weight, W_2, its end load, W_1, and the position, X, and angle, θ, of a tie rope. Then calculate the rope tension and the vertical and horizontal forces at the hinged end.

ALGORITHM DEVELOPMENT:

Step 1a: Review the problem statement. Indicate all nouns in **bold** and all verbs in *italics:*

Read in the **values for a beam length, L, its weight, W_2, its end load, W_1, and the position, X, and angle, θ, of a tie rope.** Then *calculate* the rope **tension and the vertical and horizontal forces** at the hinged end.

Formulate questions based on nouns and verbs:

Q1: How do we read in values?
Q2: How do we calculate the tension, vertical and horizontal forces?
Q3: How do we get the results out?

Step 1b: Answer the above questions.

A1: Use READ statement to read in values and assign to the variables, L, W_2, W_1, X, θ.
A2: Use assignment statement to calculate rope tension and assign to the variable, T. Then use two additional assignment statements to compute vertical forces and horizontal forces, V and H, respectively.
A3: Use PRINT statement to send the values of T, V, and H to the terminal screen.

ALGORITHM:

1. Type a prompt message on the screen to indicate order of variable input.
2. Read in L, W_2, W_1, X, θ.
3. Calculate rope tension, T.
4. Calculate V and H using formulas given.
5. Print out T, V, and H.

FLOWCHART:

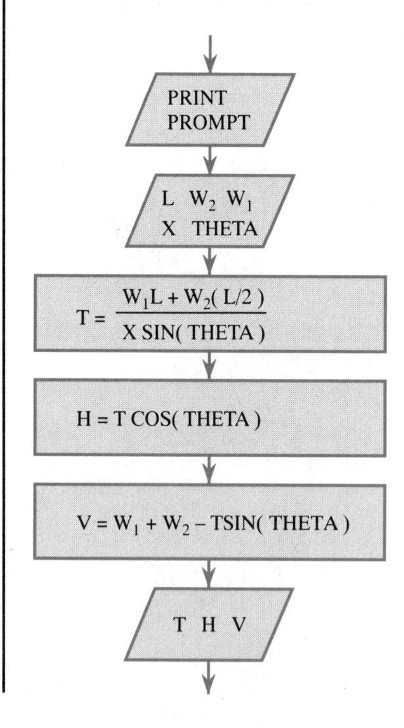

PROGRAM:
```
C     PROGRAM TO CALCULATE THE EQUILIBRIUM FORCES IN A UNIFORM
C     BEAM SUBJECT TO END LOADING AND HINGED AT ONE END. A TIE
C     ROPE IS ALSO USED TO SUPPORT THE LOADS.
C           L     = LENGTH OF BEAM
C           X     = POSITION OF TIE ROPE
C           W1    = WEIGHT OF END LOAD
C           W2    = WEIGHT OF BEAM
C           THETA = ANGLE BETWEEN ROPE AND BEAM
C           T     = TENSION IN ROPE
C           H     = HORIZONTAL FORCE AT HINGE
C           V     = VERTICAL FORCE AT HINGE
C-----------------------------------------------------------------
      REAL L
      PRINT *, 'ENTER L, W2, W1, X, THETA'
      READ *, L, W2, W1, X, THETA
      T = (W1*L + W2*L/2.0)/(X*SIN(THETA))
      H = T*COS(THETA)
      V = W1 + W2 -T*SIN(THETA)
      PRINT *, 'ROPE TENSION = ', T
      PRINT *, 'HORIZONTAL FORCE AT HINGE = ', H
      PRINT *, 'VERTICAL FORCE AT HINGE = ', V
      STOP
      END
```

Notice in this program that we have made liberal use of comment statements. At the very least, comments should come at the beginning of the program to give the reader an idea of what the program is doing. Also, it is a good idea to make a list of variables used in the program. Not only will this help in trying to read (and debug) your current programs, but it will also help you when rereading some of your older programs.

The variable for the length of the beam, L, should be a REAL variable since it is probably not a whole number. Unfortunately, according to IMPLICIT TYPING rules, L is usually an INTEGER. Therefore, we must declare L as REAL. The other variables, X, W1, W2, and THETA are already REAL, and there is no need to redeclare them.

As we mentioned once before, it is good practice to use a prompt statement before any READ. Thus, the second line of the program prints a message on the screen indicating the order of input. Note that if you do not enter the data in the expected order, you will not get the correct answer. After you type in the five numbers corresponding to the five variables, the program calculates T, V, and H according to the formulas given.

Finally, the program prints out the results with identifying information, the statement:

```
PRINT *, 'ROPE TENSION = ', T
```

will result in the message inside the single quote marks with the numerical value of the variable T. Thus, if T = 1234.567 N, the program will print on the terminal screen:

```
ROPE TENSION =    1234.567
```

The numerical value that is printed will have seven significant digits. At this point you have no control over the appearance of the number. That will have to wait until the chapter on formatting.

Exercises

SYNTAX

4.1 Locate Syntax Errors. Each of the following FORTRAN segments contains a single SYNTAX error. Find the errors and make recommendations for corrections:

a.
```
INTEGER A B C
Y=FLOAT(A)
```

b.
```
INTEGER I, J
I=-9999999
J=1,000,000
```

c.
```
INTEGER I, J
I=J=3
```

d.
```
REAL X,Y,Z
Y=2.0
Z=Y*4-3.3
X=Y**2 + Z***3
```

e.
```
REAL X,Y,Z
Y=4
Z=6.4
X=Y(Z-4.0)
```

f.
```
INTEGER I
REAL X,Y
I=3
X=2.2
Y=SIN(X)-COS(I)
```

g.
```
REAL X,Y
X=3
Y=AMAX1(ABS(X),3)
```

h.
```
REAL X,Y,SIN
X=1.349E-6
SIN=SIN(X)
Y=4.0*SIN
```

4.2 Locate Syntax Errors. The following FORTRAN programs contain several SYNTAX errors. Find the errors:

a.
```
REAL X, 6G
X=6.4
6G=2
INTEGER I
I=100,000*6G*X
J+X=0+6.4
PRINT J
STOP
END
```

b.
```
REAL=5
I=1 000 000 000 000 * REAL
J=K=I
PRINT*, I,J,K
END
STOP
```

EXPRESSIONS

4.3 FORTRAN Arithmetic. Evaluate each of the following arithmetic expressions:

a. `I=-4**2*6.4/2*MAX(16,13,2)`

b. `Z=14.2/6.82/1.01**4`

c. `L=2*3/4*5/6*7/8`
d. `X=-4**2.0/2.0+(-4.3*(SQRT(16/4.)**2))` (CAREFUL!)
e. `M=MOD(2,5)*MOD(5,2)`
f. `U=11.2-4.7*1.5+6.4-1.6/9.1`
g. `K=2+3**2*ALOG10(10.0)`
h. `V=DBLE(SNGL(1.2345678901234))` (CAREFUL!)
i. `J=MOD((14*3/MOD(13,5)),6)`

YOUR SYSTEM

4.4 Check Out Your Compiler. Generic FORTRAN is generally much more restrictive than the "enhanced" versions supplied by software distributors. For instance, FORTRAN 77 in its most austere form will allow only six characters in a variable name, whereas FORTRAN 77 sold by DEC for the VAX computer allows up to 32 characters in a variable name. Determine which of the following FORTRAN segments are valid for your system. Comments are given to help you understand what to look for. Some of these commands will work as given. Others will not work at all. Still others will work, but in a different form:

a. `REAL VERYLONGVARIABLENAME` (More than six characters allowed?)

b. (Skip lines in source code?)
```
INTEGER I,J

REAL X

REAL Y
```

c.
```
REAL A                       (Standard precision)
REAL*8 B                     (Alternate form of double precision)
REAL*16 C                    (Quadruple precision)
INTEGER D                    (Standard integer)
INTEGER*8                    (Long integer)
COMPLEX F                    (Standard complex)
COMPLEX*16 G                 (Double precision complex)
```

d. `*  THIS IS A COMMENT` (Can * replace C for comments?)

e. (Does your compiler initialize all variables to zero?)
```
PRINT*, I,J
PRINT*,X, Y, Z
```

f. `REAL JACK_IN-BOX` (Nonstandard characters allowed?)

g. `Y=FACT(13)` (Factorial function available?)

TRACING

4.5 Tracing a Program. Tracing through a program serves two purposes. First, it provides a good drill to see if you fully understand the language syntax. Secondly, it is an indispensable tool for debugging programs. With this in mind, trace through the following program segments and predict their output. Confirm your answer by running it on your system:

a.
```
PRINT*, 5.0*(4/7)+2**3
STOP
END
```

b.
```
A=11.0
J=A/2
A=J*2
PRINT*,A
STOP
END
```

c.
```
A=1
B=5
C=6
R1=(-B+(B**2-4*A*C)**(1./2.))/(2*A)
PRINT*,R1
STOP
END
```

d.
```
INTEGER I,J
REAL X,Y
I=123456789
X=I
J=X
PRINT*,J
STOP
END
```

FORmula TRANslation

4.6 Formulas. Translate the following mathematical expressions into FORTRAN code:

a. The area, A, of a circle of radius, r:

$$A = \pi r^2$$

b. The cube root of a number, X:

$$Y = X^{1/3}$$

c. The perimeter, c, of an ellipse with semiaxes a and b:

$$c = 2\pi[(\tfrac{1}{2})(a^2 + b^2)]^{1/2}$$

d. The area of a triangle, A, in terms of the lengths of its sides, a, b, and c:

$$A = [s(s - a)(s - b)(s - c)]^{1/2}$$

where $s = 0.5(a + b + c)$

e. The distance, L, of some point (x_0, y_0, z_0) form the plane $Ax + By + Cz + D = 0$:

$$L = \frac{Ax_0 + By_0 + Cz_0 + D}{\pm (A^2 + B^2 + C^2)^{1/2}}$$

PROGRAMS

4.7 Currency Exchange. Write a FORTRAN program which converts a given number of U.S. dollars to an equivalent number of Japanese yen. Assume that 140 yen equals 1$.

4.8 Car Rental Fees. At the DAVIS car rental agency the total cost to rent a car for one day depends on the class of the car and the mileage driven according to the following table. Write a program which will tell a customer the cost of renting each type of car based on the number of miles driven. Your input should consist of the customer's name and the number of miles to be driven. The output should consist of the total cost for each type of car:

Class	Daily Rental ($/Day)	Mileage Charge ($/Mile)
A (LIMO)	122	0.42
B (CORVETTE)	95	0.51
C (BMW)	79	0.46
D (PONTIAC)	60	0.29
E (CHEVROLET)	55	0.33
F (YUGO)	24	0.27

4.9 Radioactive Decay. As you know, the element plutonium is a radioactive material which is one of the principal by-products in spent fuel rods taken from nuclear reactors. In some countries plutonium is chemically removed from the rods and reused in reactors. In other countries, however, no attempt is made to recover plutonium. Rather, the radioactive materials are stored until they are no longer dangerous. But, as you will see in this exercise, storage is a very long-term process that poses many difficult problems. When the radioactive material is stored, it undergoes radioactive decay where Pu^{244} transforms into less harmful products. This process is characterized by the radioactive rate equation:

$$N = N_0 e^{-0.693t/\tau}$$

where N = number of atoms at time, t

N_0 = number of atoms at time, $t = 0$

τ = half life of plutonium

you may recall that the half life is the amount of time for half of a given amount of material to decay radioactively. The half life for plutonium is approximately 7.6×10^7 years. Write a program to calculate the amount of plutonium remaining after 100, 1000, 10,000, 100,000 years. (Note that the ratio N/N_0 represents the fraction remaining.)

4.10 Weight of Hollow Sphere. Write a program to calculate the weight, W, of a thick-walled hollow sphere of outer diameter, D, wall thickness, t, and density, δ, according to the following equations:

$$W = \delta V$$
$$V = \tfrac{4}{3}\pi(R_O^3 - R_I^3)$$
$$R_O = \frac{D}{2}$$
$$R_I = \frac{D}{2} - t$$

4.11 Trigonometric Identity. Let $A = 2.13$ and $B = 1.3$; write a short program to verify that the following identity holds true:

$$\tan(A + B) = \frac{\tan(A) + \tan(B)}{1 - \tan(A)\tan(B)}$$

4.12 Trigonometric Identity. Choose some arbitrary, nonzero value for A, and write a small FORTRAN program to verify the following trigonometric identity:

$$\tan(5A) = \frac{\tan^5(A) - 10\tan^3(A) + 5\tan(A)}{1 - 10\tan^2(A) + 5\tan^4(A)}$$

4.13 Trigonometric Identity. Choose some arbitrary value for A and write a FORTRAN program to verify the following identity:

$$\sin(\alpha) = \frac{e^{i\alpha} - e^{-i\alpha}}{2i}$$

4.14 Digit Extraction. INTEGERS in FORTRAN can have up to nine digits. Write a program which extracts each digit of a number as follows:
a. First divide the number, INT, by 10 and store the result in J1.
b. The digit in the tens place will be (INT − 10*J1). Why? To find the next digit, divide J1 by 10 and place the result in J2.
c. The second digit of the number will then be (J1 − 10*J2).
d. Extend this line of thinking to obtain all nine digits.

4.15 Rounding Decimals. Write a program which prints out a real number rounded off to two decimal places. For example, the number 123.4567 would appear as 123.45. In the next chapter you will learn about formatting output to achieve the same result, but for now try using just assignment and print statements. (*Hint:* Think about using mixed mode arithmetic.)

4.16 Doppler Effect. When a moving train is moving toward you and sounds its whistle, it has a higher pitch than when it travels away from you. This is due to the well-known DOPPLER effect given by the equation:

$$f = f_0 \frac{V + \nu_0}{V - \nu_s}$$

where f = frequency of sound sensed by observer
f_0 = frequency of sound emitted by train whistle
V = speed of sound
v_0 = speed of observer
v_s = speed of train

Write a program which will compute the pitch of a train whistle sensed by an observer in a car moving at a speed of 100 km/h and a train approaching at a speed of 120 km/h. Assume that the pitch emitted by the train whistle is set at 4000 Hz and that the speed of sound in still air is 1200 km/h. What happens if the car is stopped at a crossing gate? What happens if the car and the train are going at the same speed and in the same direction?

4.17 Wheatstone Bridge. A very common device used in many different types of laboratories is the *WHEATSTONE BRIDGE* shown here. It is a very sensitive tool which is able to measure an unknown resistance very accurately. The resistance to be measured, X, is one leg of the bridge. The other three legs are made from known resistances. The currents I_P and I_Q that flow through the points Q and P are given by:

$$I_P = \frac{X + R_1}{X + R_1 + R_2 + R_3} \quad (I_{\mathrm{TOT}})$$

$$I_Q = \frac{R_2 + R_3}{X + R_1 + R_2 + R_3} \quad (I_{\mathrm{TOT}})$$

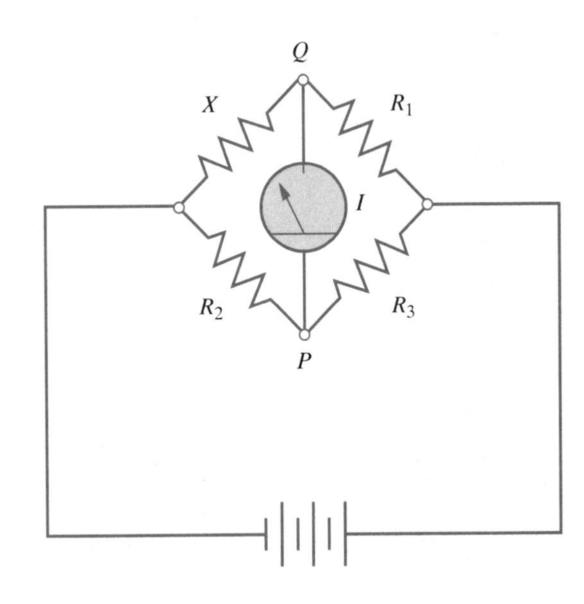

The procedure to determine the unknown resistance, X, is to use a variable resistor on one of the legs and adjust its value until the current flowing between the points P and Q is zero. In this configuration the bridge is called *balanced.* In order to achieve this result, the following must be satisfied:

$$I_Q(X) = I_P(R_2)$$

Write a program which reads in the three known resistances after the bridge is balanced and returns the value of the unknown resistor.

4.18 Resistance Strain Gauge. One application of the Wheatstone Bridge of the previous problem is the resistance strain gauge. The strain gauge is a precision resistor glued onto the surface of a part as shown here:

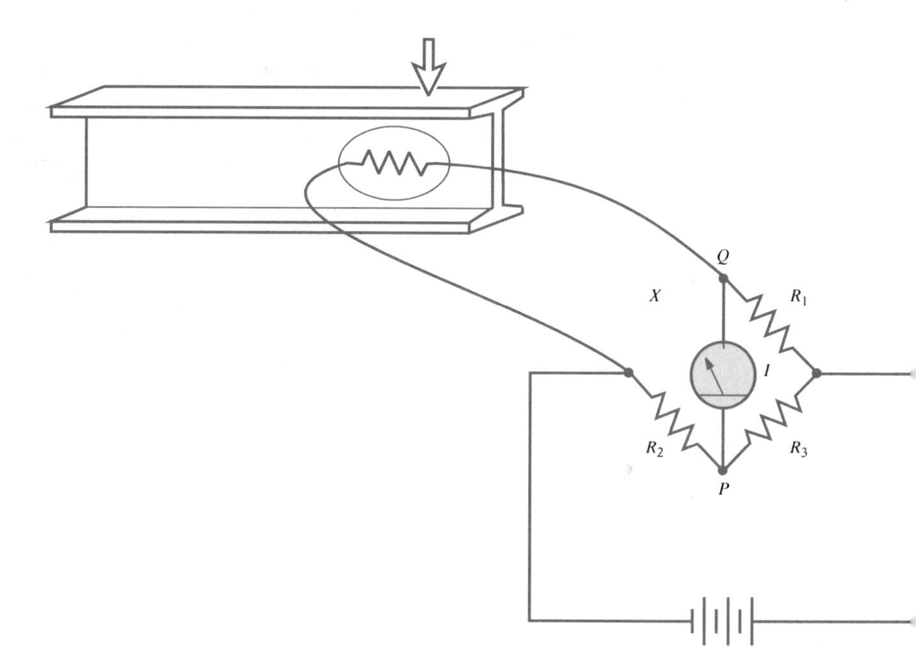

If the part is stressed, the resistance of the strain gauge will change by a small amount. To measure this change, we use the strain gauge as the fourth leg of a bridge, which is balanced twice; once when the part is unstressed, the second time when the part is loaded. Thus, for the first balancing, the resistance values reported are R_1, R_2, and R_3, whereas the second reading produces R_1', R_2', and R_3'. The corresponding unknown resistances X and X' are then calculated according to the equations of the previous problem. From this, the resistance change, $(X' - X)$, is then easily calculated. Finally, the strain in the part is calculated through the *gauge factor*, *GF*, defined as:

$$GF = \text{(resistance change)/strain}$$

Write a program to implement the process described earlier to measure strain assuming a *GF* of 2.000 Ω^{-1}.

4.19 Relativistic Mass. When a body is accelerated to a speed which approaches the speed of light, the mass of the body begins to increase according to the formula:

$$m = \frac{m_0}{[1 - (v/c)^2]^{1/2}}$$

where m = mass at relativistic speeds

m_0 = rest mass (mass at speed = 0)

v = speed

c = speed of light = 2.988×10^8 m/s

Write a computer program to use the preceding formula to compute the mass of:

a. An electron weighing 9.08×10^{-28} grams traveling at $c/2$
b. Your body weight when traveling at 99% speed of light
c. Your body weight when traveling at 99.9% speed of light

4.20 Twin Paradox Problem. Another interesting effect of relativity is known as time dilation. This postulate states that even time is modified at relativistic speeds according to the formula:

$$t_m = t_s[1 - (v/c)^2]^{1/2}$$

where t_m = elapsed time measured on moving object

t_s = elapsed time measured on stationary object

v = velocity of moving object

c = speed of light = 2.988×10^8 m/s

Assume that twins are separated at birth. One is kept on earth while the other one is placed on a rocket ship traveling to the nearest star, Alpha Centauri, which is 4.8 light-years away. Assume that the ship travels at $0.96c$ and that the trip takes 5 earth years. Write a program to calculate the difference in the apparent ages of the twins.

CHAPTER

5

FORMAT-DIRECTED INPUT AND OUTPUT AND FILES

5.1 Formatting—More than Just Pretty Print

In the previous chapter we saw how to read data into a program and how to print data to the CRT screen. But there are many occasions when we need better control over the I/O functions that the simple list-directed commands do not provide. In the first half of this chapter you will learn how to control the appearance of your output by using the FORMAT instruction. This instruction will allow you to control such things as positioning within a line and limiting the number of decimal places, and to help in the ability to align data within columns. In the second half of this chapter we will show you how to direct output to devices other than the screen.

To some students it may appear that our sole purpose in discussing format-directed I/O is to show you how to improve the appearance of the output. While that may be *one* of our objectives, it is not the only one. There are many important ideas presented in this chapter that will help improve readability. Readability of data is important in helping to understand the results of your program. It also makes it easier to spot errors. Data presented in a tabular form, for example, is always easier to read than the equivalent data presented in the list-directed format (also called *free formatting*) presented earlier. To illustrate this, compare the results of a hypothetical calculation to calculate the percent expansion of a rod as a function of temperature presented in two different ways. We print the results first with the simple free format of Chap. 4:

```
0.5000000E+01  0.1240000E+02  0.0000000E+02  0.1000000E+02
0.1273610E+02  0.2710484E-03  0.1500000E+02  0.1305540E+02
0.2507047E-03  0.2000000E+02  0.1334277E+02  0.2201124E-03
0.2500000E+02  0.1358703E+02  0.1830661E-03
```

When we print the same information with the format-directed statements to be discussed shortly, the data is immediately more readable:

TEMP (°C)	LENGTH OF BAR (CM)	EXPANSION (%)
5.0	12.400	0.000
10.0	12.736	2.710
15.0	13.055	2.507
20.0	13.343	2.201
25.0	13.587	1.831

Note that when the programmer can control the format of the data, trends are much easier to spot. For example, from the sample table you may have already observed that the rate of expansion is not a constant. In fact, the expansion rate is decreasing as a function of temperature, which may suggest an atomic level process worth looking into. At the very least, you would report this finding as noteworthy. You probably would have missed this by examining the same data printed out with the free format. Therefore, format-directed statements can do more than improve the appearance of your output. If properly used, they relieve you of the tedium of scanning reams of obscure data, thus giving you more time to focus on the science of your problem.

Of course, you must use FORMATting instructions with either an input or an output command. Therefore, we often call these an *I/O FORMAT PAIR* since the FORMAT statement is useless without the matching input or output command. In general, the input pair will have the following structure:

```
      READ 20, variable list
 20   FORMAT (list of instructions)
```

and the output pair will look like:

```
      PRINT 35, variable list
 35   FORMAT (list of instructions)
```

Notice that the READ and PRINT statements are somewhat different from the previous examples. When we wish to instruct the computer on the appearance of the data, we replace the * in the READ * or the PRINT * statements with a statement label. This statement label shows where the instructions for reading or printing are to be found. In the preceding examples we are telling the computer to go to either statement label 20 or 35 to get the instructions about the form of the data. Of course, if you are happy with free formatting, leave the * in and drop the FORMAT instruction.

The general form of the FORMAT instruction is:

```
s    FORMAT(ccc, descriptor1, descriptor2, . . . ,
          descriptorN)
```

where $\quad$ ***sl*** = statement label (Integer number up to five digits)

ccc = carriage control character

descriptor$_j$ = instructions for ***variable***$_j$

The carriage control character, CCC, is a special instruction that resets the printing head on a printer. If you use the FORMAT statement with an output command, you must include the CCC. But if you use the FORMAT statement with an input statement, you do not need the CCC.

The *FIELD DESCRIPTORS,* also called *field specifiers,* in the FORMAT statement give instructions to the computer on how the information is to be printed. Among the things that can be specified are:

Type of variable (*REAL, INTEGERs, DOUBLE, CHARACTER, etc.*)
Number of significant digits
Column in which to start printing
Floating point or exponential form (*for REALs only*)
Number of blank spaces and blank lines, and
Any text to be included

Each item in the READ or PRINT variable list must have a corresponding FIELD DESCRIPTOR. Thus, if there are five variables in the I/O list, there must be five descriptors, even if you want free format for some of them. Once you relieve the computer of the responsibility of controlling the format, you must follow through for every variable in that particular list. If you want some variables to be free format while others are to be format controlled, you must use two separate I/O statements:

```
      PRINT *, A, B
      PRINT 25, X, Y, Z
25    FORMAT (ccc, descriptor1, descriptor2,
             descriptor3)
```

In this example the values of A and B will be printed with free format, whereas X, Y, and Z will be printed according to the instructions that you specify. Although in this example we have placed the FORMAT statement after the I/O statement that refers to it, you can place it anywhere in the program. Most programmers will place all the FORMAT statements at the end of the program to make their programs more readable as in the previous example.

5.2 Format Specifiers

Several specifiers are available that you will find useful. These fall into the following general categories:

Category	Descriptor	Function
Carriage control	+, 0, 1, ⟨*blank*⟩	Vertical positioning
Numerical data	I	INTEGERS
	F, E	REAL (floating point or exponential)
	D	DOUBLE PRECISION
	G	REAL (general purpose)
Character data	A	Character variables
	apostrophe (' ')	Character strings
Spacing	X	Individual spaces
	T	Tab to column
	slash (/)	Vertical spaces
Repeat	n( )	Reuse specifier inside ( )

There are also additional edit descriptors that you will use much less frequently. We summarize these in the Appendix. The ones listed here are those that you are most likely to use for routine programming tasks.

Each descriptor in the format list must agree in type with the corresponding variable in the I/O list. Thus, a REAL variable must be matched with an appropriate descriptor for REALs, which are the F, E, D, or G statements. You cannot mix an INTEGER variable with a REAL edit descriptor, for instance. This is an example of a correct match:

```
      PRINT 27,    TIME , MILES , SPEED
                    ↑       ↑       ↑
               both |  both |  both |
               real |integer|  real |
                    ↓       ↓       ↓
27    FORMAT(ccc, spec 1, spec 2, spec 3)
```

A mismatch in the data types and the format types is one of the most common errors that occur. So, if you see an error message of the type:

INCOMPATIBLE VARIABLE AND FORMAT TYPE

double check your I/O format pairs to make sure that the variables in the I/O list and the corresponding descriptors in the format statement agree in type.

One of the biggest advantages of using format statements is that they can be used repeatedly without having to retype the instructions. Thus, if you wanted to print out a large body of information, all with the same format, this is how it can be done:

```
      PRINT 54, DIST1 , MILES1 , SPEED1
      PRINT 54, DIST2 , MILES2 , SPEED2
      PRINT 54, DIST3 , MILES3 , SPEED3
54    FORMAT(ccc, spec 1, spec 2, spec 3)
```

All three print statements can refer to the same FORMAT statement if you do not violate the rules for matching data/descriptor types. Notice that all the variables listed in the first position are REALs, those in the second position are all INTEGERs, and the third variables are again all REALs. So, if you use this device to save yourself some extra code, be careful to double check possible type mismatches.

5.3 Carriage Control

We use the carriage control character for controlling an output line, and it must come first in the list of edit descriptors. Failure to follow this rule will result in unexpected errors.

The carriage control characters have only one function: to reset a printer by moving the printing head to column 1 of a new line. For example, you use these characters to control single- or double-spacing, beginning a new page or overprinting previously typed characters. There are only four carriage control characters as summarized in the following table:

Character	Description of Function
' '	Single vertical space
'0'	Double vertical spacing
'1'	New page
'+'	No advance—Reset to beginning of current line

The most common carriage control character is the one for single-spacing, which is a blank space inside single quote marks (' '). Occasionally, you may need to use the double-space ('0'), but the other two CCC's are much less frequently seen. Check with your system operator to see if there are any special conventions adopted for your system. Some systems may have added additional CCC's beyond these basic four.

5.4 Numerical Data Specifiers

When we wish to print out data, whether REAL, INTEGER, or DOUBLE PRECISION, we usually have two primary concerns. These are the total number of spaces to be allotted and the total number of significant digits to be displayed. This is of particular concern with REAL numbers. Recall that REALs have approximately seven significant digits. But as we showed in the opening example of this chapter, it is often desirable to print only the first few. As an example, we often truncate the price of an object to two decimal places, such as $1.98 instead of $1.983450. With INTEGERs, we do not have this problem since we usually want to show all the digits. Therefore, numerical data specifiers will take the general form:

TYPE ***width . (decimals)***

where **TYPE** = a letter (*I*, *F*, *E*, *G*, or *D*) indicating type of variable

width = total width of space reserved

decimals = total number of decimal places (not needed for INTEGERs)

Details of each numerical data type will be presented in the following sections.

INTEGERS

When we print out INTEGERs, our only concern is that we leave enough space in the printed line for all the digits of the number. We need not be concerned about the number of decimal places since INTEGERs can only be whole numbers. Therefore, the form of the specifier becomes:

I*w*

where I indicates an integer number and *w* indicates the total amount of space to be reserved. To illustrate this, consider the following example:

```
      ICOUNT = 237
      JCOUNT = -14
      PRINT 33 , ICOUNT , JCOUNT
33    FORMAT('', I6 , I9)
```

In this example the first descriptor is the CCC giving the command to start a new line. In the I/O list the variable ICOUNT is the first variable, and therefore will be printed with the first edit descriptor, which is I6. The second variable in the list, JCOUNT, will be printed with the second descriptor, I9. Thus, the output page will look like this:

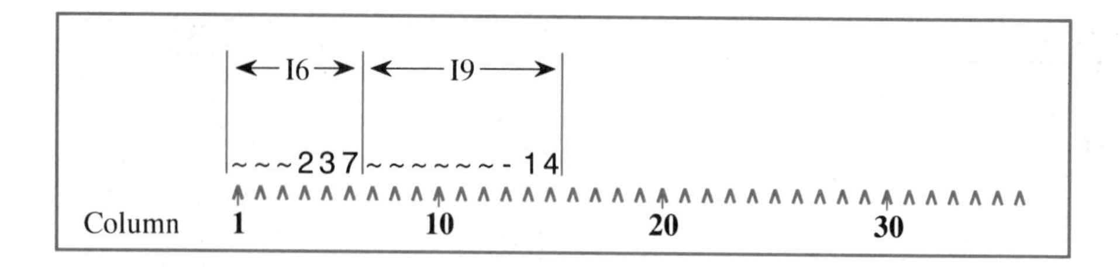

The specifier, I6, tells the computer to reserve six spaces for the numerical value of ICOUNT. Similarly, the specifier, I9, has reserved 9 spaces for JCOUNT. When the computer fills in these reserved spaces with the numerical values, the values are *right-justified.* This means that the printer places them as far to the right as possible within the field reserved for the number. If the number is smaller than the space reserved for it, the printer will leave extra blank spaces, indicated here by the symbol ~. If, however, the number is too large for the reserved space, the printer fills the field with asterisks, *, as in the following example:

```
      ICOUNT = 12345
      JCOUNT = -98765
      PRINT 98 , ICOUNT , JCOUNT
98    FORMAT('' , I5 , I5)
```

which would result in the following output:

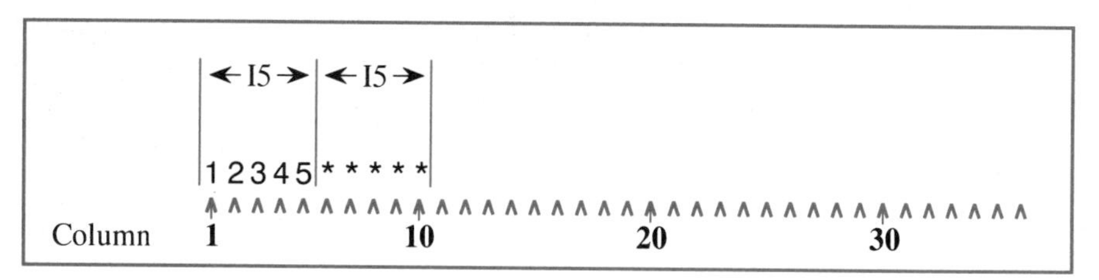

This may appear to be a strange result until you realize that even a minus sign takes up one column in the output line. Thus, the variable JCOUNT, which is equal to −98765, requires six columns for printing. However, the FORMAT specifier, I5, only allots five columns. Therefore, an overflow condition will occur.

REALS

REALs can be expressed in two different formats. The first is the *floating point* format in which we write the number in the conventional decimal notation such as 98.234. The second format is the *exponential notation* in which we express the number with an exponent, such as 0.123E+05. Each of these formats has its own descriptor. Generally, it is desirable to use the floating point format whenever possible. However, if the number is extremely large, such as 0.123E+23, or extremely small such as 0.123E−23, this is impractical. We express the floating point format in the general form:

Fw . d

where F = indicator of floating point format (decimal point required)
w = total width of field reserved for numerical value
d = desired number of decimal places

With REALs, unlike INTEGERs, you need to worry about how many decimal places to keep. If the computer stores more decimal places then specified in the FORMAT statement, the number will be rounded off. Conversely, if the computer does not keep enough decimal places, the number will be *padded* with zeros to fill out the *w.d* specification. Here is an example of how to use the F format:

```
      DIST = 12.345
      TIME = 0.00345
      VELOC = DIST/TIME
      PRINT 5 , DIST , TIME , VELOC
 5    FORMAT('' , F7.2 , F9.6 , F6.1)
```

This will produce the following output:

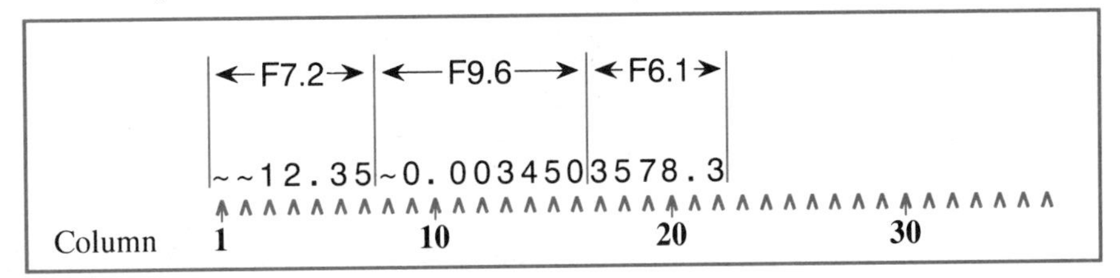

If you study the output for a moment, you will see that the computer has rounded off the value for DIST to fit it into the allotted space. The same thing has happened to VELOC, which retains only one decimal place. The variable, TIME, on the other hand, has had a zero added to it before printing.

Each field indicated in the preceding output must include spaces for three other items besides the significant digits of the number. Note that the decimal place, any negative sign, and a leading zero (zero in front of the decimal place) all require separate columns. Thus, the following rule must be obeyed when using the *Fw.d* format:

$$w \geq d + 3$$

If, for example, you tried to use a descriptor such as F12.11, you will receive a compiler error. Also, if the number is too large to fit into the space that you set up, an overflow condition will result and asterisks will be printed.

We can also express REAL numbers in scientific notation as discussed in Chapter 4. The general form for printing a REAL number in this exponential format is:

Ew . d

where E = indicates exponential format (mantissa $\times$ 10^n)

w = total width of field reserved for number

d = desired number of decimal places for mantissa

As with the F format, the E format has a special rule about how many additional spaces must be reserved. To understand how this requirement comes about, examine the form of the exponential representation of a number.

– 0 .	X X X X	E – 0 4
3 Spaces	Mantissa	4 Spaces

Besides the number of significant digits of the mantissa, the E format requires a total of seven additional spaces. Thus, the rule for the *Ew.d* format is:

$$w \geq d + 7$$

Let's reexamine the previous example, but now with the E format:

```
      DIST = 12.345
      TIME = 0.00345
      VELOC = DIST/TIME
      PRINT 5 , DIST , TIME , VELOC
 5    FORMAT('' , E12.4 , E14.6 , F6.1)
```

will produce the following output:

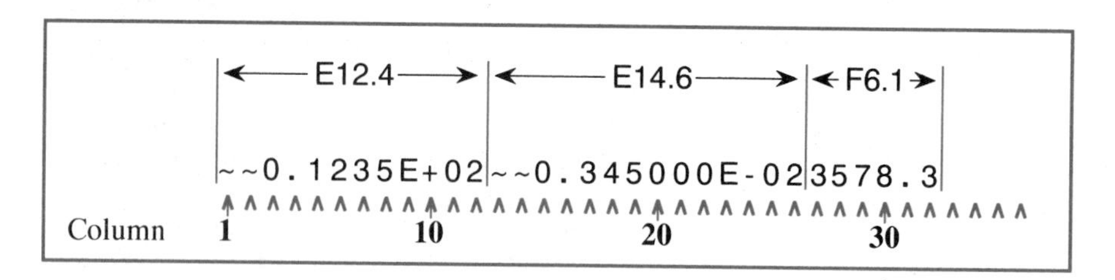

As with the F format, numbers will be rounded or zeros will be added if required to fit into a field. Also, the two different formats for REALs can be mixed as this example shows. Finally, note that the + sign indicating the sign of the mantissa and the sign of the exponent are optional. If one or both are negative, however, the printout will include the minus sign. Check with your system operators to find out what local variations exist.

We print out DOUBLE PRECISION numbers with a format very similar to that for exponential notation:

$Dw\ .\ d$

where D = indicates DOUBLE PRECISION format (e.g., 0.123D+03)
w = total width of field reserved for number
d = desired number of decimal places for mantissa

The principal difference between E and D formats is that the exponent for DOUBLE PRECISION can be significantly larger than that for SINGLE PRECISION. Therefore, you must allow for a three-digit exponent with the D format compared to two digits for E format. The following rule summarizes these requirements:

$$w \geq d + 8$$

Otherwise the D format is identical to the E format.

The final format used with REAL numbers is the *GENERAL PURPOSE* format, G, indicated by:

$Gw\ .\ d$

where G = indicates GENERAL PURPOSE format
w = total width of field reserved for number
d = desired number of decimal places for mantissa

The G format combines both the floating point and exponential formats into a single code. Recall that we prefer the F format whenever possible, but if the number is too large or too small, the E format is necessary. The problem is that sometimes you don't know the magnitude of the number to be printed. With the G format, the computer selects the format automatically. It formats a number with the *Fw.d* format if it is appropriate for a given situation but prints it out in *Ew.d* format if that is better. The computer has a specific rule to decide which format to choose. It bases this choice on the value of the exponent of the number. It compares the exponent to *d* in the *Gw.d* specification. If the exponent is between 0 and d, the computer uses the F format. Otherwise the E format will be used. To see how this works, examine the following examples:

Value	G Format	Equivalent F or E Format	Output
0.010000	G10.3	E10.3	~0.100E−01
0.100000	G10.3	F10.3	~~~~~0.100
1.000000	G10.3	F10.3	~~~~~1.000
10.00000	G10.3	F10.3	~~~~10.000
100.0000	G10.3	F10.3	~~~100.000
1000.000	G10.3	E10.3	~0.100E+04

In the first preceding example the number 0.01 is equivalent to 0.1×10^{-1}, which has an exponent of -1. Since this does not fall in the range of $0 \leq$ exponent ≤ 3, the computer selects the E format. The next four examples all have exponents between 0 and 3. Thus, the computer prints them with the F format. Finally, the last example 1000.0 or $0.1 \times 10^{+4}$ places it outside the range of 0 to 3. Thus, it reverts to the E format. Note that when we specify the *Gw.3* format, any number less than 0.1 or greater than 100.0 will be printed in scientific notation. Only if the number is within these tight limits will the machine choose the floating point format.

5.5 Character Data Specifiers

In the previous chapter we saw that you can print out a CHARACTER string by simply placing it inside quotation marks in the PRINT statement:

```
PRINT *, 'PLEASE ENTER X , Y , Z :'
```

When you begin to use FORMAT statements, to get the same effect, you simply need only to move the CHARACTER string inside as a descriptor:

```
      PRINT 21
21    FORMAT('' , 'PLEASE ENTER X , Y , Z :')
```

For the preceding trivial example, moving the STRING inside the FORMAT statement is no improvement over the list-directed example. However, if we combine numerical data output with STRINGS, we can begin to do things that we could not previously do. For example:

```
      X = 12.34
      Y = -0.025
      PRINT 34 , X , Y , X*Y
34    FORMAT('', 'X= ', F6.2,' Y= ', F6.3,
             'PROD=',F10.5)
```

will produce the following output:

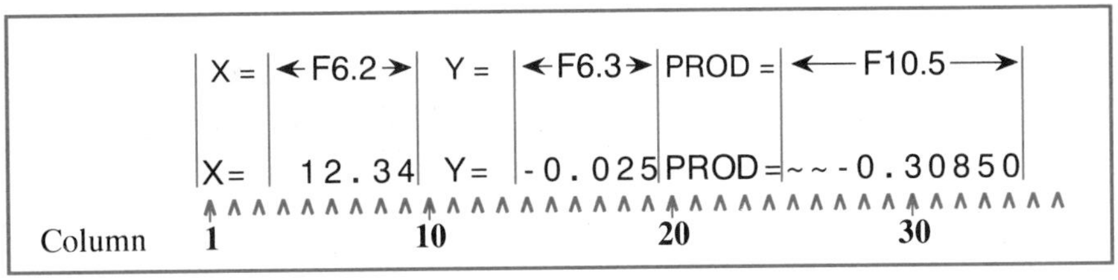

```
        | X = |←F6.2→| Y = |←F6.3→|PROD =|←— F10.5 —→|

        |X=  | 12.34| Y= |-0.025|PROD=|~~-0.30850|
        ^^^^^^^^^^^^^^^^^^^^^^^^^^^^^^^^^^^^^^^^^^
Column  1        10        20        30
```

Let's examine the descriptors within the FORMAT statement a little more carefully. We reproduce each with a brief explanation of its meaning:

' '	Carriage control character—BEGIN NEW LINE
'X= '	Character string—PRINT *X=~*
F6.2	Floating point format—PRINT OUT FIRST NUMBER AS XXX.XX
' Y= '	Character string—PRINT *~Y=~*
F6.3	Floating point format—PRINT OUT SECOND NUMBER AS XX.XXX
' PROD='	Character string—PRINT *~PROD=*
F10.5	Floating point format—PRINT OUT THIRD NUMBER AS XXXX.XXXXX

You should carefully match these descriptions with the previous printer output. Note that blank spaces inside the string quote marks produce a blank space in the output. Thus, ' Y= ' produces [blank]Y = [blank] on the output line.

We do formatting of CHARACTER variables in a way similar to that for INTEGERs, where our only concern is about the total number of spaces to reserve. The general form of the FORMAT specifier is:

Aw

where *A* = indicates CHARACTER format
w = total width of field reserved for character output

Recall that CHARACTERs can be of any length. This distinguishes them from numerical data, which has a constant length (seven significant digits for REALs, for instance). Thus, when we declared CHARACTER variables in the previous chapter, we had to specify the anticipated length. Similarly, we must tell the computer how many spaces to reserve for CHARACTER data. Consider the following simple example:

```
      CHARACTER NAME*20 , CITY*35, ZIP*7
            .
            .
            .
      PRINT 19 , NAME
19    FORMAT('', A20)
```

produces an output similar to (assuming arbitrarily that NAME = 'martin cwiakala'):

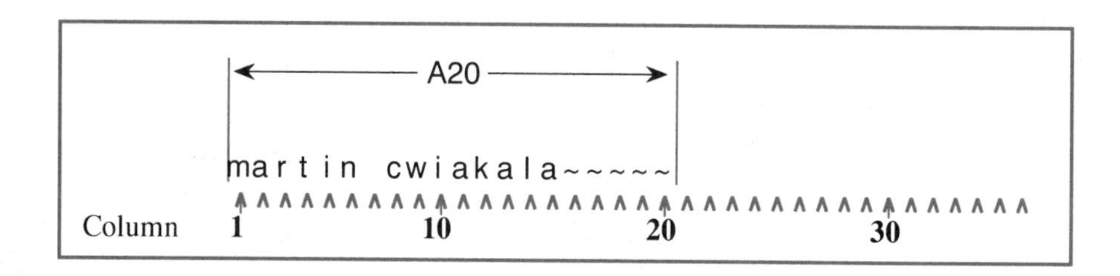

The only thing unusual about this printed line is that the CHARACTER is *left-justified,* which means that the printer pushes the string all the way to the left. Any extra spaces will appear on the right-hand side of the field. By contrast, numerical data is *right-justified.* CHARACTER data is also different from numerical data in that it is not possible to overflow a field with CHARACTERs. If the field is too small for numerical data, recall that a string of asterisks will appear. But if the field is too small for CHARACTER data, it simply truncates the extra characters as shown in the following examples:

Character Variable	Output Format	Output
BOSTON RED SOX	A1	B
BOSTON RED SOX	A10	BOSTON RED
BOSTON RED SOX	A15	BOSTON RED SOX~
BOSTON RED SOX	A20	BOSTON RED SOX~~~~~~
BOSTON RED SOX	A	BOSTON RED SOX

In the last example of the preceding table we have used an *option* that is available on most versions of FORTRAN. The simple specifier, A, without any indication of the field width, will automatically allocate just the right number of spaces for that variable. Note in the example given that there are no extra leading or trailing blank spaces.

Character output varies from system to system. So, be sure to check your local manuals to see how the computer reads and prints characters. Among the things to check for are to determine whether characters are left-justified or right-justified or whether the *default* format specifier, A, is allowed. Some systems also may not distinguish between upper- and lowercase.

5.6 Spacing Control

Several additional format specifiers are useful for spacing data output, aligning it into tables and improving the general appearance. These specifiers do not work with any data or variable in an I/O list. Rather, when the computer encounters them in a FORMAT statement, they will simply produce the spacing requested. There are three specifiers in this group:

Descriptor	General Form	Example	Function
X	nX	3X	Skip n spaces
/	/	/	Skip to next line
T	Tn	T32	Tab to column N

These format descriptors are easy to use and very effective in improving the appearance of your output. Here is an example combining all three. Assume that BASE = 12.4, HEIGHT = 9.6, and VOL = 119.04:

```
      PRINT 92, BASE , HEIGHT , VOL
92    FORMAT('',5X,F9.3,/,2X,'x',T6,F9.3,/,T6,'
                __________',/,T6,F9.3)
```

will produce the following output:

```
        ~~~~~~~~~12.400
        ~~~X~~~~~~9.600
        ~~~~~__________
        ~~~~~~~~119.040
        ^^^^^^^^^^^^^^^^^^^^^^^^^^^^^^^^^^^^^
Column  1        10        20        30
```

This may appear to be a complex example, but if you take the time to break down the descriptors one by one, you should be able to follow the process leading to the preceding printout. So, follow the descriptions given here for each specifier:

	' '	Carriage control character—BEGIN NEW LINE
(line 1)	5X	Spacing command—SKIP FIVE SPACES
(line 1)	F9.3	Floating point format—PRINT OUT FIRST NUMBER AS XXXXX.XXX
	/	Spacing command—BEGIN NEW LINE
(line 2)	2X	Spacing command—SKIP TWO SPACES
(line 2)	'x'	Character string—PRINT THE LETTER X
(line 2)	T6	Spacing command—TAB TO COLUMN 6
(line 2)	F9.3	Floating point format—PRINT OUT SECOND NUMBER AS XXXXX.XXX
	/	Spacing command—BEGIN NEW LINE
(line 3)	T6	Spacing command—TAB TO COLUMN 6
(line 3)	'__________'	Character string—PRINT __________
	/	Spacing command—BEGIN NEW LINE
(line 4)	T6	Spacing command—TAB TO COLUMN 6
(line 4)	F9.3	Floating point format—PRINT OUT THIRD NUMBER AS XXXXX.XXX

There is one additional form of the T specifier that you may see occasionally. This form tells the computer to move left or right a certain number of spaces from its current position. These instructions are:

TR*n* (move right *n* spaces from current position)
TL*n* (move left *n* spaces from current position)

Note that these two commands depend on the current position within a line, whereas the Tn command is independent of the line position.

5.7 Repeat Specifier

There are many times when you need to use the same descriptor repeatedly. A common situation where this is necessary is when all the data is of the same type and is to be printed out with the same format:

```
      PRINT 47, A , B , C , D , E , F , G , H
47    FORMAT('', F9.5 , F9.5 , F9.5 , F9.5, F9.5 ,
             F9.5 , F9.5 , F9.5)
```

As you can see, all eight variables are to be printed with the same F9.5 format. Fortunately, FORTRAN offers a shortcut to avoid repetitious use of a descriptor. All you need to do is to place the *REPEAT DESCRIPTOR* in front of the format to be repeated. Thus, the preceding FORMAT statement can be rewritten as:

```
        PRINT 47, A , B , C , D , E , F , G , H
47      FORMAT('', 8F9.5)
```

This form of the REPEAT DESCRIPTOR can be used with the I, F, E, D, G, and A formats. It cannot be used in this form with the / descriptor.

There is another form of the repeat descriptor where more complex combinations can be repeated as this example shows:

```
        PRINT 47, A , B , C , D , E , F , G , H
47      FORMAT('',F9.5,3x,F9.5,3x,F9.5,3x,F9.5,3x,F9.5,
                3x,F9.5,3x,F9.5,3x,F9.5)
```

Notice that there is a unit consisting of (..F9.5,3x,..), which repeats eight times. The idea introduced previously to place a repeat descriptor in front of the repeat unit can also be used here. However, you must place the unit inside a set of parentheses:

```
        PRINT 47, A , B , C , D , E , F , G , H
47      FORMAT('', 8(F9.5,3x) )
```

There is a slight difference between these two format statements that you probably did not catch. Note that if you write out the unit (F9.5,3x) eight times, the sequence will end in 3x. By comparison, the original format statement ended in F9.5. If you think about this, though, you will realize that the shortened form, 8(F9.5,3x) will leave three extra blank spaces at the end of the line. In effect, this is not a real difference and can be ignored.

The only exception to these rules is the / descriptor. The / mark indicates that you are finished with the current line and want the next bit of output on the following line. If you wish to skip four lines, however, the following are equivalent:

/ , / , / , /   or   ////   or   4(/)

Here are a few additional examples to show how the repeat descriptor works:

Original Format	Equivalent Format
F7.3, F7.3, F7.3	3F7.3
/ , / , / *(skip three lines)*	3(/) or ///
F7.3,I6,/,F7.3,I6	2(F7.3,I6,/)
F7.3,I6,2X,I6,F7.3,I6,2X,I6	2(F7.3,2(I6,2X))
F7.3,2X,I6,F7.3,2X,I6,F9.4,I4,F9.4,I4	2(F7.3,2X,I6),2(F9.4,I4)

5.8 Putting It All Together

Although FORMAT statements can be used with either READ or PRINT statements, they are usually used for controlling output. The reason is that if you use FORMATs with a READ statement, the data must be entered exactly as spelled out in the FORMAT statement. If you type in too many or too few zeros or spaces, the data will be read incorrectly. Therefore, we caution you to avoid formatted READ statements. The exception is when you want to read data from a data file, which we will discuss in the next few sections. In this situation you usually have no choice but to use formatted READs. Remember, though, that when a FORMAT statement is used with a READ statement, there is no carriage control character. These are limited to output on a printer.

Since we use FORMATs primarily with output functions, let's focus on them. To show how to go about setting up data in a convenient table, return to the first example of this chapter, reproduced here:

Internal Storage of Data:

TEMP:	0.5000000E+01	0.1000000E+02	0.1500000E+02	0.2000000E+02
	0.2500000E+02			
LENGTH:	0.1240000E+02	0.1273610E+02	0.1305540E+02	0.1334277E+02
	0.1358703E+02			
EXPANSION:	0.0000000E+02	0.2710484E−03	0.2507047E−03	0.2201124E−03
	0.1830661E−03			

The first item of business is to set up the headings for the table:

```
    TEMP (°C)        LENGTH OF BAR (CM)        EXPANSION (%)
column 5         column 15                 column 35
```

To do this, we need to do the following:

1. Carriage control character to reset printer
2. Express titles as character strings inside ' . . . '
3. Position with the TAB function
4. Underline with TAB LEFT (TL) function followed by character strings

The following code will do this:

```
      PRINT 23
23    FORMAT('',T5,'TEMP (oC)',T15,'LENGTH OF BAR
    1 (CM)', T35,'EXPANSION (%)',TL43,9('_'),TR2,
    1 18('_'),TR2,13('_'))
```

Here's how each of these descriptors works:

	' '	Carriage control character—BEGIN NEW LINE
(line 1)	T5	Spacing command—TAB TO COLUMN 5
(line 1)	'TEMP (oC)'	Character string—PRINT *TEMP (oC)*
(line 1)	T15	Spacing command—TAB TO COLUMN 15
(line 1)	'LENGTH..'	Character string—PRINT *LENGTH OF BAR (CM)*
(line 1)	T35	Spacing command—TAB TO COLUMN 35
(line 1)	'EXPAN ..'	Character string—PRINT *EXPANSION (%)*
(line 1)	TL43	Spacing command—TAB LEFT 43 SPACES
(line 1)	9('_')	Repeat specifier—PRINT CHARACTER STRING _ 9 TIMES
(line 1)	TR2	Spacing command—TAB RIGHT 2 SPACES
(line 1)	18('_')	Repeat specifier—PRINT CHARACTER STRING _ 18 TIMES
(line 1)	TR2	Spacing command—TAB RIGHT 2 SPACES
(line 1)	13('_')	Repeat specifier—PRINT CHARACTER STRING _ 13 TIMES

Now we can work on the data items to be printed. The best place to start is to examine one line of the desired format for the output:

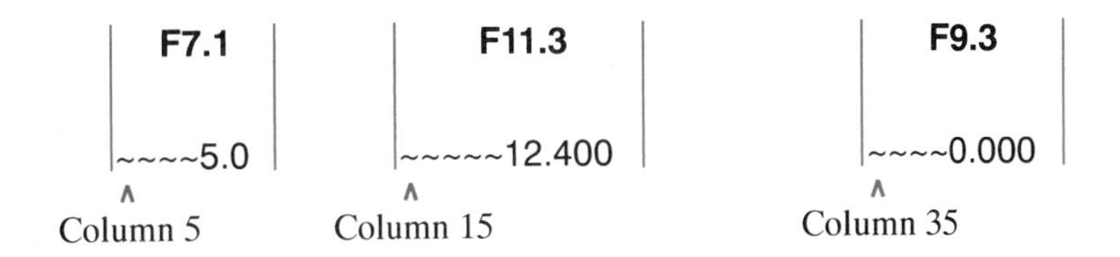

Column 5 Column 15 Column 35

The code to implement this printout is as follows:

```
      PRINT 19, TEMP , X , EXPAN
19    FORMAT('', T5, F7.1, T15, F11.3, T35, F9.3)
```

We will not discuss details of each of these specifiers since you should be able to do that for yourself at this point. We can modify this last group of instructions to print out the five sets of measurements stored in the machine. Also, we can now combine these instructions with the previously developed code to print out the heading:

```
      PRINT 23
      PRINT 19, TEMP1, X1 , EXPAN1
      PRINT 19, TEMP2, X2 , EXPAN2
      PRINT 19, TEMP3, X3 , EXPAN3
      PRINT 19, TEMP4, X4 , EXPAN4
      PRINT 19, TEMP5, X5 , EXPAN5
23    FORMAT('',T5,'TEMP (oC)',T15,'LENGTH OF BAR
     1 (CM)',T35,'EXPANSION (%)',TL43,9('_'),TR2,
     1 18('_'),TR2,13('_'))
19    FORMAT('', T5, F7.1, T15, F11.3, T35, F9.3)
      STOP
      END
```

There are two things to note in this example. (1) You may have a PRINT statement without an I/O list. (2) You may refer to the same FORMAT statement as often as necessary without having to repeat the instructions. Each time a PRINT statement references the FORMAT statement, the computer will simply reuse the instructions from the beginning. The result is the full code to produce the nice looking table from the beginning of the chapter:

TEMP (°C)	LENGTH OF BAR (CM)	EXPANSION (%)
5.0	12.400	0.000
10.0	12.736	2.710
15.0	13.055	2.507
20.0	13.343	2.201
25.0	13.587	1.831
column 5	column 15	column 35

We would be the first to admit that formatting an output is somewhat tedious. However, all the work needed is well worth the effort. You will find that as you gain experience with formats, much of the work becomes automatic. Therefore, we feel that the benefits outweigh the extra work required to create easy-to-read output.

5.9 Input and Output to Data Files

Up to now you have done all input and output through the keyboard and the CRT. When your program encountered a READ * command, you would type in the data, one item at a time. Similarly, when a PRINT * statement occurs, the computer sends output to the screen. There are many times, however, when it would be desirable to send data to a file instead. Once this is done, another program or user could access this information later. In the same vein, it is sometimes desirable to be able to read data from a file rather than to enter it from a keyboard.

A good example of the need for reading data from a data file is when you are writing a program that requires a large amount of input. Let's assume that your program needs 900 data points. Whenever you rerun the program, you must reenter the data by hand. If you need to edit the program ten times before you pull out all the bugs, you will have entered a total of 9000 data points. If instead, the data had been read from a data file, you would need to enter the data only once. An even better scheme is when the computer itself generates the data and enters it into a file for you. This way you don't need to enter the data even once!

Data files are really no different from other types of files discussed in Chap. 2. Data files may be manipulated with commands from the operating system. You can type them on your CRT screen or dump them onto the system printer(s), for example. The principal advantage of data files, however, is that you can access them from within your program. Instead of printing results to the CRT screen, you can now direct the data to an alternate output device. Mostly, this will be a data file. Other options might be a printer or magnetic tape connected to the computer.

In this section of the chapter we focus on the technique for diverting I/O from within a program to a device other than the CRT screen. To do this, there are three instructions that you must include in your program:

Instructions on opening a file
Instructions to communicate with that file
Instructions to close the file when finished

When we had the simple case of printing to the CRT, there was no need to be concerned about these things. All we had to do was to enter the command for I/O to send the data directly to the proper device. This is because the CRT has been named the *default I/O device.* Unless told otherwise the computer will always communicate through the CRT. When the system has been assigned a default device, it is easy to communicate with the system. However, if you want to use something else, like a data file, you need to do additional work. Here is a simple example of how to write to a data file named EXPER.DAT:

```
OPEN(UNIT=8, FILE='EXPER.DAT', STATUS=NEW)
WRITE(8,*) DIST, TIME, VELOC
CLOSE(8)
```

The first statement, OPEN(...), contains all the information to set up the file with the name 'EXPER.DAT', which can be called by the shorthand notation known as the *UNIT.* This UNIT number comes in handy since it is easier to refer to a single number, 8, than to use the full filename in every I/O statement. Thus, in subsequent statements when we refer to 8, we are referring to the full filename, EXPER.DAT. The final listing in the OPEN statement shows the status of the file. It will be either NEW, OLD, or SCRATCH.

The second statement, WRITE(8,*) *list,* is used to direct the output to the desired file. Remember that we have given the file the short name of 8. Thus, the WRITE statement tells the computer to send the output to UNIT #8, which has the permanent name of EXPER.DAT. The final statement of the example, CLOSE(...), simply closes the file after we finish with it.

A similar set of statements can be used to read from a data file. The primary difference from the output statement is that the file must already exist to read from it. Therefore, its STATUS will be old:

```
OPEN(UNIT=3, FILE='NOBEL.DAT', STATUS=OLD)
READ(3,*) WEIGHT, MASS, DENSIT
CLOSE(3)
```

Notice that the READ statement goes to UNIT #3, which is the shorthand notation for the file NOBEL.DAT. Also, note that the status of the file is OLD, which indicates that it is a valid file for reading. If the status had not been OLD, a compiler error would have resulted.

5.10 Opening a File

The OPEN statement contains much information about the attributes of the file to be used for I/O. The following example shows the most common form of the OPEN statement. The first two descriptive items in the parentheses are required; the remaining items are optional:

```
OPEN (UNIT= m, FILE= filename, STATUS= stat,
     ACCESS= type, FORM= form, ERR=n)
```

where UNIT = integer number used for reference

FILE = valid system filename

STATUS = NEW, OLD, SCRATCH, or UNKNOWN. If STATUS= is not used, system will assume UNKNOWN

ACCESS = SEQUENTIAL or DIRECT. If ACCESS= is not used, assumes SEQUENTIAL

FORM = FORMATTED or UNFORMATTED. Asks whether or not file is formatted. If not used, assumes FORMATTED for SEQUENTIAL files, UNFORMATTED for DIRECT files

ERR = integer number used to indicate statement label to transfer if an error has occurred

Files are either SEQUENTIAL or DIRECT ACCESS files. In a SEQUENTIAL file every item in the file must be read to get to the one that you may be interested in. DIRECT ACCESS files, on the other hand, permit you to go to the desired item if you know its position within the file. In this text we will only use SEQUENTIAL files.

The ERR=*n* statement in the OPEN command is useful because it allows you to *trap errors.* If an error occurs while attempting to access a file, control will transfer to the statement label *n*. This gives you a chance to go back and try again or to undertake another procedure.

Most of the data file work that you will need requires only the first three items in the OPEN statement. The additional specifiers are rarely necessary. So, most OPEN statements that you see will be of the form:

```
OPEN(UNIT=4, FILE= 'MINE.DAT', STATUS=OLD)
```

5.11 Closing a File

You must close all files before terminating your program. The procedure is simple by using a statement of the form:

```
CLOSE (UNIT = m, STATUS= stat, ERR= n)
```

where UNIT = integer number used for reference

STATUS = KEEP or DELETE. If STATUS= is not used, system will assume KEEP

ERR = integer number used to indicate statement label to transfer if an error has occurred

If the file in the OPEN statement has STATUS=SCRATCH, then the computer will always delete the file after you finish. Even if you use STATUS=KEEP in the CLOSE statement, the machine will delete the file. Once you declare a file to be a scratch file, you cannot save it. Most often, however, your intention will be to save the file. Therefore, you will usually declare STATUS=NEW or STATUS=OLD in the OPEN statement. In this case, when you go to CLOSE it, the computer assumes STATUS=KEEP and you do not need to restate this. Most uses of the CLOSE statement are of the form:

```
CLOSE(2)
```

For most purposes, the only time that you will run into a problem is if you attempt to write to a file before it is opened or if you try to read from a file after it is closed:

```
OPEN(UNIT=3, FILE='MINE.DAT', STATUS=NEW)
WRITE(3,*)   DIST , TIME , VELOC
CLOSE(3)
READ(3,*)    X , Y , Z
```

Once you close a file, you cannot access it again unless you use another OPEN statement. Thus, in the preceding example the third line has closed the file. After that point you cannot try to I/O to the file.

5.12 File Input and Output

The commands to transmit data to and from a file are somewhat different from the I/O statements seen thus far. The previous I/O statements, READ and PRINT, are used to control the default device. If instead we want I/O to go to another device, we need to tell the computer *which* device. Thus, in the following table you will find the simple input **READ *** statement changed to **READ**(*unit* ,*). We have changed the output statement, **PRINT ***, to **WRITE**(*unit* ,*):

Function	Input and Output to CRT Screen	Input and Output to Files
Free format input	**READ *,** ***variable list***	**READ** ***(unit ,*) variable list***
Formatted input	**READ 9 ,** ***variable list***	**READ** ***(unit ,9) variable list***
	9 **FORMAT(** ***spec list)***	9 **FORMAT(** ***spec list)***
Free format output	**PRINT *,** ***variable list***	**WRITE** ***(unit ,*) variable list***
Formatted output	**PRINT 9,** ***variable list***	**WRITE** ***(unit ,9) variable list***
	9 **FORMAT(** ***spec list)***	9 **FORMAT(** ***spec list)***

The change in the READ statement is minimal. The only difference between READ* and READ(3,*) is that the first one reads from the CRT screen and the second one reads from UNIT #3 defined in the OPEN statement.

The output statement has changed from the simple PRINT* to WRITE(3,*), but the two operate in the same way. Again, the only difference is that one writes to the CRT screen while the second one writes to UNIT #3.

Exercises

SYNTAX

5.1 Format Information Map. Create an information map of format edit descriptors. In it, include the following:

Category that each belongs to
Arguments needed
Rules governing field size, and
A simple example

5.2 Locate Syntax Errors. Find the syntax errors, if any, in each of the following program segments. Assume that UNIT=5 refers to the CRT screen:

```
a.        PRINT I , J , K
b.        READ(3,*) 'ENTER THE VALUE OF X:',X
c.        WRITE(6,*), X , Y
d.        WRITE(6,*) "THE ANSWER IS:", X
e.        WRITE(6,10) X
       10 FORMAT(I3)
f.        WRITE(6,20) X
       20 FORMAT(' ','X= ')
g.        WRITE(3,30) I , J , K , X
       30 FORMAT(4(I5,2X))
h.        WRITE(2,40)
       40 FORMAT('THE ANSWER IS: ', AMTDUE)
i.        PRINT 27, X , Y , Z ,
       27 FORMAT('+',3(F12.4),I4)
j.        READ 41, U , V , W , I , J , K
       41 FORMAT(3(F12.4, I6))
k.        PRINT(4,200) A , I , B , K
      200 FORMAT(F10.4, 3X, I4)
l.        WRITE(1,19) X , Y , I , U , V
       19 FORMAT(' ',2E12.7,2(I7,1X,2D12.4))
m.        READ(5,12) I , X , Z*Y
       12 FORMAT(' ',I4,4X,2F12.5)
```

YOUR SYSTEM

5.3 Check Out Your Compiler. The following suggestions are designed so that you can find out the limitations or extensions of your FORTRAN compilers. Run small programs to determine whether the following suggestions work. You may also need to consult the documentation for your system.

a. What is the UNIT number for CRT I/O on your system? If there are two such numbers such as 5 or 6, is there any difference in the way each behaves?
b. What is the set of valid unit numbers on your system? Do any other (aside from the CRT I/O) unit numbers have any special significance, such as tape drives, floppy disks, card readers, optical scanners, and so on?
c. How does your system respond to the following code:

```
      I = 8
      PRINT 3, I
3     FORMAT(' THIS IS A REAL NUMBER: ', F8.2)
      STOP
      END
```

d. Can you use TR or TL descriptors on your system?

e. Some systems support I$w.n$ format where w = field width and n = minimum number of digits to print. As an example, run the following segment to see if your system supports I$w.n$:

```
      I = 12
      PRINT 11,I
11    FORMAT(' ',I4.3)
```

would result in an output of 012.

f. Are there any special CCC descriptors for your system?

g. Do CCCs perform the action expected if used to print information to the CRT screen? (Remember, the CCCs are used to control a printer.) Use the following:

```
      CHARACTER*1 CC
      PRINT*, 'ENTER CC CHARACTER'
      READ 3, CC
 3    FORMAT(A1)
      X=1.0
      PRINT*,'FOR CC= ',CC
      PRINT 10, CC, X
10    FORMAT(A1,F8.2)
      STOP
      END
```

h. Some systems will allow the use of * for unit directed I/O. If that is the case, the following are equivalent. Does your system support this feature?

```
      PRINT *, X
      WRITE(*,*) X
```

i. Are CHARACTER strings printed out right or left justified?

j. Will your system allow you to drop the OPEN statement? On some compilers, either of the following will work:

```
      OPEN(UNIT=3,FILE='XXXXX',STATUS=NEW)
      WRITE(3,52)
52    FORMAT(' THIS IS A TEST ')
      CLOSE(3)
```

or

```
      WRITE(3,52)
52    FORMAT(' THIS IS A TEST ')
```

k. If your compiler allows you to drop the OPEN and CLOSE statements, what is the name of the file created?

l. What happens on your system if you forget to CLOSE a file? Write a program that reads data from a file, prints the data, but does not close the file. Now try running the program a second time. Does your system allow you to close files manually (through the operating system)?

TRACING

5.4 Tracing Program Segments. Trace through the following program segments and predict their output. For problems with formatted I/O, pay close attention to spacing:

a.
```
      X=1.5
      Y=2.56
      Z=100.01
      WRITE(6,10) X,Y,Z,Y
      WRITE(6,10) Z Y
   10 FORMAT(1X,2(F6.1),F6.2)
```
b.
```
      X=123.4567
      WRITE(2,10) X, X, X, X, X, X
   10 FORMAT(1X,2(F8.1,2(F8.2,2(F8.3))))
```
c.
```
      X1=1.0
      X2=2.0
      X3=3.0
      PRINT 100,X1,X2,X3,X3,X2,X1,X2,X3,X1
  100 FORMAT(' ',10(1X,F6.2))
```
d.
```
      X1=1.0
      X2=2.0
      X3=3.0
      PRINT 101,X1,X2,X3,X3,X2,X1,X2,X3,X1
  101 FORMAT(' ',3(1X,F6.2))
```
e.
```
      X1=1.0
      X2=2.0
      X3=3.0
      PRINT 102,X1,X2,X3,X3,X2,X1,X2,X3,X1
  102 FORMAT(' ',3(3(1X,F6.2),/))
```
f.
```
      X=1.2
      Y=1.3
      Z=1.4
      WRITE(2,10) X, Y, Z
   10 FORMAT(T10,F,6.2,T10,F6.2,T10,F6.2)
```
g.
```
      WRITE(1,2)
      WRITE(1,3)
    2 FORMAT(' ','0')
    3 FORMAT('+','/')
```
h.
```
      X=100.2
      WRITE(6,10) X
   10 FORMAT(F6.2)
```

PROGRAMS

5.5 Truncating a Real Number. Write a program which reads in a REAL number and prints out the whole number portion without using any of the intrinsic functions.

5.6 Extracting Fractional Part of Reals. Write a program which reads in a REAL number and prints out the fractional portion without using any of the intrinsic functions.

5.7 Military Time. Suppose there is a program which calculates the month, day, and year (stored in MONTH, DAY, and YEAR) and also the time in hours, minutes, and seconds (stored in HOURS, MINUTES, and SECONDS). Write the FORTRAN code to output this information in military time (24-hour basis) following this example: *The date is 10/01/90, and the time is now 15:45:37 hours.* Assume that all variable types are INTEGER.

5.8 Angular Readings. Write a program which will read in an angle reported as DEGREES, MINUTES, and SECONDS and prints it out in a decimal format as in the following example: *An angle of 22 degrees, 13 minutes, 47 seconds is equal to 22.2297 degrees.*

5.9 Tabular Output. Write a program assigning I1 = 1, I2 = 2, I3 = 3, I4 = 4, I5 = 5, and I6 = 6. Have your program print these INTEGERs in the following ways:

a. in a single row
b. in two rows containing (I1, I2, I3) and (I4, I5, I6) using two WRITE statements, but only one FORMAT statement
c. in two rows as above, but with only one WRITE statement

5.10 Common Function Table. Write a program which will print out a table containing the following X^2, X^3, $\ln(X)$, $\log(X)$, and e^x for $X = 1, 2, 3, \ldots, 10$. Use the following format:

X	X^2	X^3	$\ln(X)$	$\log(X)$	e^x
1	1	1	0.0000	0.0000	2.7183
2	4	8	0.6931	0.3010	7.3891
3	9	27	1.0986	0.4771	20.0855

5.11 Data Retrieval. Assume that you are hired as a technician to conduct a lengthy series of chemical experiments. The project is conveniently set up so that each experiment takes one day to complete. At the end of each day the data you have collected is stored in a data file called DATA.DAT in the following format:

date *technician name* *temperature* *humidity* *concentration* *activity*

Write two programs to accomplish the following:
a. Enter the data into the file at the end of a day.
b. Read the data from the file when needed.

5.12 Hardcopy Output. Rewrite Prob. 4.9 (radioactive decay of plutonium) to provide a nice looking output table of the form:

Elapsed Time (yr)	Decayed	Remaining
0	0.00001	99.99999
100	0.00113	99.99887
10000000	50.00000	50.00000

Instead of printing the data onto the CRT screen, send the output to a data file called "RADIO.DAT." Once you have run the program and generated the data file, print out the file on your system printer.

CHAPTER

CONTROL STRUCTURES

6.1 Introduction

So far we have only examined sequential execution of program instructions. As we discussed in Chap. 3, sequential execution refers to program execution in the order that it is read. This is, of course, very limited and no programmer could tolerate a language that restricted him or her to this unwieldy structure. Accordingly, all program languages have several methods of branching and looping that allow the programmer to carry out more sophisticated operations.

In this chapter we examine the *UNCONDITIONAL* and the *CONDITIONAL TRANSFER* structures in addition to *LOOPS*. However, not all the constructs will be available with all FORTRAN compilers. Therefore, we also will show you equivalent ways to set up those structures that may not be part of your compiler. Before you begin to read this chapter, refer to Chap. 3 for background information on the control structures. You will find this a useful introduction.

6.2 Unconditional Transfer—The GOTO Statement

The GOTO statement is the simplest of the control structures and it allows you to transfer or branch to a different location in the program. Note, though, that it is an *unconditional* transfer. The program has no choice. It *must* transfer to the indicated spot. The syntax of this statement is:

```
GOTO statement label
```

where *statement label* is a positive INTEGER number that points to the target of the transfer. Here is an example that you are likely to run into:

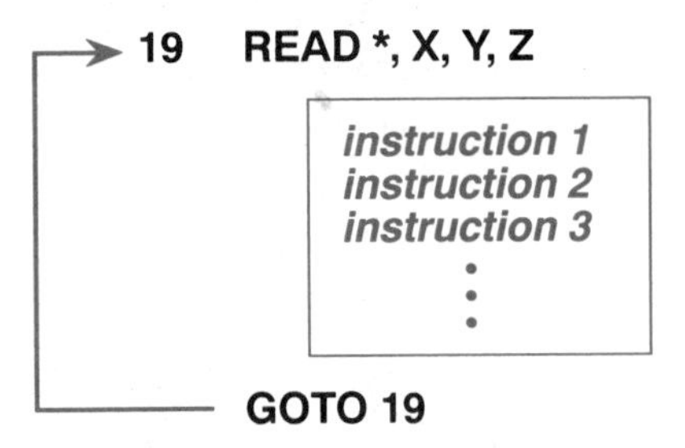

This segment will read in three numbers, assign them to the variables X, Y, and Z, and then process them according to the instructions in the block. When it comes to the line GOTO 19, it will transfer immediately to the READ statement that has the statement label 19. Notice that it does *not* transfer to the 19th line in the program like BASIC! The line with the statement label 19 can be anywhere within the program and the label has nothing to do with the line number.

The biggest problems with the use of the GOTO statement is that it can lead to *spaghetti code.* These are programs in which there are so many transfers from one end of the program to the other that it is almost impossible to follow. We will show you some examples before the end of the chapter. Our best advice is to avoid the use of GOTOs unless absolutely necessary. You will find your programs much easier to debug if you avoid them.

6.3 Conditional Transfer—The IF-THEN-ELSE Structure

The IF-THEN-ELSE structure is the most general of the IF structures, and so we will examine this one first. All the remaining structures are special cases of this general structure. As we described in Chap. 3, the IF-THEN-ELSE structure allows a conditional transfer to take place after it performs a test. Depending on the outcome, the program then executes one of two sets of instructions. The general form is:

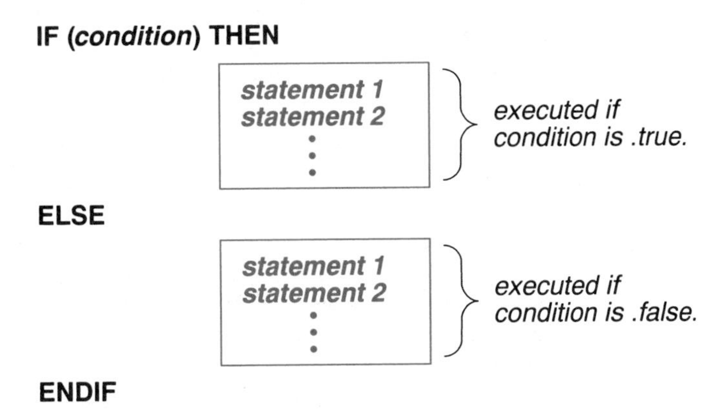

The IF-THEN-ELSE statement first evaluates the *logical condition,* which can only have two possible answers, .true. or .false.. If the answer is .true., the computer executes the instructions in the first block. But if the answer is .false., the computer executes the instructions in the second block. After completing the appropriate blocks, control jumps to the ENDIF statement.

The flowchart symbol for the BLOCK-IF structure is shown in Fig. 6.1.

FIGURE 6.1

Flowchart Structure for BLOCK-IF

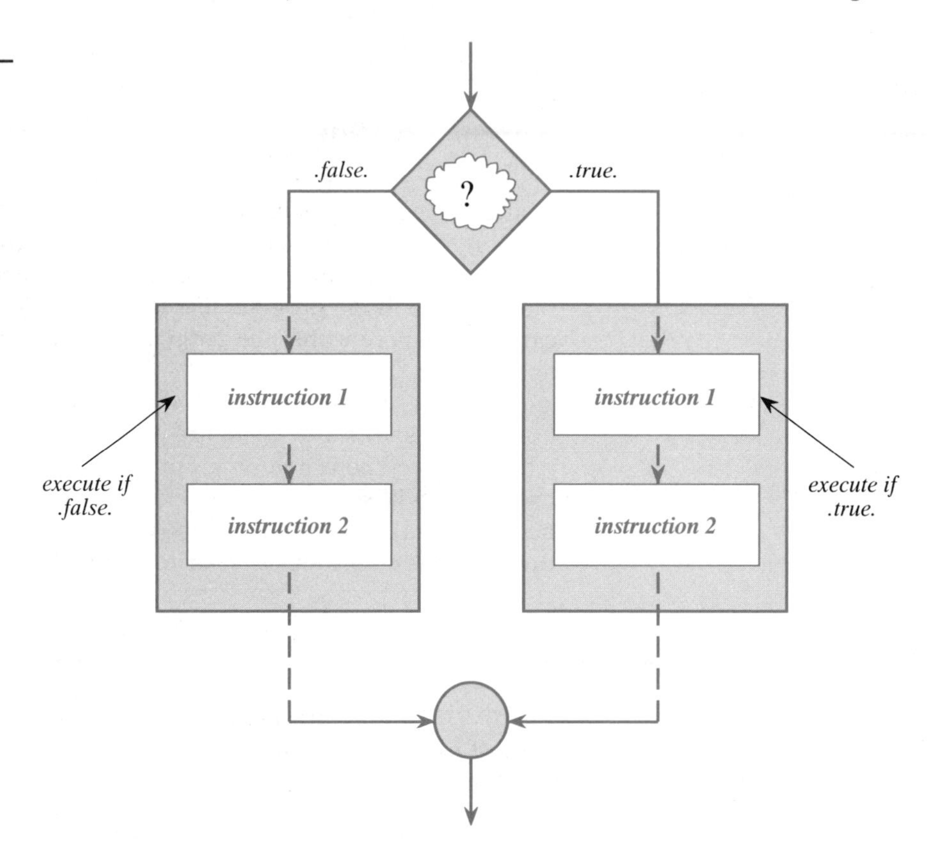

Notice in this flowchart that the *conditional* is the key to determining which way the flow will go. There can only be two possible outcomes of the conditional—.true. or .false.. There can be no ambiguity in the answer. The program does this by comparing two numerical expressions with the aid of a *RELATIONAL OPERATOR* such as:

```
( A .EQ. B )
```

In this logical expression the relational operator is .EQ., which compares the values on either side to determine if they are equal. The answer, of course, can only be true or false. The numerical expressions on either side can be of any complexity containing such things as built-in functions, mathematical functions, and so forth.

Five other relational operators are also available in FORTRAN as shown in the following table.

Operator	Math Symbol	Meaning
.EQ.	=	Equal to?
.NE.	≠	Not equal to?
.GT.	>	Greater than?
.GE.	≥	Greater than or equal to?
.LT.	<	Less than?
.LE.	≤	Less than or equal to?

Now that we have shown you the simple relational operators, here is a complete example of the IF-THEN-ELSE structure in correct FORTRAN:

```
IF ( A .EQ. B) THEN
        PRINT*, ' A & B ARE EQUAL '
ELSE
        PRINT*, ' A & B ARE NOT EQUAL '
ENDIF
```

The computer uses the .EQ. operator to compare the two numerical values stored in A and B. If they are equal, the computer prints the message "A&B ARE EQUAL." Otherwise, the message "A&B ARE NOT EQUAL" appears.

MORE ABOUT OPERATORS

The preceding table lists all simple relational operators. Although we have only shown them comparing simple variables, we also can use them to compare two complex expressions. Here are some examples of correct and incorrect usage:

Valid	Invalid	Comments
`(A.EQ.4.3)`		Variables and constants okay
`(A .EQ. 4.3)`		Spaces are ignored
`(A**2.EQ.B)`		Math operations allowed
`(SQRT(A**2).EQ.A)`		Okay, but will not give answer expected
	`(A=B)`	Must use symbolic name for =
	`(A EQ B)`	Must use periods around EQ

The only example in the table that deserves some attention is the fourth one:

```
(SQRT(A**2).EQ.A)
```

In algebra, of course, this comparison is always true since $(A^2)^{0.5} = A$ by definition. If you run this on the computer, though, you will find that the answer returned is usually .false.. To see why, here are results from an actual computer run:

$$A = 1.234567$$
$$A^2 = 1.524155$$
$$(A^2)^{0.5} = 1.234566$$

NOT EQUAL!

The condition is .false. because the computer does not store real numbers with infinite precision. In this example the computer has altered the last significant digit during the mathematical operations. So, the lesson to be learned from this exercise is: BE CAREFUL WHEN COMPARING REAL NUMBERS. Using .EQ. and .NE. are especially problematic.

There are many situations where we need *COMPOUND CONDITIONALS* that consist of two or more simple conditionals. A common example is:

$$12.3 \leq X \leq 98.0$$

Two conditions are involved in this, and both must be satisfied simultaneously. We show these here with the equivalent FORTRAN logical expressions:

$12.3 \leq X$	`( 12.3 .LE. X )`
$X \leq 98.0$	`( X .LE. 98.0 )`

Now, to have *both* conditions satisfied simultaneously, we introduce the *LOGICAL OPERATOR,* .AND.:

(*logical expression 1* .AND. *logical expression 2*)

which for the example given becomes:

```
(    12.3 .LE. X    .AND.    X .LE. 98.0    )
```

The logical operator, .AND. requires that *both* logical expressions must be .true. before it declares the whole expression .true.. We summarize this conveniently in the *truth table:*

condition 1	*condition 2*	*(condition 1 .AND. condition 2)*
.TRUE.	.TRUE.	.TRUE.
.FALSE.	.TRUE.	.FALSE.
.TRUE.	.FALSE.	.FALSE.
.FALSE.	.FALSE.	.FALSE.

There are also other logical operators that are of interest. The most important of these is the .OR. operator. This operator will return a .true. answer when *either* of the logical expressions is .true.. The only time that the overall answer is .false., is when *both* expressions are .false. simultaneously as we show in the following truth table:

condition 1	*condition 2*	*(condition 1 .OR. condition 2)*
.TRUE.	.TRUE.	.TRUE.
.FALSE.	.TRUE.	.TRUE.
.TRUE.	.FALSE.	.TRUE.
.FALSE.	.FALSE.	.FALSE.

Finally, there is the .NOT. operator, which reverses the logical value of an expression. It is different from the two previous operators in that it does not combine any two other operators. Rather, it works on a single logical value such as:

.NOT. *(logical condition)*

For example, if the logical condition were .true., the .NOT. operator would then return the value .false.. It is unlikely that you will use the .NOT. operator very often. But in a few situations it is a useful, although possibly confusing, tool.

Here is an example of the use of compound conditionals. Assume the following values for the variables already exist:

$X = 10.0 \qquad Y = -2.0 \qquad Z = 5.0$

```
IF( X * Y .LT. Z/X .OR. X/Y .GT. Z*X )

   -20.0        0.5       -5.0      50.0

        .TRUE.                .FALSE.

                   .TRUE.
```

Notice that the order in which we made the comparisons can make a difference in the results. Just as we had an hierarchy of operations for mathematical operations, we also have an hierarchy for logical operations. In addition, in examples such as the preceding one, we should be concerned about which comparison to do first—the mathematical operations or the logical ones. To help you sort out the answers to these questions, we present an updated table of the priorities of the various operators:

Priority	Math Symbol	Fortran Syntax	Meaning
1	(........)	(........)	Parentheses
2	A^B	**	Exponentiation
3	X	*	Multiplication
	$\div$	/	Division
4	+	+	Addition
	−	−	Subtraction
5	$=, \neq, <, \leq, >, \geq$	.EQ.,.NE.,.LT.,.LE.,.GT.,.GE.	Relational operator
6	$\bar{t}$	.NOT.	Logical complement
7	·	.AND.	Logical and function
8	+	.OR.	Logical or function

The program carries out any .AND. operator before any .OR. operator. Also, if there are several equivalent operators in an expression, such as several .OR.s, it evaluates them left to right. Try to follow this example to make sure that you understand the hierarchial rules:

$X = 10.0 \qquad Y = -2.0 \qquad Z = 5.0$

```
IF( X * Y .LT. Z/X .OR. X/Y .GT. Z * X .AND. Z*Y .LT. X )
   -20.0       0.5       -5.0       50.0       -10.0   10.0
        .TRUE.              .FALSE.                .TRUE.
                                         .FALSE.
                       .TRUE.
```

In the following table we list several compound conditionals. Trace through them to see if you can duplicate the indicated answers. Assume that the values of X, Y, and Z, are:

$X = 10.0$

$Y = -2.0$

$Z = 5.0$

Compound Conditional	Logical Value
`(X.EQ.Y.OR.X/Y+Z.EQ.0.0.OR.Y.GE.Z)`	.FALSE.
`(X.NE.Y.AND.Y.NE.Z.AND.X.NE.Z)`	.TRUE.
`(.NOT.(X.EQ.Y.OR.X.Y+Z.EQ.0.0).AND.Y.LE.Z)`	.TRUE.
`(X.GE.Z.AND..NOT.((Z*Y.LE.X).OR..NOT.(X.EQ.Y)))`	.FALSE.

If you are having trouble with the expressions given in the example, help yourself by placing parentheses around individual comparisons. Then evaluate each logical expression, and you will find that the problem is greatly reduced.

Fortunately, few problems, if any, ever require logical expressions as complex as the preceding ones. Usually, the most complex one you will see looks like this:

```
IF( (...) .AND. (...) ) THEN
```

In addition, any compound conditional as complex as the preceding examples should be rewritten in several simpler steps. As the complexity of these expressions increases, the likelihood of making an error increases rapidly—so, try to keep compound conditionals to a minimum.

6.4 Special Forms of the IF-THEN-ELSE Statements

The flowchart shown in Fig. 6.1 is the most general of the IF-THEN-ELSE statements and is always correct to use. There are several situations, however, where we can simplify this structure to save ourselves some extra lines of code. The special cases that we will look at in the next few pages are:

SINGLE ALTERNATIVE IF Statements
ONE LINE IF Statements
NESTED IF-THEN-ELSE Structures
SELECT-CASE Structure

As you become more familiar with FORTRAN, you will automatically switch between these several different types of IF structures. At the beginning, however, you probably will give in to the temptation to use only the full-blown IF-THEN-ELSE structure. We encourage you to resist this temptation and try to use some of these simpler structures whenever possible.

SINGLE ALTERNATIVE IFs

The general structure of the IF-THEN-ELSE structure is shown in Fig. 6.1. We sometimes call this structure the *DOUBLE ALTERNATIVE IF* structure since there are two sets of instructions for the program from which to choose. You may have guessed already that the *SINGLE ALTERNATIVE IF* structure is identical with the DOUBLE ALTERNATIVE structure, except that one of the alternatives is to do nothing. Figure 6.2 shows the flowcharts for both the DOUBLE and SINGLE ALTERNATIVE IF constructs.

In the SINGLE ALTERNATIVE structure there are no instructions on one side of the conditional. In this example nothing is to be done if the logical expression is *.false.*. You could just as easily have set up the SINGLE

FIGURE 6.2

Flowchart Symbols for SINGLE and DOUBLE ALTERNATIVE IF Structures

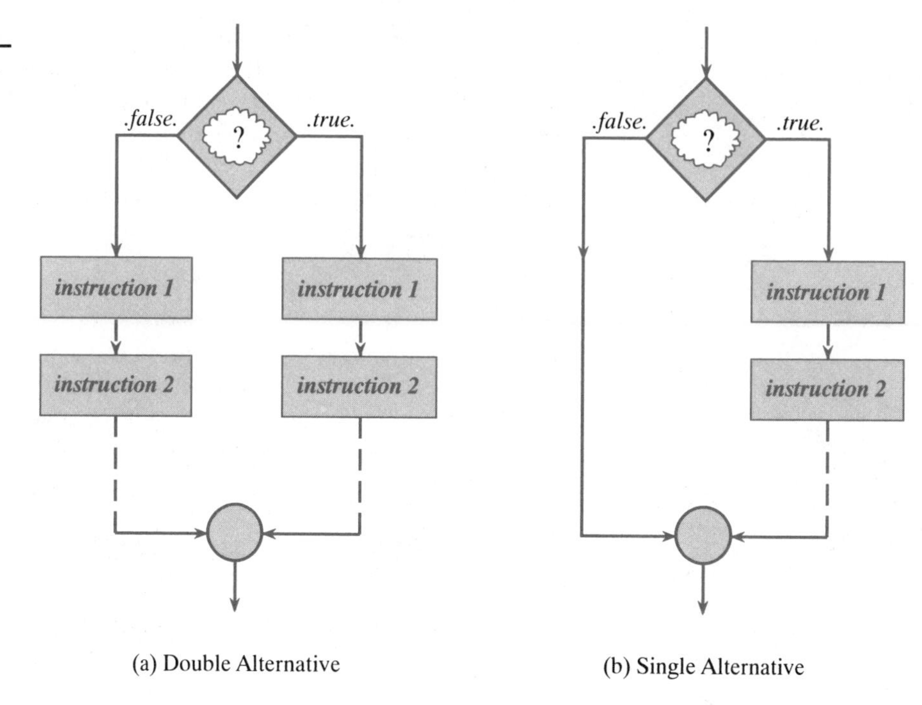

(a) Double Alternative (b) Single Alternative

ALTERNATIVE so that it does nothing on the *.true.* side, but it is better to set up the SINGLE ALTERNATIVE as indicated. The reason for this will be apparent shortly.

When we convert the SINGLE ALTERNATIVE into FORTRAN, we obtain the following:

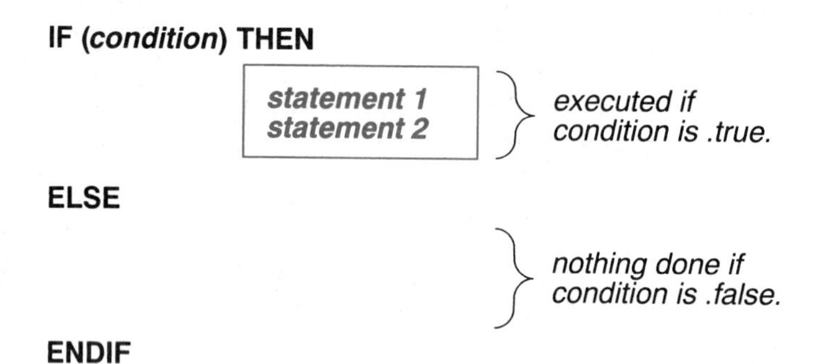

Notice that the ELSE statement serves no purpose here since it is obvious where the *.true.* block ends. Remember, in the DOUBLE ALTERNATIVE we needed the ELSE statement to set the two blocks apart. But for the SINGLE ALTERNATIVE, we don't need it anymore. So, let's drop it altogether:

IF (*condition*) THEN

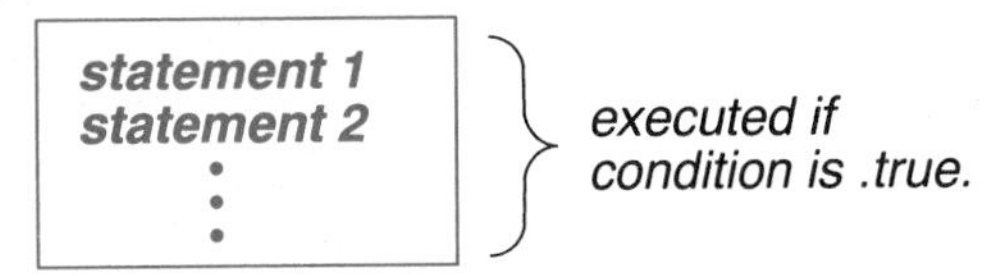

ENDIF

The benefit of doing this is simply a reduction of one line of code. This may not seem like much until you write a program with 100 such structures. A savings of 100 lines of code is noteworthy. In addition, the structure of the SINGLE ALTERNATIVE IF is easier to follow than its parent. So, a side benefit of this structure is a slight improvement in readability of the program. This should make the program a little bit easier to debug.

ONE LINE IF STATEMENT

The ONE LINE IF statement is a special case of a special case. Namely, it can be used when the SINGLE ALTERNATIVE IF structure has only one line to execute if the logical expression is true. The flowchart for the ONE LINE IF is shown in Fig. 6.3.

In the SINGLE ALTERNATIVE IF structure the *.true.* block can contain as many executable lines as you wish. The SINGLE LINE IF structure on the other

FIGURE 6.3

Flowchart Symbols for SINGLE ALTERNATIVE and SINGLE LINE IF Statements

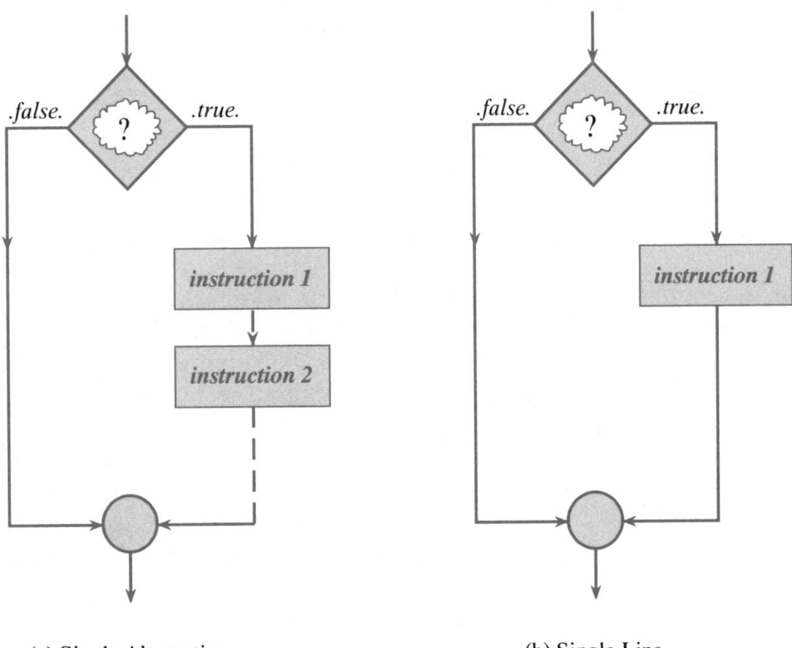

(a) Single Alternative

(b) Single Line

hand, can have only one line in the *.true.* block. Under this very restrictive circumstance the BLOCK-IF structure can be reduced to a single line:

```
IF (condition )     instruction 1
```

This is the simplest, but also the most restrictive of the IF structures. Note that we have succeeded in eliminating the THEN, ELSE, and ENDIF statements and that the executable instruction must be a *single line.* The following figure illustrates the flow of the logic:

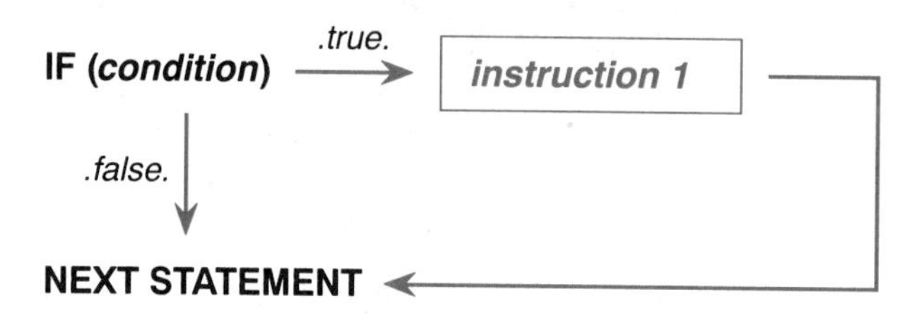

If the logical expression is *.true.*, the single statement on the same line executes. After this statement executes, control then passes to the next line in the program. If, however, the logical expression is *.false.*, control passes immediately to the next line.

The SINGLE LINE IF statements are very easy to use, but don't abuse them since they lead to programs with logic that is difficult to follow. Beginning programmers tend to use single-line IF statements with GOTO, almost to the exclusion of the other, more powerful structures. Try to avoid this trap and use the other IF-THEN-ELSE structures whenever possible.

NESTED BLOCK-IF STRUCTURES

The block of instructions within a BLOCK-IF structure may have any complexity you want. In fact, among the most common types of complex structures is the *NESTED BLOCK-IF* structure, in which a BLOCK-IF resides within another BLOCK-IF as Fig. 6.4 shows. With the NESTED BLOCK-IF structure, you can perform tests that have more than two choices. The SINGLE ALTERNATIVE structure offers only one choice. The DOUBLE ALTERNATIVE offers two choices and NESTED BLOCK-IF offers three choices. Examine the figure carefully, and you will see that there are indeed three choices.

To demonstrate how this works, let's try to write a program segment to test for whether a number is positive, negative, or zero. Of course, in this problem there are three possible outcomes. So, the following might be a reasonable algorithm:

1. Check to see if the number is > zero; if it is, then print "number is positive."

FIGURE 6.4

NESTED BLOCK-IF Structure

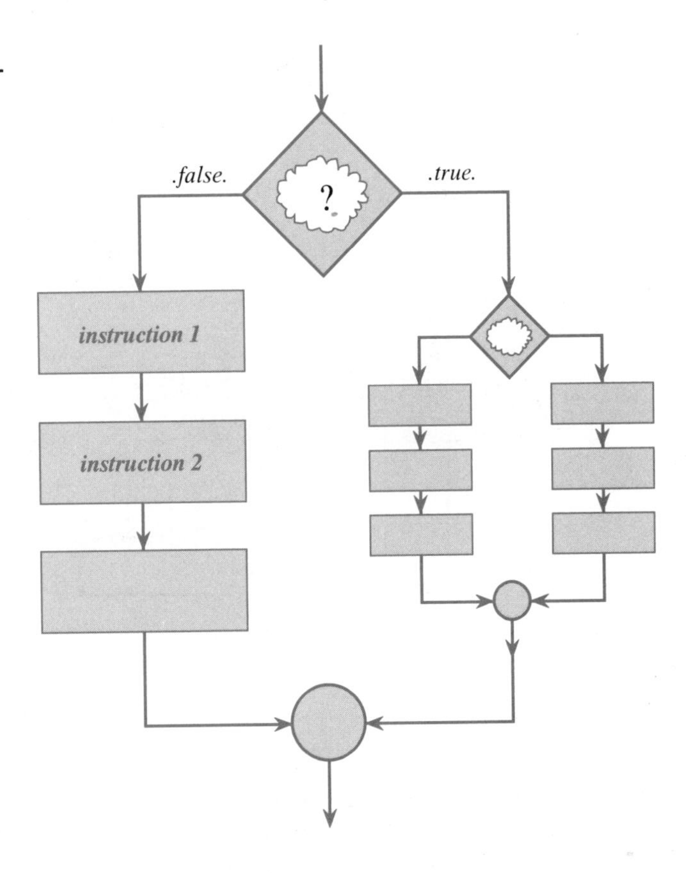

2. If the first test fails, then the number can be zero or negative.
3. Check to see if number < zero; if it is, then print "number is negative."
4. If both tests fail, then the number is zero.

This simple algorithm contains two tests. The first checks to see if the number is greater than zero. The program executes the second test only if the first one fails. Once we fill in the instructions, the flowchart looks like the one in Fig. 6.5.

Try a few numbers to see if you understand how this flowchart operates. It is good practice when tracing a program or flowchart to use *MAGIC BULLETS* or data that is likely to show errors in the logic. Here the most likely *MAGIC BULLET* is the number zero, which is neither positive nor negative. So, be sure to include this number in your list of the ones to try.

SELECT CASE

The SELECT-CASE construct is a very simple way to handle many ALTERNATIVE IF structures. In effect, it is simply an extension of the NESTED BLOCK-

FIGURE 6.5

Flowchart for Determining Sign of a Number

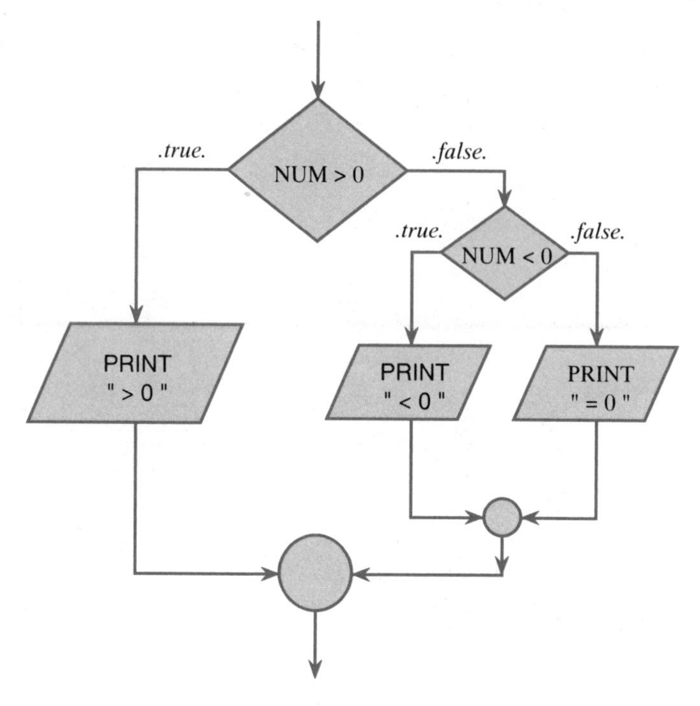

IFs to even more alternatives. Unfortunately, it is not part of the FORTRAN 77 standard, and few compilers support it as an option. Therefore, if your compiler does not have this structure, you may want to skip to the next section. The CASE construct, though, will be included in the next FORTRAN standard and will probably appear in most compilers shortly after that. So, this is a good place to discuss this topic.

Unlike the NESTED BLOCK-IFs, the SELECT-CASE structure performs only a single logical comparison. We introduced this in Chap. 3 and repeat it in Fig. 6.6.

The idea of the SELECT CASE is a very simple one. The single expression to be evaluated can be numerical, character, or logical data. The answer then

FIGURE 6.6

The SELECT CASE Structure

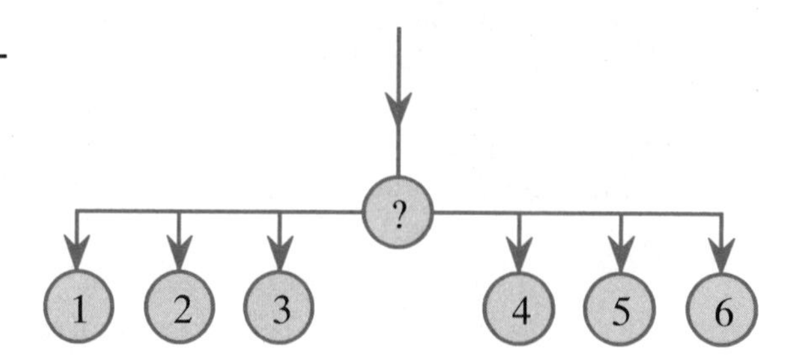

determines which direction the logic flows. For example, we might have a list such as the following from which to choose:

IF X = 0 THEN PRINT "ATTEMPTED DIVISION BY ZERO"
IF X = 1 THEN EVALUATE X*Y*Z
IF X = 3 THEN EVALUATE X/Y/Z
IF X = 4 THEN EVALUATE X*Y/Z
IF X = 7 THEN STOP

The SELECT-CASE construct can be used to do this in the following way:

```
READ *, X
SELECT CASE (X)
      CASE (0)
            PRINT *, 'ATTEMPTED DIVISION BY ZERO'
      CASE (1)
            PRINT *, X*Y*Z
      CASE (3)
            PRINT *, X/Y/Z
      CASE (4)
            PRINT *, X*Y/Z
      CASE (7)
            STOP
END SELECT
```

Once you generate the value of X, the computer uses the statement SELECT CASE to compare X with one of the values listed. If it finds a specified CASE, it carries out the instructions listed there. Thus, if you enter a value of 4 for X, the machine will print out the numerical value of X*Y/Z. One requirement of the SELECT-CASE structure is that the value entered must match one of the CASES.

The general form of the SELECT-CASE structure is:

SELECT CASE (*expression*)

CASE (*selector 1*)

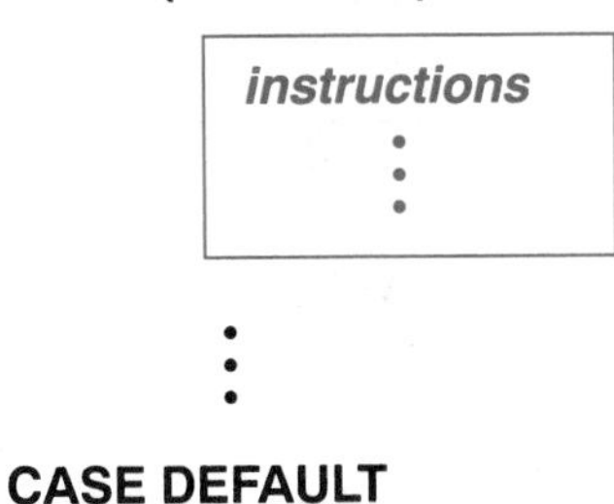

CASE DEFAULT

instructions — *optional*

END SELECT

The expression may give a numeric value, a character string, or a logical value, provided that the corresponding selectors in the CASE(..) statement are of the same type. Thus, for example, you may have an INTEGER expression, but you cannot have a character selector such as:

```
SELECT CASE (I)
   .
   .
   CASE ( 'TUESDAY' )
   .
   .
   .
   .
END SELECT
```

The special instruction CASE DEFAULT is a catchall section that is optional and contains instructions on what to do if the value of the expression is not among those listed. To show how this works, we will modify the previous example to handle any illegal values of X, which we will define as any value not equal to 0, 1, 3, 4, or 7:

```
10  READ *, X
    SELECT CASE (X)
         CASE (0)
               PRINT *, 'ATTEMPTED DIVISION BY ZERO'
         CASE (1)
               PRINT *, X*Y*Z
         CASE (3)
               PRINT *, X/Y/Z
         CASE (4)
               PRINT *, X*Y/Z
         CASE (7)
               STOP
         CASE DEFAULT
               PRINT *, 'ILLEGAL VALUE, REENTER:'
               GOTO 10
    END SELECT
```

In the event that X is not an allowed number, the CASE DEFAULT intercepts it, prints out a message to reenter, and then transfers control to the input statement. If you try to do this with NESTED BLOCK-IFs (and we will shortly), you will find it a much more complex structure. The SELECT-CASE structure is a much better way to do multiple selections than to use NESTED BLOCK-IFs. If your compiler does not support this feature, however, you must use the BLOCK-IFs as discussed in the next section.

ELSE-IF STRUCTURES

The final special case of BLOCK-IF structures that we will discuss is the ELSE-IF. This is another way of carrying out the multiple alternative structure that we examined in the previous sections. The ELSE-IF structure is a NESTED BLOCK-IF that economizes on the number of ENDIF statements. This makes the structure somewhat easier to trace and debug. To show how this works, here is the problem from the last section written in a full-blown NESTED BLOCK-IF structure:

```
10  READ*,X
    IF(X.EQ.0) THEN
        PRINT *,'ATTEMPTED DIVISION BY ZERO'
    ELSE
        IF(X.EQ.1) THEN
            PRINT *, X*Y*Z
        ELSE
            IF(X.EQ.3) THEN
                PRINT *, X/Y/Z
            ELSE
                IF(X.EQ.4) THEN
                    PRINT *,X*Y/Z
                ELSE
                    IF(X.EQ.7) THEN
                        STOP
                    ELSE
                        PRINT *,'ILLEGAL VALUE. REENTER:'
                        GOTO 10
                    ENDIF
                ENDIF
            ENDIF
        ENDIF
    ENDIF
```

Of course, a segment such as this is tedious to construct and even more difficult to debug. An additional problem with this approach is that it is not clear where each block ends. If you draw lines around each block as we have done, though, you should have no trouble. In addition, having all the ENDIFs bunched together at the end is not very helpful, not to mention inelegant. So, FORTRAN has a way to improve the situation. What we will do is the following:

1. Eliminate all but one of the ENDIFs.
2. Combine the two lines
```
ELSE
  IF (...) THEN
```
3. Replace with
```
ELSEIF (...) THEN
```

This is how we can rewrite the preceding program with this new structure:

```
10  READ*,X
    IF(X.EQ.0) THEN
          PRINT *,'ATTEMPTED DIVISION BY ZERO'
    ELSEIF(X.EQ.1) THEN
          PRINT *, X*Y*Z
    ELSEIF(X.EQ.3) THEN
          PRINT *, X/Y/Z
    ELSEIF(X.EQ.4) THEN
          PRINT *,X*Y/Z
    ELSEIF(X.EQ.7) THEN
          STOP
    ELSE
          PRINT *,'ILLEGAL VALUE.REENTER: '
          GOTO 10
    ENDIF
```

The ELSEIF structure is a considerable improvement over the NESTED BLOCK-IF structure. Besides the reduction in lines of code from 23 to 15, the ELSEIF structure is significantly easier to follow. The general form of the ELSEIF structure is as follows:

```
IF (condition 1 ) THEN

        block 1

ELSEIF (condition 2 ) THEN

        block 2

ELSEIF (condition 3 ) THEN

        block 3
          .
          .
          .
ELSE

        block N

ENDIF
```

Keep in mind that there is only one ENDIF statement to service all the IF statements. Therefore, when any one of the IF statements becomes true, control will transfer to the end after completing the instructions in the appropriate BLOCK. Thus, it is possible for only one IF statement, the first one, to be evaluated.

6.5 LOOP Structures

The third type of control structure that we will discuss is the LOOP. As we discussed in Chap. 3, there are three forms of the loop in common usage as summarized here:

COUNTED LOOP
CONDITIONAL LOOPS: REPEAT UNTIL
DO WHILE

The COUNTED LOOP is one where the loop executes a predetermined number of times and the variables controlling the loop cannot be altered during the loop execution. The CONDITIONAL LOOPs, on the other hand, lack a predetermined stopping point, and the loop alters the variables controlling it.

We review the three loop structures in this chapter. The COUNTED LOOP is the only one supported by FORTRAN 77 and the new standard. Commercial compilers, however, generally support one or both of the conditional LOOPs. Therefore, we include these in our discussion. Also, for those of you whose compilers do not support these extensions, we will show you alternate ways of doing loop execution.

COUNTED LOOPS

The COUNTED LOOP is the most widely used loop among the high-level languages. As we discussed in Chap. 4, these loops have a very rigid structure. They execute for a predetermined number of iterations and the variables controlling the loop cannot be altered once the loop begins. In FORTRAN we call these DO LOOPs, and they have the flowchart symbol, as shown in Fig. 6.7.

The DO LOOP uses a *LOOP CONTROL VARIABLE* that controls the number of times that the loop executes. As the flowchart in Fig. 6.7 shows, we use a diamond-shaped box to mark the beginning of the loop. Usually, you place in the box the starting and stopping values for the loop with the step size. For example, if you wanted the variable L to go from 1 to 100 in steps of 1, the numbers 1, 100, and 1 would be placed in the box. After we inform the computer that a loop is about to begin, it determines the number of loops based on the starting, stopping, and step size values. The computer then carries out the instructions inside the body of the loop. After each *iteration* through the loop, the LOOP CONTROL VARIABLE increases by the step size. Finally, the computer runs a test to see if the LOOP CONTROL VARIABLE is less than the stop value. If it is, then the loop executes. Otherwise, the loop will stop and control will pass to the next step after the end of the loop.

FIGURE 6.7

Flowchart Symbol for DO LOOPs

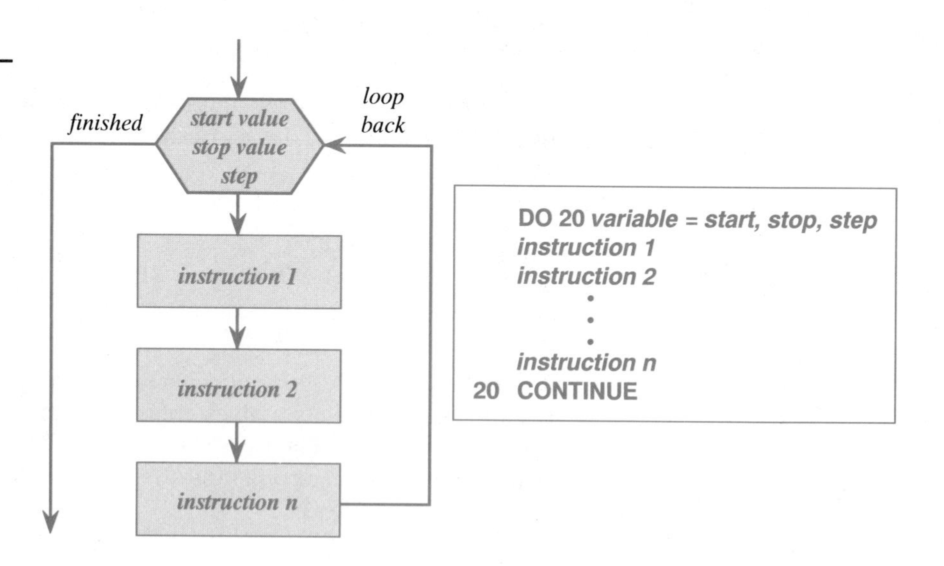

The general form of the corresponding FORTRAN statements is shown in the figure and summarized as follows:

```
          DO label    LCV = start, stop (,step)

                            instruction 1
                            instruction 2

    label     CONTINUE
```

The DO statement marks the beginning of the loop and contains all the information necessary to set up the loop. It shows where this loop ends by using the *statement label.* Of course, the statement label marking the end of the loop must correspond to the one used in the DO statement. Also, each statement label must be unique.

The DO statement also contains the information on the LOOP CONTROL VARIABLE (LCV). The program assigns the starting value to this variable and after each cycle through the loop increases the value by the *step* size. Before the loop can recycle, however, the computer checks to see if the new value of the LCV exceeds the *stop* value. If it does, then the loop stops and control passes to the next line after the end of the loop.

The (*label* CONTINUE) statement marks the end of the loop. By itself, though, it does nothing. It is a contrived statement to avoid confusion. Any executable statement, such as PRINT can be used in its place. But if you use such a statement, it is not always clear whether the program should execute the statement every time. So, to avoid this confusion, we use this do nothing statement.

The computer has complete control of the loop and handles all the tasks associated with it. These include:

- *Initialize* the LCV to the start value.
- *Increment* the LCV by the step value each time through the loop.
- *TEST* the LCV to see if it exceeds the stop value.
- Decide when to terminate the loop.

You don't need to do anything yourself. The computer does all this automatically. The only pitfall you might encounter occurs if you attempt to modify the LCV while within the loop. The LCV belongs to the computer. You may use it for calculations, but you may not change it.

Here is an example of a DO LOOP in action:

```
      DO 100      ICOUNT = 1, 10, 1
           PRINT *, 2 * ICOUNT
100   CONTINUE
      PRINT *, ICOUNT
```

This is a simple loop, which will print out the even numbers from 2 to 20. An interesting point about loops is that the LCV increments inside the loop until it *exceeds* the final value, 10 in this example. Of course, the first integer number that exceeds 10 is the number 11. Therefore, the PRINT statement outside the DO LOOP will print the final value of ICOUNT as 11. If we ran this program on a computer, this is what the output would look like on a terminal screen:

```
$ RUN PROG          <CR>
     2
     4
     6
     8
    10
    12
    14
    16
    18
    20
    11
FORTRAN STOP
$
```

Notice that we can *use* the LCV inside the loop, but we cannot *modify* it. For example, if you tried something like this, you would receive a compiler error:

```
      DO 10 ICOUNT = 1, 10, 1
            ICOUNT = ICOUNT + 1
   10 CONTINUE
```

This kind of error is sometimes not so obvious, especially when you begin to use subroutines. We will give you an example in the chapter on debugging of what can happen if the compiler allowed you to change the LCV inside a loop. The results can be disastrous. So, be careful when using the LCV inside loops.

The *start, stop,* and *step* values need not always be constants. In fact, it is usually advisable to use variables instead. Thus, consider the following:

```
      READ*, ISTART, ISTOP, ISTEP
      DO 30 ICOUNT = ISTART, ISTOP, ISTEP
            PRINT *, 2 * ICOUNT
   30 CONTINUE
      PRINT *, ICOUNT
```

With this structure, we can enter the values for ISTART, ISTOP, and ISTEP at execution time. For example, if we type in 1, 10, and 1, we will simply obtain the previous results. If we run the segment a second time and type in new values of 5, 9, and 2, we obtain a different set of output data: 10, 14, 18, and 11. The first three values correspond to 2 × 5, 2 × 7, and 2 × 9 and the last value printed is ICOUNT after the loop stops or 11. Be careful in tracing through this example that you increment properly because in this second run we specified an increment of 2.

The STEP variable is optional. If you do not include it, the computer assumes that you want a value of 1. But if you include it, the machine will give you what you ask for.

The LOOP CONTROL VARIABLES can be either REALs or INTEGERs, although we recommend INTEGERs. The reason is that the computer stores INTEGERs precisely in its memory while it stores REALs with some imprecision. Thus, a loop with REAL LCVs may execute one too many or one too few times because of this imprecision. Since this effect is not easily predictable, we advise you to use INTEGER LCVs exclusively.

Here are some examples of correct and incorrect usage of the DO statements:

Correct	Incorrect	Comments
DO 10 I = 1, 10		Step is optional (assumed = 1)
DO 20 I = J, 10		Mixing variables, constants okay
DO 30 I = 10, 1, −1		Decreasing index okay
	DO 40 I = 1.0, 5.0, 0.1	Mixed mode
	DO 50 I = 1, 10, I	Subtle attempt to modify LCV
	DO 60 I = 10, 1	Loop does not converge

NESTED DO LOOPS

The body of the DO LOOP can have any complexity desired. The simple examples previously shown contain only sequential instructions. Yet, it is possible to put other branching and looping instructions in the body. One very common structure in programming is to *NEST* one loop inside another. You will see this frequently with arrays and complex I/O. In block form here is how we do it:

```
      DO 10 I = ISTART, ISTOP, ISTEP
            DO 20 J = JSTART, JSTOP, JSTEP     } Body of
20          CONTINUE                           } DO 10 Loop
10    CONTINUE
```

FIGURE 6.8

Flowchart of NESTED DO LOOPs

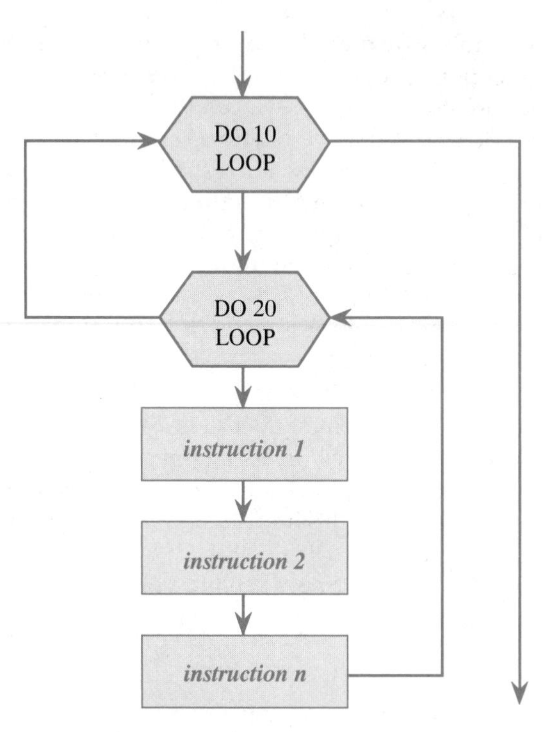

Figure 6.8 shows what the flowchart of a nested DO LOOP might look like.

The flowchart is useful for seeing the flow of the logic. Notice, for example, that the innermost loop lacks an outlet except through the outer loop. Also, the inner loop must cycle to completion before the outer loop increments even once. Finally, note that the inner loop resides completely within the outer loop.

This last point is one that often creates confusion, so we should spend some time discussing it. Here are a few loops that are *PROPERLY NESTED:*

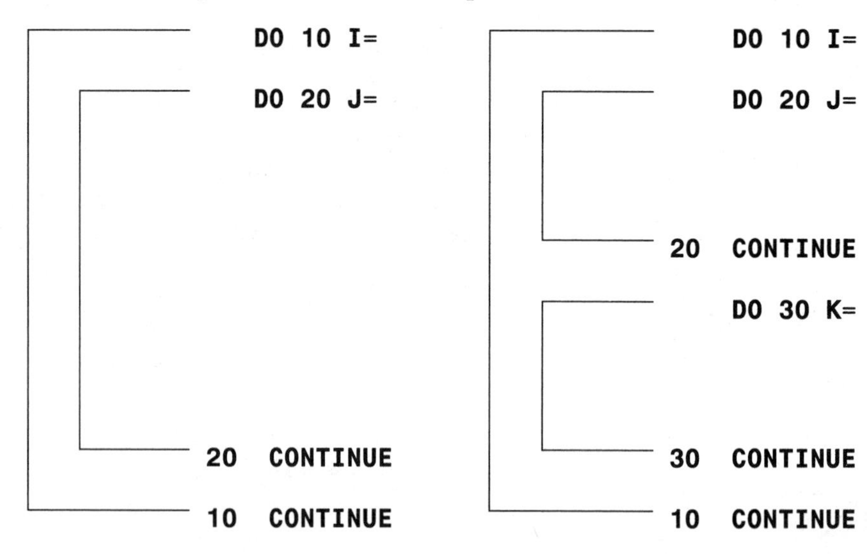

An important note about properly nested loops is that they each have a different LCV. If we tried to use the same variable for two nested loops, the compiler will interpret this as an attempt to modify the value of the LCV. This is strictly forbidden. Now here are some improperly nested loops:

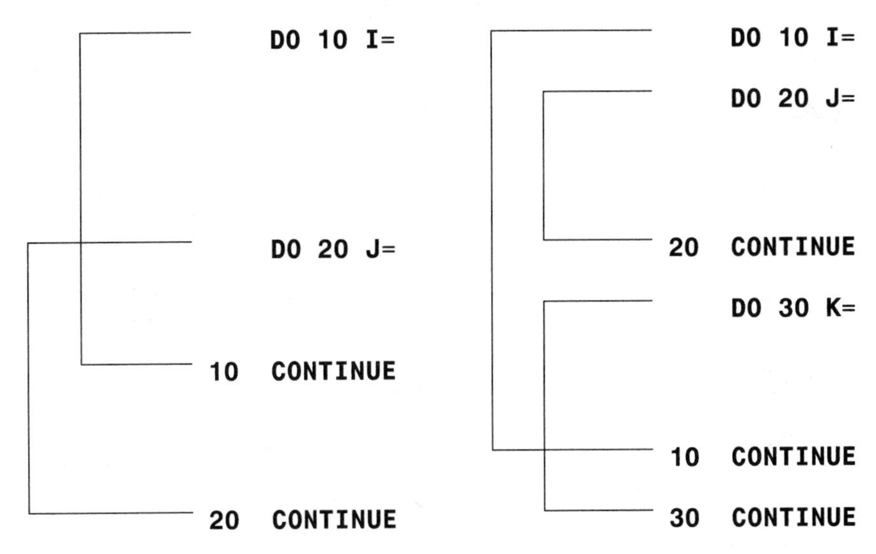

Improper nesting of loops is an easy error to catch since compilers recognize this error very easily. Also, since the beginning and end of DO LOOPS are clearly marked, it is easy to spot nesting errors.

Here is an example of how to use nested loops to generate a simple multiplication table. At this point we don't have the sophistication yet to produce a nice square table. But at least this little program will generate the data.

```
      READ *, I , J
      DO 10    OUTER = 1, I
            DO 10 INNER = 1, J
                  PRINT*, INNER*OUTER
   10 CONTINUE
```

Suppose we enter 2 for I and 3 for J. The nested loops tell us that the innermost loop will execute more rapidly than the outer loop. Here is a trace table for the variables:

OUTER	INNER	OUTPUT
1	1	1
1	2	2
1	3	3
2	1	2
2	2	4
2	3	6

Notice that the variable OUTER remains fixed while the variable INNER goes through its range. After the inner loop finishes, OUTER increases by one and then the inner loop begins all over. Be sure that you understand how these nested loops execute, since you will see them repeatedly when we get to arrays.

Did you notice in the previous example that we combined both CONTINUE statements into one? The compiler will allow this if there are no program lines between the two CONTINUE statements:

Allowed:

```
      DO 10 I=
         DO 20 J=
            .
            .
20    CONTINUE
10    CONTINUE
```

```
      DO 10 I=
         DO 10 J=
            .
            .
10    CONTINUE
```

Not allowed:

```
      DO 10 I=
         DO 20 J=
            .
            .
20    CONTINUE
      PRINT*, ...
10    CONTINUE
```

```
      DO 10 I=
         DO 10 J=
            .
            .
10    CONTINUE
      PRINT*,...
```

This device saves you a few extra lines, and in longer programs it may save you hundreds of lines. So, it is well worth using. If you try to set up your loops so that there are no program lines between the CONTINUE statements, you can use this device frequently.

DO WHILE LOOPS

The *DO WHILE LOOP* is available on many FORTRAN compilers, but there are still a few that do not support it. Therefore, if you are one of the unlucky ones, skip to the later section that will show you how to generate an equivalent structure using IF-THEN-ELSEs and GOTOs.

The DO WHILE structure is a form of a conditional loop as we discussed in Chap. 3. The loop will execute indefinitely until a particular condition becomes .false.. Of course, to start the loop, the condition must initially be .true.. Obviously, the condition must somehow change from .true. to .false.. This is a stark difference from the DO LOOP, where the computer does not allow us to change the control variable. For the DO WHILE loop, the control variable *must* change. Otherwise we will be trapped in an infinite loop.

The DO WHILE loop should be used when you do not know in advance how often to execute the loop. A good example is when we want to read in some data from an experiment. Generally, you do not know how many data items there will be. Thus, you set up a loop so that if a data item has a specific value, this will mark the end of the data. This will trigger the loop to stop.

To demonstrate this, let's assume that we are reporting the weights of a laboratory full of rabbits. Since rabbits multiply so fast, we never know in advance how many there will be. Therefore, we set up the loop to read in the weights continuously. When one of the weights is greater than 500 pounds (lb), the loop will stop. We sometimes call this special value a *SENTINEL* value, and it is essential for situations like this. Here is the loop in FORTRAN:

```
DO WHILE (WEIGHT.LE.500.0)
     READ*, WEIGHT
          .
          .
          .
END DO
```

When the loop begins, the computer does not know the value of WEIGHT and probably initializes it to zero. Therefore, the first conditional is .true. and the loop will begin. It then reads in a weight value and does something with it according to the body of the loop. After it finishes, it will return to the conditional statement to check to see if you entered a value greater than 500 lb. Since few rabbits weigh 500 lb, this condition will always be .true. and the loop will continue. When you have entered all the data, you type in a weight of something like 99999 and the loop will end. Since a weight of 99999 exceeds 500, the condition is now .false. and the loop will exit.

There is no special flowchart symbol for the DO WHILE loop. Instead, we use a conditional structure with GOTO type symbols. Figure 6.9 shows the previous problem in a flowchart form.

FIGURE 6.9

Flowchart Showing the DO-WHILE LOOP

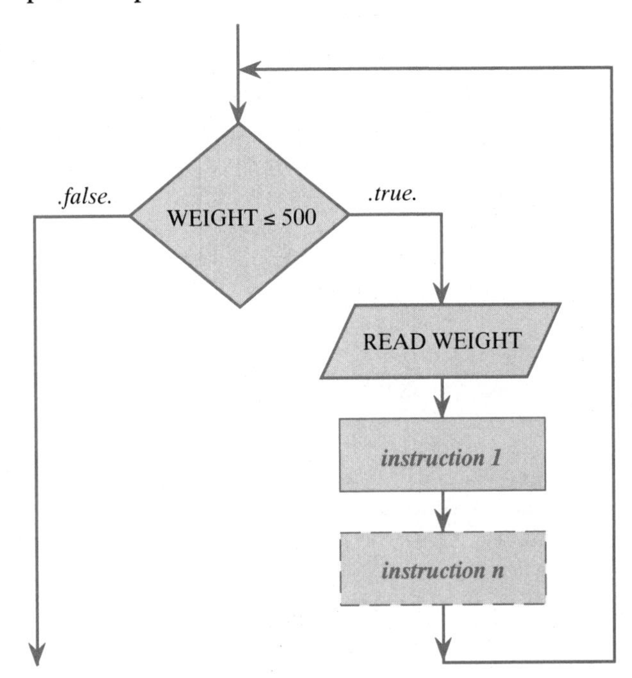

The logical condition must satisfy the rules discussed earlier in this chapter. Also, each loop must end with an END DO statement and begin with the DO WHILE statement followed by the conditional.

Some compilers have slightly different statements for this loop, such as WHILE DO, END WHILE, and so forth. So, check your manual to find out the precise form. Hopefully, you will find this construct in your compiler. If not, then you must use the structure discussed in the next section.

ALTERNATE FORM OF DO WHILE

If your system does not support the DO WHILE construct, you must set up your programs with other structures that we have already discussed, namely, the IF-THEN-ELSE and the GOTO statements. To see how to do this, look at the flowchart in Fig. 6.9, and you will see that the following code accomplishes the desired function:

```
10  IF(WEIGHT .LE. 500.) THEN
          READ*, WEIGHT
          instruction 1
          instruction 2
                .
                .
                .
          instruction n
          GOTO 10
    ENDIF
```

In this program segment the computer first tests to see if WEIGHT is within range. If it is, then the program will read the next value and process it. The GOTO statement at the end then transfers control back to the beginning of the IF statement for a recheck. Thus, we have converted a conditional statement into a loop.

This structure should be avoided if possible, however, because of the GOTO. GOTOs eventually cause serious problems in debugging programs. So, if at all possible, use the DO WHILE structure.

REPEAT UNTIL LOOPS

The REPEAT UNTIL LOOP is similar to the DO WHILE loop in that both are conditional loops. The difference is that the REPEAT UNTIL LOOP performs the conditional test at the end of the loop. In effect, it asks the question "Do another?". If the answer is yes, then control transfers to the top of the loop. Otherwise the loop terminates.

We can modify the previous loop to enter weights of rabbits, as in Fig. 6.10, to see how this works.

FIGURE 6.10

Flowchart Showing the REPEAT UNTIL LOOP

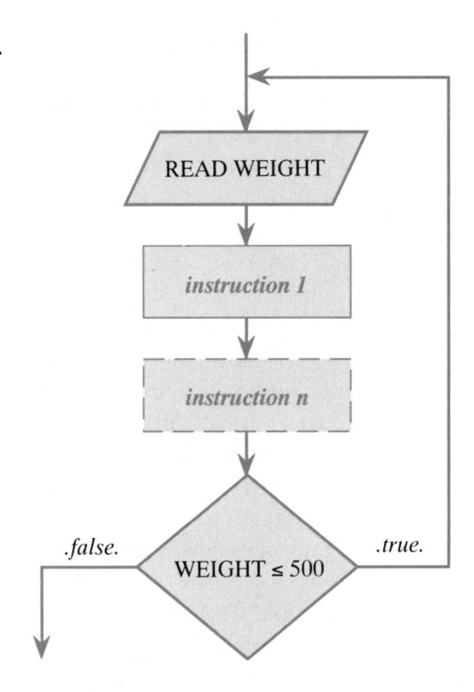

If your compiler supports the REPEAT UNTIL structure, the FORTRAN code would look like this:

```
REPEAT
      READ *, WEIGHT
      instruction 1
      instruction 2
          .
          .
          .
      instruction n
UNTIL (WEIGHT.GT.500.)
```

The conditional has changed slightly from the previous structure. In the DO WHILE structure, we tested to see if WEIGHT $\leq$ 500. In the REPEAT UNTIL structure we test to see if WEIGHT > 500. In most situations you can use the REPEAT UNTIL and the DO WHILE structure interchangeably. The only thing that you must watch for is the change in the conditional.

6.6 Algorithms

In this section we give four examples that demonstrate the control structures of this chapter. The problems that we show are examples of language structures (LS) for performing specific tasks. With only slight modifications, you can use them in your own programs.

The four examples that we present are:

ONE LINE IF statement:	Odd or even number?
BLOCK-IF structure:	Function evaluation over a range
COUNTED LOOP:	Factorial of a number
CONDITIONAL LOOP:	Evaluating an infinite series

All four of these LSs will be used in the exercises at the end of the chapter. Therefore, when you develop your algorithms to solve the assigned problems, you can stop the refinement process when you get to one of these structures. Simply incorporate them into your algorithm.

The first structure that we present is a method to see if a number is an even or odd integer. This process is so trivial that we don't give it much thought. Yet, it is an important element in many scientific calculations. One way to find out if a number is even or odd is to divide it by 2 and then multiply it by 2 and see if you get the same number back. This seems strange—$(N/2) \times 2$ is always equal to N, right? In algebra the answer is yes, but not if we use the rules of integer division! Remember, when we do division of integers, the computer drops any remainder. Here are some examples of the $(N/2) \times 2$ formula:

N	N/2	(N/2)*2	N=(N/2)*2
1	0	0	NO
2	1	2	YES
3	1	2	NO
4	2	4	YES
5	2	4	NO

This is a simple idea. When N is even, then $(N/2) \times 2 = N$, but if N is odd, then the equality does not hold. This is a perfect setup for a logical condition. We now set up the test:

IS (N/2) * 2 = N ?

If the condition is .true., then N is even. If the condition is .false., then N is odd.

LS FOR ODD OR EVEN NUMBER

PROBLEM STATEMENT: Read in an integer, N, and determine if it is odd or even?

ALGORITHM DEVELOPMENT:

Step 1a: Review the problem statement. Indicate all nouns in **bold** and verbs in *italics:*

Read in the **Integer, *N*,** and *determine* if it is odd or even.

Formulate questions based on nouns and verbs:

Q1: How do we read in a number?
Q2: How do we determine if the number is odd or even?

Step 1b: Construct answers to above questions:

A1: Use TERMINAL INPUT command.
A2: Check to see if $N = (N/2)*2$
If true, then N is even.
If false, then N is odd.

ALGORITHM:

1. Read in number, assign to variable, N
2. Compare (N/2)*2 to N to see if they are equal
3. If equal, print message "EVEN"
4. If not equal, print message "ODD"

SEGMENT:

```
READ *, N
IF( (N/2)*2 .EQ. N) PRINT*,'EVEN'
IF( (N/2)*2 .NE. N) PRINT*,'ODD'
```

or

```
READ *, N
IF( (N/2 *2) .EQ. N) THEN
        PRINT *, 'EVEN'
ELSE
        PRINT *, 'ODD'
ENDIF
```

FLOWCHART:

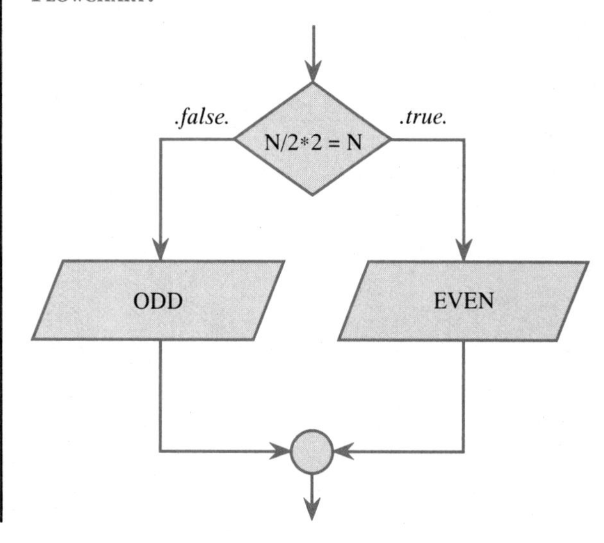

In this example we have shown two different FORTRAN versions of the algorithm. The first uses two consecutive ONE LINE IF statements and the second uses the BLOCK-IF structure. Although the ONE LINE IF statements result in a shorter program segment, two lines versus five, we still prefer the BLOCK-IF. The two ONE LINE IFs in this example do not present a problem, but in more complicated programs they may result in more complicated logic. Therefore, we favor the BLOCK-IF structure, even if it involves more code.

The second LS that we will show is the evaluation of a mathematical function over a dual region. There are many physical problems where the equations governing the behavior will differ depending on the region of operation. The example that we show you here is behavior of a diode. A diode is an important electrical component that allows current to flow in one direction but not in the reverse direction. When you apply a voltage, V, in the forward direction (called forward biasing), the diode allows current, i_D , to flow, as the curve in Fig. 6.11 illustrates. But when we reverse the voltage (called reverse biasing), very little current flows. Figure 6.11 shows both the theoretical curve and the experimentally measured curve.

In the forward bias region (positive voltage) there is good agreement between the theory and the experiment. But in the reverse bias region (negative voltage) there is a voltage, V_R, where the diode breaks down and then no longer blocks the current. Thus, there are two regions where the equations describing the current-voltage characteristics are different:

$$i_D = \begin{cases} I_0(e^{kV} - 1) & V > V_R \\ -I_0(e^{-kV} - 1) & V \leq V_R \end{cases}$$

where I_0 and k are diode characteristics. The way that we handle this is to check first to see if the value of V exceeds V_R. If it does, then we use the first formula. Otherwise we use the second formula. Notice that this is a good situation to use the IF-THEN-ELSE structure since there are two alternatives and a simple test to perform. The two alternatives are the two equations and the simple test is to decide in which voltage region we are.

FIGURE 6.11

Voltage (V), Current (I) Characteristics of a Diode

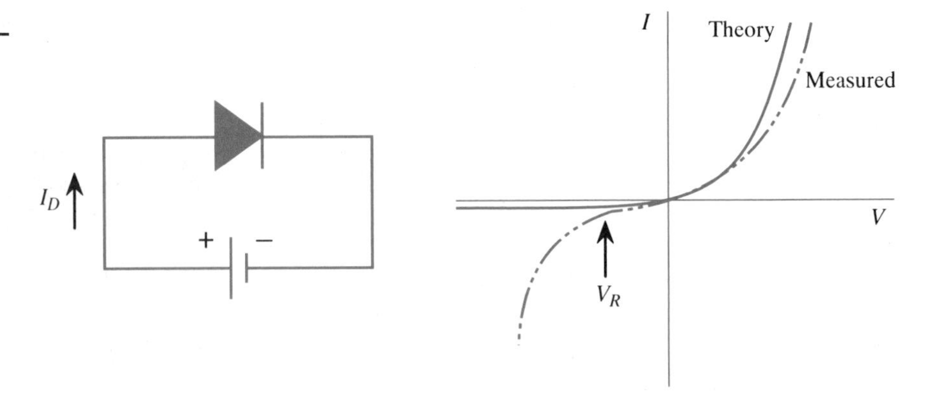

Although the algorithm shown in the LS contains only the IF-THEN-ELSE structure, you can also set it up with two ONE LINE IF statements. However, we have avoided this structure and will continue to avoid it to be consistent with our comments in the last example:

LS FOR DUAL RANGE FUNCTIONS

PROBLEM STATEMENT: Calculate the diode current, i_D, for a given applied voltage, V, if the characteristics are given by:

$$i_D = \begin{matrix} I_0(e^{kV} - 1) & V > V_R \\ -I_0(e^{-kV} - 1) & V \leq V_R \end{matrix}$$

where I_0 and k are diode characteristics.

ALGORITHM DEVELOPMENT:

Step 1a: Review the problem statement. Indicate all nouns in **bold** and verbs in *italics:*

Calculate the **DIODE current, i_D,** for a given **applied voltage, V,** if the characteristics *are given* by equations.

Formulate questions based on nouns and verbs:

Q1: How do we calculate diode current?
Q2: How do we apply the two equations?
Q3: Where do we get the constants, V, V_R, I_0, and k?

Step 1b: Construct answers to above questions:

A1: Use equations to calculate i_D for a given V
A2: See if V is greater than or equal to V_R:
If true, then use first equation.
If false, then use second equation.
A3: Use **TERMINAL INPUT command** to enter required constants.

ALGORITHM:

1. Read in constants, assign to variables, *V, VR, I*0 and *K*
2. Compare V to VR to see if they V ≥ VR
3. If equal, use Eq. #1
4. If not equal, use Eq. #2

SEGMENT:

```
REAL IO, K
READ *, V, VR, ID, IO, K
IF( V .GE. VR) THEN
      ID = IO * ( EXP( K*V ) - 1 )
ELSE
      ID = -IO * ( EXP(-K*V) - 1 )
ENDIF
PRINT *, 'DIODE CURRENT =', ID
```

FLOWCHART:

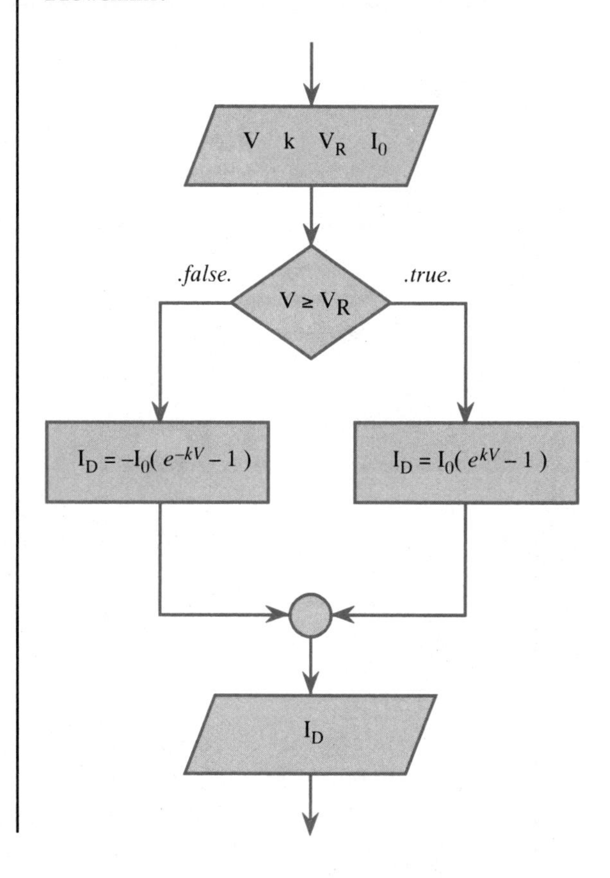

This algorithm can be easily modified to evaluate any split function. Simply replace the V_R in the conditional with the value where the splitting occurs. Also, replace the two functions with the appropriate functions that you wish to evaluate. To extend this idea to more than two regions is a simple process and is the subject of Prob. 11 at the end of the chapter.

The third algorithm that we will examine is the calculation of the factorial ($N!$). This is a very common function that we use often in engineering and science, especially when we need to use statistical analyses. The factorial function is defined by

$$N! = (N)\,(N-1)\,(N-2)\cdots(3)\,(2)\,(1)$$

This function involves the multiplication of all numbers from 1 to N inclusive. For example, if $N = 9$, the 9! equals:

$$9! = (9)\,(8)\,(7)\,(6)\,(5)\,(4)\,(3)\,(2)\,(1) = 362880$$

The way we break the problem down so that we can program it is to take partical products. If you think about it, you will realize that when you calculate 9! by hand, you multiply only *two* numbers at a time, though we write it algebraically as 9 multiplications. Thus, we start with $9 \times 8 = 72$. Then we multiply 72 times the next number in the sequence, or 7 and obtain 504. Then we multiply 504 by the next number, 6, and obtain $504 \times 6 = 3024$. We repeat this process until we reach the last number, 1.

We give the partial product a variable name, FACT. Notice that after each multiplication, we replace the value of FACT by the new value. Note also that the first value of FACT is 9 here, or N in the general case. This algorithm is developed in the following LS:

LS FOR FACTORIALS

PROBLEM STATEMENT: Calculate factorial, $N!$, where

$$N! = (N)*(N-1)*(N-2)*\cdots(3)\,(2)\,(1)$$

ALGORITHM DEVELOPMENT:

Step 1a: Review the problem statement. Indicate all nouns in **bold** and verbs in *italics:*

Calculate the **factorial, $N!$,** where $N! =$ (equation)

Formulate questions based on nouns and verbs:

Q1: How do we calculate the factorial?
Q2: Where do we get N?

Step 1b: Construct answers to above questions:

A1: Use equation to calculate $N!$
A2: Use **TERMINAL INPUT command** to enter N

ALGORITHM REFINEMENT:

Step 2a: Review previous answers to see if refinement is needed. Answer A1 is not sufficiently detailed. Construct new questions with nouns in **bold** and verbs in *italics:*

Q3: How do we apply equation?

Step 2b: Construct answers to above question:

A3: Assign FACT = N. Then multiply FACT by the next number in the sequence and assign this new value to FACT. Repeat this process until all numbers in the sequence are used.

ALGORITHM:

1. Read in N.
2. Assign N to variable, FACT.
3. Loop: from L = N − 1 to 2
   Assign FACT * L to FACT.
4. Print out results.

SEGMENT:

```
      INTEGER FACT
      READ*, N
      FACT = N
      DO 10 L=N-1,2
            FACT=FACT*L
   10 CONTINUE
      PRINT *, FACTORIAL =', FACT
```

FLOWCHART:

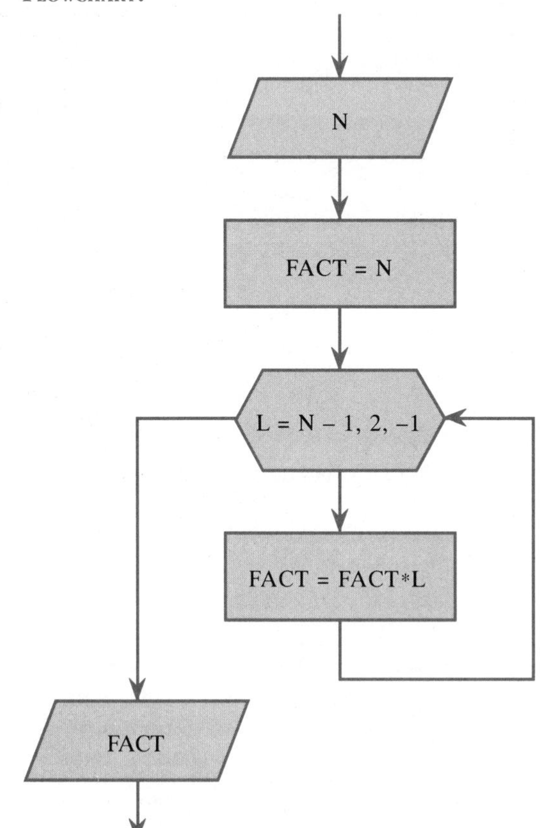

One key element of the algorithm for the factorial that students often forget is the *initialization* of the variable, FACT. If you forget the statement, FACT = N, before the loop, the machine will set FACT to zero when the program calls it. Thus, the answer will always be zero unless you remember to initialize it yourself.

The final LS that we will investigate here is the SUMMATION OF AN INFINITE SERIES. This is an important algorithm since it contains several key ideas that you will use in many of your programs. The title is misleading since we can never carry the summation of an infinite series to its conclusion, for by definition this will take an infinite amount of time. For practical reasons, we need to truncate it when we have a sufficiently accurate approximation. To demonstrate, consider the following series:

$$A = 1 + \frac{1}{2^3} + \frac{1}{3^3} + \frac{1}{4^3} + \cdots$$

If we expand this series, we have:

$$\begin{aligned} A &= 1 + 0.125000 + 0.037037 + 0.015625 + \cdots \\ &= 1.177662 \end{aligned}$$

The true value of the series is approximately 1.201. So, only a few terms have given us an accuracy of better than 2 percent. Yet, for many applications this would not be nearly good enough. Accordingly, we may need to calculate the series to a much better approximation. From the following table, you can see that this series gives approximations within 1 percent with only six terms. But to get better than 0.1 percent, we need to go to 16 terms. Here is what the series approximation is for different numbers of terms:

Number of Terms	Approximation	Error (%)
1	1.000000	20.1
2	1.125000	6.3
3	1.162037	3.2
4	1.177662	1.9
5	1.185662	1.3
6	1.190292	0.9
7	1.193207	0.6
8	1.195160	0.5
9	1.196532	0.4
10	1.197532	0.3

If we knew in advance how many terms to include in the series, the program segment would be simple:

```
      DO 10 K=1,N
             SUM=SUM+1/K**3
   10 CONTINUE
```

where N = the desired number of terms. But, what if we don't know the number N? Don't forget that we know the value of N only if we expand the series ourselves. Instead, what we would like is a program segment that calculates the approximation until it is very close to the true value. Sometimes 1 percent is close enough, but other times we may need 0.00001 percent.

One way to achieve this goal is to examine the contribution of each term to the series:

Term	Series	This Term	Contribution (%)
1	1.000000	1.000000	100
2	1.125000	0.125000	11.11
3	1.162037	0.037037	3.19
4	1.177662	0.015625	1.33
5	1.185662	0.008000	0.67
6	1.190292	0.004630	0.39
7	1.193207	0.002916	0.24
8	1.195160	0.001953	0.16
9	1.196532	0.001372	0.11
10	1.197532	0.001000	0.08

Notice that as the series approaches the true value, each additional term contributes a smaller and smaller amount to the series total. So, for example, when we reach the tenth term, its contribution is only 0.001, which is only 0.08 percent of the total.

The contribution of each additional term is one possible condition that we can use to terminate a series. Assume that we can set the condition that if a term contributes something less than 0.01 percent, then we can stop the calculation. Of course, you can enter any cut-off value that you wish. This will give you some control over the final accuracy of the result. Some series, like this one, may require many terms to satisfy this condition. Other series may satisfy the condition in only a few iterations.

The key here is that we do not know beforehand how many loops will be run since this will depend on how rapidly the series converges. Also, the quantity that will decide when to stop the loop changes inside the loop. Here, this quantity is the percentage change in the term contribution. The following LS will show you how to set up this process:

LS FOR SUMMATION OF INFINITE SERIES

PROBLEM STATEMENT: Calculate the series approximation for A:

$$A = \frac{1}{1^3} + \frac{1}{2^3} + \frac{1}{3^3} + \frac{1}{4^3} + \cdots$$

ALGORITHM DEVELOPMENT:

Step 1a: Review the problem statement. Indicate all nouns in **bold** and verbs in *italics:*

Calculate the **series approximation** for A.

Formulate questions based on nouns and verbs:

Q1: How do we calculate the series total?
Q2: How do we know when to stop the approximation?

Step 1b: Construct answers to above questions:

A1: Use loop to add terms $1/n^3$
A2: When the new term contributes less than a predetermined amount, we stop the calculation N

ALGORITHM REFINEMENT:

Step 2a: Review previous answers to see if refinement is needed. Answer A2 is not sufficiently detailed. Construct new questions with nouns in **bold** and verbs in *italics*:

Q3: How do we determine the contribution of each new term?
Q4: Where do we get the value for stopping the calculation?

Step 2b: Construct answers to above questions:

A3: Calculate $1/n^3$ for the term. Take ratio of term value to previous series total; this is the contribution.
A4: Use TERMINAL INPUT command to enter variable eps.

ALGORITHM:

1. Read in eps (desired accuracy).
2. Initialize variables, Sum = 0.0, N=1, Contrib =1.
3. CONDITIONAL LOOP: DO WHILE (Contrib >eps)
   Term value = $1/N^3$.
   Increase Sum by Term value.
   Contribution = Term value/Sum.
   Increment N by 1.
4. Print out Sum.

SEGMENT:

```
DATA N,SUM,CONTRIB/1, 0.0, 1.0/
READ*, EPS
DO WHILE (CONTRIB.GT.EPS)
      TERM=1.0/N**3
      SUM=SUM+TERM
      CONTRIB=TERM/SUM
      N=N+1
END DO
PRINT*, SUM
```

FLOWCHART:

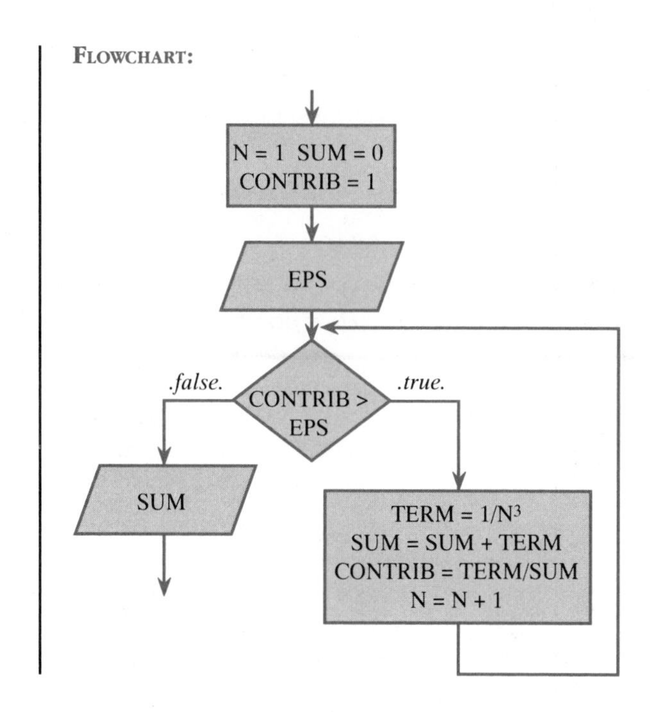

You should try to trace through the program segment and the flowchart with the series given to make sure you understand how it works. Also, we leave it up to you to convert the flowchart into a program segment using the BLOCK-IF structure instead of the DO WHILE structure.

This example is an important one since many students find the idea of evaluating an infinite loop intimidating. So, study this example carefully before you try any of the exercises.

6.7 Putting It All Together

We can apply the methods of this chapter to many interesting and important problems. To show this, we explore a small program to calculate the binding energy of an ionic crystal. An ionic crystal (in one dimension), Fig. 6.12, has

FIGURE 6.12

Schematic Drawing of One-Dimensional Ionic Crystal

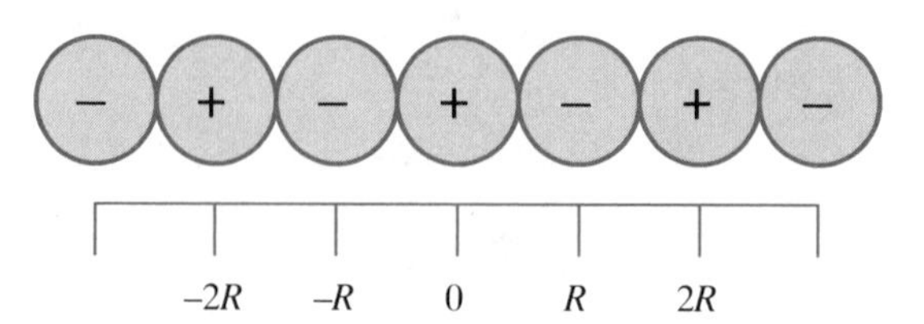

alternating electrical charges at regular distances from each other. We arbitrarily place a positive charge at the origin. There will then be negative charges at $\pm R$, $\pm 3R$, $\pm 5R$, and so forth. Also, there will be positive charges at $\pm 2R$, $\pm 4R$, $\pm 6R$, and so forth. Because electrical charges interact, we will have both attractive and repulsive forces on the charge at the origin. For example, the positive charge at the origin tries to repel all the other positive charges, but the negative charges tend to pull them back. Thus, there is a net attractive force that holds this structure together.

We start the discussion by looking at how two charges interact. Then we will extend this idea to many charges arranged in a periodic structure like the previous one. The energy of two interacting (and only two) charges, Z_1 and Z_2, contains a term, U, defined by:

$$U = \frac{-Z_1 Z_2}{\text{separation distance}} = \frac{-Z_1 Z_2}{R}$$

If the two charges have opposite signs, U is positive and the two will attract. But if the two charges have the same sign, U is negative and the charges will repel. Now, if we add the other charges to give us the preceding alternating sequence, the term, U, becomes:

$$U = \alpha \frac{Z^2}{R}$$

where α is the MADELUNG constant, and considers the contribution of all the charges. The following alternating series describes this constant:

$$\alpha = 2\left[1 - \frac{1}{2} + \frac{1}{3} - \frac{1}{4} + \frac{1}{5} - \cdots\right]$$

The term inside the brackets comes from the alternating repulsive and attracting charges. The constant 2 in front of the brackets comes from the contributions on either side of the origin.

The task here is to evaluate the alternating series inside the brackets. Notice that the series does not converge very quickly. In fact, to obtain the final value to an accuracy of 0.001 requires over 1000 terms. Just calculate individual components of the series by hand, and you will see how slowly it approaches the final value (4 log 2 or 1.204).

There are two issues to consider before we convert this series into a program. First, we must develop a general formula for any term in the series. By inspection, you should see that each term comes from:

$$(-1)^{i+1} \frac{1}{i}$$

Thus, if $i = 3$, the term would be $(-1)^4/3$. This is a convenient format since we want to sum the series made from these terms. We saw this process once before in our previous example of the summation of an infinite series. The shorthand notation for this series is:

$$\alpha = 2 \sum_i (-1)^{i+1} \frac{1}{i}$$

and we carry the series out to infinity. We approximate the series by calculating each term, starting with $i = 1$ and adding the value of the term to a variable, SUM. Then we repeat this process for subsequent terms in the series:

	(initial value of approximation)	SUM = 0.0
$i = 1$	Term = $(-1)^2/1 = +1.000$	SUM = 1.000
$i = 2$	Term = $(-1)^3/2 = -0.500$	SUM = 0.500
$i = 3$	Term = $(-1)^4/3 = +0.333$	SUM = 0.833
$i = 4$	Term = $(-1)^5/4 = -0.250$	SUM = 0.583

We can do this calculation with a loop in which i is the loop control variable. We then use i to compute the value of the individual terms before adding it to the variable SUM. The only issue still unresolved is when to stop the computation. An easy way to do this is to check the value of the term, and stop if it is an acceptably small value. For example, we might be satisfied if α is computed down to terms of the order of 0.01. But other times we may want terms down to 0.0001. Therefore, a better approach would be to make the cutoff a variable, eps, and read it in at execution time.

Notice from this discussion that we need a loop. But we don't know in advance how many times to run it. This implies that we need a conditional loop (DO WHILE). This loop will stop whenever (term < eps). Thus, our program must contain the following:

Input section	a prompt to enter a value for eps
Process section	conditional loop to calculate term and add this value to the variable, SUM
	check to see if term < eps
	keep track of how many terms computed so far
Output section	value of SUM
	how many terms needed for given accuracy

Here then is the algorithm:

EXAMPLE PROGRAM FOR CALCULATING BINDING ENERGY OF IONIC CRYSTAL

PROBLEM STATEMENT: Evaluate the following alternating series for all terms greater than a value, eps, read in at execution time:

$$\alpha = 2(1 - \tfrac{1}{2} + \tfrac{1}{3} - \tfrac{1}{4} + \tfrac{1}{5} - \cdots)$$

EXAMPLE PROGRAM OF IONIC CRYSTAL (cont'd)

ALGORITHM DEVELOPMENT:

Step 1a: Review the problem statement. Indicate all nouns in **bold** and all verbs in *italics:*

Evaluate the following alternating **series** for all **terms** greater than a **value, eps,** read in at execution time.

Formulate questions based on nouns and verbs:

Q1: How do we evaluate a series?
Q2: How do we calculate each term?
Q3: How do we know when to stop the approximation?
Q4: What value do we want to print out?

Step 1b: Construct answers to above questions:

A1: Use a loop to evaluate each term in the series and add that value to the variable SUM.
A2: Each term is given by (−1)**(*i* + 1)/*i*, where *i* is the number of the term.
A3: Whenever the absolute value of any term is less than eps, we may stop.
A4: Want the approximation and the number of terms.

ALGORITHM:

1. Read in the value for eps.
2. Calculate first term = 2.
3. Initialize SUM to term; i=2.
4. Loop: (while ABS(term) > eps).
   term = (−1)**(i + 1)/i
   add term to SUM
   increment i
5. Print out SUM and i.

FLOWCHART:

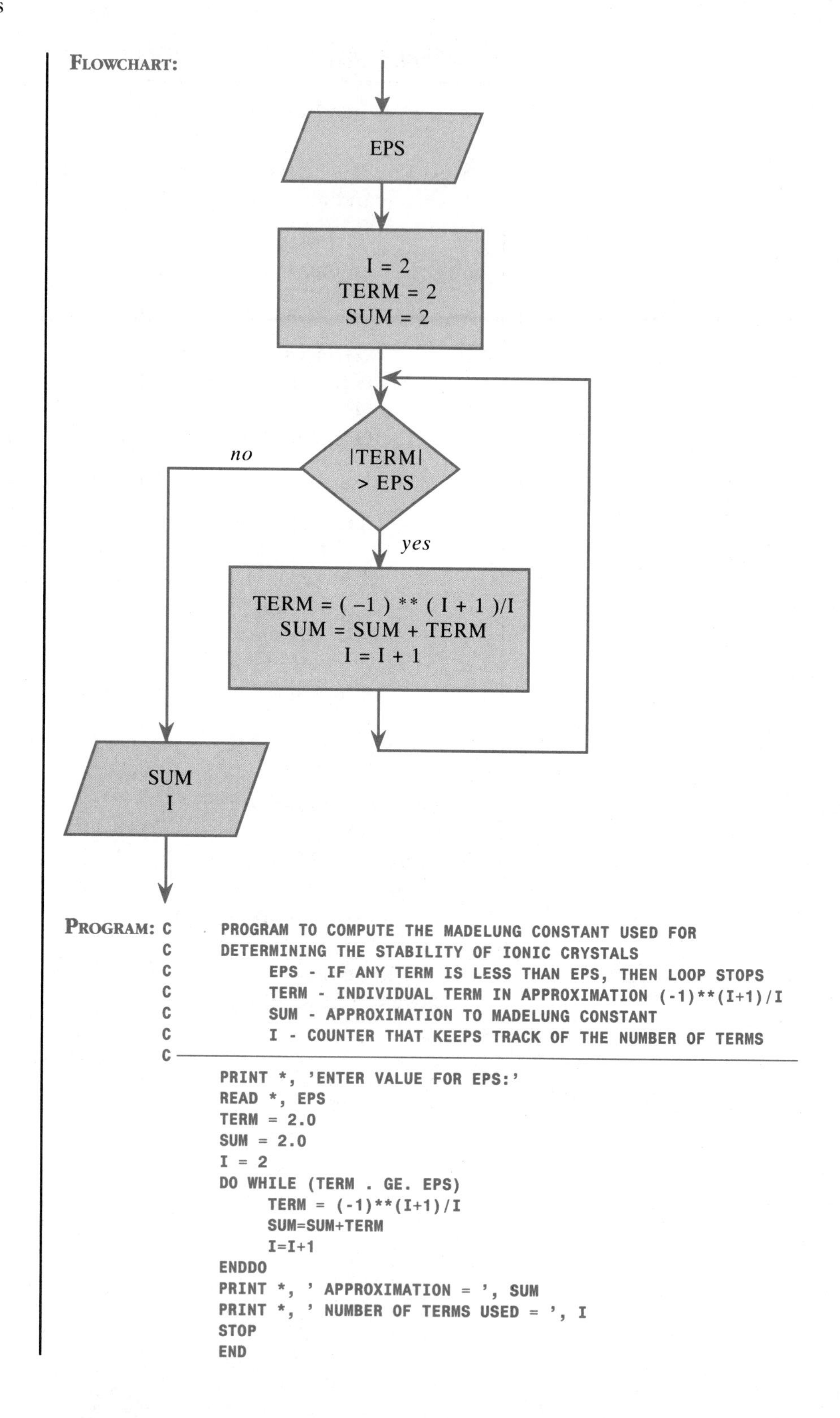

PROGRAM:

```
C       PROGRAM TO COMPUTE THE MADELUNG CONSTANT USED FOR
C       DETERMINING THE STABILITY OF IONIC CRYSTALS
C            EPS - IF ANY TERM IS LESS THAN EPS, THEN LOOP STOPS
C            TERM - INDIVIDUAL TERM IN APPROXIMATION (-1)**(I+1)/I
C            SUM - APPROXIMATION TO MADELUNG CONSTANT
C            I - COUNTER THAT KEEPS TRACK OF THE NUMBER OF TERMS
C ------------------------------------------------------------------
        PRINT *, 'ENTER VALUE FOR EPS:'
        READ *, EPS
        TERM = 2.0
        SUM = 2.0
        I = 2
        DO WHILE (TERM . GE. EPS)
             TERM = (-1)**(I+1)/I
             SUM=SUM+TERM
             I=I+1
        ENDDO
        PRINT *, ' APPROXIMATION = ', SUM
        PRINT *, ' NUMBER OF TERMS USED = ', I
        STOP
        END
```

Before we give you the flowchart and the program, there are a few points that we want to make. First, note that we have set the first term and sum to 2 outside the loop. This is because the loop stops whenever the absolute value of the term is small. Note that if we did not assign a value to the term outside the loop, the loop would never execute. This is because the term would be equal to zero. Therefore, we evaluate the first term outside and then start the loop on the second term in the series.

The final point to note about the algorithm is that we must use the absolute value of the term to decide when to stop the loop. Otherwise a negative value for the term would stop the loop, even though the magnitude of the number is significant.

One final word about this program, you could write it with a COUNTED LOOP. But if you did, you would need a very large number of iterations because the loop converges so slowly. It would be easy to guess incorrectly and not include enough terms. Also, you would need a method to get out of the loop when any term's magnitude is less than eps. This would require a GOTO instruction, which we are trying to avoid. For these two reasons, we prefer the CONDITIONAL LOOP.

Exercises

SYNTAX

6.1 Locate Syntax Errors. Each structure contains syntax errors. Find the errors and make corrections:

a.
```
IF(X.LE.Y) THEN STOP
```
b.
```
IF(X.EQ.Y) .AND. (U.NE.V) STOP
```
c.
```
IF(X.EQ.Z.OR.Y) GOTO 12
```
d.
```
IF(X = Y) PRINT*,'THEY ARE EQUAL'
```
e.
```
IF(X*Y .EQ. Y*X)
THEN PRINT*,'EQUAL'
ENDIF
```
f.
```
IF(X.NGT.Y) THEN
PRINT*,'LESS'
ELSEIF(X.LE.Y) THEN
PRINT*,'GREATER'
ELSEIF
PRINT*,'EQUAL'
ENDIF
```
g.
```
IF(.NOT.(A.LT.B) THEN
PRINT*,'LESS THAN'
ELSE
PRINT*,'GREATER THAN'
ENDIF
```

h.
```
IF(X.GE.Y) THEN IF(X.EQ.Y) THEN
PRINT*,'EQUAL'
ELSE
PRINT*,'LESS'
ENDIF
```

6.2 Syntax Errors. Each LOOP structure contains syntax errors. Find the errors and make corrections:

a.
```
      DO 10 I=1,5,I
```

b.
```
      DO 20, J=I,K,L
```

c.
```
      DO 30 L=1,M,
```

d.
```
      DO 40 M=1,I**2,-1
```

e.
```
      DO WHILE(X ≤ 0.0)
      PRINT*,X
      ENDDO
```

f.
```
      DOWHILE(X.LE.Y)
      X=X+0.1
      PRINT*,X
      Y=X
      ENDDO
```

g.
```
      DO 10 I=1,100
      DO WHILE(X.LE.I)
          PRINT*,X*I
          X=X+0.1
   10 CONTINUE
      ENDDO
```

h.
```
      DO 10 I=K,L,M
      DO 10 K=I,M,L
          PRINT*, I,K,L,M
   10 CONTINUE
```

YOUR SYSTEM

6.3 Check Out Your Compiler. The following suggestions are designed so that you can find out the limitations or extensions of your FORTRAN compiler. Run small programs to determine whether the following suggestions work. You may also need to consult the documentation for your system.

a. Some compilers will always execute a loop at least once, even if the structure of your loop tells it otherwise. Try the following:

```
      DO 10 I=1,0
          PRINT*, I
   10 CONTINUE
```

b. See if your compiler supports the DO WHILE extension. Some compilers use the DO WHILE structure. Others may use the WHILE(..)DO structure. Try both:

```
      DO WHILE (X.LE.1.0)
          PRINT*,X
          X=X+1
      ENDDO
```

```
      WHILE(X.LE.1.0) DO
          PRINT*,X
          X=X+1
      ENDWHILE
```

c. See if your compiler supports the SELECT-CASE structure by running the following:

```
CHARACTER*1 A, B, C, D, ANS
PRINT*,'SELECT: A, B, C, D'
READ *,ANS
SELECT CASE(ANS)
     CASE('A')
          PRINT*, 'A SELECTED'
     CASE('B')
          PRINT*, 'B SELECTED'
     CASE('C')
          PRINT*, 'C SELECTED'
     CASE('D')
          PRINT*, 'D SELECTED'
     CASE DEFAULT
          PRINT*, 'SOMETHING WENT WRONG'
END SELECT
```

d. Most FORTRAN compilers will not allow you to transfer into the middle of a DO loop. Try the following code to see if yours catches this problem:

```
    READ*,N
    IF(N.NE.0) GOTO 10
    DO 20 I=1,10
10        PRINT*,'I=',I
20  CONTINUE
```

e. Using REAL values to control a DO loop is generally not a good idea because REALs are stored imprecisely. Run both segments on your system and compare results.

```
     SUM=0.0
     DO 100 X=0.0,1.0,0.0001
          SUM=SUM+X
100  CONTINUE
     PRINT*, SUM

     ISUM=0
     DO 100 I=1,10000
          ISUM=ISUM+I
100  CONTINUE
     PRINT*,ISUM/10000.0
```

TRACING

6.4 Tracing Program Segments. Debugging control structures is nearly impossible without manual tracing. Therefore, to give you some practice tracing, go through the following segments and predict the output of each:

a. How many lines will this senseless program print?

```
      INTEGER OUTER, INNER, MIDDLE
      DO 100 OUTER = 2,8,2
            DO 100 MIDDLE =OUTER,2,1
                  DO 100 INNER = 1,MIDDLE,2
 100  PRINT*,OUTER, MIDDLE, INNER
```

b. What is the output?

```
      DO 90  I=5,17,3
            II=I/5
            DO 99 III=7,II,4
                  IF(I.EQ.I/III*III) GOTO 90
  99        CONTINUE
            PRINT*, I, II, III
  90  CONTINUE
```

c. What is the output? (assume that ALPHA=200.598641)

```
      READ*, ALPHA
      RSUM=10.0
      SFACT=1.0
      DOWHILE (ALPHA.GT.10.0)
          ALPHA=SFACT*ALPHA/RSUM
          SFACT=SFACT+1
          IALPHA=ALPHA
          ALPHA=IALPHA
          RSUM=RSUM-2
      ENDDO
      IF(ALPHA.LE.5) THEN
            ALPHA=ALPHA*2.0
      ELSEIF(ALPHA.GT.7) THEN
            ALPHA=ALPHA**2*2.0
      ELSE
            ALPHA=SQRT(ALPHA)
      ENDIF
      PRINT*, ALPHA
```

d. What is the output?

```
      IMPLICIT INTEGER (A-Z)
      DATA P,R,C,E/2,5,3,10/
      DO 10 I=1,P
            DO 20 J=1,5,2
                  R=R+(-1)**I*J
                  C=C-R
```

```
20          CONTINUE
10    CONTINUE
      IF(R.GT.0.AND.C.GT.0) THEN
            DO WHILE(E.NE.0)
                 P=E*E
                 E=E-3
            ENDDO
      ELSEIF(R.LT.0.AND.C.LT.0) THEN
            R=R/C+C/R
            C=(C-R)*(R-C)
      ELSE
            R=((R-C)*(C+R))
            C=(C/R)*10+20
      ENDIF
```

PROGRAMS

6.5 Converting Flowcharts. Convert the following flowcharts into well-structured FORTRAN programs. Assume that all conditionals have the *.false.* to the left and *.true.* to the right:

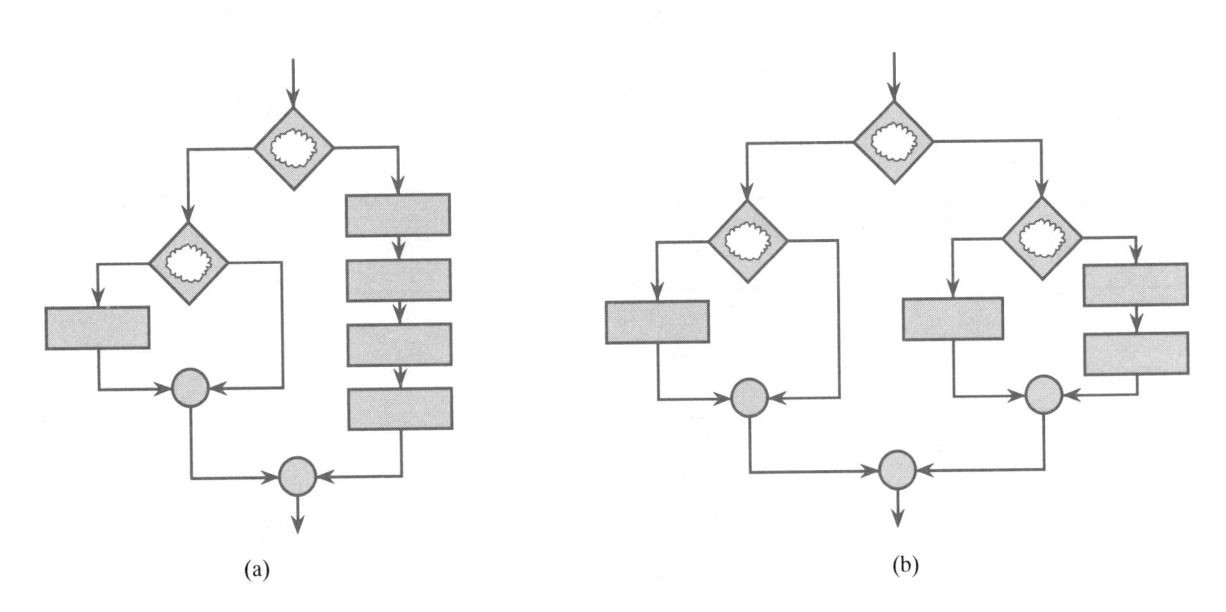

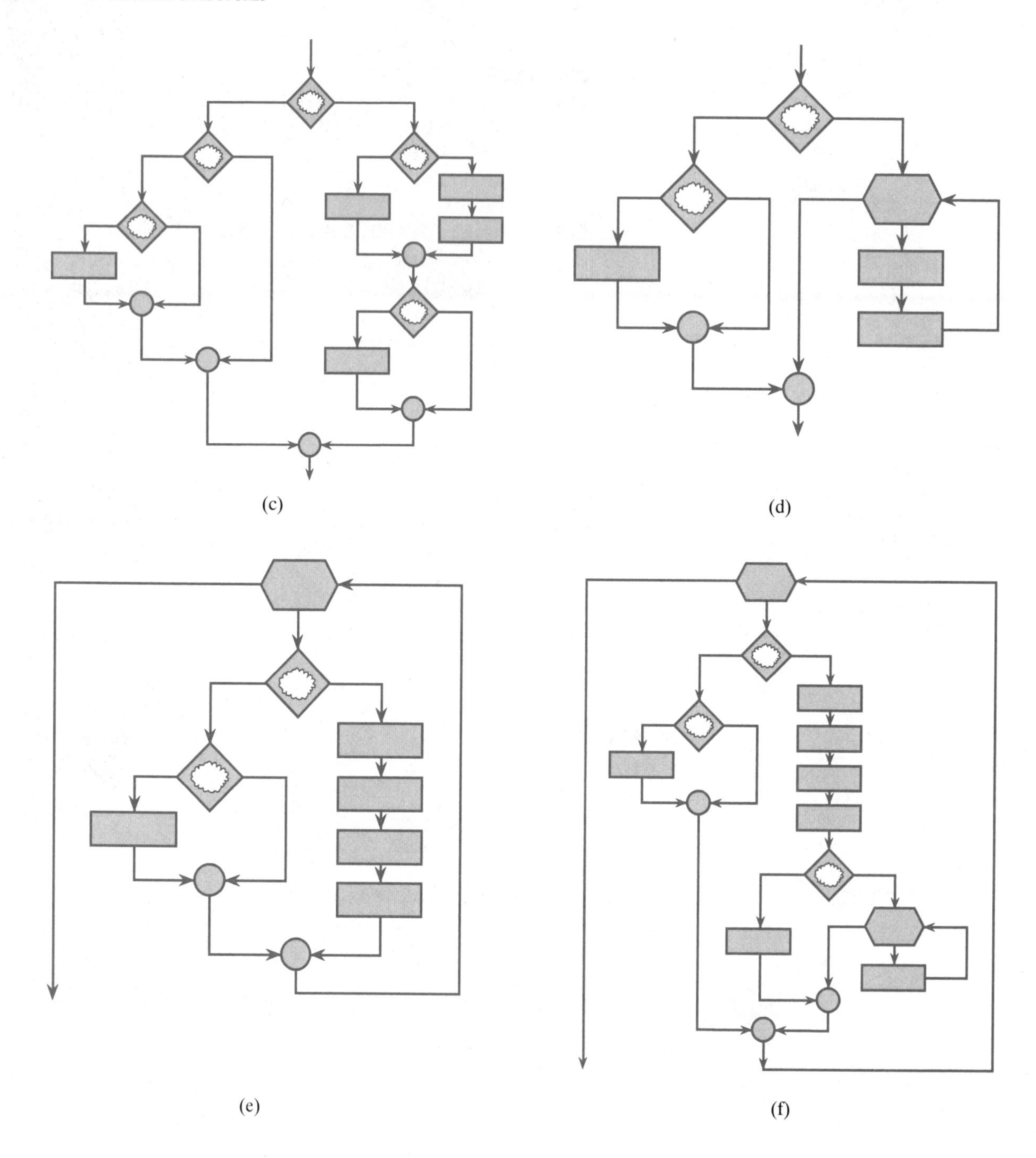

(c) (d)

(e) (f)

6.6 Program Segments. Write small FORTRAN program segments for each of the following:

a. Read in values for A, B, and C and print their sum. Repeat this procedure until all values of A, B, and C are negative.
b. Read in three INTEGER values and determine if *all* are odd or *all* even.
c. Read in a REAL number and repeatedly divide it by 2 until it is smaller than 0.001. Print out the result after every five divisions.
d. Read four INTEGERs into I, J, K, and L. Print out appropriate messages if any of the following conditions occur:

    any two numbers are equal
    any three numbers are equal, and
    all four numbers are equal

e. Read in two INTEGERs and find the sum of all the numbers between these two numbers.
f. Read in two INTEGERs and find the sum of all *odd* integers between these two numbers.
g. Read in two INTEGERs and find the product of all the numbers between these two numbers. This is different from the previous problem since you must be careful to skip over the number zero.
h. Read in three INTEGERs, I, J, and K. Use these data to print out a table consisting of I rows, each of which is J columns wide with characters chosen from a list based on the value of K as follows:

    K < 0—character to use is *
    K = 0—character to use is #
    K > 0—character to use is $

i. Read in a series of numbers and keep track of the running total and the number of data items. Stop collecting data when a value < 0 is entered. Then calculate the average and report it.
j. Read in a series of numbers and find the largest and smallest. Stop reading data when a negative value is entered.
k. Read in a dollar amount and a monthly interest rate. Calculate the interest earned each month and the total amount on deposit. Terminate the program when the initial deposit has doubled.
l. Read in the radius, R, of a circle centered at the origin. Then read in coordinate pairs, (X,Y) and determine if that point lies within the circle. Use the condition, that if:

$$(X^2 + Y^2)^{0.5} < R$$

then the point is within the circle. Terminate the program the first time that $(X^2 + Y^2)^{0.5} > 2R$.

6.7 Tabular Data Calculation. Write a program to calculate the values of Y, where Y is given by:

$$Y = \frac{1}{X} - 4.3 \log(X) + X^4$$

for values of X between -1.0 and $+10.0$ in increments of 0.01. Also, include output statements to print out the values of X and Y for every nth X value (e.g., every tenth value for $n = 10$). The value of n is to be read in at execution.

6.8 Evaluating *e*. The value of e = 2.718282 can be approximated by the infinite series:

$$e = \sum_{n=0}^{\infty}\left(\frac{1}{n!}\right) \approx \left(\frac{1}{0!}\right) + \left(\frac{1}{1!}\right) + \left(\frac{1}{2!}\right) + \left(\frac{1}{3!}\right) + \cdots$$

The factorial function, $n!$, is the product of integers from 2 to n and $0! = 1$ by definition. Write a program to approximate the value of e for five terms in the series. Then modify it to compute the approximation for m terms in the series, where m is read in at execution time.

6.9 Refinement of Computation of *e*. One problem with the approach of the previous problem is that you never know how many terms to use for the approximation. One way that has proven to be very successful is to have the series terminate automatically when each new term adds little to the approximation. For example, the 13th term is $1/13!$ or 1.6059×10^{-10}, which is insignificant compared to all the other previous terms. Therefore, you should modify your program for Prob. 6.8 to allow for termination of the series when any term is less than *eps*, which is a variable that you read in.

6.10 Prime Numbers. Write a program to determine if a number is *PRIME*. A PRIME number is one which is divisible only by itself and 1. Use the following algorithm:

a. Successively divide N by all INTEGERs lying between 2 and $N/2$.
b. With each division, check for a remainder.
c. If there is no remainder for a given division, then the number is not a prime. Stop the process at this point.
d. Print out a message in either case (PRIME or NONPRIME).

6.11 Calculating the Inverse Tangent. Your computer may not have the $\tan^{-1}$ function built into it. In situations like this, therefore, you must develop your own program to calculate this function. A useful approximation that you can use is:

$$\tan^{-1}(X) = \begin{cases} X - \dfrac{X^3}{3} + \dfrac{X^5}{5} - \dfrac{X^7}{7} + \cdots & -1 < X < 1 \\[2ex] \dfrac{\pi}{2} - \dfrac{1}{X} + \dfrac{1}{3X^3} + \dfrac{1}{5X^5} + \cdots & X \geq 1 \\[2ex] \dfrac{-\pi}{2} - \dfrac{1}{X} + \dfrac{1}{3X^3} + \dfrac{1}{5X^5} + \cdots & X \leq -1 \end{cases}$$

Be sure to include in your program a means for allowing the series to truncate automatically.

6.12 Area Under a Curve. The simplest way to calculate the area under a curve is shown in the following figure:

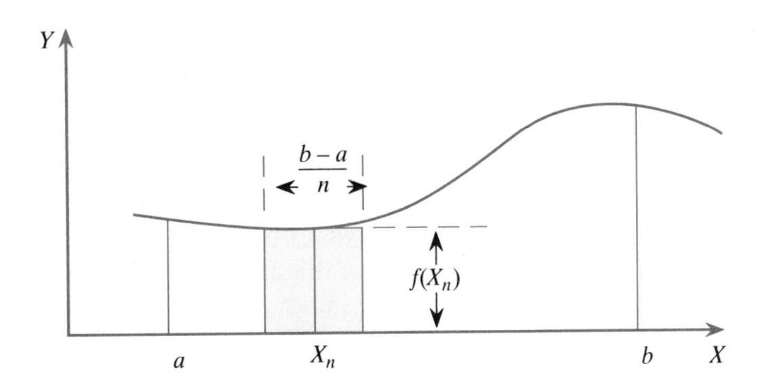

This method is the least accurate method, but the easiest to program. We break the curve into n equal rectangles. The width of each rectangle is $(b - a)/n$. The height of each rectangle will vary and is given by $F(X)$, where X is the midpoint of the box and $F(X)$ is the value of the function at that point. Thus, the area, A_n, of the small rectangle is:

$$A_n = F(X_n)\frac{(b - a)}{n}$$

If you add up all the small areas, you will have an approximate value for the area under the curve, provided n is large. Write a program to implement this scheme for the equation:

$$F(X) = \exp(-5.0X)\,\sin(20.0X)$$

over the range $a = 1.0$ to $b = 2.0$.

6.13 Fibonacci Series. A famous sequence is the Fibonacci series given by:

1 1 2 3 5 8 13 21 34 . . .

This series is purported to appear in many naturally occurring phenomena. For example, successive rows of sunflower seeds duplicate the series. It also describes a population explosion, among rabbits, for example. The first two numbers in the series are 1 and 1. All the additional terms of the series are the sum of the two previous terms. Thus, the ninth term, 34, is the sum of the seventh and the eighth terms, or $13 + 21$. Write a program to calculate the first n terms of the series.

6.14 Population Explosion. Write a program to simulate a population explosion. Start out with a single bacteria cell that can produce an offspring by division every 4 h. The new cell must incubate for 24 h before it can divide. The parent cell meanwhile will continue to divide every

4 h. Assume that any new cells will follow this pattern. How many cells will you have in 1 day, 1 week, and 1 month if none of the new cells dies? Now assume a life span of 4 days and repeat the preceding calculations.

6.15 Numerology. Many people place a lot of faith in the study of numbers. They believe that they can predict your future if they know one of your vital statistics, such as your social security number (SS#). They base their method on reducing your number to a single digit number by adding all the digits together. For example, if your SS# is 123-45-6789, the sum of the digits is 45. Since this is still a two-digit number, the process needs to be repeated. The result (4 + 5) is 9. Write a program to carry out this unusual addition process for any general number such as a phone number, body weight, and so on.

a

CHAPTER

7 THE ART OF DEBUGGING—PART I

7.1 Debugging—A Matter of Survival

Survival—it's what debugging is all about. A computer can be a wonderful tool for increasing productivity, accuracy, and organization of a project. It can also be a recurring nightmare if you are not serious about your debugging practices (Yes—Freddy Lives!). As you will discover, novice programmers spend most of their time debugging code instead of generating it. Locating and correcting errors is what debugging is all about.

One difference between an experienced programmer and an inexperienced one is his or her sophistication in debugging programs. Beginners fall into the trap of relying on trial and error. Experienced programmers, on the other hand, systematically analyze their programs to find the fault. One area where these differing approaches is apparent is in the turnaround time. A novice, using trial and error, makes dozens of changes an hour with little overall improvement in the program. Experienced programmers, meanwhile, tend to make just a few changes per day. For these lucky people, sometimes a single change is all it takes. But is it luck or skill that gives the experienced programmers the edge? Which category do you fall into?

Everyone writes programs with errors. Writing a program that runs correctly the first time is rare and certainly not the rule—even for experts. Yet, it is possible to write programs with **few** errors that lend themselves to easy debugging. The key idea to achieving this is the knowledge of a few simple debugging tools and the experience to apply them. This chapter presents methods that allow you to gain experience in a way that may help you to bypass the traditional *"trial by fire"* approach. There will be some of that too, but, hopefully, we can show you some tricks to make it a little less painful.

In this book we divide the subject of debugging into two chapters. This chapter provides an approach for you to follow to start debugging programs at an early stage. In a later chapter we will present more advanced ideas that will require more advanced knowledge of FORTRAN. We feel very strongly that you should begin to develop a technique for debugging early so that you don't develop bad habits. We think that it is a mistake to wait until you have learned all the FORTRAN syntax. That is why we are presenting this chapter at this time.

There are three types of errors that you will encounter. These are:

Syntax errors
Run-time errors
Logic errors

Syntax errors result from an incorrect entry that violates a FORTRAN *syntax* rule. Examples might include missing a comma in a PRINT command, putting text in the wrong column (e.g., a continuation marker), or making a typing mistake. Syntax errors are the simplest ones to detect because the compiler can find most of them and point them out to you.

Run-time errors are those that occur during the execution of the program. For example, entering data that would result in a divide by zero operation would result in a run-time error. Another example of a run-time error is when you might forget to define properly a variable that results in incorrect output. This type of error is more elusive than the simple syntax error. But you can usually detect these easily by using a feature known as the list-file generation.

The third type of error, logic errors, deals with errors in the algorithm and the method that you proposed to solve your problem. This type of error usually causes incorrect output. The program will execute, but the results are invalid. This type of error is the most difficult type to detect and often requires manual tracing of the program. This is why we have included tracing problems after each chapter. Without this practice, you would be unprepared to locate logic errors.

In this chapter debugging without the use of a *debugger* will be presented. A debugger is a software package used to aid in the detection of program bugs and is available on many different systems. Unfortunately, the operation of a debugger depends on the specific software package. Therefore, it is difficult to talk about general methods of debugging based on these many different packages. By contrast, the methods presented in this chapter use features such as list-files and available FORTRAN commands, which are more general and make debugging practices less system-dependent. Accordingly, you will be able to use these methods even if a debugger does not exist on your system. In addition, the methods that we show you in this chapter will help you write better programs which are inherently more resistant to bugs to begin with.

7.2 Syntax Errors

Syntax errors are the easiest of the three types of errors to locate, if you know how to look for them, and if you use the tools made available here. Recall Fig. 1.8 from Chap. 1 that shows the life of a program. You will find that syntax errors occur during compilation, the process of turning source code into run-time code. When you compile a program (see Fig. 7.1), the compiler will return error messages concerning the syntax errors that it has located. These are typically written to the CRT screen. Most compilers today return reasonable, understandable messages with a minimum of computer jargon. Some, though, will return an error number that must be looked up in a manual to describe the error. If you have such a system (usually a small PC with limited disk storage), make a list of common errors you encounter for quick reference while you work. The following proverb summarizes this basic idea:

> **PROVERB 1:**
> **NUMBERS, NUMBERS, NUMBERS!**
> **If your compiler returns only error numbers, make a list describing the meaning for the most common errors.**

As we develop additional ideas in this chapter, we will also highlight them as we have for the first proverb. A total of 11 proverbs are presented in this chapter. Although not every student will need all 11, we are confident that many of them will greatly improve your debugging abilities.

Let's now examine a typical situation of a first draft of a program that contains a few minor syntax errors. To be consistent with Chap. 1, we show

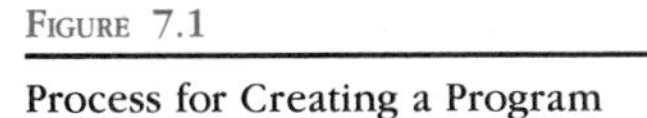

FIGURE 7.1

Process for Creating a Program

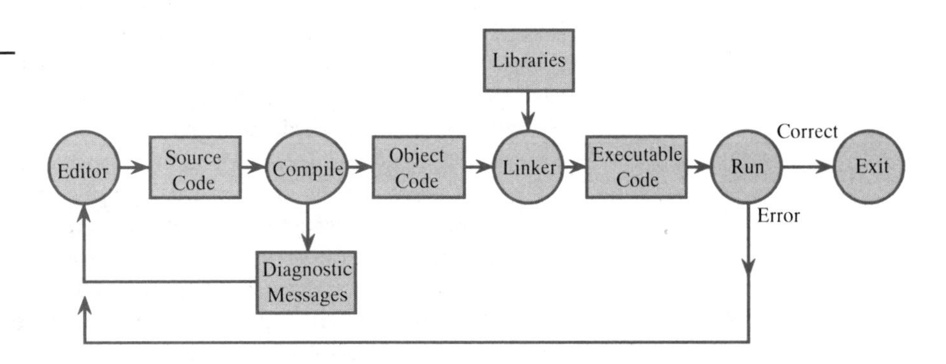

the EMACS editor commands used to create the program. Also, we include the command to compile the program. These commands may be different for your system, but the program itself will always be the same:

```
$ EMACS DUMMY.FOR;1                <CR>
EMACS EDITOR DUMMY.FOR, VERSION 1
      REAL IN,CM
      PRINT 'ENTER LENGTH IN INCHES'
      READ IN
      CM=IN*2.54
      PRINT IN ' INCHES = ' CM
      STOP
      END
<CTRL>X<CTRL>S
<CTRL>X<CTRL>C
FILE SAVED IN DUMMY.FOR;1
```

Once you have used the editor to create the program, the next step is to submit it to the compiler. But before compiling, it is a good idea to first type the program onto the screen. The reason for doing this is to catch errors that may not be obvious during editing. One example of such an error is a program segment that is accidently typed twice or typed out of order. It is often very easy to catch such errors by simply displaying the program on the screen. Of course, the compiler may detect such errors, but then again, it may not. So a preliminary reading of the program is a good idea.

The next step is to submit your program to the FORTRAN compiler. In addition to converting your program into machine language, one of the compiler's jobs is to find and report simple syntax errors. There are usually displayed on the CRT screen during the compiling process. In some cases though, there may be so many errors that they cannot fit onto a single screen. In this case, the error messages can be sent to a file for easy reference or printout. We will discuss how to do this shortly.

When we compile the program above, the compiler finds three syntax errors. As we will seee shortly, the compiler reports these errors in a format that will help you correct the faulty program. Unfortunately, the compiler cannot make any of the corrections for you. You must do this for yourself. But the compiler messages are a useful aid in guiding you to the needed corrections.

The following dialogue illustrates the three syntax errors found by the compiler for the above program:

```
$ TYPE DUMMY.FOR          <CR>
    REAL IN,CM
    PRINT 'ENTER LENGTH IN INCHES'
    READ IN
    CM=IN*2.54
    PRINT IN ' INCHES = ' CM
    STOP
    END
$ FORT TEST
      %FORT-F-MISSDEL,      Missing operator or delimiter symbol
                            in module TEST$MAIN at line 2
      %FORT-F-IOINVFMT,     FORMAT specifier in error
                            [READ IN          ] in module TEST$MAIN at line 3
      %FORT-F-IOINVFMT,     Format specifier in error
                            [PRINT IN ' ] in module TEST$MAIN at line 5
      %FORT-F-ENDNOOBJ,     DISK$USERFILES:[MACWIAKALA]TEST.FOR;2 completed
                            with 3 diagnostics—object deleted
```

In response to the command TYPE DUMMY.FOR, the computer prints out the file sent to the compiler. It is a good idea to see the program on the screen next to the error messages, if the entire program can fit. After we submit the program to the compiler, it will return a series of messages. The general format adopted for these messages is:

Error #1 acronym—*details*
Error #2 acronym—*details*
Error #3 acronym—*details*

.
.
.

Summary statements

The acronym used to describe the error can be used when you need to go to the compiler documentation to find out more information. Greater detail about the type of error made can be found in the manuals under the acronym reported in the message. Usually, however, the short messages in the *details* section of the error description is sufficient to correct the error.

From the messages reported in the example, you can see that the compiler detected three errors in lines 2, 3, and 5. Let's look carefully at the first message to see how much information is contained in it:

%FORT-F-MISSDEL, Missing operator or delimiter symbol in module TEST$MAIN at line 2

This message tells us an operator or a delimiter might be missing from line 2. An operator is a mathematical operations such as +, −, *, / or **. A delimiter is a separator such as a comma. Now, reexamine line 2 of the program to see what is missing:

```
PRINT ' ENTER LENGTH IN INCHES'
```

Of course, the PRINT statement is incorrect. It should be PRINT *, *'string'*. The comma after the PRINT statement is required and the compiler reported it as a missing delimiter.

The second and third errors are similar, so we will examine them together. The messages tell us that the format specifier is the origin of both errors. But, wait a minute—there is no format specifier! So, how can we have an error with a statement that doesn't exist? Fortunately, there is a simple answer. Here are the two lines:

```
(line 3)    READ IN
(line 5)    PRINT IN 'INCHES = ' CM
```

The compiler is interpreting the arguments (variable I/O list) of the READ and PRINT statements as the format specifiers. The programmer has not given the compiler any indication of the format to be used. So, the machine had to search. Remember, format specification (either free or controlled) comes immediately after the READ or PRINT statement. What it found in this location was not a valid format specifier (either * or statement label), so it reported a FORMAT SPECIFIER error. This lesson is well worth noting:

> PROVERB 2:
> **THINGS MAY NOT BE AS THEY SEEM!**
> **The actual errors are often not reported accurately by the compiler.**

This type of error occurs often. The compiler will report one type of error, but the real error is somewhere else. In the previous example the error was that the free format specifier, *, was missing. The error messages did not show this, though. So, don't be surprised if you find errors that are different from those detected by the compiler. Read the error messages very carefully, but also reexamine the faulty lines yourself to find the real error.

In the following program we have made the necessary corrections to lines 2, 3, and 5 by adding "*" in the appropriate places. Besides the errors detected by the compiler, we also found missing commas separating items in the I/O list of line 5. The compiler did not find these because it misinterpreted the list as a FORMAT SPECIFIER:

```
REAL IN,CM
PRINT *,'ENTER LENGTH IN INCHES'     (added *)
READ *,IN                            (added *)
CM=IN*2.54
PRINT *,IN,' INCHES = ',CM           (added * & ,)
STOP
END
```

The preceding syntax errors are simple ones that the compiler is able to detect. You should be aware that different compilers may return different error messages and even different numbers of errors. One way to get a "feel" for your compiler is to create simple programs with known errors. For example, you should create programs in which you mistype commands, leave out commas, leave out arguments to commands, and so on, and note the error messages generated. Start with a simple program that runs correctly, then systematically introduce errors. Be careful to start with a program that is running correctly to begin with before you introduce your selected known errors. This exercise will provide you with some understanding regarding how your compiler handles various errors. The experience gained can save you much time later on. For instance, the example program had only a single kind of error, a missing * to indicate a free format. Yet, the compiler reported the same error two different ways, which depended on the type of argument following the I/O command:

> **PROVERB 3:**
> **DON'T BE AFRAID TO EXPERIMENT!**
> **Gain experience by making deliberate errors and see how your compiler reports them.**

7.3 Run-Time Errors

Because the compiler reports no syntax errors does not mean that you have thoroughly debugged your program. More elusive errors may await you. The computer may detect run-time errors and logic errors during the execution of the program. Run-time errors can be thought of as syntax errors that have gone undetected during compilation. Sometimes the compiler does not detect all the errors present. Most compilers do a good job of detecting syntax errors, but sometimes a few will remain. Here's an example of a program that is syntactically and logically correct, yet gives the wrong answer:

```
REAL IN CM
PRINT *,'ENTER LENGTH IN INCHES'
READ *,IN
LET CM=IN*2.54
PRINT *,IN, ' INCHES = ', CM
STOP
END
```

When you go to execute the program, a strange result occurs. Suppose that we enter a value of 2.5 after the 'ENTER LENGTH IN INCHES' prompt (line 2 of the program). Follow the dialogue carefully and compare it to the example program. See if you can tell what's wrong before we show you the problem:

```
$ RUN TEST2                        (command to run
                                   program)

ENTER LENGTH IN INCHES             (prompt from program)
     2.5      <CR>                 (value that you
                                   enter)
2 INCHES = 0.0000000E+00 CM        (answer reported
                                   from program)

FORTRAN STOP                       (indicates end of
                                   program)
$                                  (indicates system
                                   ready for command)
```

Something is obviously wrong. We entered a value of 2.5 in, but the program read it as 2 in. Also, the program calculated the length in centimeters to be zero. This program compiled with no errors and what's more, the logic is correct! Both errors are linked to extra blank spaces that caused three lines of the program to do something other than what we intended:

```
(line 1)   REAL IN CM
(line 3)   READ *,IN
(line 4)   LET CM=IN*2.54
```

Look carefully at line 1. FORTRAN ignores the space between IN and CM. Therefore, the compiler interprets the variable, INCM, to be REAL, not the intended IN and CM. Thus, when the computer executes line 3, the computer assigns the number typed into the variable IN, which is an INTEGER, not REAL as intended. Therefore, the computer truncates the value of 2.5 to 2 before assignment. This explains why the program reports the value of 2.5 as 2.

The second run-time error was that the program calculated CM to be 0, not the expected 6.35. It turns out that the same type of problem is present. The compiler interprets the LET statement (commonly used by BASIC programmers) as part of a variable name LETCM. So, the equation LET CM=IN*2.54 calculated a value of (2*2.54) and assigned it to the variable LETCM, not to CM. Thus, when it came time to print out the value of CM, the computer had never assigned a value to the variable, and so it reported a value of zero. It should be noted at this point that some computers will initialize an undefined variable to zero while others will not. We explored this problem once before in an exercise for Chap. 5. If your compiler does not initialize variables to zero, then when the computer executes the program, it might return a different number each time you run the program. Once we make the necessary changes, the corrected program should be:

```
REAL IN , CM                                ( add comma )
PRINT *,' ENTER LENGTH IN INCHES '
READ *, IN
CM = IN * 2.54                              ( LET CM → CM )
PRINT *, IN , ' INCHES =' , CM
STOP
END
```

Note that the errors which caused the program to give an incorrect answer were not detected by the compiler. This is because the errors did not violate any of the syntax rules of the language. They are errors nonetheless.

The type of error just illustrated causes great frustration for many new FORTRAN programmers, although there is a simple way to avoid it. This method involves the use of a compiler feature known as LIST-FILE GENERATION. A compiler list-file is a file generated by the compiler that lists errors with other information related to compilation of the program. Most compilers have this feature. Here is an example of how to generate and interpret a list-file:

```
$ FORT/LIST DEMO.FOR              (selects lis option)
$ TYPE DEMO.LIS                   (types out LIST-FILE)

1           REAL IN CM
2           PRINT *,'ENTER LENGTH IN INCHES'
3           READ *,IN
4           LET CM=IN*2.54
5           PRINT *,IN, ' INCHES = ', CM
6           STOP
7           END
```

Symbolic Name	Storage Class	Attributes
CM	DYNAMIC	REAL*4
IN	DYNAMIC	INTEGER*4
LETCM	DYNAMIC	INTEGER*4
INCM	DYNAMIC	REAL*4

```
0000 ERRORS [<.MAIN.>F77 Rev. Rev. 21.0]

$
```

To create a LIST-FILE, you select an option when you send the source code to the compiler for translation. On most systems the command is FORT/LIS *filename.* This will automatically create a new file that has an extension similar to XXXXX.LIS. When you request a printout of this newly created file, something similar to that shown above is found. The LIST-FILE usually has the following format:

Program line #1
  (any syntax errors for line #1)
Program line #2
  (any syntax errors for line #2)
  .
  .
  .
Program line last
  (any syntax errors for line #last)
List of program variables and attributes

The LIST-FILE will contain each line of the program with embedded syntax errors. This format is useful because it immediately draws your attention to lines containing errors. In the preceding example there are no syntax errors. Thus, we print only the seven program lines. What is of interest to us, though, is the list of the program variables and attributes at the bottom of the LIST-FILE. This table shows you how the compiler interpreted your program, the variable names, and the data types assigned to each. Notice in this table, for instance, that there are four variables listed, whereas the program typed in contained only two. There are two new variables listed, LETCM and INCM, which do not belong here. Also, the compiler treats the variable IN as INTEGER, not as REAL as we intended. If you had inspected this table, you would quickly see that something was wrong; which deserves closer attention.

Using the LIST-FILE feature, you should easily detect errors such as:

- Missing commas in a declarative statement: example REAL IN CM
- Mistyping a variable name: Common mistakes are to use the letter O for the number 0, or the letter I for the number 1. Examples would be LO for L0, or H1GH for HIGH.
- Incorrect type declaration: An example would include using the variable LENGTH to store a real number, but then forgetting to declare it as real.

By reviewing the LIST-FILE generated by the compiler, you can quickly locate mistakes like those previously outlined. The LIST-FILE also serves as a useful tool when requesting help from an aid or an instructor since it includes the source code, errors, and other useful data. This leads us to two new proverbs:

> **PROVERB 4:**
> **LET THE COMPUTER DO THE WORK!**
> **Let the computer put together in one place, the LIST-FILE, all the information you need to debug your programs.**

and a closely related corollary:

> **PROVERB 5:**
> **HELP YOUR INSTRUCTOR HELP YOU!**
> **When requesting help from an instructor or aid station, have a copy of the LIST-FILE of your program available.**

On occasion, the run-time errors you encounter will result in the program stopping before completion. This type of error is typically a math error, such as divide by zero. The process of locating the source of this error usually involves adding PRINT statements to help in the diagnosis. The lowly PRINT statement can be one of your best friends for help in debugging. The best way to use PRINT statements is to attempt to break your program down into smaller and smaller modules until you isolate the problem. To see how this helps, consider the following simple program, which converts Centigrade temperatures into Fahrenheit:

```
C    PROGRAM NAME: TEMP.FOR
C
C    PURPOSE:  PRINTS TEMPERATURE IN K,C AND
C                   F FOR K RANGING FROM 0 to 375
C                   IN STEPS OF 0.5.
C
     START=0
     STOP=375
     STEP=0.5
     DO 10 K=START,STOP,STEP
           C=K-273
           F=9/5*C+32
           PRINT *,'K=',K,'C=',C,'F=',F
10         CONTINUE
     STOP
     END
```

When you try to run this program, a run-time error occurs and a strange looking error message appears. It seems from the error message that there is an attempt to divide by zero somewhere in the program:

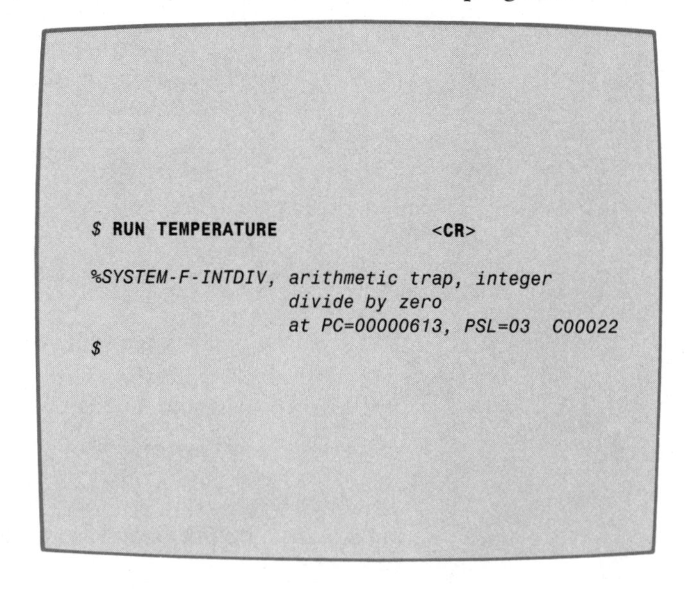

The error generated by this program is somewhat perplexing since there is only one division, and it involves nonzero constants (9/5). So, to find the source of this strange error, let's follow our own advice and create a LISTFILE. By doing this (and you should try it yourself), you will find that the compiler treats K as an integer. But, how could this cause a divide by zero error? To find out, we suggest that you add PRINT statements at key locations in the program. A good approach is to place a PRINT statement at the beginning and end of key junctions of the program to isolate first the section of the program that is generating the error. In this program we add PRINT statements before and after each calculation. We use the PRINT statement before the calculation to print out the variables used and to identify the particular equation, thus verifying the input portion of the expression. The PRINT statement following the calculation tells us if the computer successfully evaluated the equation, and if so, what the value was, thus verifying the process:

PROVERB 6:
USE PRINT STATEMENTS LIBERALLY!
Use PRINT statements before and after each section to isolate problems. Print variables used and calculation results.

So, let's add PRINT statements to our program and see what happens:

```
C    PROGRAM NAME: TEMP.FOR
C
C    PURPOSE: PRINTS TEMPERATURE IN K,C and F FOR
C             K RANGING FROM 0 to 375 IN STEPS OF 0.5.
C
     START=0
     STOP=375
     STEP=0.5
     PRINT *,'K LOOP; START , STOP ,
   1 STEP: ',START,STOP,STEP
     DO 10 K=START,STOP,STEP
           PRINT *,'BEFORE CALC OF C, K=' , K
           C=K-273
           PRINT *,'AFTER CALC OF C, C=', C
           F=9/5*C+32
           PRINT *,'K=',K,'C=',C,'F=',F
10   CONTINUE
     STOP
     END
```

In the previous example we have added three new PRINT statements. The first statement appears before the loop and shows the values of the loop control variables before we start. The other two PRINT statements are inside the loop and show the values of K before calculating C and the value of C after the computation. When we execute this program, we don't get very far before the program again crashes:

```
$ RUN TEMP                    <CR>

K LOOP: START, STOP, STEP: 0.0000  375.00  0.5000

%SYSTEM-F-INTDIV, arithmetic trap, integer
                  divide by zero
                  at PC=00000613, PSL=03 C00022

$
```

Notice that the only line of output was the PRINT statement before the DO LOOP and that the values of START, STOP, and STEP are correct. But at this point the program attempts to divide by zero before it gets to the next PRINT statement. Of course, there is only one line between the two PRINT statements, so this must be the location of the error:

```
                              .
                              .
                              .
      PRINT *,'K LOOP: START , STOP , STEP: ', START ,
    1 STOP , STEP
      DO 10 K=START , STOP , STEP
         PRINT *,'BEFORE C CALC, K= ', K
                              .
                              .
                              .
```

The program printed out the first PRINT statement, but didn't reach the second. Therefore, we have isolated the problem to the statement, DO 10 K=START, STOP, STEP. Now we must focus on what is wrong with this statement. After a short study, you should see that there is a problem with mixed mode. The computer is attempting to perform an internal division to calculate the number of times to execute the loop. Because K is an INTEGER by default, the step size that we intended to be 0.5 was truncated to 0. That zero step size resulted in the divide by zero error. Simply changing K to a REAL is all you must do to get the program running:

```
REAL K                          ( add declaration statement )
START = 0
STOP = 375
STEP = 0.5
DO 10 K = START , STOP , STEP
      .
      .
      .
```

At this point the program is running, but is it correct? We use the term *validation* to check the correctness of our output. You check to see that your program is returning correct answers. For the program shown, we validate the program by taking a line of output and seeing whether we can manually calculate the same results. For more complex programs, other methods of validation will be necessary. These will be covered in following chapters.

Here are the first two lines of the printout from the program once we insert the statement REAL K:

K = 0.0000   C = −273.0000   F = −241.000
K = 0.5000   C = −272.5000   F = −240.500

Now check the output manually for the first few values of K:

	Program Result		Manual Result		
K	C	F	C	F	Comments
0.0	−273.0	−241.0	−273.0	−459.4	C calculation okay F value wrong
0.5	−272.5	−240.5	−272.5	−458.5	C calculation okay F value wrong

As you can see, something is still wrong with the program. The value calculated for F is incorrect. To find out the cause of this latest problem, you need to reinsert the diagnostic PRINT statements and examine the printout to find the faulty line:

```
$ RUN TEMP                <CR>

K LOOP: START, STOP, STEP: 0.0000  375.00  0.5000
BEFORE CALC OF C, K=0.0000
AFTER CALC OF C, C=-273.00
K=0.0000  C=-273.00  F=-241.00

BEFORE CALC OF C, KI=0.5000
AFTER CALC OF C, C=-272.50
K=0.5000  C=-272.50  F=-240.50
```

Because of the diagnostic PRINT statement, we have isolated the problem to the following program line:

```
F = 9/5*C + 32
```

By now you should see that this is again a problem with mixed mode. Although we defined the variables F and C as REALs, the division of 9 by 5 involves two INTEGERs. In INTEGER arithmetic, of course, 9/5 = 1, not the 1.8 that we expected. To correct this, one or both of the numbers in 9/5 should be changed to a REAL.

There are two valuable lessons to be learned from this exercise. The next two proverbs summarize these points:

> **PROVERB 7:**
> **ALWAYS VALIDATE PROGRAMS!**
> **Trace through your program manually with a few sets of data to double check the program results.**

and

> **PROVERB 8:**
> **VALIDATE EXPRESSIONS!**
> **If an expression is correct when manually checked, but incorrect when the program runs, check for integers or mixed mode.**

As you have seen with two simple programs, some FORTRAN errors can be difficult to track down. By using the compiler LIST-FILE and additional

PRINT statements, you stand a good chance to find even the most elusive errors. With this background of syntax and run-time errors, it is now time to start considering the most difficult error to detect, the logic error.

7.4 Logic Errors

Logic errors deal with errors in the algorithm development. To decide that a logic error is present, you must first establish that no syntax errors or run-time errors are present. Once you eliminate these errors, you will use a method called *tracing* to find logic errors.

Tracing is the process of manually executing the commands that the computer performs. You are in effect the computer. Tracing is tedious work, so good organizational skills are needed. Therefore, we must discuss methods for keeping track of the values stored in variables, how a loop executes, and how conditional transfers work. Each method is presented using a program that works correctly, and then used to solve a few problems that have logic errors.

VARIABLE TABLES

Variables represent areas in memory. When you first use a variable, the computer establishes a space for it in memory. Each time the variable name appears on the left-hand side of an assignment statement, the computer updates the value stored for that variable. Keeping track of the variables used and their values requires a method of organization. One method for doing this is to create a variable table.

A variable table is a list of every variable used in the program. As the program assigns values to the variables, they are entered into the table. The current value of any variable is the last value entered. If you have a compiler with the LIST-FILE feature, you have most of the work already done for you. Most LIST-FILEs report all the variables used in the program. This can be the start of your variable table. To illustrate how to use variable tables, consider the following program, which performs several simple operations:

```
1   A=1
2   B=2
3   A=A+B
4   B=B*A
5   C=A**2+B
6   C=A+B+C
7   PRINT *,'A,B & C HAVE VALUES ',A,B,C
8   STOP
9   END
```

A ______________________________

B ______________________________

C ______________________________

The line numbers in the preceding program are only for reference. We list in the table the three variables used in the program. As each variable assumes a value, the variable table is updated. For example, after executing lines 1 and 2, the trace table would look like this:

A	1
B	2
C	

When the computer executes line 3, it retrieves the values for A and B from the table, adds them, and then assigns the result to the target variable A. The trace table now looks like this:

A	1 , 3
B	2
C	

Completing the trace results in the following variable table and output, we have:

A	1 , 3
B	2 , 6
C	15 , 24

Thus, when the PRINT statement in the program executes, the output will contain the final values of each variable. The new assignment statements overwrite the previous values, such as A = 1. The old values are lost forever. The only record of the previous contents of each variable is the variable table that you have just created. You will find these useful to help track down faults in program logic:

> PROVERB 9:
> **KNOW YOUR VARIABLES!**
> **Use a variable table to keep track of the value of each variable when performing a trace.**

TRACING AND DO LOOPS

Before trying to trace through a loop, you must do some preliminary work. The first step is to print out the program and *highlight* the loop bodies. One method of doing this is by drawing boxes around the body of each loop. Another popular method is to *indent* the loop body. The following program is an example of nested loops without any highlighting. Its purpose is to print out a table of lengths in feet and inches, with the corresponding measure in total inches:

```
C   PROGRAM NAME:     LENGTH.FOR
C
C   PURPOSE:
C   GENERATE A TABLE OF FT-IN TO TOTAL INCHES FOR
C   LENGTHS RANGING FROM 0'-0" TO 12'-0" IN
C   1 INCH  INCREMENTS.
C
      INTEGER FT,IN,INTOT
      DO 20 FT=0,12
      DO 10 IN=0,12
      INTOT=FT*12+IN
      PRINT *,FT,' FEET AND ',IN,' INCHES = ',INTOT,'
    1 IN'
 10   CONTINUE
 20   CONTINUE
      STOP
      END
```

FT ______________________________

IN ______________________________

CM ______________________________

We now show the same programs with the loop structures highlighted using both methods. The indentation method is strongly suggested even if you prefer the line method, only because you can do it while entering the program. By using the indentations, you prepare for the likelihood of tracing. If you still prefer the line method, having the indents makes it much easier to locate tops and bottoms of loops:

```
C
C     LOOPS HIGHLIGHTED WITH BOXES
C
            INTEGER FT,IN,INTOT
      ┌──── DO 20 FT=0,12
      │ ┌── DO 10 IN=0,11
      │ │   INTOT=FT*12+IN
      │ │   PRINT *,FT,' FEET AND ',IN, '
      │ │ 1 INCHES = ',INTOT,' IN'
      │ └── 10  CONTINUE
      └──── 20  CONTINUE
            STOP
            END
```

```
C
C     LOOPS HIGHLIGHTED BY INDENTATIONS
C
          INTEGER FT,IN,INTOT
          DO 20 FT=0,12
               DO  10  IN=0,11
                      INTOT=FT*12+IN
                      PRINT *,FT,' FEET AND ',IN,
     1                ' INCHES = ',INTOT,' IN'
   10          CONTINUE
   20     CONTINUE
          STOP
          END
```

Before beginning the trace, review the action and rules governing DO LOOPS. The following are among the actions performed during the execution of the loop:

- Upon entry, the DO variable is initialized to the start value.
- The computer checks the DO LOOP control variable to see if it is beyond the stopping value. For step sizes greater than 0, the DO variable would be larger than the stopping value. For step sizes less than 0, the DO variable would be less than the stopping value. If the loop variable is within the range, then control transfers to the line following the end of the loop. If not, the loop body executes.
- After the loop body, the DO variable increments by the step size.

These steps can be summarized with the following code using an IF-THEN, GOTO, and the accumulation LP. The example illustrated here contains a positive step size:

```
C     X     = THE DO VARIABLE
C     START = INITIAL VALUE
C     STOP  = FINAL VALUE
C     STEP  = STEP SIZE
C ___________________________________
          DO 3  X = START , STOP , STEP
                         .
                         .
                         .
                     BODY OF LOOP
                         .
                         .
                         .
      3   CONTINUE
```

```
C
C     AN ALTERNATE WAY OF DOING THE SAME THING
C
C ______________________________________________
            X = START
        5   IF (X . GT . STOP) THEN
                  GOTO 10
            ENDIF
                  .
                  .
                  .
            BODY OF LOOP
                  .
                  .
                  .
            X=X+STEP
            GO TO 5
       10   CONTINUE
```

The reason for this review is to remind you that DO LOOPs increment their DO variables beyond the final value. This is important if the program accesses the DO variable after the DO LOOP finishes.

A rule also worth noting concerning DO LOOPs is that the control values cannot change during the execution of the loop. This fact is useful if variables define the loop control values (START, STOP, AND STEP). Since these values cannot change during the looping, they can be noted above the loop in pencil for easy reference during the trace of the loop. Tracing a DO LOOP takes a bit of patience since some variables change frequently. The following variable table shows the printout for the previous problem:

Variable Table	Output
FT 0, 1, 2, . . .	0 FEET AND 0 INCHES = 0 IN
IN 0, 1, 2, 3, 4, 5, 6, 7, 8, 9, 10, 11, 12, 0, 1, . . .	0 FEET AND 1 INCHES = 1 IN
INTOT 0, 1, 2, 3, 4, 5, 6, 7, 8, 9, 10, 11, 12, 13, . . .	0 FEET AND 2 INCHES = 2 IN
	0 FEET AND 3 INCHES = 3 IN
	0 FEET AND 4 INCHES = 4 IN
	0 FEET AND 5 INCHES = 5 IN
	0 FEET AND 6 INCHES = 6 IN
	0 FEET AND 7 INCHES = 7 IN
	0 FEET AND 8 INCHES = 8 IN
	0 FEET AND 9 INCHES = 9 IN
	0 FEET AND 10 INCHES = 10 IN
	0 FEET AND 11 INCHES = 11 IN
	1 FEET AND 0 INCHES = 12 IN
	1 FEET AND 1 INCHES = 13 IN
	. . .

Upon entering the DO 20 LOOP, FT is 0. The program then checks to see if it exceeds 12, which it doesn't, so the body of the loop executes. Entering the DO 10 LOOP, IN is initially 0 and is checked to see if it exceeds 11. The body of the loop executes; INTOT is calculated and printed to the terminal screen. At the bottom of the loop IN is incremented by the default step size of 1 and control transfers to the top of the loop. This process repeats until IN increments to 12. At this value the DO 10 LOOP stops. This brings control to the bottom of the DO 20 LOOP where FT increments to 1 and control transfers to the top of the loop. Since the value of FT does not exceed 12, the body of the loop executes, reentering the DO 10 LOOP. This process repeats until FT increments to 13. The final values of FT and IN will be 13 and 12, respectively, beyond the bottom of the loops.

Writing down the steps that the computer is performing is one good method of documenting your trace—it is very important if you should request help from your instructor or aid station. Here is a written description of the trace that helps to describe the function of these double nested loops:

```
Enter DO 20, init. FT=0, is it > 12?, FALSE so execute body
Enter DO 10, init. IN=0, is it > 11?, FALSE so execute body
INTOT=FT*12+IN, INTOT=0
Print FT,IN,INTOT
Bottom of loop 10, increment IN by 1, goto top of loop
IN=1, is it > 11?, FALSE so execute body
INTOT=FT*12+IN, INTOT=1
Print FT,IN,INTOT
Bottom of loop 10, increment IN by 1, goto top of loop
IN=2, is it > 11?, FALSE so execute body
INTOT=FT*12+IN, INTOT=2
Print FT,IN,INTOT
Bottom of loop 10, increment IN by 1, goto top of loop
IN=3, is it > 11?, FALSE so execute body
INTOT=FT*12+IN, INTOT=3
Print FT,IN,INTOT
...
Bottom of loop 10, increment IN by 1, goto top of loop
IN=12, is it > 11?, TRUE skip past loop
Bottom of loop 20, increment FT by 1, goto top of loop
FT=1, is it > 12?, FALSE so execute body
Enter DO 10, init. IN=0, is it > 11?, FALSE so execute body
INTOT=FT*12+IN, INTOT=12
Print FT,IN,INTOT
...
```

As we stated earlier, tracing can be a tedious task. But by using variable tables and documenting your trace, you are much less likely to make mistakes. When you do make mistakes, you or someone else will be able to find them more easily:

> PROVERB 10:
> **DOCUMENT YOUR TRACE!**
> **Document your trace by writing down the commands and actions performed by the computer.**

TRACING IF-THEN STRUCTURES

Like DO LOOPs, IF-THEN statements are structured commands (language structure or LS). Whenever you deal with tracing programs that have LSs, the first step is to highlight the structures. As with DO LOOPs, IF-THENs can be illustrated by drawing lines connecting the bottom to the top, or by indentations. The following program is used to solve the quadratic equation $ax^2 + bx + c = 0$.

IF-THEN structures pose a problem not found with DO LOOPs. With a DO LOOP, it is always obvious where the loop ends since the DO statement contains a statement label marking the end. With IF-THEN structures, however, all structures end with ENDIF statements and there is nothing to distinguish one ENDIF from another. Therefore, you must decide where the structure terminates based on its context. This takes some getting used to:

> PROVERB 11:
> **INDENT, INDENT, INDENT!**
> **Use the indentation method to display the bodies of structured commands.**

As with the DO LOOPs, the indentation method of displaying a LS is superior because you can do it while entering the program. Lines following the IF-THEN statements should be indented by some fixed amount. If your system accepts tabs, indent by an additional tab, which is usually five to seven spaces. Once you have entered the last line in the structure (ENDIF for IF-THENs, CONTINUE for loops), indent backward one tab stop. We will show you the difference between programs without indentation and those with indentation:

```
REAL A,B,C,X1,X2,RADCL,RL,IM
PRINT *,'FOR THE EQUATION A*X**2+B*X+C=0'
PRINT *,'ENTER VALUES FOR A,B, and C'
READ *,A,B,C
RADCL=B**2-4*A*C
IF (A.EQ.0.0) THEN
X1=-C/B
PRINT *,'EQUATION ENTER WAS LINEAR'
PRINT *,'THE ROOT IS ',X1
ELSE
```

```
IF (RADCL.LT.0.0) THEN
RL=-B/(2*A)
IM=SQRT(-RADCL)
PRINT *,'SOLUTION IS COMPLEX'
PRINT *,RL,' + I',IM
PRINT *,RL,' - I',IM
ELSE
IF (RADCL.EQ.0.0) THEN
X1=-B/(2*A)
PRINT *,'REPEATED ROOTS ARE PRESENT'
PRINT *,'ROOT IS ',X1
ELSE
X1=(-B+SQRT(RADCL))/(2*A)
X2=(-B-SQRT(RADCL))/(2*A)
PRINT *,'TWO REAL ROOTS ARE PRESENT'
PRINT *,'X1 = ',X1,' AND X2 =',X2
ENDIF
ENDIF
ENDIF
STOP
```

As you can see, with no method of breaking the blocks up, this program is almost impossible to follow. Look now at what happens when we indent the different blocks to highlight where each one belongs. The following illustrates the same program highlighted using the indentation method to display IF-THEN structures:

```
REAL A,B,C,X1,X2,RADCL,RL,IM
PRINT *,'For the equation A*X**2+B*X+C=0'
PRINT *,'Enter values for A,B, AND C'
READ *,A,B,C
RADCL=B**2-4*A*C
IF (A.EQ.0.0) THEN
   X1=-C/B
   PRINT *,'Equation enter was linear'
   PRINT *,'The root is ',X1
   ELSE
   IF (RADCL.LT.0.0) THEN
         RL=-B/(2*A)
         IM=SQRT(-RADCL)
         PRINT *,'Solution is complex'
         PRINT *,RL,' + I',IM
         PRINT *,RL,' - I,IM
         ELSE
```

```
            IF (RADCL.EQ.0.0) THEN
                   X1=-B/(2*A)
                   PRINT *,'Repeated roots are present'
                   PRINT *,'Root is ',X1
                   ELSE
                   X1=(-B+SQRT(RADCL))/(2*A)
                   X2=(-B-SQRT(RADCL))/(2*A)
                   PRINT *,'Two real roots are present'
                   PRINT *,'X1 = ',X1,' and X2 =',X2
                   ENDIF
            ENDIF
      ENDIF
  STOP
```

Most programmers prefer the indentation method for isolating blocks since it can be done easily in the editor used to enter the program. You may use any number of spaces for indentation, although most people use the ⟨TAB⟩ key. Remember that the idea is to *visually* break up the code to make it easier to follow. Therefore, the actual number of spaces is unimportant.

A few programmers however, prefer to use the line method. In this case though, you cannot draw the lines with the editor. These are drawn in after the program is printed out. Locating the bodies of nested IF-THENs is straightforward when using the line method. The steps involved are:

- Check that the number of IF-THEN statements equals the number of ENDIFs. If it does not, the program will not compile without errors. *Do not* confuse the single alternative form (IF (. . .) STATEMENT) with the IF-THEN form.
- Start at the top of the program and work your way down looking for ENDIFs. When you find an ENDIF, draw a line back to the first available IF-THEN (one that does not already have a line connected to it).
- After you have drawn the line, continue searching for ENDIFs past the one you have just found. Repeat this process until you match all ENDIFs with IF-THENs.
- To assign ELSEs (which are optional and may, or may not be present), start at the top of the program and work your way down looking for ELSEs. When you find one, draw a line from it to the IF-THEN-ENDIF line. Repeat this step for all remaining ELSEs.

If you prefer the line method, it is a good idea to outline loops on one side of the program and IF-THENs on the other side for step 4. By doing this, you will find that the program will appear less cluttered, thus making it easier to read and trace. The quadratic formula program is illustrated here using the line method:

```
REAL A,B,C,X1,X2,RADCL,RL,IM
PRINT *,'For the equation A*X**2+B*X+C=0'
PRINT *,'Enter values for A,B, and C'
READ *,A,B,C
RADCL=B**2-4*A*C
IF (A.EQ.0.0) THEN                                   1
X1=-C/B
PRINT *,'Equation enter was linear'
PRINT *,'The root is ',X1
ELSE
IF (RADCL.LT.0.0) THEN                               2
RL=-B/(2*A)
IM=SQRT(-RADCL)
PRINT *,'Solution is complex'
PRINT *,RL,' + I',IM
PRINT *,RL,' - I,IM
ELSE
IF (RADCL.EQ.0.0) THEN                               3
X1=-B/(2*A)
PRINT *,'Repeated roots are present'
PRINT *,'Root is ',X1
ELSE
X1=(-B+SQRT(RADCL))/(2*A)
X2=(-B-SQRT(RADCL))/(2*A)
PRINT *,'Two real roots are present'
PRINT *,'X1 = ',X1,' and X2 =',X2
ENDIF
ENDIF
ENDIF
STOP
```

In this program we number each IF-THEN structure for ease of documentation during the trace. Before performing a trace on the program, review the actions of an IF-THEN statement. The general form of an IF-THEN is as follows:

```
IF (Boolean expression) THEN
      (Statements to execute if .TRUE.)
ELSE (optional)
      (...Statements to execute if .FALSE.)
ENDIF
```

The Boolean expression is a combination of relational and logical operators that when evaluated, return a value of *.TRUE.* or *.FALSE.* If the Boolean expression has a *.TRUE.* value, then the statements between IF-THEN and ELSE (or ENDIF if no ELSE is present) are executed. Control then transfers to the line following the ENDIF statement. If the Boolean expression is *.FALSE.,* then statements between ELSE and ENDIF execute. If the ELSE is not present, control immediately transfers to the line following the ENDIF.

To illustrate the process of tracing a program with IF-THEN statements, we use the preceding program. Values assigned to A, B, and C will be 1, 0, and −9, respectively. Those inputs correspond to the quadratic equation, $X^2 - 9 = 0$, which has roots $X = +3$ and $X = -3$. The variable table and trace are as follows:

Variable	Value	Trace
A	1	PRINT Equation
B	0	PRINT Prompt for A, B, C
C	−9	READ A, B, C = 1, 0, −9
X1	3	RADCL=B**2−4*A*C, 36
X2	−3	IF-THEN 1, A=0?, FALSE goto ELSE1
RADCL	36	IF-THEN 2, RADCL<0?, FALSE, goto ELSE2
RL		IF-THEN 3, RADCL=0?, FALSE goto ELSE3
IM		X1=(−B+SQRT(RADCL))/(2*A), 3
		X2=(−B−SQRT(RADCL))/(2*A),−3
		PRINT two real roots present
		PRINT X1,X2
		ENDIF for IF-THEN 3
		ENDIF for IF-THEN 2
		ENDIF for IF-THEN 1
		STOP

As you can see from this trace, we did not use all the variables in the trace table since we defined some for special cases (complex solution, etc.). Programs with nested IF-THENs tend to look confusing and complex. When tracing them, you will find that the number of commands actually executing is small compared to the amount of code present.

EXAMPLE OF LOGIC ERROR

Let us now use some ideas illustrated earlier to debug a program containing a logic error. We designed this program to produce the following sequence of fractions:

$$\frac{2}{1}, \frac{3}{2}, \frac{5}{3}, \frac{8}{5}, \ldots$$

The numerator of each fraction is the sum of the numerator and denominator of the previous fraction. The denominator of the new fraction is then assigned to the numerator of the previous fraction. The following program prints out the first 40 terms of this sequence using real arithmetic. The output contains the term number (2/1 is term number 1, e.g.), with the numerator, denominator, and value of the fraction:

```
      REAL NUM,DEN,VAL
      INTEGER TERM
      NUM=2
      DEN=1
      DO 10 TERM=1,40
            VAL=NUM/DEN
            PRINT *,TERM,'# ',NUM,'/',DEN,' = ',VAL
            NUM=NUM+DEN
            DEN=NUM
 10         CONTINUE
      STOP
      END
```

When we execute this program, the computer prints the following output on the CRT screen:

```
1# 2.00000/1.00000 = 2.00000
2# 3.00000/3.00000 = 1.00000
3# 6.00000/6.00000 = 1.00000
4# 12.0000/12.0000 = 1.00000
5# 24.0000/24.0000 = 1.00000
6# 48.0000/48.0000 = 1.00000
7# 96.0000/96.0000 = 1.00000
```

As you can see from the output, the numerators and denominators are equal, except for the first term. The first step to find the bug is to do a trace on the program (if no syntax or run-time errors can be found). The following shows a listing of the variable table and the trace analysis for the first few fractions:

Variable	Values
NUM	20, 30
DEN	10, 30
VAL	20
TERM	1, 2

NUM = 2

DEN = 1

Enter DO LOOP, assign TERM=1

Is it >40? No, then execute loop

VAL = NUM/DEN = 2.0

PRINT TERM, NUM, DEN, VAL

NUM=NUM+DEN=3.0

DEN=NUM=3.0

We can carry out this trace further, but it is not necessary since the error has already occurred. After the first iteration through the loop, the expected values of each variable and the actual values already differ as summarized here:

Variable	Expected Value	Actual Value
NUM	3.0	3.0
DEN	2.0	3.0
VAL	1.5	1.0

We copied the expected values from the problem statement. For the term, 3/2, we expect the denominator to be 2.0 and the numerator to be 3.0. The program, however, returns the values of 3.0 and 3.0. Clearly, then, the problem lies with the variable, DEN. The problem is that the statement that updated the numerator destroys the value of the previous numerator (which we need to define the new denominator):

```
      REAL NUM,DEN,VAL,PRENUM,PREDEN
      INTEGER TERM
      NUM=2
      DEN=1
      DO 10 TERM=1,40
            VAL=NUM/DEN
            PRINT *,TERM,'# ',NUM,'/',DEN,' = ',VAL
            PRENUM=NUM
            PREDEN=DEN
            NUM=PRENUM+PREDEN
            DEN=PRENUM
10          CONTINUE
      STOP
      END
```

One method of solving this is to use additional variables to store the previous numerator and denominator. Using those values, we can then calculate the new numerator and denominator. The corrected program includes these two new variables, PRENUM and PREDEN. Once we correct the logic error, the output is:

```
1# 2.00000/1.00000 = 2.00000
2# 3.00000/2.00000 = 1.50000
3# 5.00000/3.00000 = 1.66667
4# 8.00000/5.00000 = 1.60000
```

```
5# 13.0000/8.00000 = 1.62500
6# 21.0000/13.0000 = 1.61538
7# 34.0000/21.0000 = 1.61905
8# 55.0000/34.0000 = 1.61765
9# 89.0000/55.0000 = 1.61818
.
.
.
40# 2.679143E+08/1.655801E+08 = 1.61803
```

As you can see from this output, the program is now working correctly. You will find that there is no substitute for tracing to remove logic errors. You must know what the output of your program should look like and be willing to trace through the program by hand to see that the logic is correct. This approach will work as we described here for relatively simple programs, such as those you have written so far. For more complex programs, you will need more advanced procedures. We will explore these in Chap. 11, which deals with advanced methods of debugging.

7.5 Summary

In this chapter we presented the topic of debugging or removing errors from FORTRAN programs that can be used on most systems. We discussed three types of errors. These were syntax errors, run-time errors, and logic errors. Syntax errors are mistakes in grammar such as misspelling a command, or leaving out a comma. This type of error normally appears during compilation. Run-time errors are those errors detected when the program executes. Typical errors would be math errors such as a divide by zero. Another example might be an incorrect answer due to the mistyping of a variable name. The final type of error discussed was the logic error. This is due to faulty algorithm development. Typically, the only way to detect this error is to trace the program manually. Throughout this chapter key points were recorded as debugging proverbs. It is worthwhile to recall these in one place for easy reference:

- Proverb 1 — If your compiler returns only error numbers, make a list describing the meaning for the most common errors.
- Proverb 2 — The actual errors are often not reported accurately by the compiler.
- Proverb 3 — Gain experience by making deliberate errors and see how your compiler reports them.
- Proverb 4 — Let the computer put together in one place, the LIST-FILE, all the information that you need to debug your programs.
- Proverb 5 — When requesting help from an instructor or aid station, have a copy of the LIST-FILE of your program available.

Proverb 6	Use PRINT statements before and after each section to isolate problems. Print variables used and calculation results.
Proverb 7	Trace through your programs manually with a few sets of data to double check results.
Proverb 8	If an expression is correct when manually checked, but incorrect when the program runs, check for integers and mixed mode.
Proverb 9	Use a variable table to keep track of the value of each variable when performing a trace.
Proverb 10	Document your trace by writing down the commands and actions performed by the computer.
Proverb 11	Use the indentation method to display the bodies of structured commands.

Exercises

SYNTAX

7.1 If-Then Blocks. Use computer generated LIST-FILEs to locate the syntax errors in the following program segments:

a.
```
      IF(A .GE. B) THEN Z=Y
      END IF
```
b.
```
      REAL A
      INTEGER B
      B=A
      IF(A > B) PRINT *,'A IS REAL'
```
c.
```
      REAL A , B , C
      IF(A.GT.B) GOTO 10
      IF(A.GT.C) THEN
      X=5
      ELSE
   10 X=10
      END IF
```
d.
```
      REAL A, B, X
      IF(A.GE.B) THEN
      X=A
      ELSE
      IF(4*A.GE.B) THEN
      X=B
      ELSE
      X=A*B
      ENDIF
      PRINT*,X
```

e.
```
      IF(A.LT.B*2) THEN
 10         A=ABS(A-B)
      ELSE IF(A.GT.B*2) THEN
            B=ABS(B-A)
      ENDIF
      GOTO 10
```
f.
```
      IF(X.GT.Y.OR.Z)THEN AVE=(X+Y)/2
      ELSEIF(A.NGT.X+Y)PRINT,' NO GOOD'
      ENDIF
```

7.2 Loop Structures. Use program-generated LIST-FILEs to locate the syntax errors in the following program segments:

a.
```
      DO 10 I=1,15
            J=I**2
            I=I+1
 10   PRINT*, J, I
```
b.
```
      DO 10 I=1,10
            DO 20 J=1,I
                  K=I*J
 10         CONTINUE
 20   CONTINUE
```
c.
```
      DO 10 I=1,J
            X=1.3*J**2
            DO 10 I=1,3
                  Y=4.5*I
 10   CONTINUE
```
d.
```
      DO 10 X=X0+DX,XF,DX
 10   ANS=ANS+SIN(X)
      WRITE(3,10) ANS*DX
 10   FORMAT(1X,' THE ANSWER IS ', F10.5)
```
e.
```
      DO 10 I=1,10
 100        N=N+I
 10   CONTINUE
      IF(N.LT.95) GOTO 100
```
f.
```
      DO 10 I=10,0,-2
            N=2*I
            PRINT*,N
            I=I-2
            PRINT*,I
 10   CONTINUE
```
g.
```
      DO 10 I=10,1
            K=I**2/K
            PRINT*,I,K
 10   K=2*K
```

h.
```
      DO10 I=I1,I2,I1+I2
            AVE=(I1/I2*I2)**0.5
      DO 20 I=1,10
10    PRINT*,I*AVE
      DO WHILE(1.GE.I1)
           AVE=AVE+I*2
           IF(I.EQ.AVE) PRINT*,'OK'
      ENDDO
20    I=I+1
```

YOUR SYSTEM

7.3 Check Out Your Compiler. Write small FORTRAN program segments to check out how your compiler responds to the following potential errors:

a. What is the maximum and minimum values that can be assigned to default INTEGER and REAL variables on your computer? What type of error message is generated if you exceed any of these values?
b. Is any error message returned when you READ a REAL number into an INTEGER variable or an INTEGER number into a REAL variable?
c. Are any error messages reported for any of the following conditions?
   REAL number raised to an INTEGER power
   INTEGER number raised to a REAL power
   Logarithm of an INTEGER
   Logarithm of a negative REAL
d. Run the following segments through your compiler to see what errors, if any, that it returns:

```
      DO 10 I=1,10,0
            PRINT *,I
10    CONTINUE
```

```
      DO 10 X=1,10,0
            PRINT *,X
10    CONTINUE
```

e. Run these segments also:

```
      ISTRT=1
      ISTOP=10
      ISTEP=0
      DO 10 I=ISTRT,ISTOP,ISTEP
            PRINT *,I
10    CONTINUE
```

```
      STRT=1
      STOP=10
      STEP=0
      DO 10 X=STRT,STOP,STEP
            PRINT *,X
10    CONTINUE
```

TRACING

7.4 DO LOOPs. Trace through the following programs and predict their output. Verify your results by entering and executing the program:

a.
```
      SUM=0.0
      DO 10 X=1,5,0.5
            SUM=SUM+X
   10 CONTINUE
      PRINT*,SUM,SUM*X
      STOP
      END
```

b.
```
      Y=2.0
      Z=4
      DO 10 X=Y,Z**2,2*Y
            Z=Z-1
            Y=Y+1
   10 CONTINUE
      PRINT *,Y
      STOP
      END
```

c.
```
      DO 10 I=1,10,1
   10 IF(I/7.0-I/7.EQ.0) PRINT*,I
      STOP
      END
```

d.
```
      X=1.0
      DO WHILE(X.LE.100.0)
           PRINT *,X
           X=(X-1)**2+2.0
      END DO
```

e.
```
      DO 10 I=1,3
            DO 10 J=1,4
   10 PRINT*, I,J
      STOP
      END
```

f.
```
      DO 10 I=1,3
            DO 10 J=1,4
   10 CONTINUE
      PRINT *, I, J
      STOP
      END
```

7.5 BLOCK IF Structures. Trace through the following program segments and predict their output. Verify your results by entering and executing the programs for A = 2.7:

a.
```
      X=4.1
      Y=-7.9
      I=38
      J=7
      K=A
      IF(X*Y.LT.2.AND.Y.GT.7) THEN
            PRINT*,'CASE 0'
      IF(9*K+2.EQ.J.AND.I/J.LE.5) THEN
            PRINT*,'CASE 1'
      ELSE IF(K*2.LT.100) THEN
            PRINT*,'CASE 2'
      ELSE IF(X.LT.10.OR.Y.LT.100) THEN
            PRINT*,'CASE 3'
      ENDIF
      ENDIF
```

b.
```
      X=20.0
      Y=0.0001
      DO 10 I=10,5,-1
        IF(X*I.GT.100.)THEN
            Z=X*Y
        ELSE IF(Y.LT.0.001)THEN
            Y=Y*10
        ENDIF
   10 CONTINUE
        PRINT*, Y
```

c.
```
      A=4.2
      B=9.6
      C=2.7
      D=4.1
      IF(A.GE.B) THEN
      IF(A.GE.C) THEN
      IF(A.GE.D) THEN
            PRINT*,A
      ELSE
            PRINT*,D
      ENDIF
      ELSE
      IF(C.GE.D) THEN
            PRINT*,C
      ELSE
            PRINT*,D
      ENDIF
      ENDIF
      ELSE
```

```
      IF(B.GE.D) THEN
            PRINT*,B
      ELSE
            PRINT*,D
      ENDIF
      ELSE
      IF(C.GE.D) THEN
            PRINT*,C
      ELSE
            PRINT*,D
      ENDIF
      ENDIF
      ENDIF
```

d.

```
      OPEN(UNIT=3,FILE='IN.DAT')
      READ(3,*) K,L
      IF(K.EQ.L**2) THEN
      IF(K.EQ.L) THEN
      IF(K.NE.0) THEN
      PROD=K/L**2
      PRINT*, 'PROD=',PROD
      ELSE
      DO 10 N=1,4
      DO 10 M=1,N
      DO 10 I=M,N
      PROD=PROD*I**4
   10 CONTINUE
      PRINT*,'PROD=',PROD
      ENDIF
      ELSE
      ENDIF
      DO 20 N=I,L,L
      K=K**L
      DO 20 M=N,K,2
      PRINT*,'K=',K
   20 CONTINUE
      ELSE
      PRINT*,'OUT OF BOUNDS'
      ENDIF
      CLOSE(3)
      STOP
      END
   DATA FILE: 4, 3
   (IN.DAT)
```

RUN TIME & LOGIC ERRORS

7.6 When the following program is executed, the computer gives an error message:

> Error Message: Computation terminated at line 05
> during second iteration.

What went wrong?

```
T=2.0
I=2
DO WHILE(I.GT.-10)
      A=X*B/T**2
      T=-T
      P=B*A/(I-1)
      I=I-1
ENDDO
```

7.7 A problem requires that a program should read in a number less than −1.0 or greater than 1.0. What is logically wrong with the following program written to accomplish this?

```
READ*, X
IF(X.GT.-1.0) THEN
      PRINT*,' NUMBER NOT ACCEPTABLE'
ELSE IF(X.LT.1.0) THEN
      PRINT*,' NUMBER NOT ACCEPTABLE'
ENDIF
```

7.8 The following program segment reads in a real number and if it is very large, prints it with the E format; otherwise it prints the number with the F format. What is logically wrong?

```
    PRINT*, 'ENTER A REAL NUMBER: '
    READ*, R
    IF(R.GT.0.1E7.OR.R.LT.-0.1E7) THEN
          PRINT 10, R
    ELSE IF(R.LT.0.1E-7.OR.R.GT.-0.1E-7) THEN
          PRINT 20, R
    ELSE
          PRINT 20, R
    ENDIF
10  FORMAT(' ',E10.3)
20  FORMAT(' ',F16.7)
```

7.9 What is wrong with this program segment designed to compute the squares of the numbers between 1 and 10?

```
    N=10
    DO 20 I=1,N
          J=N**2
          PRINT*, J
20  CONTINUE
```

7.10 The following program was designed to approximate the sum of the infinite series

$$1 + \tfrac{1}{2} + \tfrac{1}{4} + \tfrac{1}{8} + \tfrac{1}{16} + \cdots$$

Find and correct the logical error:

```
      SUM=1.0
      DO 10 I=2,100
10    SUM=0.5**I
      PRINT *, SUM
```

7.11 This program to compute the factorial, $N!$, of a number, N, is flawed. Find and correct the errors. The factorial of a number is given by:

$$N! = N(N-1)(N-2)(N-3)\cdots(3)(2)(1)$$

```
      PRINT *,' ENTER N'
      READ*, N
      DO 10 I=2,N
10    FACT=FACT*I
```

7.12 The first two numbers of the famous Fibonacci series are 1 and 2. Succeeding terms of the series are given by the sum of the previous two terms. Hence, this series begins as 1, 2, 3, 5, 8, 13, ... The following program was designed to compute the first 100 terms. What went wrong?

```
      INTEGER F1,F2,F3
      F1=1
      F2=2
      DO 10 I=3,100
         F3=F1+F2
         F2=F3
         F1=F2
10    PRINT*, F3
```

7.13 Rewrite the last program in Sec. 7.4 so that it uses only one additional variable instead of two (PRENUM and PREDEN).

7.14 The following program calculates the value of the constant $e = 2.71828\ldots$ using:

$$e = 1 + \tfrac{1}{1!} + \tfrac{1}{2!} + \tfrac{1}{3!} + \tfrac{1}{4!} + \cdots$$

The program is designed to stop when the difference between two successive approximations is less than 0.01. Find the logic error:

```
      SUM=1.0
      PREV=0.0
      DO 10 X=1.0,10000.0
            FACT=FACT*X
            SUM=SUM+1.0/FACT
            IF(SUM-PREV.LT.0.01.OR.PREV-SUM.LT.0.01)
            GOTO 20
10    CONTINUE
20    PRINT 11,SUM
11    FORMAT(' ','SERIES APPROXIMATION = ',F9.6)
```

7.15 The following program is syntactically correct, but when executed, it generates the error:

Error message: Arithmetic trap, Integer divide by zero
Locate the logic error and fix it:

```
          REAL INDEX IRATIO
          PI=3.1415926178
          DO 100 INDEX=2*PI, -PI/2., -PI/7.
               ICOUNT=ICOUNT+1
               AREA=PI*RADIUS**2.
               RADIUS=300.0
               ARC=INDEX*RADIUS
               CIRCUM=PI*RADIUS*2
               IRATIO=CIRCUM/ARC
               PRON=AREA/ICOUNT
    100   CONTINUE
          PRINT*, PRON, AREA,ICOUNT
```

7.16 The following program tests the validity of this approximation:

$$\frac{1}{1 - X} = 1 + X + X^2 + X^3 + \cdots$$

The program computes the series for values of X, n, and epsi, which you read in at execution time. N is the desired number of terms to include in the approximation. Epsi is the allowed error in the approximation. Identify the logic errors in this program. Pay particular attention to the run-time error that results when $X = 1.0$. Suggest ways to get around the problem that you identify:

```
      DO 20 I=1,2
            READ*,X,N,EPSI
```

```
      DO 10 I=1,N
            APPROX=APPROX+X**I
10    CONTINUE
      LHS=1/(1-X)
      PRINT*,'LEFT HAND SIDE= ',LHS
      IF(ABS(LHS-APPROX).LE.EPSI) THEN
            PRINT*,' THE APPROX IS VALID FOR X= ',X
      ELSE IF(ABS(LHS-APPROX).GT.EPSI) THEN
            PRINT*,' THE APPROX IS INVALID FOR X= ',X
      ENDIF
20    CONTINUE
      STOP
      END
```

CHAPTER

ARRAYS

8.1 The Need for Arrays

Scientists and engineers often work with large amounts of data. For example, we may run an experiment in which there are several thousand data points that we need to process. With the techniques that we have presented so far, this would be a very difficult task. To demonstrate this, let's focus on a much simpler task: to write a program to read in ten numbers and print them out in reverse order. One solution is the following:

```
READ *, X1, X2, X3, X4, X5, X6, X7, X8, X9, X10
PRINT *, X10, X9, X8, X7, X6, X5, X4, X3, X2, X1
```

This will work, of course, but it's not too elegant. Nor is it very practical. If we wanted to use the program to do the same thing for 11 numbers, we would need to rewrite the entire program. A program that runs for only a specific set of numbers is not very useful.

Now, let's make the task a little more challenging. Your task is to expand the program to read in 100 numbers, find their average, and print the numbers out in reverse order with the average. If we duplicate the approach of the last example, our program would look like this:

```
READ *, X1,X2,X3,X4,X5,X6,X7,X8,X9,X10,X11,X12,X13,
        X14,X15,X16,X17,X18,X19,X20,X21,X22,X23,
        X24 ...
AVG = (X1+X2+X3+X4+X5+X6+X7+X8+X9+X10+X11+...)/100
PRINT *, X100,X99,X98,X97,X96,X95,X94,X93,X92,X91,
         X90,X89,X88,X87,X86,X85,X84,X83,X82,X81,
         X80,X79 ...
```

where the ... in each statement indicates that you must type out all 100 variables *explicitly.* No shortcuts are allowed here. If you use this approach, you must type in all 100 variables—three different times! We humans understand what ... means, but the computer doesn't. So, you have to type everything out.

"Okay," you might say, "this is a tedious task, but certainly possible." So, if we haven't convinced you yet that we need a better way, then how about modifying the program again for 1,000 numbers, 10,000 numbers, or a million numbers? By now you should see that the direct method is not adequate for many of our needs.

Mathematics has solved this problem for us by introducing the idea of the *subscripted variable.* To find the average of a list of numbers, for example, mathematicians would use this special notation:

$$\text{Average} = \bar{X} = \frac{1}{N}\left(\sum_{i=1}^{i=N} X_i\right) = \frac{1}{N}(X_1 + X_2 + X_3 + \cdots + X_N)$$

where N = number of items in list

X_i = ith item in the x list

This expression is a simple way to show the summation of the items in the list. The subscript i attached to the variable X shows that we should go the ith position in the list to obtain the numerical value stored there. Graphically, Fig. 8.1 shows how we might represent this.

So, when our summation comes to the terms, $X_7 + X_8 + X_9 + X_{10}$, the machine adds the numbers stored in the appropriate locations:

$$\cdots + X_7 + X_8 + X_9 + X_{10} + \cdots$$
$$\cdots + 3.7 - 0.2 + 11.9 + 6.1 + \cdots$$

The advantage of representing data in this way is that we need only *one* variable, X, to store a large quantity of data. We can then manipulate all that information by referring to the subscript. Notice that the mathematical shorthand for summing all the terms sets up the subscript, i, as a number that changes from $i = 1$ to $i = N$. As we will see shortly, this subscript will be under the control of the program and will make our work much easier.

FIGURE 8.1

Schematic Representation of Storage of Array Elements

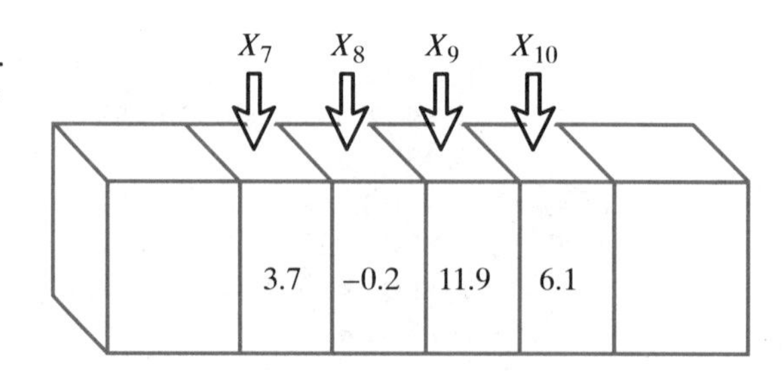

Although we have not yet presented the procedure for using arrays, you might like to see how we compute the sum of 100 numbers with and without arrays:

Without Arrays:

```
SUM= X1+X2+X3+X4+X5+X6+X7+X8+X9+X10+X11+X12+X13+X14
     +X15+X16+X17+X18+X19+X20+X21+X22+X23+X24+X25
     +X26+X27+X28+X29+X30+X31+X32+X33+X34+X35+X36
     +X37+X38+X39+X40+X41+X42+X43+X44+X45+X46+X47
     +X48+X49+X50+X51+X52+X53+X54+X55+X56+X57+X58
     +X59+X60+X61+X62+X63+X64+X65+X66+X67+X68+X69
     +X70+X71+X72+X73+X74+X75+X76+X77+X78+X79+X80
     +X81+X82+X83+X84+X85+X86+X87+X88+X89+X90+X91
     +X92+X93+X94+X95+X96+X97+X98+X99+X100
```

With Arrays:

```
      DO 10 I = 1 , 100
            SUM=SUM+X(I)
 10   CONTINUE
```

The second program segment is equivalent to the mathematical shorthand notation for the summation operation. Notice that the subscript, (I), is under the control of the DO LOOP. Therefore, each time it goes through the loop, the value of I changes and we reference a different position within the X list. In succession, we will add X(1), X(2), X(3), . . . , X(99), X(100) to the variable SUM. The idea is no more complicated than that. So, although you do not yet know how to set up and use arrays efficiently, you should at least appreciate the economy of code that you can achieve. The difference is even more dramatic if you rewrite the previous examples to add 1000 numbers together.

8.2 The DECLARATION Statement for Arrays

Before you can begin to use an array within a program, you must first *declare* it. The DECLARATION statement comes at the beginning of the program before any executable statements. Here are some examples:

```
REAL X(100)
INTEGER MEAN(10), AVG(10)
```

These DECLARATION statements contain the following information about each array:

- Name of array
- Type of array (REAL, INTEGER, etc.)
- Size of array

The first statement, REAL X(100), tells the compiler that the array X will have 100 *elements* or components and that all 100 will be REAL. Similarly, the second DECLARATION statement tells the compiler that the arrays MEAN and AVG will each have ten elements, all INTEGERs.

The general form of the DECLARATION statement is:

```
TYPE name ( lower limit : upper limit )
```

where

TYPE is the data type (REAL, INTEGER, CHARACTER, etc.) for the array
name is the name of the array (limited to six letters/characters)
lower limit is the lowest subscript to be used
upper limit is the highest subscript to be used

The lower limit/upper limit pair must be INTEGERs that tell the computer how many memory spaces to reserve for the individual components. Usually, though, arrays will begin with a lower limit of 1. Here you may omit the lower limit and only specify the upper limit. Some examples of correct and incorrect usage are listed in the following table:

Valid	Invalid	Comments
`REAL X(1:10)`		Okay, but 1 is unnecessary
`REAL X(10)`		Equivalent to previous example
`REAL X(10) , Y(20)`		Multiple arrays okay
`INTEGER AMT , MEAN(10)`		Okay to mix arrays with single-valued variables
	`REAL X(N)`	Variable size array not allowed
	`REAL X(10.0)`	INTEGERs only for limits
	`INTEGER X(5:-5)`	Wrong order of limits

Usually, we think of components as starting with the first element. But there are some problems when we may wish to start with the zero element instead of the first one. Therefore, DECLARATION statements give us that flexibility. For instance, in the first example REAL X(−5:5) tells the computer that there will be 11 components running from X(−5) to X(5).

The first invalid example in the table is an interesting one since it is something that we would very much like to do. For example, what if you wrote a program to analyze data from an experiment but you don't know in advance how many data points there will be. You might be tempted to set up your program to use a variable size array as follows:

Forbidden:

```
REAL X(N)
      .
      .
      .
READ *, N
```

In this example we tried to enter the number of data points, N, at execution time and then use this value to set up the array. Unfortunately, FORTRAN does not allow us to do this. There is a compromise:

Allowed:

```
PARAMETER (N=30)
REAL X(N)
      .
      .
      .
```

This example introduces the PARAMETER statement that allows us to set up constants whose values cannot change during the program. The PARAMETER statement in the example assigns a value to the variable, N, which the compiler then uses to declare the array.

This is not exactly what we wanted since we would have to modify the PARAMETER statement before execution time to match the desired number of data points. It is useful, though, when we have many arrays to declare and want a simple way to change all of them with a minimum of fuss. Here is an example of how we might do this:

```
PARAMETER (I=10, J=20, K=150)
REAL X(I), Y(I), Z(I), TOT(I), AVG(I)
INTEGER INT(J), NEW(J), VOL(K)
```

Using the PARAMETER statement concentrates all changes in one location so that you will not forget any during editing for the new data set. This is especially useful in very long programs, when the number of changes may number in the hundreds.

8.3 Manipulating Arrays

As we showed earlier, we can access individual elements of an array through its subscript. Assume that we have the array VOLTS(I) of Fig. 8.2 that stores voltage measurements as a function of time. The subscript I represents time. I = 1 for example, represents the first time increment; I = 2 represents the second increment, and so on.

There are several questions we need to address in our discussion:

- How do we store numbers into the array elements?
- How do we retrieve the numbers?
- How do we use arrays in assignment statements?

Storing data into array elements is a simple process: You only need to use the desired array element as the target of an assignment statement:

```
VOLTS(4) = 2.9
VOLTS(5) = 3.6
```

In a similar way, we can print out data stored in the array element by again referring to the element:

```
PRINT *, VOLTS(4), VOLTS(5)
```

which will print the numbers 2.9 and 3.6 to the CRT screen on the same line. Input and output of arrays is an important topic and we will have more to say about that shortly.

Manipulation of arrays is also a simple matter. We treat array elements just like the single-valued variables that we discussed in Chap. 5. For example, if we wanted to add the four voltages in the preceding list and compute the average, we could use the following code:

```
AVG = (VOLTS(3) + VOLTS(4) + VOLTS(5) + VOLTS(6))/4.0
```

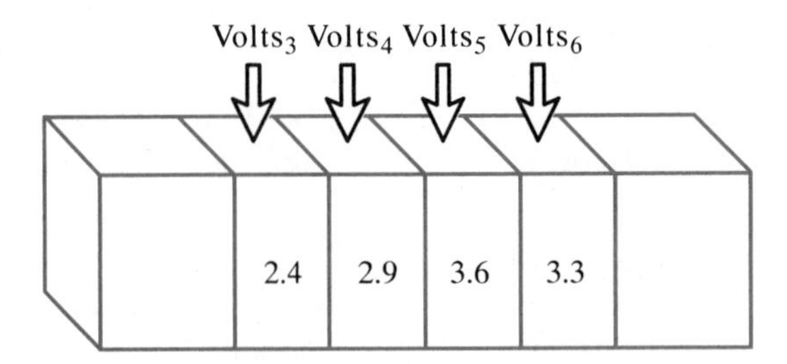

FIGURE 8.2

Storage of Voltage Readings as a Function of Time

Note that this is no different from storing the voltages in single-value variables VOLTS1, VOLTS2, VOLTS3, and VOLTS4. You might ask, then, why go to all the bother? The answer is that the use of arrays opens up a new way to manipulate the data that single-valued variables cannot match. This economy involves the use of a DO LOOP that controls the index or subscript of the array:

```
      DO 10  I = 3,6
             SUM = SUM + VOLTS(I)
10    CONTINUE
      AVG=SUM/4.0
```

Let's trace through this program segment to see how it matches the previous expression for computing the average. The DO LOOP has the index, I, take on the values 3 to 6 inclusive. Thus, when I = 3, the computer adds the numerical value of VOLTS(3) to SUM. Next, I = 4 and the computer then adds the value of VOLTS(4), and so forth, until the loop finishes. Finally, the program divides the value of SUM by 4 to obtain the average. This is shown in the following table:

Process	I	VOLTS(I)	SUM	AVG
START LOOP	3	VOLTS(3)=2.4	2.4	
INCREMENT I	4	VOLTS(4)=2.9	5.3	
INCREMENT I	5	VOLTS(5)=3.6	8.9	
INCREMENT I	6	VOLTS(6)=3.3	12.2	
FINISH LOOP				
AVG=SUM/4.0				3.05

For the specific example chosen, this is not much of an improvement over the single assignment statement. But the DO LOOP approach with the arrays offers the opportunity to do the same process with many more data points. Here is how we would rewrite the loop:

```
      REAL VOLTS(1000)
      READ*, N
      READ*, (data points )
      DO 10 I=1,N
             SUM = SUM + VOLTS(I)
10    CONTINUE
      AVG=SUM/N
```

Notice that we have made the loop perform the summation as a general loop running from 1 to N, where the user enters N at execution time. Also, we have modified the formula to compute the average for the general case. This is a common occurrence in programming, so study this structure well.

We have set up the DECLARATION statement in the previous example for many more data points than we will ever need. This is common practice when we do not know the exact number of points to be entered at execution time. It is permissible to make the array larger than you really need. In the DECLARATION statement you are merely promising the compiler that the number of array elements *will not exceed the number declared.* The compiler permits you to use less than the number declared. This is particularly useful in situations where you do not know the exact number of data points until execution.

Finally, the process for reading the data points into the array is missing from the preceding example. There are shorthand methods for doing this, which we will discuss shortly.

Let's now expand on this example and compute the deviation of each voltage from the average. We define the deviation as the difference between a particular voltage and the average. Thus,

deviation voltage 1 = VOLTS(1) − AVG
deviation voltage 2 = VOLTS(2) − AVG
.
.
.

This can be done with the following algorithm:

- Read in all the data points.
- Compute the average.
- Subtract the average from each voltage.

We now set up a second array, DEV(I), which contains the deviations corresponding to each voltage reading. Thus, DEV(4) corresponds to the deviation of VOLTS(4) from the average. Here is the code:

```
      REAL VOLTS(1000), DEV(1000)
      READ*, N
      READ*, (data points )
      DO 10 I=1,N
            SUM = SUM + VOLTS(I)
   10 CONTINUE
      AVG=SUM/N
      DO 20 I=1,N
            DEV(I) = VOLTS(I) - AVG
   20 CONTINUE
```

The code for computing the average is the same as that previously shown. The program then uses the variable, AVG, to compute the deviation of each voltage. Notice that the computation of the average must be completed before you can compute the deviations. Otherwise the program is straightforward.

You should note that DO LOOPs and processing of arrays are almost inseparable. Arrays, by definition, contain large amounts of data that can be processed easily by changing the subscript. This is most efficiently done with DO LOOPs. So, whenever using arrays, you should begin to think of loops as the means to manipulate the data.

8.4 Input and Output of Arrays

The simplest way to enter arrays or to print them out is to refer to them individually, just as you would for single-valued variables:

```
READ *, X(1), X(2)
PRINT *, X(100), X(99)
```

Of course, this method is very inefficient and ignores the shortcuts inherent in the use of arrays. Also, the preceding method is not feasible when there are many data points for I/O.

A much better way is to combine DO LOOPs with the arrays, just as we did in the last section. Here is how you would read in 100 data points into the array X and then print them out in reverse:

```
    REAL X(100)
    DO 10 I=1,100
          READ *, X(I)
10  CONTINUE
    DO 20 J=1,100
          PRINT *, X(101-J)
20  CONTINUE
```

This is no different from the ideas of the last section. The subscripts, I and J, are under the control of the DO LOOPs and control storage or retrieval of the data. One new thing introduced here is the use of an INTEGER expression for calculating the array element. In the example we refer to the element, X(101 − J), where the expression 101 − J follows the rules of INTEGER arithmetic. You must be careful when referring to an array element since the computer will always truncate the result to an INTEGER, even if you use REAL variables.

The input and output of arrays occurs so frequently in the preceding form that programmers have developed two shorthand methods for doing the same process. These shorthand methods result in a significant reduction of code.

The distinguishing fact about the I/O of arrays is that DO LOOPs are almost *always* used. Therefore, the shorthand method involves the simplification of the loop to a structure known as the IMPLIED DO LOOP:

EXPLICIT DO LOOP	Equivalent IMPLIED DO LOOP
DO 10 I=1,10 READ *, A(I) 10 CONTINUE	READ *, (A(I), I=1,10)

Since we know that the I/O of an array will involve a DO LOOP, we condense the loop down to the single one-line statement shown. Note that we still have the LOOP CONTROL VARIABLE (I=1,10), but now we include it inside parentheses in the READ statement. The general form of the IMPLIED DO LOOP is as follows:

```
READ *,
PRINT *,     (array (LCV), LCV = start, finish, step )
```

The rules for the LOOP CONTROL VARIABLE, LCV, in an IMPLIED DO LOOP are identical with those for the general DO LOOP that gave rise to this special form. The LCV should be an INTEGER and the start, finish, and step values can be constants, variables, or expressions. Thus, for the general problem that we introduced in the previous section for averaging, an unknown number of data points would become:

```
      REAL VOLTS(1000), DEV(1000)
      READ*, N
      READ*, ( VOLTS(I) , I=1,N )
      DO 10  I=1,N
             SUM = SUM + VOLTS(I)
10    CONTINUE
      AVG=SUM/N
      DO 20  I=1,N
             DEV(I) = VOLTS(I) - AVG
20    CONTINUE
      PRINT *, ( DEV(I) , I=1,N )
```

Notice that both the input and output (I/O) of the arrays use IMPLIED DO LOOPs with the variable, N, which the user enters at execution time. This makes the program general enough so that you can use it for almost any data set up to 1000 points.

Even the IMPLIED DO LOOP can be further simplified since you will enter or print most arrays in ascending order, X(1), X(2), X(3), and so on. Therefore, even the LCV is redundant since almost all arrays will be entered from beginning to end. So, a third form of I/O for arrays exists:

EXPLICIT DO LOOP	IMPLIED DO LOOP	Short Form
DO 10 I=1,10 READ *, A(I) 10 CONTINUE	READ *, (A(I), I=1,10)	READ *, A

The third form implies that every element of the array is to be read in. Looking at the READ statement, it is not obvious that A is an array. But, of course, the DECLARATION statement tells the compiler how to treat the variable, A. Thus, if you have a code like this:

```
REAL A(50)
  .
  .
  .
READ *, A
```

the computer will enter the first 50 numbers into A(1), A(2), A(3), and so forth. You should limit your use of this form of the I/O statement since the array *must* be filled with data before any other processing can proceed. Thus, if you want to set up the array for 50 elements but use only 10 of them, then this form of the I/O statement should be avoided.

8.5 Formatting of Array Output

Because of the large quantities of data that arrays can output, formatting may present some problems. You can usually solve these problems with the *repeat specifier*. Let's assume that we have an array with 100 elements and want the array to output 5 elements per row with an (F7.2, 3x) format such as:

```
xxxx.xx   xxxx.xx   xxxx.xx   xxxx.xx   xxxx.xx
xxxx.xx   xxxx.xx   xxxx.xx   xxxx.xx   xxxx.xx
xxxx.xx   xxxx.xx   xxxx.xx   xxxx.xx   xxxx.xx
```

The format specifier (F7.2, 3x) repeats five times on a line before moving to the next line. This can be done with the following:

5(F7.2, 3X), /, 1X

Recall that the slash specifier, /, tells the computer to finish the current line and go to the next line. The 1x that follows is the carriage control character. Thus, the preceding statements will use the (F7.2, 3x) five times and then move on to the next line. Now to use this for every line, we need to repeat this whole group of instructions 20 times. As you might guess, we can do this with the repeat specifier again:

20(5(F7.2, 3X), /, 1X)

which we can incorporate into the final output format pair:

```
     PRINT 7, (A(I), I=1,100)
   7 FORMAT(' ', 20( 5(F7.2, 3X), /, 1X) )
```

As we have shown you a few times already, arrays are often of an unknown size. This presents an additional problem for printing out the numbers. For example, the previous problem used the format, (5(F7.2, 3x), /, 1x), 20 times since we knew that there were 100 data items with five items per line. But when we do not know in advance the number of items to print out, we have a special problem. Fortunately, the solution is an easy one:

```
     PRINT 7, (A(I), I=1,N)
   7 FORMAT(' ', (5(F7.2, 3X), /, 1X) )
```

The only difference between this format statement and the previous one is the omission of the number 20. If the I/O command statement has more variables than the format statement has specifiers, then some specifiers will be reused. Thus, the descriptor, (5(F7.2, 3X), /, 1X), correctly tells the computer how to print out the first five values. But what will it do with the sixth value? Well, the computer will just use the same format repeatedly until it prints all the data. In effect, the computer determines how often to repeat the specifier.

8.6 The DATA and PARAMETER Statements

In Sec. 8.2 we introduced you to the PARAMETER statement. You will find this to be a useful statement when declaring arrays whose size is likely to change. For example, you may write a program to process 100 data points as follows:

```
   REAL VOLTS(100), I(100), IMPED(100), RESIST(100)
   INTEGER TIME(100), COUNTS(100), SIZE(100)
```

If you now wish to change the program for 1000 data points, you must use your editor and change each array dimension from 100 to 1000. An easier way is to use the PARAMETER statement when you first set up your program:

```
PARAMETER (N=100)
REAL VOLTS(N), I(N), IMPED(N), RESIST(N)
INTEGER TIME(N), COUNTS(N), SIZE(N)
```

Now, when you want to increase the size of the arrays, you only need to make one change in the PARAMETER statement. You still must use the editor, but the changes are far fewer.

This device is especially useful for changing variables scattered throughout your program. For example, you might have DECLARATION statements for the arrays, DO LOOP variables to process the data, and I/O statements:

```
      PARAMETER (N=100)
      REAL VOLTS(N), I(N), IMPED(N), RESIST(N)
      INTEGER TIME(N), COUNTS(N), SIZE(N)
                  .
                  .
                  .
      READ *, (VOLTS(K), K=1,N)
                  .
                  .
                  .
      DO 10  L=1,N
                  .
                  .
                  .
10    CONTINUE
                  .
                  .
                  .
      PRINT *, (IMPED(M), M=1,N)
```

If you change the variable in the PARAMETER statement, you will change all sections of the program that use the variable, N. Thus, the size of the arrays, the number of data items read in and printed out, and the number of iterations of the loop will change. You only need to make one change. If you had to make all changes individually, you might easily miss one and create additional work for yourself locating it. So, it is good practice to get into the habit of using the PARAMETER statement for variables that are likely to change.

The general form of the PARAMETER statement is:

PARAMETER *(variable1=value , variable2=value , ...)*

All you must do is to give values to each variable in the PARAMETER list separated by commas. You should place the PARAMETER statement before any executable statements. Also, it usually appears even before any DECLARATION statements since several variables in the PARAMETER statement may be used to set the size of arrays.

Once a variable appears in the PARAMETER statement, it cannot change inside the program. Although we think of variables, like N, as those that can take on any values, the PARAMETER statement converts N from a variable into a constant. Thus, you cannot change it, either deliberately or accidentally, to another value.

Another useful structure is the DATA statement. It is useful to assign *initial values* to a variable. Whereas the PARAMETER statement assigns permanent values, the DATA statement assigns temporary values. One way of assigning temporary values to variables is with the simple assignment statements:

```
 VOLTS = 5.3
RESIST = 10000.0
CAPICT = 0.000035
```

This can be done a little more economically with the DATA statement:

```
DATA VOLTS, RESIST, CAPICT/5.3, 10000.0, 0.000035/
```

The DATA statement first lists each variable followed by the intended values set off by two slash marks, /:

```
DATA var 1, var2,... /value 1, value 2, ... /
```

In this form the DATA statement is not much of an improvement over the simple assignment statements. Yet, there are two situations where the DATA statement offers an advantage. The first situation is when several variables are all given the *same* initial value:

```
DATA A, B, C, D, E, F / 1.0, 1.0, 1.0, 1.0, 1.0, 1.0/
```

which can be shortened to:

```
DATA A, B, C, D, E, F / 6*1.0 /
```

The expression 6*1.0, which is equivalent to 1.0, 1.0, 1.0, 1.0, 1.0, 1.0 contains the initial values for the variables. The * in the expression does not suggest multiplication. Instead, it indicates repetition of the value to follow. Similarly, if we had written /3*0.0, 3*1.0/, the first constant, 0.0, repeats three times, then the second constant repeats three times.

The second situation where DATA statements are useful is when you want to initialize an array. For example, the direct way to set all elements of an array to a value of 1.0 is:

```
      DO 10  I=1,100
             A(I) = 1.0
   10 CONTINUE
```

You can do the same thing with an IMPLIED DO LOOP and a DATA statement:

```
   DATA (A(I), I=1,100) / 100 * 1.0 /
```

The expression for the values for assignment, 100*1.0, have the same meaning as before. The IMPLIED DO LOOP, (A(I), I=1,100), is equivalent to the IMPLIED LOOP that we used for I/O operations.

An important point that we should emphasize here is that you can only use IMPLIED DO LOOPs with READ, WRITE, PRINT, and DATA statements only! You cannot use them for doing any calculations, for example.

8.7 Syntax Summary

The following tables, by way of example, should summarize most of the rules for the syntax related to arrays. You may assume that we have made the following declarations:

```
   REAL A(10), B(10,10), C(10)
   INTEGER K(10)
```

Valid	Invalid	Comments
REAL A(5), B(10,10)		Okay to mix one-dimensional and two-dimensional arrays
A(I+1)=A(I)		Okay to use arrays in assignment
A(K(5))		Okay since K(5) is an INTEGER
IF(A(3).GT.A(5))		Okay to use arrays in conditionals
PRINT *, X, Y, A(5)		Okay to mix arrays with other data
PRINT *, A(1)+A(2)		Math in PRINT statement okay
PRINT *,(A(I), C(I), I=1,10)		Two arrays in IMPLIED DO okay
DATA (A(I), I=1,10)/10*1.0/		Initializes all elements to 1.0
DATA A /10*1.0/		Equivalent to previous example
	A(A(5))	Subscript must be INTEGER
	A(4.0*I)	Mixed mode gives REAL
	PRINT *, (A(I),I=1,100)	Subscript out of bounds
	SUM=SUM+(A(I), I=1,10)	IMPLIED DO LOOPs for I/O only
	PARAMETER (I=7) ⋮ I=I+1	Variable in PARAMETER cannot change value

8.8 Algorithm Primitives for One-Dimensional Arrays

Arrays are an important programming feature that you will use often. It is hard to envision any program for engineers and scientists that will not use an array. Therefore, we devote a significant amount of time here to present several useful algorithm primitives. These will include:

Searching for MAX/MIN value
Sorting an array
Summing elements of an array
Adding vectors
Scalar (dot) product of two vectors

We will use these primitives in the numerical methods sections to follow. More importantly, they are useful for demonstrating how to use arrays.

SEARCHING A LIST

We often need to scan a list of numbers to find the maximum (or minimum) value in the list. This is often a preliminary step before scaling or normalizing the data. For example, assume that we have the following short list of numbers:

45.4 19.8 35.9 51.0 28.3

We can see immediately by inspection that 51.0 is the largest value. If we now wish to normalize the data, we divide each by that maximum value. This is an important step before trying any analysis, such as graphing the data.

Notice that scanning the list to find the largest value is an automatic process for humans. We don't give it much thought. But to write a program to do something similar, we must think much more carefully about details of the procedure.

To simulate the process, we need to create a temporary storage location for the largest value in the list. We call this BIG:

BIG 0

45.4 19.8 35.9 51.0 28.3

The first thing we do is to assume that the first number in the list is the biggest value:

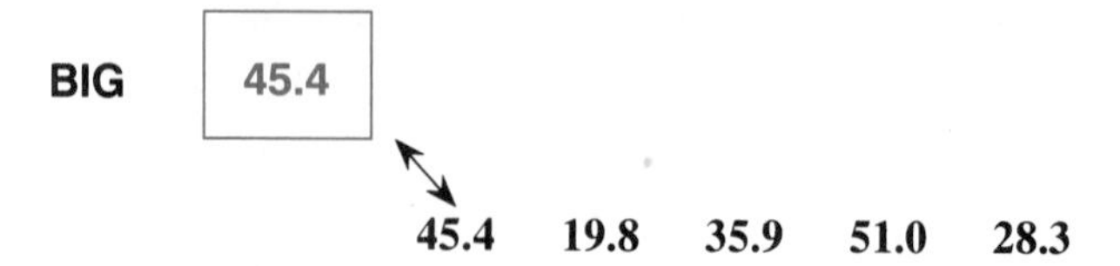

Next, we examine the second number to see if it is larger than the value that we have temporarily assigned to BIG. If this number is bigger than BIG, we reassign it to BIG:

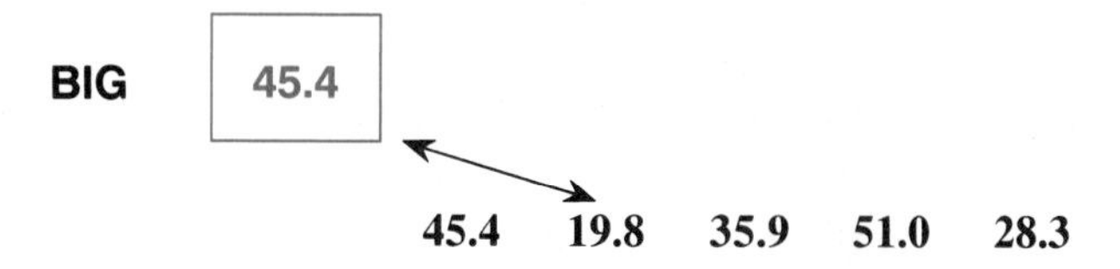

Since the second number, 19.8, is smaller than BIG, we do nothing. We repeat this process until we have examined all numbers in the list:

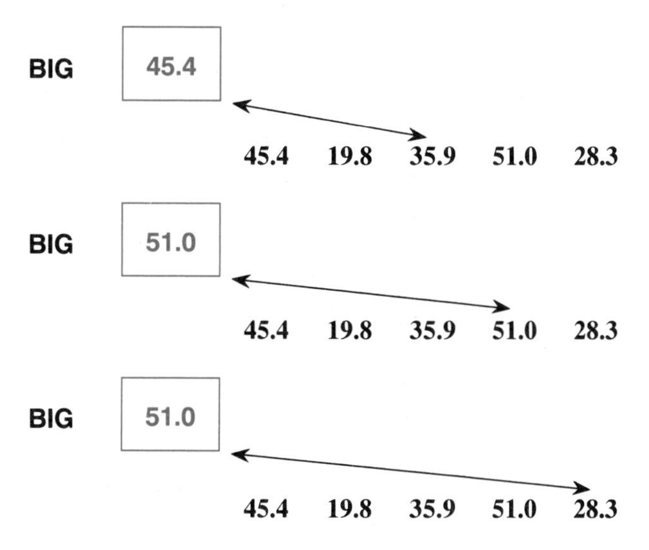

With the aid of these figures, you should see that we can carry out the process with a loop. For example, the first iteration of the loop compares BIG to the first number; the second iteration compares BIG to the second number and so on. Also, we can represent the individual numbers as elements of an array. For example, when comparing the second number to BIG, we can do that this way:

```
IS NUM(2) > BIG ?
```

Finally, you should note that we went through the list of numbers in ascending order, first element, second element, and so forth. This suggests that the subscript referring to each element can be the same variable as the LOOP counter. In general, the process will look like this:

```
LOOP: I = 1 to 5 (or N for general case)
    Compare NUM(I) to BIG
    Switch if needed
END LOOP
```

Notice that each number is stored as an element of the array, NUM. Thus, when the loop counter, I, has the value of I = 1, then the first number in the list, NUM(1), will be compared to BIG. Then, as the loop counter increments to the next value, I = 2, the second number in the list, NUM(2) is compared. This continues for all numbers in the list. In the simple example shown above, only 5 values were given. But in the general case, there will be N numbers in the list. Therefore, it is a good idea to set up the program segments for the general case of N numbers.

The following LS can find the largest value in a list of numbers, and with only modest changes, can also find the smallest value in a list. For brevity however, we will only show you how to find the largest value. We leave the conversion of the program into a MIN search up to you as an exercise.

LS FOR MAX/MIN SEARCH

PROBLEM STATEMENT: Read in a list of numbers and determine the largest (smallest) value.

ALGORITHM DEVELOPMENT:

Step 1a: Review the problem statement. Indicate all nouns in **bold** and verbs in *italics:*

Read in a list of **numbers** and *determine* the largest (smallest) **value.**

Formulate questions based on nouns and verbs:

Q1: How do we read in a list of numbers?
Q2: How do we determine the largest value?

Step 1b: Construct answers to the above questions:

A1: Use **IMPLIED DO LOOP**
A2: Assume BIG = first value; scan remaining values; if any value bigger than Big, then assign it to BIG. Otherwise continue to search.

ALGORITHM:

1. Read in N, which is number of items in list.
2. Use IMPLIED DO LOOP to read in A(1) to A(N).
3. Assign A(1) to BIG.
4. LOOP: from A(2) to A(N)
   is A(I) > Big?
      True: BIG ← A(I)
      False: continue search.
5. Print out value of BIG.

FLOWCHART:

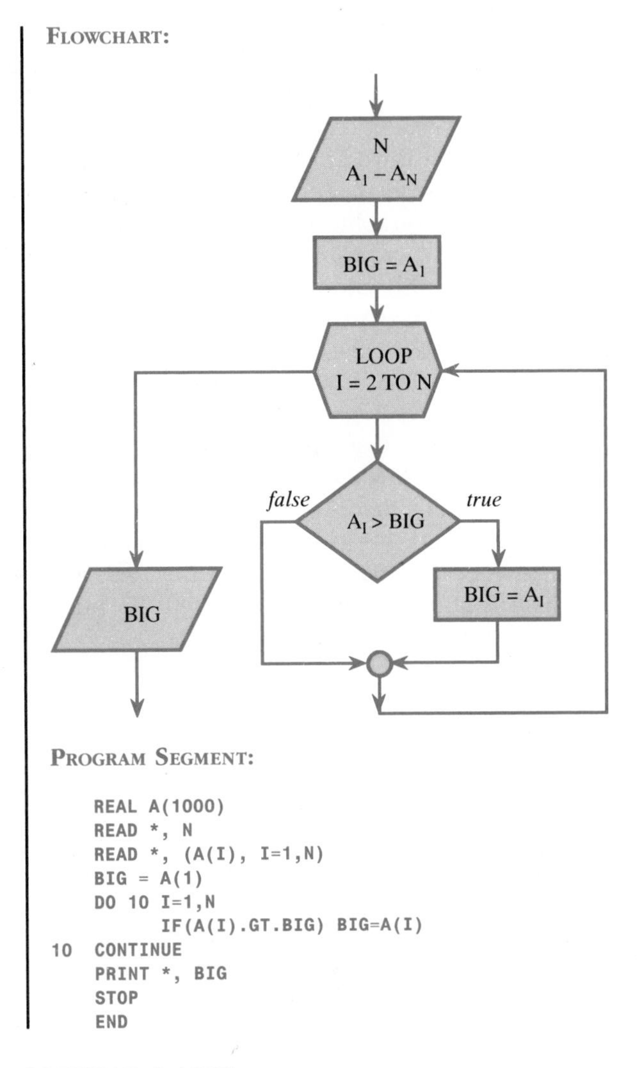

PROGRAM SEGMENT:

```
      REAL A(1000)
      READ *, N
      READ *, (A(I), I=1,N)
      BIG = A(1)
      DO 10 I=1,N
            IF(A(I).GT.BIG) BIG=A(I)
10    CONTINUE
      PRINT *, BIG
      STOP
      END
```

SORTING A LIST

Statistical analysis of data often requires sorting an array. For example, if you want to find the *median* grade of an exam, you must first put the grades in order. The median value represents the point separating the grades into the top half and the bottom half. Usually, the average value and the median value are different. We will discuss this subject in more detail in a later chapter.

There are many different sorting algorithms. The one that we show here is the MAX/MIN sort. The MAX/MIN sort is probably the easiest of the sorting techniques to understand, but, unfortunately, it is also very inefficient. However, it is more important at this point that you understand how to set up the solution rather than concentrate on efficiency. There will be plenty of time for that later.

The MAX/MIN sort is a natural extension of the MAX/MIN search procedure just presented. The sorting procedure uses the search procedure to locate the maximum value and to move it to the top of the list. Then the search repeats, except that the search begins with the second item in the list. When it finishes, a third search begins, but it starts with the third item. The process repeats on smaller and smaller lists until all numbers are in descending order. Figure 8.3 graphically illustrates this.

We use the MAX/MIN search algorithm of the previous example with one minor modification. Instead of conducting the search from A(1) to A(N), we conduct the search from A(L) to A(N). The variable, L, is the current position in the list for which the maximum value is being sought. For example, L = 1 for the first search when we scan A(1) to A(N). The second search has L = 2 and the search goes from A(2) to A(N), and so forth. It should be apparent that L will be another LOOP counter. Thus, the search procedure will be placed inside a loop.

Notice that this new language structure (LS) contains two previous LSs, namely, the MAX/MIN search LS and the LS for switching two numbers. As promised at the beginning of this text, we will strive to build a library of useful routines, many of which will be built upon previous routines. This is a good example of that approach:

LS FOR MAX/MIN SORT

PROBLEM STATEMENT: Read in a list of numbers and put them into descending order.

ALGORITHM DEVELOPMENT:

Step 1a: Review the problem statement. Indicate all nouns in **bold** and verbs in *italics:*

Read in a list of **numbers** and *put* them into descending **order.**

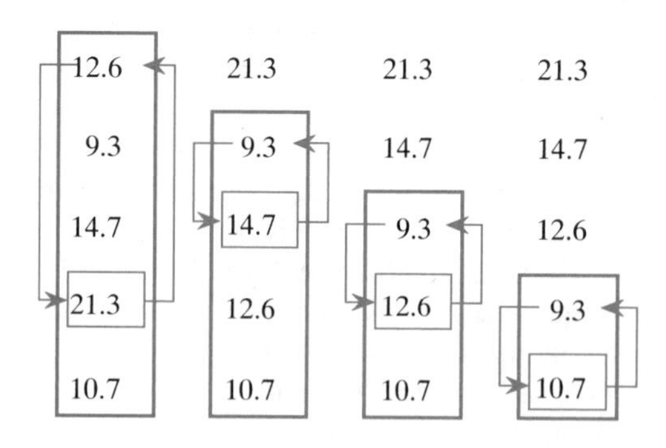

FIGURE 8.3

Process for Sorting a List

Formulate questions based on nouns and verbs:

Q1: How do we read in a list of numbers?
Q2: How do we put them into descending order?

Step 1b: Construct answers to the above questions:

A1: Use **IMPLIED DO LOOP.**
A2: Search the list for largest value (use MAX/MIN SEARCH LS):
Move largest to top.
Repeat for reduced list until all numbers are ordered.

ALGORITHM:

1. Read in N, which is the number of items in the list.
2. Use IMPLIED DO LOOP to read in A(1) to A(N).
3. LOOP: from L=1 to N
   Use MAX/MIN SEARCH LS to locate BIG and position, I; switch A(I) and A(L); assign A(L) to BIG.
4. Print out ordered list with IMPLIED DO LOOP.

FLOWCHART:

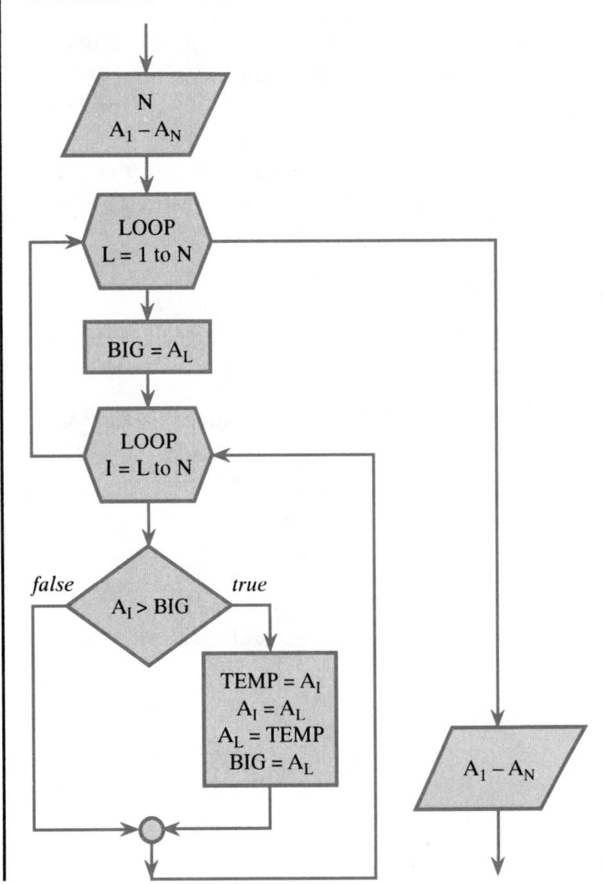

PROGRAM SEGMENT:

```
      REAL A(1000)
      READ *, N
      READ *, (A(I), I=1,N)
      DO 10 L = 1, N-1
            BIG = A(L)
            DO 10 I=L,N
                  IF(A(I).GT.BIG) THEN
                          TEMP=A(I)
                          A(I)=A(L)
                          A(L)=TEMP
                          BIG=A(L)
                  ENDIF
10    CONTINUE
      PRINT *, (A(I), I=1,N)
```

SUMMING ELEMENTS IN A LIST

At the beginning of this chapter we started our discussion of arrays by presenting the problem of adding a string of numbers to obtain the average or mean:

$$\text{Average} = \bar{X} = \frac{1}{N}\left(\sum_{i=1}^{i=N} X_i\right) = \frac{1}{N}(X_1 + X_2 + X_3 + \cdots + X_N)$$

This is a common operation in many fields of data analysis and statistics, among others. The solution to the problem is not so difficult. We set up a variable called SUM that will hold a temporary value representing a partial sum of numbers previously added.

If we want to add the string of numbers shown here, we start with the variable SUM, whose initial value is zero:

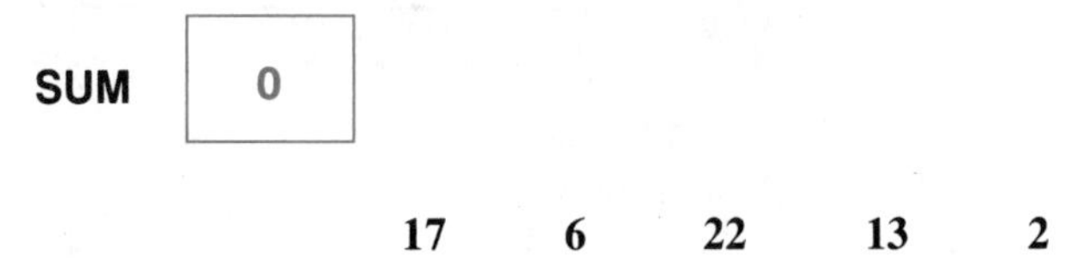

Now we add the first element to the value in SUM:

Then we add the second element:

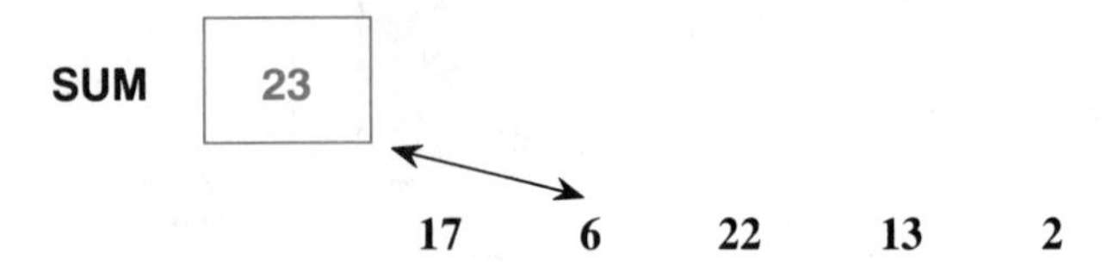

Next, we add the third, fourth, and fifth elements until all elements are added:

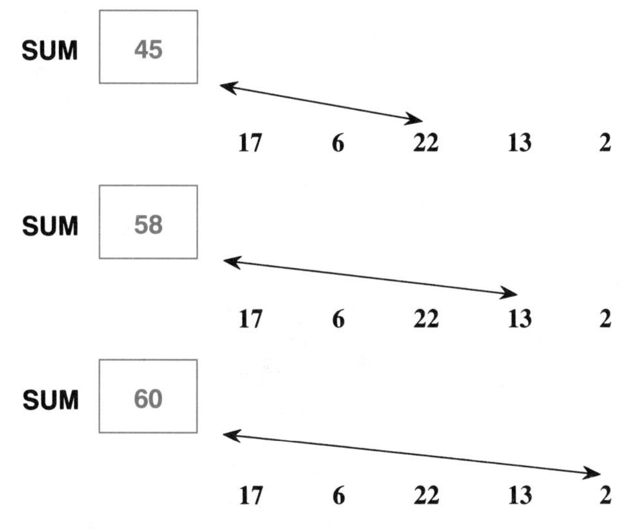

Again, we see that a LOOP is needed to duplicate the previous process in a program. Each element in the list will be an array element and we refer to the element by its position in the list. To add the array element to the total, SUM, we add the old value of SUM to the value of the element and put the result back into SUM. We do this in FORTRAN with:

```
SUM = SUM + A(I)
```

The algorithm and program segment using this approach are shown here:

LS FOR SUMMING ELEMENTS OF AN ARRAY

PROBLEM STATEMENT: Read in a list of numbers and sum them.

ALGORITHM DEVELOPMENT:

Step 1a: Review the problem statement. Indicate all nouns in **bold** and verbs in *italics:*

Read in a list of **numbers** and *sum* them.

Formulate questions based on nouns and verbs:

Q1: How do we read in a list of numbers?
Q2: How do we sum them?

Step 1b: Construct answers to the above questions:

A1: Use **IMPLIED DO LOOP.**
A2: Add one element at a time to a temporary total.

ALGORITHM:

1. Read in N, which is the number of items in the list.
2. Use IMPLIED DO LOOP to read in A(1) to A(N).
3. Set dummy variable, SUM, to zero.
4. LOOP: from I = 1 to N
   Add A(I) to SUM.
5. Print out SUM.

FLOWCHART:

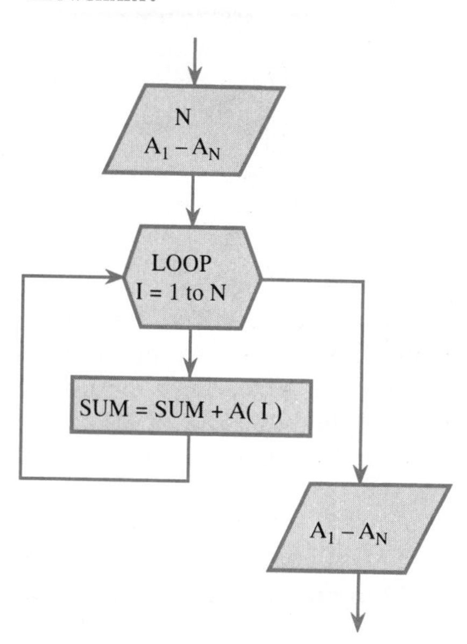

PROGRAM SEGMENT:

```
      REAL A(100)
      READ *, N, (A(I), I=1,N)
      SUM=0.0
      DO 10 I=1,N
            SUM=SUM+A(I)
10    CONTINUE
      PRINT*, SUM
      STOP
      END
```

ADDING VECTORS

Vector quantities are those quantities that possess both magnitude and direction. Both physicists and engineers use them frequently in a variety of applications. For example, a force, which is a vector, that acts on a body will change not only the acceleration of the body, but also the direction of the acceleration. Thus, if a force acts in the X direction, then the acceleration in that direction also changes.

Since vectors have a direction, it is easier to work with them by breaking the vectors into their spatial components. A force, F, acting in an arbitrary direction can be broken into its F_X, F_Y, and F_Z components acting parallel to the X, Y, and Z directions, respectively. This idea is an important one since it allows us to find the net result of many vector quantities acting at once. Consider the simple example of two forces of equal magnitude acting on a body, Fig. 8.4. In the first example the two forces are opposed (pointing in the opposite direction). As a result, the net force on the body is zero. The result will be very different if the two forces act at 90° to one another.

To find the net force on the body, you must add the components for all forces. A common way to represent the components is to list the components within parentheses:

$$\mathbf{F} = (F_X, F_Y, F_Z)$$

To add two forces $\mathbf{F}^1$ and $\mathbf{F}^2$, you simply add the components:

$$\begin{aligned}\mathbf{F}_{\text{net}} &= \mathbf{F}^1 + \mathbf{F}^2 \\ &= (F_X^1 + F_X^2, F_Y^1 + F_Y^2, F_Z^1 + F_Z^2)\end{aligned}$$

To write a program to add two vectors, we must first write each vector as an array, usually with three elements each. Thus, the custom is to represent the force vector, $\mathbf{F}^1$, by the array F1(I). The element, F1(1), represents the X component, F1(2) represents the Y component, and F1(3) is the Z component. A second vector, F2(I), also will have three components. With this background, it is then a simple matter to add the two vectors to find the *resultant*. Note that the resultant is also a vector and must be an array with the same number of elements as the two vectors:

```
RESULT (1) = F1 (1) + F2 (1)
RESULT (2) = F1 (2) + F2 (2)
RESULT (3) = F1 (3) + F2 (3)
```

or for the general case:

```
RESULT (J) = F1 (J) + F2 (J)
```

FIGURE 8.4

Illustration of Addition of Vectors

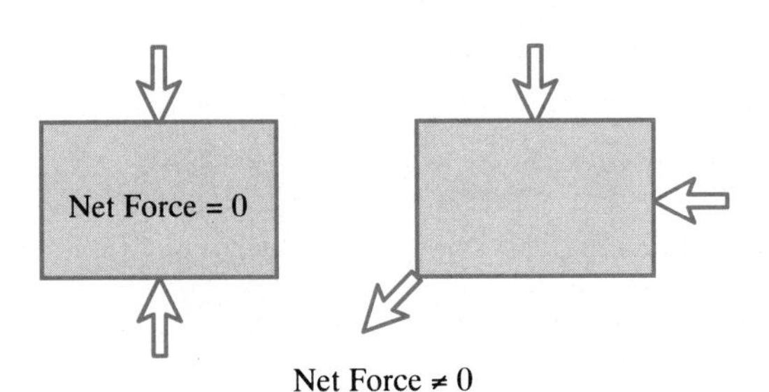

The general formulation is what we will use for the program since we can set up a LOOP with J as the LCV running between 1 and 3:

LS FOR ADDING VECTORS

PROBLEM STATEMENT: Read in two vectors and determine the resultant.

ALGORITHM DEVELOPMENT:

Step 1a: Review the problem statement. Indicate all nouns in **bold** and verbs in *italics:*

Read in two **vectors** and *determine* the **resultant.**

Formulate questions based on nouns and verbs:

Q1: How do we read in a vector?
Q2: How do we determine the resultant?

Step 1b: Construct answers to the above questions:

A1: Use an **IMPLIED DO LOOP** to read in the components of each vector and store it as an element of an array.
A2: Add the same element (first, for example) of each vector array and store the result in the resultant array element.

ALGORITHM:

1. Use IMPLIED DO LOOP to read in F1(1) to F1(3) and F2(1) to F2(3).
2. LOOP: for I = 1 to 3
   RESULT(I) = F1(I) + F2(I).
3. Use IMPLIED DO LOOP to print out RESULT(1) to RESULT(3).

FLOWCHART:

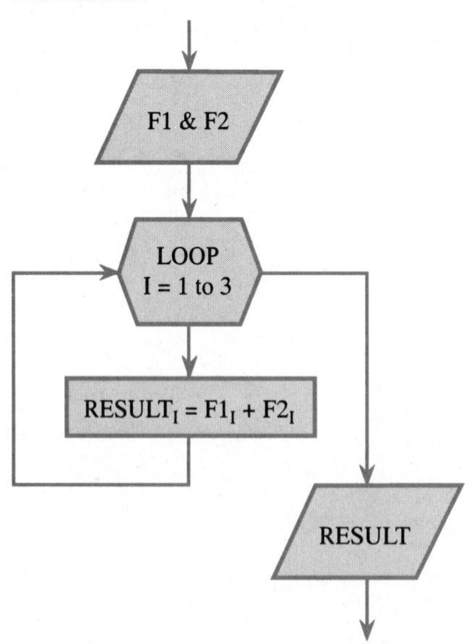

PROGRAM SEGMENT:

```
      REAL F1(3), F2(3), RESULT(3)
      READ *, (F1(I),I=1,3), (F2(I),I=1,3)
      DO 10 I=1,3
            RESULT(I)=F1(I)+F2(I)
10    CONTINUE
      PRINT *, (RESULT(I),I=1,3)
```

It is an easy matter to modify this program to add more than two vectors. In fact, most problems will involve several vectors acting simultaneously. All you must do is to add additional vectors in the DECLARATION statement and include them in the assignment statement inside the DO LOOP. We leave this as an exercise.

DOT PRODUCT OF TWO VECTORS

The *DOT PRODUCT* is an important tool in vector analysis that you will encounter frequently. Mathematically, we define it by:

$$\mathbf{F}^1 \cdot \mathbf{F}^2 = \sum_{i=1}^{i=3} F_i^1 F_i^2 = F_X^1 F_X^2 + F_Y^1 F_Y^2 + F_Z^1 F_Z^2$$

We have seen the summation operation before when we summed all elements of an array, A(I). In this situation, however, we perform the summation on the product of array elements. This requires only a very small modification to our SUMMATION OF ELEMENTS LS as we illustrate in the following example:

LS FOR DOT PRODUCT

PROBLEM STATEMENT: Read in two vectors and take the DOT PRODUCT.

ALGORITHM DEVELOPMENT:

Step 1a: Review the problem statement. Indicate all nouns in **bold** and verbs in *italics:*

Read in two **vectors** and *take* the **DOT PRODUCT.**

Formulate questions based on nouns and verbs:

Q1: How do we read vectors?
Q2: How do we take the DOT PRODUCT?

Step 1b: Construct answers to the above questions:

A1: Use **IMPLIED DO LOOP** to read in the components of each vector and assign them to array elements.
A2: Multiply corresponding elements of each array together. Repeat this for all three elements and add the results.

ALGORITHM:

1. Use IMPLIED DO LOOPs to read in F1(1) to F1(3) and F2(1) to F2(3).
2. Set dummy variable, DOT, to zero.
3. LOOP: from I = 1 to 3
   Add F1(I)*F2(I) to DOT.
4. Print out DOT.

FLOWCHART:

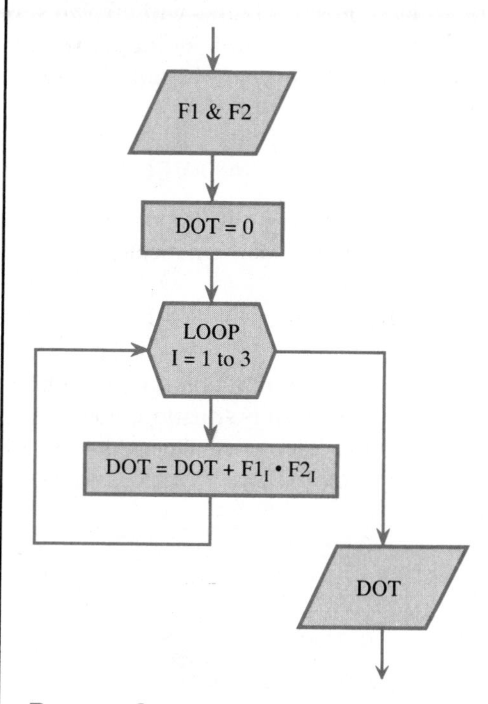

PROGRAM SEGMENT:

```
      REAL F1(3), F2(3)
      READ *, (F1(I), I=1,3), (F2(J), J=1,3)
      DOT=0.0
      DO 10 I=1,3
            DOT = DOT + F1(I) * F2(I)
10    CONTINUE
      PRINT *, DOT
```

The DOT PRODUCT is sometimes also called the *scalar product.* A scalar quantity is something that has magnitude but does not have a direction. The mass of an object is an example. Notice in the example algorithm that DOT is a scalar, or a single-valued variable. It has no components, and therefore is not an array. Be sure to contrast this with the addition of two vectors when the result is itself a vector.

8.9 Two-Dimensional Arrays

One-dimensional arrays, which were the subject of the previous sections, represent lists of data. Engineers and scientists, though, are more likely to work with tables of data. These cannot be easily represented by the simple lists described by one-dimensional arrays. Therefore, we need to explore the two-dimensional array, which is the computer programming analog to tabular data.

In mathematics we handle tabular data with the double-subscripted variable:

$$X_{ij}$$

Two subscripts are present, showing the row and column in the table, as in Fig. 8.5, where the computer stores the data. In Chap. 4 we used an illustration of how to store and retrieve such data.

We manipulate the data stored in the table by controlling the two subscripts. With the simple lists of the previous sections, we had only to control one subscript. With tables, we have two variables to worry about. As we will see shortly, this often means that nested DO LOOPs will be used routinely.

When using tabular data in FORTRAN, we only need to extend some ideas of the one-dimensional array. This applies to the declaration, I/O, and manipulation of the arrays. These extensions are usually straightforward.

To use two-dimensional arrays, you must first declare the array with a type statement. The difference from the one-dimensional array is that you

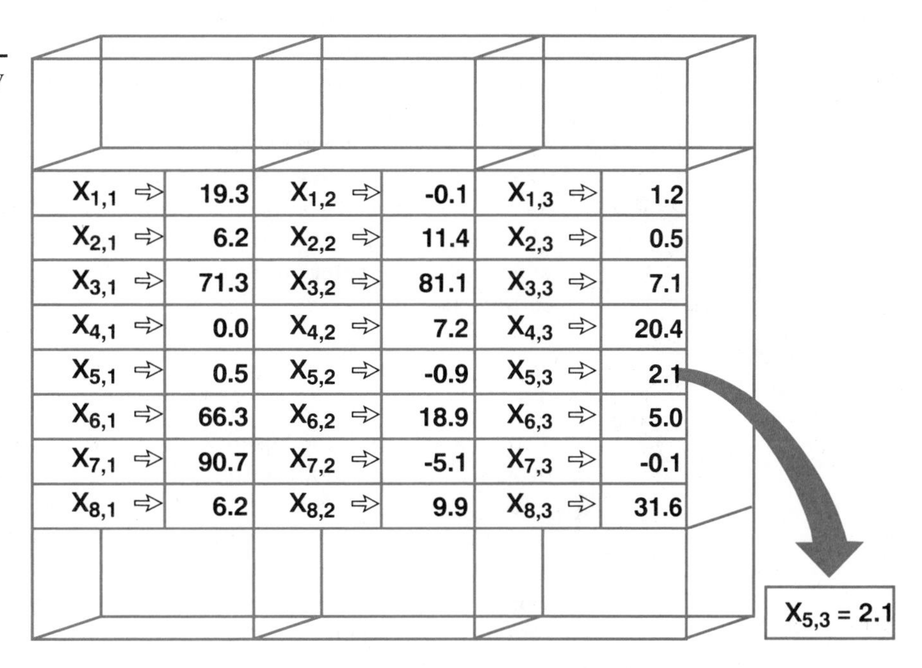

FIGURE 8.5

Storage of a Two-Dimensional Array as a Table

must now specify the number of rows and columns needed to set up your table:

```
TYPE name(row lower limit: row upper limit,
        column lower limit: column upper limit)
```

where

TYPE is the data type (REAL, INTEGER, CHARACTER, etc.) for the array
name is the name of the array (limited to six letters/characters)
row lower limit is the lowest subscript used for rows
row upper limit is the highest subscript used for rows
column lower limit is the lowest subscript used for columns
column upper limit is the highest subscript used for columns

The lower limit on the row and column subscripts is usually 1, in which case we omit it from the DECLARATION statement. Typical declarations will then look like this:

```
REAL STRESS(100,100), TEMP(150,20)
INTEGER INTENS(50,50)
```

The first DECLARATION statement sets up an array, STRESS, which is a table of 100 rows and 100 columns, and a second array, TEMP, which is 150 rows long and 20 columns wide. All data stored in these two arrays will be REAL. The second DECLARATION statement sets up a 50 × 50 table in which all data are INTEGERs.

The size of the arrays in the DECLARATION statement is a limit to make sure that the compiler reserves enough memory space for all the data. Thus, there is no error if you do not use all the allotted space.

You must be careful when using two-dimensional arrays to remember which subscript refers to the row and which refers to the column. In the preceding statements we use the order of:

name (row, column)

Thus, for the table in Fig. 8.5, the value of X(2, 3) is 0.5, whereas the value of X(3, 2) is 81.1. So, you must be careful to remember the ordering of the subscripts.

Manipulating two-dimensional arrays is no more complicated than manipulating one-dimensional arrays, except that we have to do a lot more of it. Tables hold much more data than lists, and therefore you should expect more required processing. If we want to add two numbers from the table, we use the subscripts of each to show their respective positions within the table. Thus, based on the X table in Fig. 8.5:

```
VALUE = X (6,1) - X (4,3)
      =   66.3  -  20.4
      =        45.9
```

You will find frequently that we need to use nested DO LOOPs to manipulate all the data within a table. One LOOP controls the column subscript while the other LOOP controls the row subscript. We explore several examples of this in our ALGORITHM PRIMITIVES.

Input and output of two-dimensional arrays follow the ideas set forth with one-dimensional arrays. The general form of printing out a table is:

```
      DO 10 I=1,10
            DO 10 J=1,10
                  PRINT *, A(I,J)
10    CONTINUE
```

These statements will cause the computer to print the data one item per line since we have omitted any formatting. Notice that the LOOP CONTROL VARIABLE, J, changes more rapidly than the LCV, I. Therefore, the table would be printed out by rows. The computer would print all elements in the first row first, followed by the second row, and so forth:

```
A(1, 1)     row 1; column 1
A(1, 2)     row 1; column 2
A(1, 3)     row 1; column 3
   .              .
   .              .
   .              .
A(1, 10)    row 1; column 10
A(2, 1)     row 2; column 1
   .              .
   .              .
   .              .
```

This is not a very pretty output; most people would find this data hard to read. Also, it wastes paper. So, to improve the appearance of the output, we can use the IMPLIED DO LOOP:

```
      DO 10 I=1,10
            PRINT *, (A(I,J), J=1,10)
10    CONTINUE
```

The IMPLIED DO LOOP after the PRINT statement is the same as the one we discussed with one-dimensional arrays. The only pitfall you are likely to encounter is that the variable controlling the IMPLIED DO LOOP must follow the same rules for all DO LOOPs. Specifically, you must make sure that the variable does not attempt to modify the other LCVs. Here, the IMPLIED DO LOOP uses the variable J, whereas the outer LOOP uses the variable I. There is no conflict, and so the compiler will permit it.

The previous statements will produce a neater output. The IMPLIED DO LOOP will print out the ten values on a single line. Thus, each time the outer DO LOOP executes a full row of the array will appear on the CRT screen:

```
A(1,1)  A(1,2)  A(1,3)  A(1,4)  A(1,5)  A(1,6)  A(1,7)  A(1,8)  A(1,9)  A(1,10)
A(2,1)  A(2,2)  A(2,3)  A(2,4)  A(2,5)  A(2,6)  A(2,7)  A(2,8)  A(2,9)  A(2,10)
A(3,1)  A(3,2)  A(3,3)  A(3,4)  A(3,5)  A(3,6)  A(3,7)  A(3,8)  A(3,9)  A(3,10)
  .
  .
  .
```

Recall that the IMPLIED DO LOOP is a shorthand notation for the full-blown DO LOOP. So, for the previous example we condensed the inner DO LOOP into an IMPLIED LOOP. We were left with one full-blown LOOP and one IMPLIED LOOP. A natural question might then be, why can't we do it again? Why not condense the remaining DO LOOP into another IMPLIED LOOP? There is nothing to prevent us from doing this, and here is the result:

```
PRINT *, ( ( A(I,J) , I=1,10 ), J=1,10 )
```

Notice that the form of the NESTED IMPLIED DO LOOP is:

(implied do loop,* LCV = *start* , *stop* , *step)

where the inner IMPLIED DO LOOP obeys the rules that we discussed before. When using this structure, you must be sure that the two LCVs do not conflict. They must both be different variables and the innermost one must not attempt to change the value of the outermost one. In this structure the innermost LOOP will execute to completion before the outermost LOOP increments to the next value. Thus, the NESTED IMPLIED DO LOOPs behave just like any other pair of nested LOOPs.

Finally, the simplest form of the NESTED IMPLIED DO LOOP is to use the full array name:

```
PRINT *, A
```

If you have previously defined A as a two-dimensional array, the compiler understands this statement to mean that you want the *entire* contents of the array printed. Although you may not have used all elements in the array, the preceding statement will print them. The only question that remains with this form is whether the computer prints the data by rows or columns. By convention, the computer will print out the entire array by columns:

```
A(1, 1)  A(2, 1)  A(3, 1)  A(4, 1)  A(5, 1)  A(6, 1)  A(7, 1)  A(8, 1)  A(9, 1)  A(10, 1)
A(1, 2)  A(2, 2)  A(3, 2)  A(4, 2)  A(5, 2)  A(6, 2)  A(7, 2)  A(8, 2)  A(9, 2)  A(10, 2)
A(1, 3)  A(2, 3)  A(3, 3)  A(4, 3)  A(5, 3)  A(6, 3)  A(7, 3)  A(8, 3)  A(9, 3)  A(10, 3)
   .
   .
   .
```

With any of the IMPLIED DO LOOPs, you can use a FORMAT statement. Generally, it is a good idea; otherwise, the computer will mix E and F formats as the need arises. So, to make your tables easy to read, use FORMATted output.

8.10 Algorithm Primitives for Two-Dimensional Arrays

All the algorithms that we previously presented for one-dimensional arrays are equally valid for two-dimensional or higher arrays. Only a few minor modifications are necessary. Based on our previous discussions, we will show you how to do the following with two-dimensional arrays:

Search for MAX/MIN value in a table
Summing elements in a row of a table
Summing elements in a column of a table

In addition, we introduce two new *algorithm primitives* that are limited to two-dimensional arrays:

Matrix addition
Matrix multiplication

Where appropriate, we rely heavily on our previous discussions and development of *algorithms.*

SEARCHING A TABLE

The method for searching a table is identical to the one for searching a list. All we must do is to treat each row in the table as a list and search that list for the maximum or minimum value. After we finish with that row, we start over with the second row, and so forth. We previously solved the problem of scanning a list. With that algorithm, this is what the variable BIG will look like after we finish scanning the first row:

After Searching Row 1:

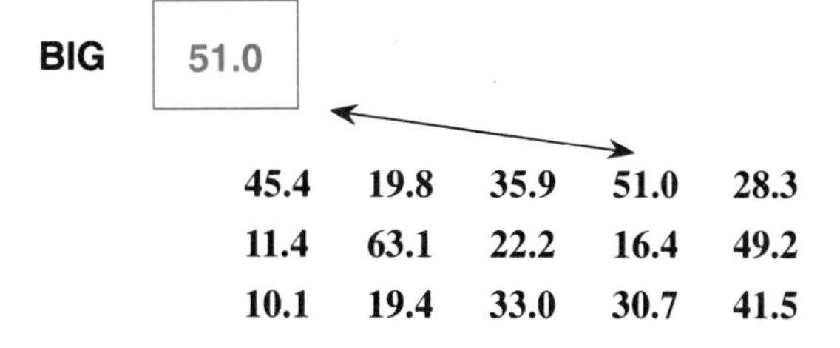

After Searching Row 2:

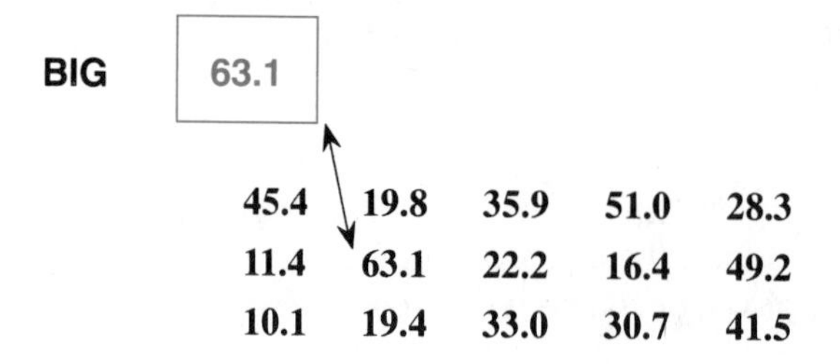

After Searching Row 3:

BIG 63.1

45.4	19.8	35.9	51.0	28.3
11.4	63.1	22.2	16.4	49.2
10.1	19.4	33.0	30.7	41.5

Since the largest value is in the second row, the algorithm did not find any larger value in the third row. Thus, the program made no exchange. Otherwise the search executes as before. To carry out this search, we need to make only a few changes to our algorithm. The biggest change is that wc must add an outer LOOP that selects the row for the search. The inner LOOP will be the previous search algorithm with a two-dimensional array for the one-dimensional array of the previous example:

LS FOR MAX/MIN SEARCH OF TABLE

PROBLEM STATEMENT: Read in a table of numbers and determine the largest (smallest) value.

ALGORITHM DEVELOPMENT:

Step 1a: Review the problem statement. Indicate all nouns in **bold** and verbs in *italics:*

Read in a table of **numbers** and *determine* the largest (smallest) **value.**

Formulate questions based on nouns and verbs:

Q1: How do we read in a table of numbers?
Q2: How do we determine the largest value?

Step 1b: Construct answers to the above questions:

A1: Use **NESTED IMPLIED DO LOOP**

A2: Assume BIG = first value:

Scan remaining values in first row; if any value is bigger than Big, then assign it to BIG. Otherwise continue to search the next row.

ALGORITHM:

1. Read in M and N for size of table (M × N).
2. Use NESTED IMPLIED DO LOOP to read in A(1, 1) to A(M, N).
3. Assign A(1, 1) to BIG.
4. LOOP: for L = 1 to M
5. LOOP: from I = 1 to N
   is A(L,I) > Big?
   True: BIG ← A(L,I)
   False: continue search.
6. Print out value of BIG.

FLOWCHART:

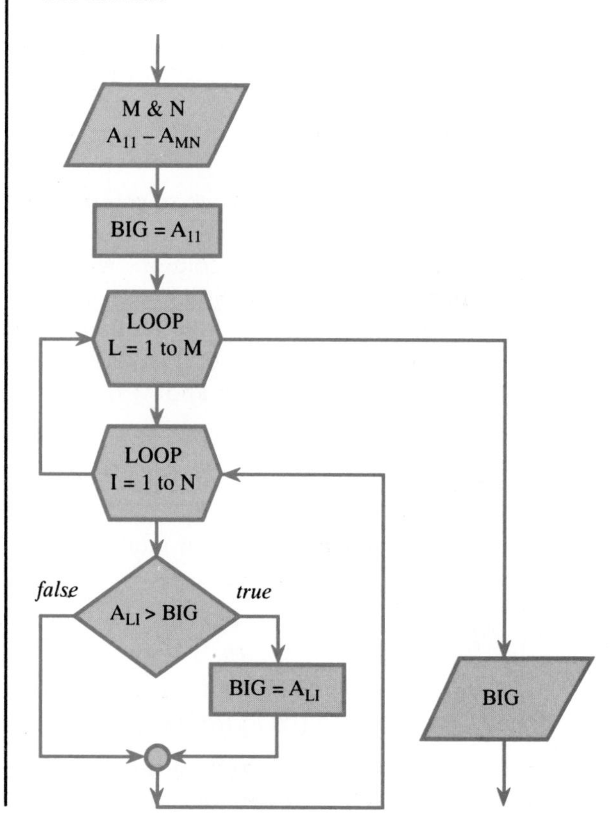

PROGRAM SEGMENT:

```
      REAL A(100,100)
      READ *, M,N,((A(I,J),J=1,N), I=1,M)
      BIG = A(1,1)
      DO 10 L=1,M
            DO 10 I=1,N
                  IF(A(L,I).GT.BIG) BIG=A(L,I)
10    CONTINUE
      PRINT *, BIG
      STOP
      END
```

SUMMING A ROW OR COLUMN

We previously showed you the algorithm for summing a list of numbers. This can be used with only a minor change to sum a column or row within a table. The row or column can be thought of as a list embedded within the table. Therefore, we only have to point to the desired list and apply the previous algorithm:

SUM: 162.3

45.4	19.8	35.9	51.0	28.3
11.4	63.1	22.2	16.4	49.2
10.1	19.4	33.0	30.7	41.5

In this example we have selected the second row. Therefore, we use the summing algorithm to sum the elements of A(2, I). Similarly, if we want to sum the elements in the third column, we submit the list A(I, 3) to the summing algorithm. Notice that in either case that I will be a variable under control of the summing algorithm, while the other index remains constant. Thus, in the first example, we will add A(2, 1), A(2, 2) . . . A(2, 5) and in the second example we will add A(1, 3), A(2, 3) and A(3, 3).

LS FOR SUMMING ROW OR COLUMN IN A TABLE

PROBLEM STATEMENT: Read in a table of numbers and sum the elements in a selected row or column.

ALGORITHM DEVELOPMENT:

Step 1a: Review the problem statement. Indicate all nouns in **bold** and verbs in *italics:*

Read in a **table of numbers** and *sum* the elements in a **selected row or column.**

Formulate questions based on nouns and verbs:

Q1: How do we read in a table of numbers?
Q2: How do we select the row (column) for adding?
Q3: How do we sum the elements in that row (column)?

Step 1b: Construct answers to the above questions:

A1: Use **NESTED IMPLIED DO LOOP** to read in a two-dimensional array.
A2: Read in an INTEGER for the selected row (column).
A3: Use SUMMATION LS to calculate the sum of a list.

ALGORITHM:

1. Read in M and N, which indicate the size of the table.
2. Use NESTED IMPLIED DO LOOP to read in A(1, 1) to A(M, N).
3. Read in L, which indicates row # (column #) for summing.
4. Set dummy variable, SUM, to zero.
5. LOOP: from I = 1 to N (assuming summation of a row)
   Add A(L,I) to SUM.
6. Print out SUM.

FLOWCHART:

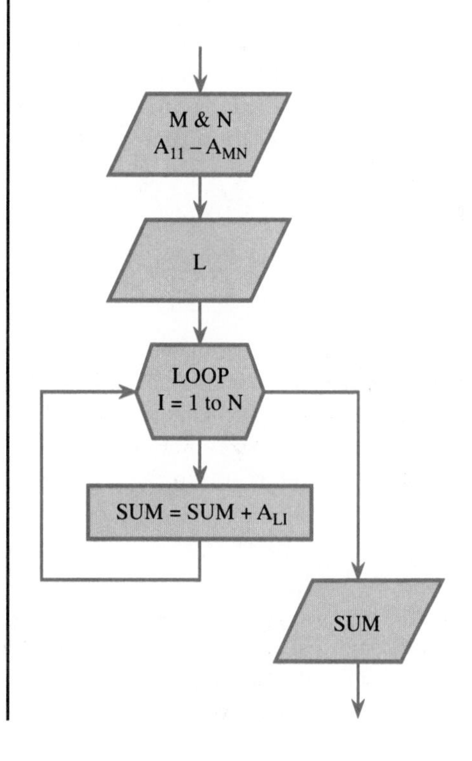

PROGRAM SEGMENT:

```
      REAL A(100,100)
      READ *, M,N,((A(I,J), I=1,M), J=1,N)
      READ *, L
      SUM=0.0
      DO 10 I=1,N
            SUM=SUM+A(L,I)
   10 CONTINUE
      PRINT *, SUM
```

In the preceding example we have summed across a given row. If you want to sum down a column, you must interchange the subscripts and modify the DO LOOP as follows:

Row Summation	Column Summation
DO 10 I = 1,N SUM=SUM+A(L,I) 10 CONTINUE	DO 10 I = 1,M SUM=SUM+A(I,L) 10 CONTINUE

MATRIX ADDITION

Matrices are a common mathematical tool used by engineers and scientists to solve a variety of problems. A matrix is a M × N table of data and is directly analogous to the two-dimensional array that we have been discussing. A common mathematical operation that we perform on matrices is addiiton. Here is a simple example:

$$\begin{bmatrix} 1 & 3 & 5 \\ 4 & -1 & 7 \\ 3 & 2 & 0 \end{bmatrix} + \begin{bmatrix} 2 & 4 & 8 \\ 0 & 3 & 9 \\ 5 & 5 & 2 \end{bmatrix} = \begin{bmatrix} 1+2 & 3+4 & 5+8 \\ 4+0 & -1+3 & 7+9 \\ 3+5 & 2+5 & 0+2 \end{bmatrix}$$

To do *matrix addition,* we simply add corresponding elements of each matrix. Of course, the two matrices must be the same size, and they need not be square as in this example. Also, note that the result of the addition is itself a matrix of the same size. Mathematically, if we wish to add matrix A to matrix B and store the result in matrix C, we can write it in a more convenient form:

$$C_{ij} = A_{ij} + B_{ij}$$

Thus, in the previous example we computed the term C_{12} to be $A_{12} + B_{12}$, or 7. The mathematical notation is very convenient since it is identical with the notation that we will use in our program. Each variable, I and J, will be under the control of DO LOOPs, which will run from 1 to M and 1 to N, respectively:

LS FOR ADDING TWO MATRICES

PROBLEM STATEMENT: Read in two matrices and add them.

ALGORITHM DEVELOPMENT:

Step 1a: Review the problem statement. Indicate all nouns in **bold** and verbs in *italics:*

Read in two **matrices** and *add* them.

Formulate questions based on nouns and verbs:

Q1: How do we read in a matrix?
Q2: How do we add two matrices?

Step 1b: Construct answers to the above questions:

A1: Use **NESTED IMPLIED DO LOOP** to read a single matrix into a two-dimensional array.
A2: For each element, add corresponding terms in each matrix.

ALGORITHM:

1. Read in M and N, which indicate the size of the matrix.
2. Use two NESTED IMPLIED DO LOOPs to read in matrix A(1, 1) to A(M, N) and B(1, 1) to B(M, N).
3. LOOP: from I = 1 to M
4. LOOP: from J = 1 to N
   C(I, J) = A(I, J) + B(I, J).
5. Use the NESTED IMPLIED DO LOOP to print out C(1, 1) to C(M, N).

FLOWCHART:

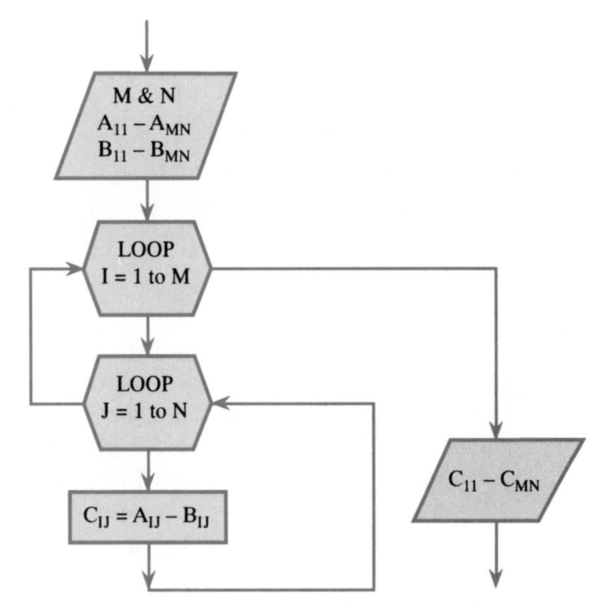

PROGRAM SEGMENT:

```
      REAL A(100,100), B(100,100), C(100,100)
      READ *, M, N, ((A(I,J), I=1,M), J=1,N),((B(I,J),I=1,M),J=1,N)
      DO 10 I = 1,M
            DO 10 J = 1,N
                  C(I,J) = A(I,J) + B(I,J)
10    CONTINUE
      PRINT *, (C(I,J), I=1,M), J=1,N)
```

Notice that we have set up the program segment for matrix addition for the general case of a $M \times N$ matrix up to 100×100. It is unlikely that you will ever encounter a matrix larger than this.

The approach that we use here is something that you may want to use in your programs. Set up the arrays in the DECLARATION statements for the largest conceivable number of data items. Then set up the I/O and the manipulation of the arrays to use only a small portion of this memory space.

MATRIX MULTIPLICATION

Another common mathematical procedure with matrices that you will need is the multiplication of two matrices. Multiplication of two matrices is more complex than the simple addition of the previous example:

$$\begin{bmatrix} 1 & 3 & 5 \\ 4 & -1 & 7 \\ 3 & 2 & 0 \end{bmatrix} \times \begin{bmatrix} 2 & 4 & 8 \\ 0 & 3 & 9 \\ 5 & 5 & 2 \end{bmatrix} = \begin{bmatrix} 2+0+25 & 4+9+25 & 8+27+10 \\ 8+0+35 & 16-3+35 & 32-\ 9+14 \\ 6+0+0 & 12+6+0 & 24+18+0 \end{bmatrix}$$

The value of the element in the product matrix is the sum of the row of the first matrix times the column of the second matrix:

$$[\ \ ] \times [\ \ ] = [\ \ ]$$

Mathematically, we can write the multiplication of A and B as:

$$C_{ij} = \sum_k A_{ik} B_{kj}$$

This may seem like a complex notation, so we should explain it before we proceed any further. The easiest way to explain it is with an example. Let's assume that the A and B matrices are 3×3 square matrices. Now if we want to calculate the element $C_{2,3}$ of the product matrix, we expand the summation (Note: k = number of columns in A = number of rows in B = 3):

$$C_{2,3} = \sum_{k=1}^{k=3} A_{2,k} B_{k,3} = A_{21}B_{13} + A_{22}B_{23} + A_{23}B_{33}$$

$$= (4)\,(8) + (-1)\,(9) + (7)\,(2) = 37$$

Examine the notation carefully. You should see that the algorithm multiplies the second row of A, indicated by $A_{2,k}$, by the third column of B, indicated by $B_{k,3}$. After the computer calculates all products, it adds their sum to $C_{2,3}$. Notice that the indices of the C element, 2 and 3, are the same as the index of the row in A and the index of the column of B.

Now we have to convert this into a program segment. Notice that for the general element, C_{ij}, the value involves a summation over only one variable, K. This means that we can do this part of the task with a single LOOP. The variable K is equal to the number of columns in the A matrix, and therefore is known. A portion of the algorithm might look like this:

```
For given values of i and j (say, for example, i = 2, j = 3);
LOOP: for k = 1 to 5 assuming A has five columns
   add A(2,k)*B(k,3) to SUM
when finished with loop, assign SUM to C(2,3)
```

Next, notice that this only gives us a single value of the product matrix, C(2,3), for example. Therefore, we need to repeat the preceding procedure for every element of the C matrix. Since the matrix is a two-dimensional array, we need to repeat the procedure for all values of i and j. This tells us that we need two more loops:

```
LOOP: for i = 1 to M (all rows)
  LOOP: for j = 1 to N (all columns)

        LOOP: for k = 1 to M
        add A(i,k)*B(k,j) to SUM
        when finished, assign SUM to C(i,j)
```

The LOOP inside the box is similar to the earlier example of the general case for C(I, J). You should trace through this example by hand to make sure that you can generate the results given in the example:

LS FOR MULTIPLYING TWO MATRICES

PROBLEM STATEMENT: Read in two matrices and multiply them.

ALGORITHM DEVELOPMENT:

Step 1a: Review the problem statement. Indicate all nouns in **bold** and verbs in *italics:*

Read in two **matrices** and *multiply* them

Formulate questions based on nouns and verbs:

Q1: How do we read in a matrix?
Q2: How do we multiply two matrices?

Step 1b: Construct answers to the above questions:

> **A1:** Use **NESTED IMPLIED DO LOOP** to read a single matrix into a two-dimensional array.
>
> **A2:** For each element of the product matrix:
> Multiply row of A matrix by column of row of B matrix.
> Add together all products and assign to element of C matrix.

ALGORITHM:

1. Read in M and N, which indicate the size of the matrix A.
2. Use NESTED IMPLIED DO LOOPs to read in matrices A and B.
3. LOOP: from I = 1 to M
4. LOOP: from J = 1 to N
5. LOOP: from K = 1 to M
   C(I, J) = C(I, J) + A(I, K) * B(K, J).
6. Use NESTED IMPLIED DO LOOP to print out C(1, 1) to C(M, N).

FLOWCHART:

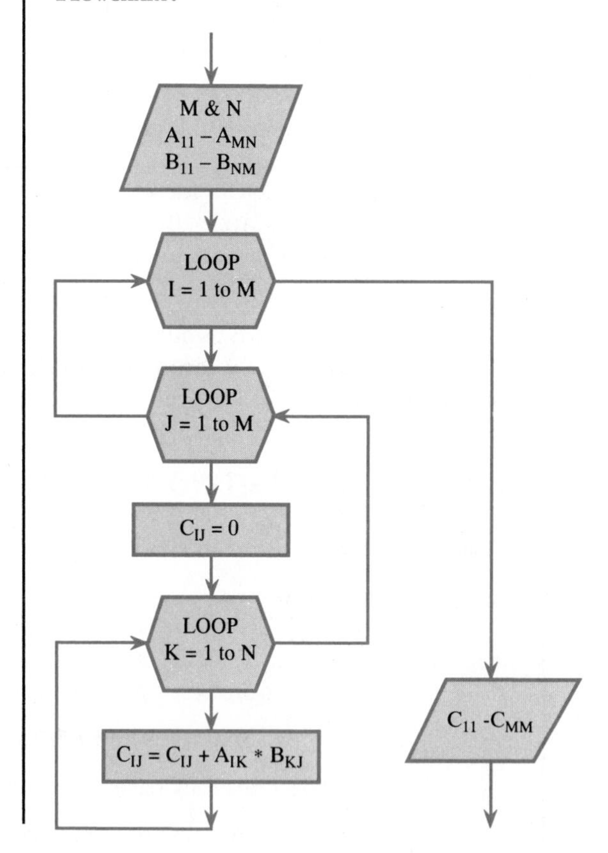

PROGRAM SEGMENT:

```
    REAL A(100,100), B(100,100), C(100,100)
    READ *, M, N, ((A(I,J), I=1,M), J=1,N),((B(I,J),I=1,N),J=1,M)
    DO 10 I = 1,M
          DO 10 J = 1,M
                C(I,J)=0.0
                DO 10 K = 1,N
                      C(I,J) = C(I,J) + A(I,K) * B(K,J)
10  CONTINUE
    PRINT *, (C(I,J), I=1,M), J=1,M)
```

This example is a very important one for two reasons. First, it is the first problem where we needed triple nested DO LOOPs. Second, the mathematics behind this solution is more complex than the previous examples. Therefore, you should review this problem very carefully to make sure that you understand all important features presented.

8.11 Putting It All Together

There are many engineering problems that can be treated by a simple averaging process. In this process we mathematically divide a component into small elements, and the properties of that element are the average of itself and its neighbors. The mathematics is no more complicated than taking the average of three values. We will review one such example here.

If you place one end of a solid insulated bar in 100°C water and the other end of the bar in 0°C water, heat will flow from the hot end to the cold end. To calculate the temperature at any intermediate point, we must break the bar into many small elements as Fig. 8.6 shows. The leftmost segment of the bar is always at 100°C and the rightmost segment of the bar is at 0°C. We can determine the temperature at the intermediate points by taking the average of the temperatures at that point and the two nearest neighbors. For example, in the bar shown in Fig. 8.6 the temperature of the second element is (100 + 0 + 0)/3. The three numbers for the averaging are the temperatures of the first, second, and third elements, respectively. Thus, the temperature in the second element is 33°C (after rounding any decimal fractions). Similarly, the temperature in the third element is (0 + 0 + 0)/3 or 0°C. Notice that we used the old temperature of each element to calculate the new temperatures. That is why we used 0 for the second element instead of 33 for the third element. We can express this process mathematically as:

$$T_i^{\text{new}} = \frac{T_i^{\text{old}} + T_{i-1}^{\text{old}} + T_{i+1}^{\text{old}}}{3}$$

With this equation, you should be able to duplicate the preceding profile labeled as the first iteration.

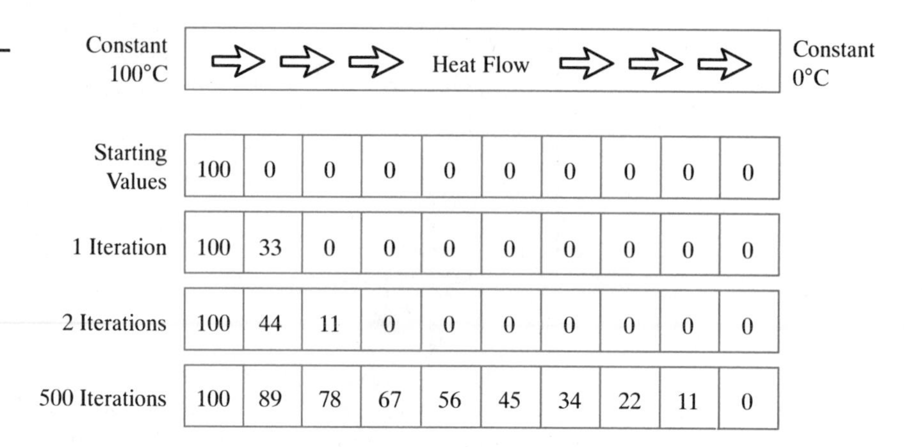

FIGURE 8.6

Iterative Process for Determining Temperature Profile Along a Bar

A single pass with this formula is not sufficient. Look at the results after the first iteration. Do they seem reasonable? Would you expect 80 percent of the bar to be at 0°C? The problem is that we have not gone far enough, and the previous process should be repeated until the temperature distribution in the bar stabilizes. We call this ***equilibrium***.

In the preceding temperature profiles we show the temperatures after 2 and 500 iterations. Clearly, after the second iteration there has been a significant change in the distribution. But after the 500th iteration, there is negligible change from the 499th iteration (not shown). Therefore, it is reasonable to assume that the bar is now in equilibrium with the heat source at one end and the heat sink at the other end. If you plot the data, you will find that the temperature profile is a straight line between 100 and 0°C, which seems like a very reasonable result.

Our program to calculate these profiles must take the following facts into account:

- The temperature in the first element will always be 100°C and the temperature in the last element will always be 0°C.
- Temperatures in the intermediate elements will be calculated as follows:

  For each element: New $\text{temp}_i = 1/3(\text{old temp}_i + \text{old temp}_{i-1} + \text{old temp}_{i+1})$, starting from the second element and going to the $N - 1$ element.
- Check to see if the rod is in equilibrium. If *all* points meet the condition (new $\text{temp}_i -$ old $\text{temp}_i <$ some small value, eps), then equilibrium has been established, and the calculation should stop.
- After the entire new temp profile has been calculated, copy the new temp values into the old temp array.
- After equilibrium has been reached, print out the final temperature profile.

A key idea here is the check for equilibrium. We read in a value for eps at execution time. We use this to see if the temperature of any element has changed by more than eps. If it has, then the bar has not reached equilibrium and the calculation should repeat. A reasonable value for eps might be 0.1°C. A smaller value for eps does not change the final temperature profile by any appreciable amount.

When constructing the algorithm, note that we need a COUNTED LOOP to obtain the new temperature profile from the old. Since we know in advance how many points to calculate, the COUNTED LOOP is the preferred method. The determination of equilibrium also uses a COUNTED LOOP. Again, we know in advance how many calculations to make.

A reasonable thing to do would be to combine these two LOOPs into one. While we are calculating the new temperature, we also can check to see if the value has changed by more than eps. If it has, then we can *set a flag*. A flag is a variable that indicates whether or not a condition is met. For example, we might set the flag to 1 if equilibrium is established, but 0 otherwise. We can then use this flag value to control whether to repeat the calculation. Notice that this is a good candidate for a CONDITIONAL LOOP. Therefore, the skeleton of our algorithm should look like this:

```
CONDITIONAL LOOP: DO WHILE ( FLAG = 0 ?)
        Set FLAG to 1
        COUNTED LOOP: Element 2 to Element (N−1)
                NEW_i = 1/3 (OLD_i+OLD_i−1+OLD_i+1)
                if NEW_i−OLD_i > eps, then reset FLAG to 0
        END COUNTED LOOP
        COUNTED LOOP: Element 2 to Element (N−1)
                Copy NEW_i into OLD_i
        END COUNTED LOOP
END CONDITIONAL LOOP
Print out final temperature profile (OLD_i)
```

Pay special attention to how we handled the FLAG. When starting the CONDITIONAL LOOP, we check its value. If it is zero, then we know that the bar has not reached equilibrium and we can proceed. Once inside the LOOP, we reset FLAG to 1. Thus, we temporarily assume that equilibrium will be reached during this pass through the LOOP. As we calculate the new temperature profile inside the COUNTED LOOP, we check to see if the bar is in fact at equilibrium. If we find *that any* element changes too much, we set FLAG to zero. Thus, when we finish with the LOOPs, FLAG will again be 0 when the CONDITIONAL LOOP next checks it. But, if no element changes by more than eps, then FLAG will be unchanged. Some students may prefer to use a LOGICAL variable, EQUIL, in this situation. This is what that might look like:

CONDITIONAL LOOP: DO WHILE (EQUIL= .false. ?)
  Set EQUIL=.true.
  COUNTED LOOP: Element 2 to Element (N−1)
    $NEW_i = 1/3\ (OLD_i + OLD_{i-1} + OLD_{i+1})$
    if $NEW_i - OLD_i$ > eps, then reset EQUIL=.false.
  END COUNTED LOOP
  COUNTED LOOP: Element 2 to Element (N−1)
    Copy NEW_i into OLD_i
  END COUNTED LOOP
END CONDITIONAL LOOP
Print out final temperature profile (OLD_i)

If the use of the LOGICAL variable makes more sense to you, then use that. But, if you feel more comfortable with the numerical approach, then stick to that. Note, however, that the idea in both cases is the same. Based on these ideas, we can then develop the following flowchart, as shown in Fig. 8.7, for this problem.

The flowchart shows that the CONDITIONAL LOOP uses an IF statement at the beginning to examine the value of FLAG. But, when we convert this into

FIGURE 8.7

Flowchart for Calculating Temperature Distribution in a Bar

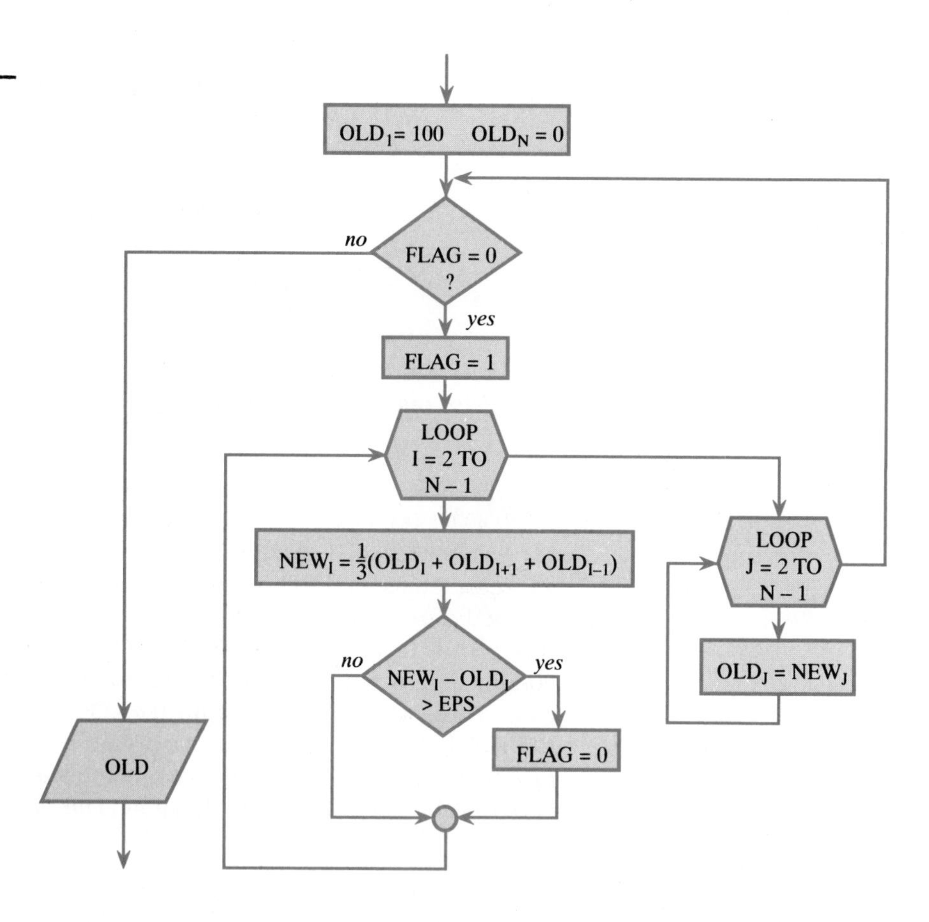

FORTRAN, we use the more convenient DO WHILE structure. Remember, if your compiler does not support this feature, you must rewrite this program with the block IF/GOTO combination discussed in previous chapters.

For this problem, we set up the arrays to have only 10 elements. If we were trying to do a more refined calculation, we would need many more elements, 100 to 1000 would be preferred. We have left the arrays at 10 elements, though, to make it easier to trace. Here then is the final program:

```
C      PROGRAM TO CALCULATE TEMPERATURE DISTRIBUTION
C      IN A SOLID, INSULATED ROD. THE TEMPERATURE AT
C      EACH POINT IS THE AVERAGE OF ITSELF AND ITS TWO
C      NEAREST NEIGHBORS. THE FIRST AND LAST POINTS,
C      THOUGH ARE HELD CONSTANT AT 100 AND 0
C      RESPECTIVELY. THEREFORE, THE LOOPS TO CALCULATE
C      THE NEW TEMPERATURES SKIP OVER THESE POINTS.
C      EQUILIBRIUM IS CHECKED TO SEE IF THE PROCESS
C      SHOULD BE REPEATED. THIS IS DETERMINED BY
C      EXAMINING ALL THE ELEMENTS BEFORE AND AFTER
C      AVERAGING. IF THE CHANGE IS LARGER THAN SOME
C      VALUE, EPS, THEN THE BAR IS NOT IN EQUILIBRIUM,
C      AND THE CALCULATION IS REPEATED.
C           OLD(I) = TEMPERATURE PROFILE OF BAR
C           NEW(I) = TEMPORARY SCRATCH SPACE FOR
C                    CALCULATING NEW VALUES OF OLD(I)
C           EPS    = VALUE READ IN AT EXECUTION TIME
C                    TO CHECK FOR EQUILIBRIUM
C-----------------------------------------------------------
       REAL NEW(10), OLD(10)
       INTEGER FLAG
       PRINT*, 'ENTER EPS (TO CHECK FOR EQUILIBRIUM)'
       READ*, EPS
       FLAG=0
       OLD(1)=100.0
       OLD(10)=0.0
       DOWHILE(FLAG.EQ.0)
            FLAG=1
            DO 10 I=2,9
               NEW(I)=(OLD(I)+OLD(I-1)+OLD(I+1))/3.0
               IF(ABS(NEW(I)-OLD(I)).GT.EPS) FLAG=0
10          CONTINUE
            DO 20 J=2,9
               OLD(J)=NEW(J)
20          CONTINUE
       ENDDO
       PRINT 25, (OLD(K), K=1,10)
25     FORMAT(10(F6.2,2X))
       STOP
       END
```

The DO 10 LOOP calculates the new temperature profile for the elements 2 to 9. Note that it skips over the first and last element since these are at a constant temperature and cannot change. While the program is calculating the new profile it checks for equilibrium. If any of the points change by a large amount, the value FLAG changes to zero. This will be used later by the DO WHILE statement to see if the calculation should be repeated.

After the program calculates the new temperature profile, it copies the temporary values stored in *new* into the original array, *old*. We do this with DO 20 LOOP. Finally, at some point the profile will not change significantly, and FLAG will still equal 1. At this point the CONDITIONAL LOOP stops, and the final profile will be printed. Notice that we have used an IMPLIED DO LOOP with a repeat specifier in the FORMAT statement to print the results.

Exercises

SYNTAX

8.1 Locate Syntax Errors. Find and correct the syntax errors in each of the following segments:

a.
```
      INTEGER I
      REAL ARRAY(I)
```

b.
```
      REAL A(1,10)
      DO 10 I=1,10
   10 A(I)=I**2
```

c.
```
      INTEGER I(10)
      DO 10 I=0,9
   10 I(I+1)=10**2
```

d.
```
      REAL SIN(10)
      X=3.14159/10
      DO 10 I=1,10
   10 SIN(I)=SIN(I*X)
```

e.
```
      REAL A[10]
      DO 10 I=0,9
   10 A(I+1)=I**2
```

f.
```
      DOUBLE PRECISION X(100)
      DATA X/100*2.0/
      DO 10 I = 1, INT(X(I))
               PRINT *, X(I)*I
   10 CONTINUE
```

YOUR SYSTEM

8.2 Check Out Your Compiler. Try the following suggestions with your compiler to find the limitations and extensions available:

a. Usually, the allowed subscripts for an array are j and k (constants), m and n (variables), m+k, m−k, k*m, j*m+k, j*m−k. But some FORTRAN compilers will allow you to use others. Try a subscript such as m**k, j+k*m, m/n, m**n, and so forth.

b. Does your compiler allow you to use the element of an array as a subscript of another array. Try this:

```
      INTEGER II(10)
      J=42342
      DO 10 I=1,10
   10 II(I)=RAN(J)*10+1          (RAN is the random
      PRINT *, II(II(II(1)))      number generator)
```

c. It is sometimes desirable to begin a vector with a lower bound other than 1, such as A(−1). Most FORTRAN compilers will allow this. Does yours?

d. Standard FORTRAN 77 requires that we use INTEGERs for array subscripts. See if your compiler has an extension to permit REAL subscripts:

```
      REAL I, A(5)
      DO 10 I=1,5
10    A(I)=I
```

e. How many subscripts of an array will your system permit? Enter the following DECLARATION statement with a progressively increasing number of subscripts until you get an error message.

```
      REAL A(2,2,2,2,2,2,2)
```

f. Does your system initialize an array to zero?

```
      REAL A(5,5)
      PRINT*, ((A(I,J), I=1,5), J=1,5)
```

g. What happens if you try to change the value of a variable named in a PARAMETER statement?

```
      PARAMETER (X=5.2)
      READ *, X
      PRINT *, X
```

h. What is the limit to the size of an array that you can use?

i. What happens when you refer to an array element in a program in which you forgot to declare it?

j. If you make reference to an array element that is out of bounds, either a compiler or a run-time error will occur. If the compiler does not detect the error, you may get strange results before your program crashes. Run this program to see how your system responds:

```
      REAL A(10)
      DO 10 I=1,1000
            A(I)=0.0
            PRINT *, I, A(I)
10    CONTINUE
```

TRACING

8.3 Tracing Program Segments. Trace through the following program segments and predict their outputs:

a.
```
      REAL II(100)
      DO 10 I=1,100
10    II(I)=I*2
      PRINT*, II(II(II(II(II(1)))))
      STOP
      END
```

b.

```
      REAL A(10)
      DO 10 I=1,10
   10    A(I)=I**2
      DO 20 I=2,10,2
   20 PRINT*, A(I/2+1)
      STOP
      END
```

c.

```
      REAL A(3,3), B(3), C(3)
      DATA A/1,2,3,4,5,6,7,8,9/
      DATA B/1,2,3/
      DO 10 I=1,3
            C(I)=0.0
            DO 10 J=1,3
   10 C(I)=C(I)+A(I,J)*B(J)
      WRITE(5,100) C
  100 FORMAT(1X,3(F6.1, 1X))
      STOP
      END
```

d.

```
      REAL A(3,3), B(3,3), C(3,3)
      DATA A/1,2,3,4,5,6,7,8,9/
      DATA B/1,2,3,4,5,6,7,8,9/
      DO 10 I=1,3
            DO 10 J=1,3
                  C(I,J)=0.0
            DO 10 K=1,3
   10 C(I,J)=C(I,J)+A(I,K)*B(K,J)
      WRITE(5,100) C
  100 FORMAT(1X,3(F6.1, 1X))
      STOP
      END
```

e.

```
      REAL X(2), Y(2)
      DATA X, Y/1, 2, 3, 4/
      DATA NDIM,Z,P/2,2.0,0.0/
      DO 10 I=1,NDIM
            TERM=1
            DO 20 J=1,NDIM
                  IF(I.NE.J) TERM=
     1            TERM*(Z-X(J))/
     1            (X(I)-X(J))
   20       CONTINUE
            TERM=TERM*Y(I)
            P=P+TERM
   10 CONTINUE
      PRINT *, P
      STOP
      END
```

f.
```
      REAL MAT(10,10)
      DATA ((MAT(I,J),I=1,3),J=1,4)
     1    /1,2,3,4,5,6,7,8,9,10,11,12,
     1     13,14,15,16,17,18,19,20/
      DO 15 I=1,2
            DO 15 J=1,4
   15 MAT(I+1,J)=MAT(I+1,J)+MAT(I,J)
      PRINT *,((MAT(I,J),I=1,3),J=1,3)
      STOP
      END
```

g.
```
      REAL X(10,10)
      DATA N,X/10,3,5,7,9,11,10,8,6/
      DO 10 I=2,10
            L=I-1
            M=M-1
            DO 10 J=1,M
   10 X(I,J)=X(L,J+1)-X(L,J)
      PRINT *, X
      STOP
      END
```

h.
```
      INTEGER A(10)
      DO 10 I=1,9,2
            A(I)=I**2
   10 CONTINUE
      DO 20 I=2,8,2
      A(I)=A(I-1)-A(I+1)
   20 CONTINUE
      PRINT 6, (A(I), I=1,10)
    6 FORMAT(' ',2(I7,3X),/)
      STOP
      END
```

8.4 What Does It Do? Tracing can be a tedious task. You can reduce the work required if you try to see what the program is doing while you trace it. Usually, after tracing through a small portion of the data, you will see what is happening and can skip to the final answer. To train you to do this, try to explain in words what the following segments are doing. Make up any data that you will need.

a.
```
      REAL MAT(10,10)
      OPEN(UNIT=10,FILE='DATA.IN')
      OPEN(UNIT=11,FILE='DATA.OUT')
      DO 5 I=1,3
    5 READ(10,*) (MAT(I,J), J=1,4)
      DO 15 I=2,3
            DO 15 J=1,4
   15 MAT(I,J)=MAT(I-1,J)+MAT(I,J)
      DO 10 J=1,4
```

```
 10   WRITE(11,*) (MAT(I,J), I=1,3)
      STOP
      END
```

b.

```
      REAL PIC(10,10)
      DO 10 I=2,9
       DO 10 J=2,9
        SUM=0.0
        DO 20 K=I-1,I+1
         DO 20 L=J-1,J+1
 20       SUM=SUM+PIC(K,L)
 10   PIC(I,J)=SUM/9
```

c.

```
      INTEGER PRIME(1000), LIMIT
      PRINT*, 'ENTER THE LIMIT'
      READ*, LIMIT
      PRIME(1)=2
      K=1
      DO 20 N=3,LIMIT
            FLAG=0
            I=1
            DO WHILE(PRIME(I).LE.SQRT(REAL(N)).AND.
            FLAG.EQ.0)
                IF(MOD(N,PRIME(I)).EQ.0) FLAG=1
                WRITE(5,100) N,PRIME(I),I
                I=I+1
            END DO
            IF(FLAG.EQ.0) THEN
                K=K+1
                PRIME(K)=N
                PRINT*, N, 'IS A PRIME NUMBER'
            ENDIF
 20   CONTINUE
100   FORMAT(1X,T10,I5,2X,I5,2X,I5)
      STOP
      END
```

PROGRAMS

8.5 Transpose of a Matrix. Write a program that reads in a matrix, A, of order $n \times m$ and determines its transpose. The transpose of a matrix is an $m \times n$ matrix where the columns are the rows from A and the rows are the columns from A. Thus, if:

$$A = \begin{bmatrix} 12 & 5 & 2 \\ 1 & 2 & 3 \end{bmatrix} \qquad \text{TRANS(A)} = \begin{bmatrix} 12 & 1 \\ 5 & 2 \\ 2 & 3 \end{bmatrix}$$

8.6 Scalar Multiplication of a Matrix. A matrix, A, can be multiplied by a scalar quantity, k, by multiplying every element in the matrix by the constant:

$$k \times \begin{bmatrix} 12 & 5 & 2 \\ 1 & 2 & 3 \end{bmatrix} = \begin{bmatrix} k \times 12 & k \times 5 & k \times 2 \\ k \times 1 & k \times 2 & k \times 3 \end{bmatrix}$$

Write a program to perform the scalar multiplication for any size $n \times m$ matrix and any constant.

8.7 Matrix Subtraction. Use the program for MATRIX ADDITION in the text to perform MATRIX SUBTRACTION of [A] − [B]. First, use the scalar multiplication process of the previous problem to multiply [B] by −1. Then add the resulting matrix to [A] to complete the subtraction process.

8.8 Orthogonal Vectors. Two vectors are *orthogonal* if they are perpendicular to each other. Use the DOT PRODUCT algorithm to write a program to determine if two vectors are orthogonal.

8.9 Determinant of a 3 × 3 Matrix. Given the 3 × 3 matrix, A, the determinant, det(A) is defined by:

$$\begin{vmatrix} a_{11} & a_{12} & a_{13} \\ a_{21} & a_{22} & a_{23} \\ a_{31} & a_{32} & a_{33} \end{vmatrix} = a_{11}(a_{22}a_{33} - a_{23}a_{32}) - a_{12}(a_{21}a_{33} - a_{23}a_{31}) + a_{13}(a_{21}a_{32} - a_{22}a_{31})$$

Write a program to read in a 3 × 3 matrix, and calculate the determinant.

8.10 Three-Point Curve Smoothing. Experimental data always has "noise" associated with it. There are many techniques to reduce this noise, but none is easier to use than the three-point method. Here is some experimental data before we remove any of the noise:

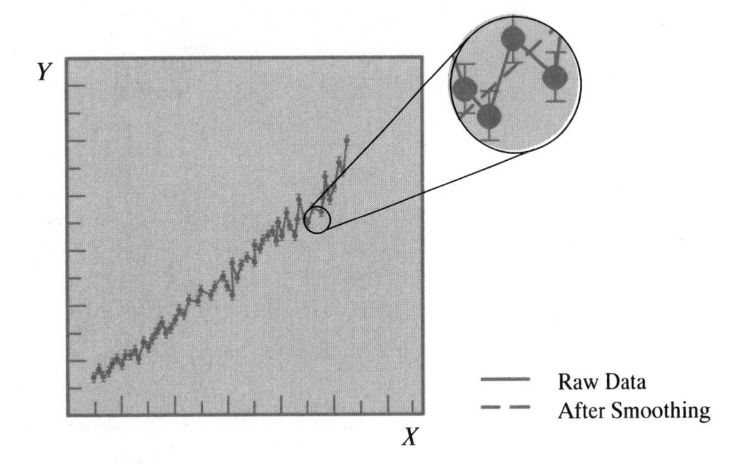

The process for smoothing is to replace each data point, Y(I), by the average of itself plus its two nearest neighbors, Y(I + 1) and Y(I + 1):

$$Y(I) = (Y(I - 1) + Y(I) + Y(I - 1))/3$$

Let's examine the three points highlighted:

Point	Before Smoothing		After Smoothing	
⋮			⋮	
	562			
234	541		558	
235	571	SPREAD=	556	SPREAD=
236	556	30	569	13
	580			
⋮			⋮	

There are two data points where you must be careful, namely, Y(1) and Y(FINAL) since they have only one neighbor. For example, when you try to smooth Y(1), you must use Y(0), which does not exist. Therefore, we use a different formula for these two special cases:

```
Y(1)     = (2*Y(1)+Y(2))/3
Y(FINAL) = (Y(FINAL-1) + 2*Y(FINAL))/3
```

Write a program to perform three-point smoothing for an arbitrary number of points.

8.11 General Method for Smoothing. The method described in Prob. 8.10 for three-point smoothing can only be used when the data points are evenly spaced. A more general procedure for unevenly spaced data is to replace each data point, Y(I), by:

$$Y(I) = Y(I - 1) + \frac{Y(I + 1) - Y(I - 1)}{X(I + 1) - X(I - 1)}(X(I) - X(I - 1))$$

Write a program that reads in a set of 100X and Y data pairs and smooths the data. If the data is evenly spaced (X(I) − X(I + 1) equals a constant), use the formula from the previous problem. But if the data is unevenly spaced, use the formula given here.

8.12 Five-Point Smoothing. Sometimes a smoothing process spread over more data points is more desirable than the three-point process of Prob. 8.10. Modify the three-point algorithm to average over five points:

$$Y(I) = (Y(I - 2) + Y(I - 1) + Y(I) + Y(I + 1) + Y(I + 2))/5$$

There are now four special cases: Y(1), Y(2), Y(FINAL − 1), and Y(FINAL). Be sure to take care of these special cases in your program.

8.13 Two-Dimensional Smoothing. The ideas in the two previous problems can be extended to smooth two-dimensional arrays. An example might be a photographic image sent with a FAX machine since noise is inevitable in transmissions of this type. The picture is made from PIXELS (picture elements) that are too small to see with the naked eye. If you look closely at a TV image, though, you can see the individual PIXELS. By varying the intensity and color of each PIXEL, you create the image. The simplest image is black and white created from the grey scale. The usual grey scale ranges from 0 (white) to 255 (black). To smooth the image, a PIXEL is replaced by the average of its eight nearest neighbors:

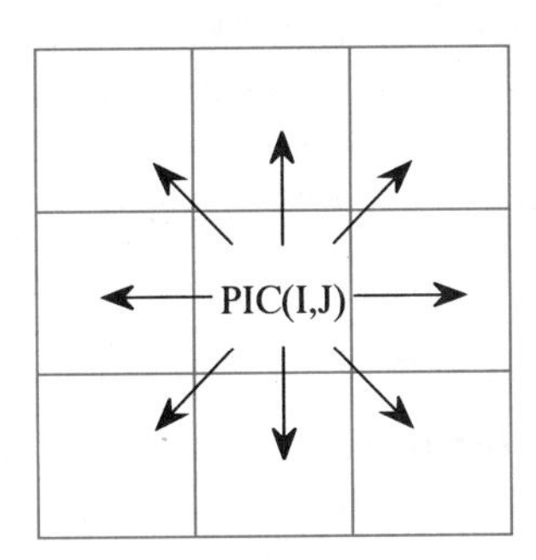

$$\text{PIC(I,J)} = \frac{\sum \text{(EIGHT NEIGHBORS)}}{8}$$

Write a program to use the preceding process to smooth a digitized image. Once again, there are special cases: corners and edges. In these cases you should smooth over only the existing elements. For example, corners have only three neighbors and edges have only five.

8.14 Gradient Sharpening of Images. When an image is digitized (PIXEL assigned an INTEGER grey scale value), the sharpness of the image may suffer. One useful method to restore the sharpness employs a gradient technique. Briefly, these techniques involve the rate of change in the grey scale. As you get close to an edge in an image, the grey scale changes rapidly. We can use this fact to find the edges and then enhance them by other methods. The key is to locate the edge.

The ROBERTS gradient technique creates a new image array, B, by assigning to any point, B(I, J), the value of the gradient at the same point in A:

$$\begin{aligned} \text{B(I, J)} &= \text{GRAD(A(I, J))} \\ &= ((\text{A(I, J)} - \text{A(I + 1, J + 1)})^2 + (\text{A(I + 1, J)} - \text{A(I, J + 1)})^2)^{1/2} \end{aligned}$$

This technique has disadvantages in that edges of objects sometimes are blurred. This can be overcome somewhat by using the following condition:

$$B(I,J) = \begin{cases} \text{GRAD}(A(I,J)) & \textit{if } GRAD(A(I,J)) > T \\ A(I,J) & \textit{Otherwise} \end{cases}$$

where T is a threshold value read in at execution time. With a properly chosen value of T, you can emphasize significant edges without destroying the characteristics of smooth backgrounds.

Write a program to read in the grey scale values of an image array of size 100 × 100 from a data file. Create a sharpened image using the preceding technique. Be careful of special cases such as the bottom and rightmost edges.

8.15 Enlarging an Image. A three-dimensional holographic image is stored in a 10 × 10 × 10 integer array. Because three-dimensional arrays take up so much memory space, they are often squeezed to reduce storage requirements. When we want to examine these images, we decompress them by a simple expansion algorithm.

The image, S, is to be enlarged to the 19 × 19 × 19 array, L. The transformation begins by copying the smaller array into the larger array as follows:

$$L(2I - 1, 2J - 1, 2K - 1) \leftarrow S(I,J,K) \qquad (I,J,K = 1 \text{ to } 10)$$

Note that the new array contains only elements in positions with three *odd* indices. For the moment, all other elements are blank. The second step is to fill in the blank elements by linear interpolation as follows:

$$L(I,J,K) = (S(I-1,J,K) + S(I+1,J,K))/2 \qquad (I \text{ even}; J,K \text{ odd})$$

$$L(I,J,K) = (S(I,J-1,K) + S(I,J+1,K))/2 \qquad (J \text{ even}; I,K \text{ odd})$$

$$L(I,J,K) = (S(I,J,K-1) + S(I,J,K+1))/2 \qquad (K \text{ even}; I,J \text{ odd})$$

$$\begin{aligned} L(I,J,K) = (&S(I-1,J-1,K) + S(I-1,J+1,K) \\ &+ S(I+1,J-1,K) \\ &+ S(I+1,J+1,K))/4 \qquad (K \text{ odd}; I,J \text{ even}) \end{aligned}$$

$$\begin{aligned} L(I,J,K) = (&S(I-1,J,K-1) + S(I-1,J,K+1) \\ &+ S(I+1,J,K-1) \\ &+ S(I+1,J,K+1))/4 \qquad (J \text{ odd}; I,K \text{ even}) \end{aligned}$$

$$\begin{aligned} L(I,J,K) = (&S(I,J-1,K-1) + S(I,J+1,K-1) \\ &+ S(I,J-1,K+1) \\ &+ S(I,J+1,K+1))/4 \qquad (I \text{ odd}; J,K \text{ even}) \end{aligned}$$

$$\begin{aligned} L(I,J,K) = (&S(I-1,J-1,K-1) + S(I+1,J-1,K-1) \\ &+ S(I-1,J+1,K-1) + S(I-1,J-1,K+1) \\ &+ S(I+1,J+1,K-1) + S(I+1,J-1,K+1) \\ &+ S(I-1,J+1,K+1) \\ &+ S(I+1,J+1,K+1))/8 \qquad (\text{all even}) \end{aligned}$$

Write a program to read the original data file from data file IMAGE.IN, create the expanded array, and send it to data file, IMAGE.OUT.

8.16 Pyramidal Compression of an Image. A relatively recent technique developed for image processing of binary images (only 1 or 0 value allowed) is known as pyramidal processing. These operations are most efficiently performed on parallel processing machines, but can also work on conventional computers. The process is quite simple. The original image must be square and the total number of pixels must be a multiple of 4 (16, 64, 256, ...). Operations are performed on the image to produce another smaller image, which has only one-fourth the number of PIXELS. For example, if the original image on level 1 has 64 pixels (8 × 8), the image on pyramid level 2 will have 16 PIXELS (4 × 4). Level 3 will have 4 pixels (2 × 2), and so on, until the final image has just 1 pixel. Therefore, any pixel on a given level is associated with 4 pixels on the preceding level:

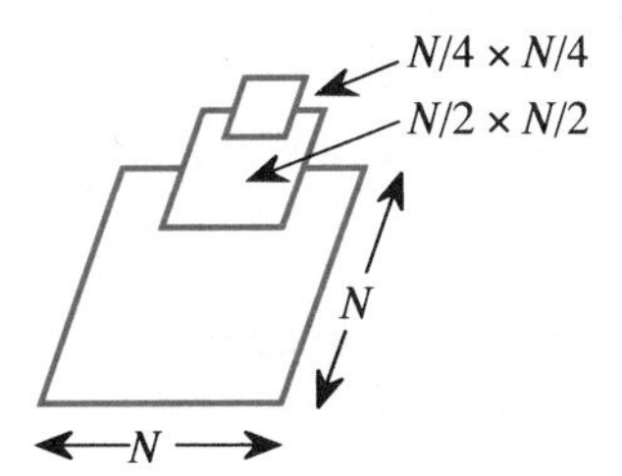

Two of the simpler operations are known as **AND**ing up and **OR**ing up. In ANDing up, 4 adjacent pixels to be compressed are compared. The value for the new pixel is based on the following conditions:

- If all pixels are 1s, the single pixel on the next level will be a 1.
- If any of the 4 pixels is a 0, then the pixel on the new level is 0.

The ORing up process is the logical opposite of the ANDing up function:

- If all pixels are 0s, the single pixel on the next level will be a 0.
- If any of the 4 pixels is a 1, then the pixel on the new level is 1.

ANDing is a kind of smoothing operation, whereas ORing is a sharpening technique. Notice that the ANDing up procedure is very restrictive. In 3 out of 4 cases, the value sent up is zero. The only time that a value of 1 is sent up is when *all* four values are simultaneously 1. By the same reasoning, ORing is much less restrictive, since 3 out of 4 times, the value sent up is 1. Here is an example of how these processes work:

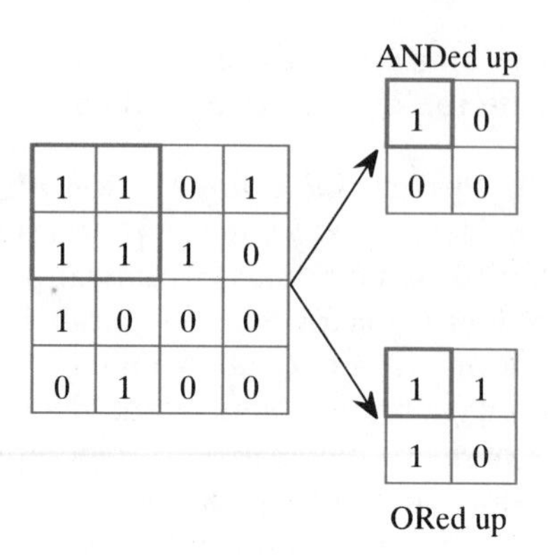

Write a program to read in a 64×64 binary image from a data file and compress the image according to the preceding process. Your program should ask the user for the degree of compression ($2\times, 4\times, 8\times, 16\times$, etc.) and whether the user wants the ANDing or ORing function. Output should go to a data file.

8.17 Heat Flow Down a Rod. At some time, t_0, the temperature of a uniform rod is 0°C. Instantaneously, the temperature of one end of the rod is raised to 100°C while the other end is held fixed at 0°C. The sides of the rod are not insulated, and so heat is lost out the sides. It has been determined that the heat lost per unit of time, Q_i, by a small element of the rod is given by:

$$Q_i = K * (\text{TEMP}_{i-1} - 2*\text{TEMP}_i + \text{TEMP}_{i+1}) - H*(\text{TEMP}_i)$$

where TEMP_i = temperature of the ith element in the rod
K, H = material constants

The first term represents heat flow down the rod, and the second term represents heat lost out the sides. After a small time increment, Δt, the temperature within an element is given by:

$$\text{TEMP}_i^{\text{new}} = \text{TEMP}_i^{\text{old}} + C * Q_i * \Delta t$$

This process is shown schematically here:

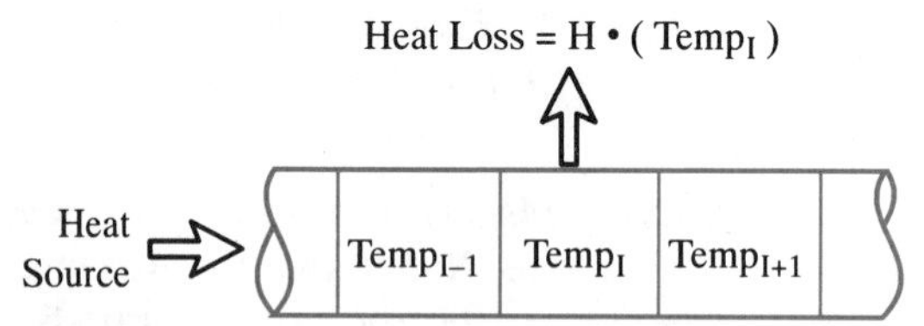

The rod is mathematically broken into many small slices of equal size. We may then assume that the temperature of each block is given by the preceding equations. In a way this is comparable to the smoothing of data or images of the previous problems. Write a program that will calculate the temperature at 1-s intervals until a final time, TF. Enter all necessary information, TF, K, C, and H, and the number of elements (no more than 100) at execution time.

8.18 Temperature Distribution in a Cooling Fin. A two-dimensional problem in heat flow is the cooling fin shown here:

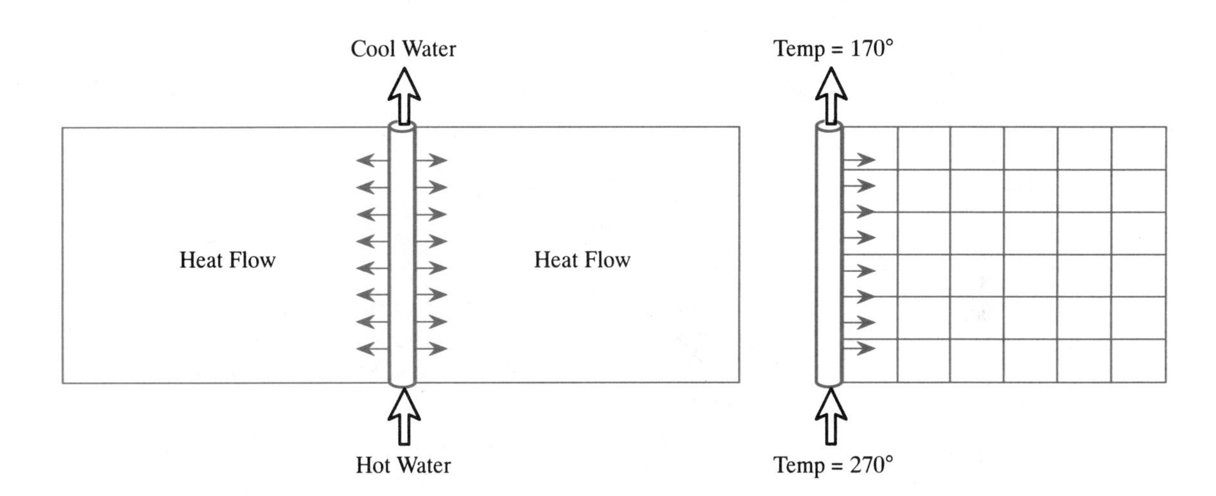

The hot water enters at the bottom through a tube running up the center of the fin. Heat is extracted by the fins and the water exits at the top at a lower temperature. In problems of this sort the temperature distribution may lead to uneven stresses, called thermal stresses, throughout the fin. To calculate these stresses, we must first find the temperature distribution in the fin. To do this, we break up the fin into small area elements, which we can make any size. We can then make the following assumptions:

- The temperature in the pipe varies linearly along its length.
- Elements along the outside edges always remain at ambient temperature, 80°F.
- Elements that touch the pipe always remain at the pipe temperature.
- Elements in the interior of the fin have a temperature that is the average of its eight neighbors.

The calculation starts out with all elements at ambient temperature, except those touching the tube. All elements then change according to the last assumption. This process will then repeat until no element changes by more than 1°. Here are the first two iterations for the temperatures given:

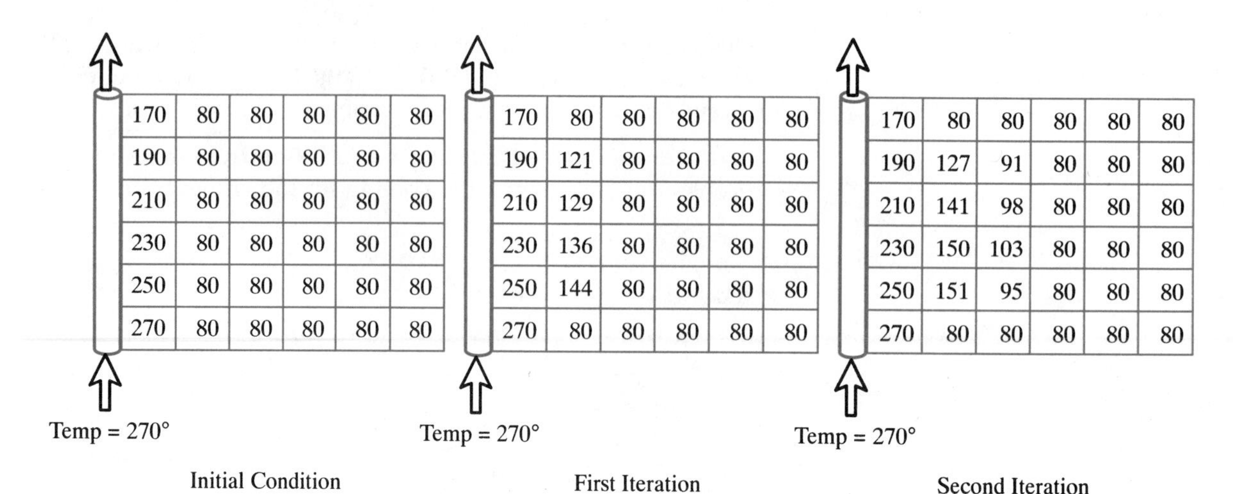

Write a program that reads in the entrance and exit water temperatures, the ambient temperature, the number of elements, and the condition for terminating the calculations. Your program should then calculate the temperature distribution.

8.19 Machining Operation. A three-dimensional array of size 100 × 100 × 100 is used to represent a solid part. Each element has either a value of 0 or 1. A value of 1 indicates that the volume element is filled. A value of 0 indicates that the volume element is empty. Each element represents 0.125 in^3.

a. Write a program that reads in the array using IMPLIED DO LOOPs and then calculates the volume of the part.
b. Now imagine a machining operation in which a hole is to be drilled into the solid. This causes many of the small volume elements to be set to zero where they once had a value of 1. Modify your program so that the center and radius of the hole are read in. Your program then examines the volume and decides which elements are to be removed. Recalculate the volume of the body and the amount of the material machined away.

8.20 Barcodes. Barcodes have become a common sight on items in stores. They are also used by the post office to help speed the mail. One type of barcode, shown here, can be read by an optical scanner that can distinguish between tall bars and short bars. Since the scanner can distinguish between only two states (tall and short), everything must be written in this binary code. Therefore, we refer to these as 1 and 0, respectively. The code is constructed as follows:

- The code always starts with 1, 1 (two tall bars).
- The value will be 16 digits long and expressed in base 2. The preceding example is the value 39 or 0000000000100111 in base 2.

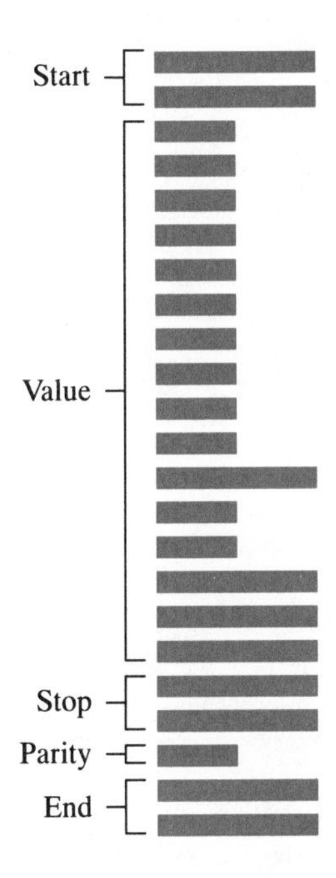

- The next two digits are 1, 1 (two tall bars) that indicate the end of the number.
- The next digit is the parity digit. Parity is set to 0 if the total number of 1s in the value is even. If the number of 1s is odd, parity is set to 1. This acts as a double check on the number. If the parity does not match the value read by the scanner, then a warning is sounded by the machine. For the previous value, 0000000000100111, there are four 1s. Therefore, parity = 0.
- The code must then end with 1, 1 (two tall bars).

Write a program to read in a value up to 65535, convert it into a base 2 number, and then construct the barcode. The final barcode can be printed out in numerical form such as 1100000000001001111011 for the preceding example.

CHAPTER

SUBPROGRAMS

9.1 Modularity—The Key to Programming Success

If we could ask ten professional programmers the question, "What is the best strategy to use to write a complex program?" all ten probably would answer "Modularize!" In effect, the professionals are telling us that by breaking down the complex problem into several smaller problems, we stand a much better chance of being successful. A "divide-and-conquer" approach is the key to programming success.

What do we mean by "modularize"? The answer can mean different things to different programmers. But most agree that to modularize a complex task is to break it down into individual, well-focused ***subtasks.*** If you concentrate on solving one small subtask at a time, the overall complex task is much simpler to solve. There is nothing unique about this approach. We do it all the time in our personal lives. How often have you heard, "I can only do one thing at a time."? Well, that's what modularization is all about—concentrating on one thing at a time.

In this chapter we do not learn any new ideas about programming, other than how to repackage it differently. We do break down long programs into several smaller subprograms, all glued together by A MAIN Program, whose primary function is to oversee all subtasks.

The best way for us to introduce the need for modularization is to introduce a simple problem. Suppose that a laboratory instructor wants to divide 12 students into two groups, one with 7 students and the other with 5 students. The number of possible combinations, C, that can occur is:

$$C = \frac{12!}{5!\,7!} = 792$$

If we generalize this to a class of N students and groups of I and $N - I$, then the number of combinations is:

$$C = \frac{N!}{I!\,(N - I)!}$$

A program to calculate the value of C would require three separate loops to calculate $N!$, $I!$, and $(N - I)!$ as follows:

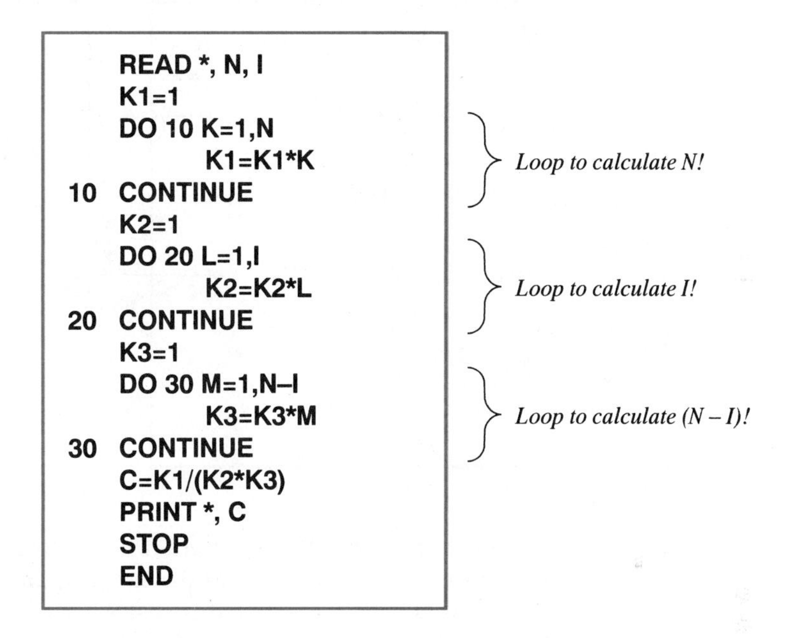

```
      READ *, N, I
      K1=1
      DO 10 K=1,N
            K1=K1*K
   10 CONTINUE
      K2=1
      DO 20 L=1,I
            K2=K2*L
   20 CONTINUE
      K3=1
      DO 30 M=1,N-I
            K3=K3*M
   30 CONTINUE
      C=K1/(K2*K3)
      PRINT *, C
      STOP
      END
```


Although this program works, it is an awkward structure. We use the loops for calculating the factorial three times with only small changes. The problem becomes even more acute when you calculate the probability, P, of being dealt a royal flush (A, K, Q, J, 10 of the same suit) in a game of poker:

$$P = \frac{\dfrac{4!}{1!\,3!}\;\dfrac{4!}{1!\,3!}\;\dfrac{4!}{1!\,3!}\;\dfrac{4!}{1!\,3!}\;\dfrac{4!}{1!\,3!}}{\dfrac{52!}{5!\,47!}}$$

There are a total of 18 factorials to calculate, although that can be reduced by inspection to fewer than five. The point, however, is that complex calculations often use the same mathematical functions repeatedly. This is especially true in statistical analysis as we show with this example. Thus, you would need to write the code for the factorial 18 times (or 5 times if you see the shortcut). What a waste of time!

Fortunately, FORTRAN has a solution for such unnecessary repetition. A subprogram can be set up outside the MAIN program to calculate the factorial, *N*!, for any value of *N*. The subprogram worries about the details of how to calculate *N*!, whereas the MAIN program worries about sending the right values to the subprogram and what to do with the results.

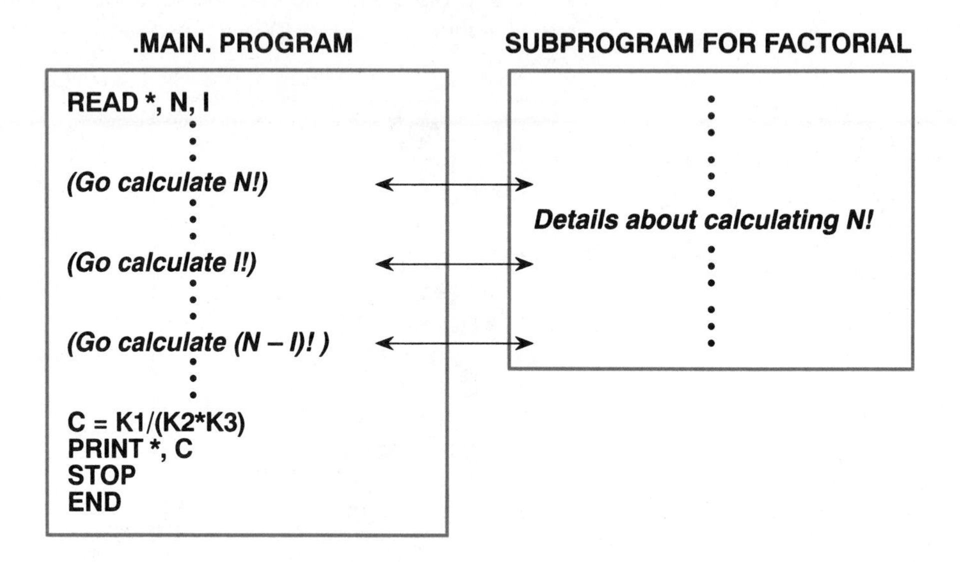

Every time the MAIN program needs to calculate the *factorial,* it sends the number *N*, *I*, or $(N - I)$ to the subprogram and lets it do the calculation. One advantage of this method is that we write the subprogram only once. If done correctly, it can calculate the factorial for any number sent to it. Notice that the MAIN program controls what to send and what to do with the results. It leaves all details of the calculation to the subprogram.

There are several major advantages to breaking down programs into smaller subprograms:

- You can concentrate on the small task assigned to the subprogram, making the big problem much easier to solve.
- If the MAIN program uses the subprogram calculation several times, you can reuse the code in the subprogram as often as needed.
- Subprograms are *portable.* This means they can be saved in a library that you create and use in other programs. Thus, the subprogram that you write for the factorial here can be used later for a different problem.
- Subprograms make it easier for you to debug your programs. A 20-line subprogram is much easier to test than a 200-line program with the same code embedded in it.

9.2 Function Subprograms

The simplest type of subprogram is the FUNCTION SUBPROGRAM. Its purpose is to calculate a *single numerical* answer outside the MAIN program and return the result. You have already seen many examples of FUNCTIONs when you used statements like:

```
Y = SQRT(X)
```

This is an example of an INTRINSIC or BUILT-IN FUNCTION for determining the square root of the number stored in X.

The execution of the preceding statement is probably so automatic that you didn't give it much thought when you used it. But it is important that we review the process for summoning the FUNCTION and how to transfer the data back and forth. The statement, SQRT(X), is a CALLing statement. It tells the computer that you want to use the black box for calculating the square root. It then transfers the value of X to the appropriate black box. When the black box finishes, SQRT(X) returns the answer and assigns it to the variable, Y.

Even the extensive list of INTRINSIC functions found on most mainframe computers is not complete, and so there are many functions that are missing. The factorial, $N!$, is one of these. Therefore, we need a way to construct the missing black boxes. These will be CALLed exactly as we did in the SQRT example. The difference is that we now need to construct a FUNCTION SUBPROGRAM that has the syntax:

```
type FUNCTION name (list of variables)
      ┌──────────────────────────┐
      │ SUBPROGRAM INSTRUCTIONS  │
      └──────────────────────────┘
RETURN
END
```

The first line of the subprogram must state that this is a FUNCTION and must give it a name. The name is the same as the one to be used in the CALLing statement in the MAIN program. This name is also a variable within the subprogram to which the desired value will be assigned. Following the name is an optional list of variables, separated by commas, which the FUNCTION needs for the calculation.

Since the name of the FUNCTION will be used as a variable, we sometimes need to declare the type (REAL, INTEGER, CHARACTER, and so forth). This typing is optional, though, if the *implicit* typing rules are sufficient. Thus, the following two statements are equivalent:

```
INTEGER FUNCTION IFACT( N )
          or
FUNCTION IFACT ( N )
```

Since the variable IFACT is implicitly INTEGER, we do not need to redeclare it. In the following exercises we always assume implicit typing. This option is only available with REAL or INTEGER data. With any other type, CHARACTER, DOUBLE PRECISION, LOGICAL, or COMPLEX, the type declaration is required in the FUNCTION statement. The body of the FUNCTION SUBPROGRAM is identical with any other program. It contains DECLARATION and ASSIGNMENT statements, loops, and branches. This is where the calculations are done. At the end of the subprogram are the RETURN and END statements. The computer uses the RETURN statement at execution time to return control back to the module that called it. The compiler uses the END statement to locate the end of the module.

So, let's return to our previous example of the number of combinations of groups of students of size I and N − I. Here, though, we replace the code that was repeated three times with a FUNCTION. Since the factorial function does not exist on most compilers, we must write our own FUNCTION SUBPROGRAM:

```
      READ *, N, I
             .
             .
             .
      K1 = IFACT ( N )                    .MAIN. PROGRAM
      K2 = IFACT ( I )
      K3 = IFACT (N-1)
             .
             .
             .
      C = K1/(K2*K3)
      PRINT *, C
      STOP
      END
```

```
      FUNCTION IFACT ( L )
           IFACT = 1
           DO 10 I = 1 , L                .FACT. MODULE
                 IFACT = IFACT * I
 10        CONTINUE
      RETURN
      END
```

The FUNCTION immediately follows the end of the MAIN program. We have drawn boxes around it here only to highlight it. The computer will know where the MAIN ends and the SUBPROGRAM begins by their corresponding END and FUNCTION statements.

When the MAIN program comes to the ASSIGNMENT statement:

```
K1 = IFACT ( N )
```

the computer will seek out the function called IFACT. It transfers N to the subprogram and returns the value for N! to the variable K1. When the program comes to the next CALL:

```
K2 = IFACT ( I )
```

the same thing will happen, except that now we transfer the value of I down and return the result to K2. It is important to observe here that we use the same FUNCTION SUBPROGRAM three times, but each time with different input data.

The FUNCTION SUBPROGRAM, here called IFACT, computes the value of the variable within the parentheses in the CALLing statement. Note that the subprogram uses what we call a *DUMMY VARIABLE,* L in this example. The value transferred to L is used for the computation of the factorial. To BASIC programmers this will seem very strange. If we are trying to calculate N!, then how does N get down to the subprogram? The FUNCTION SUBPROGRAM has no N in it, and the transfer of the data is not obvious unless you understand the purpose of the DUMMY variable.

Variables in FORTRAN are *LOCAL.* They exist only in the module in which they appear explicitly. Thus, if we had two variables with the same name within a MAIN and a FUNCTION, they would not be the same. In BASIC, where variables are *GLOBAL,* they would have the same value:

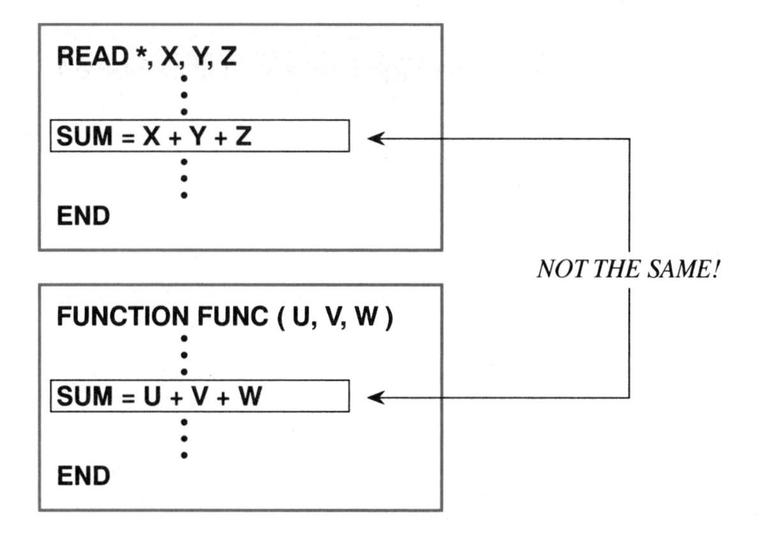

The LOCAL feature is very useful because it allows different sets of input data to use the same function repeatedly and still protect the results from our previous calculations. Thus, the primary way that data is sent back and forth is through the argument list. This is a controlled path and provides good protection for variables in the different modules. Later on, we will consider an alternate way to transfer data with the COMMON BLOCK.

The following figure schematically shows the primary way to transfer data:

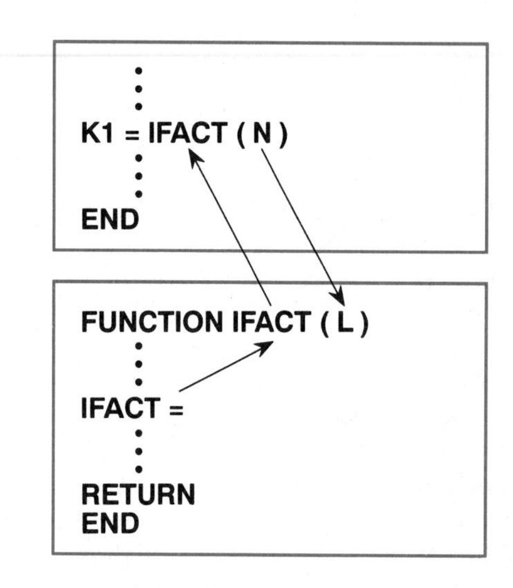

The variable inside the parentheses in the MAIN program CALLing statement goes to the subprogram although the two variables do not have the same name. This feature makes it easy to transfer different values to the subprogram. The result of the calculation returns to the MAIN program through the name of the FUNCTION. Notice first, though, that the subprogram uses the FUNCTION name as a variable to store the result. Thus, the name, IFACT in this example, occurs at least three different times:

- In the CALLing statement in the MAIN
- In the name of the FUNCTION SUBPROGRAM
- As a variable within the subprogram

An important point about subprograms is that the type of corresponding variables in the MAIN program and the subprogram must match. Thus, in our example the variables sent down to the subprogram are INTEGERs, as are the dummy variables set up to receive them. Similarly, IFACT is implicitly INTERGER in both modules. You must pay special attention to the type of the variables since it is easy to create accidentally a mismatch. Also, you must be careful that the argument list in the CALLing statement and the FUNCTION statement match in the number of variables and in their order. If you send down four variables, you must have four variables to receive them. Finally, there is a one-to-one correspondence between the position of the variables within the two lists. The third variable in one list will be sent to the third variable in the other list, for example.

Assume that we wish to send three variables to a FUNCTION called ADD. The first two variables, X and Y, are REAL, but the third is an INTEGER:

```
         real   real   integer
           ↖      ↑      ↗
Y = ADD (X,     Y,     J)
```

Then the FUNCTION statement must have the variables match in number, order, and type:

```
                real   real   integer
                  ↖      ↑      ↗
FUNCTION ADD ( A,      X,     I)
```

Note that X from the MAIN program will be assigned to A in the FUNCTION, not to X in the FUNCTION, even though they have the same variable name. Position within the list determines the assignment, not the similarity of the variable names!

If necessary, you may need to redeclare some variables in the FUNCTION. For example, if we had used W instead of the variable I, we would use the following:

```
FUNCTION ADD ( A, B, W )
INTEGER W
       .
       .
       .
RETURN
```

To highlight the local nature of the variables, we present here a variable table and a trace of the example. Note that there are two separate tables corresponding to actions of each module:

.MAIN.

N 12

I 7

FACTN 12!

FACTI 7!

FACTNI 5!

C 12!/7!/5!

READ N,I = 12,7	
FACTN=FACT(N);	Transfer 12 to FACT
	Value returned = 12!
FACTI=FACT(I);	Transfer 7 to FACT
	Value returned = 7!
FACTNI=FACT(N−I);	Transfer 5 to FACT
	Value returned = 5!

C=FACTN/(FACTI*FACTNI)=792

.FACT.

L 12,7,5

FACT 1*1*2 *12

1*1*2 *7

1*1*2 *5

I 1,2,12

1,2, 7

1,2, 5

FACT(12);	Assign 12 to L
FACT=1.0;	Assign 1.0 to FACT

Enter DO LOOP, assign I=1

FACT = 12*11*10*9*8*7*6*5*4*3*2*1

Return value of 12!

FACT(7);	Assign 7 to L
FACT=1.0;	Assign 1.0 to FACT

Enter DO LOOP, assign I=1

FACT = 7*6*5*4*3*2*1

Return value of 7!

FACT(5);	Assign 5 to L
FACT=1.0;	Assign 1.0 to FACT

Enter DO LOOP, assign I=1

FACT=5*4*3*2*1

Return value of 5!

FUNCTION SUBPROGRAMs are simple to use, but only for those applications where a single numerical value is sufficient. Therefore, they are the preferred method to calculate mathematical functions. For more sophisticated functions, we need the subroutine.

VALID EXAMPLES

The following tables should help you to understand the syntax rules and method of passing data for FUNCTION SUBPROGRAMs:

Calling Statement	Subprogram	Comments
Y = TEST (I, J, K)	FUNCTION TEST (L, M, N)	Variable names need not match
X=MIX(A, B, 3.2)	FUNCTION MIX(X,Y, Z)	Transfer of constant → variable okay
PRINT*, MA(U,V)	FUNCTION MA(A, B)	Okay to use FUNCTION like a variable
Z=F(X, 2*Y)	FUNCTION F(S,T)	Okay to do math in CALLing statement
A=FACT(5)/FACT(3)	FUNCTION FACT(N)	Okay to have multiple calls to FUNCTION
A=FACT(FACT(3))	FUNCTION FACT(N)	Okay to have nested calls provided no type mismatch

INVALID EXAMPLES

Calling Statement	Subprogram	Comments
W=F(I, J, X)	FUNCTION F(I, J, K)	Type mismatch (X → K)
Y=TEST(I, J, 2)	FUNCTION TEST(I, J, 2)	Cannot transfer constant → constant
X=MIX(A, B, C)	FUNCTION MIX(U,V,W, X)	Number mismatch (3 → 4)
MATH(X,Y)=X+2*Y	FUNCTION MATH(A, B)	FUNCTION may not be a target of assignment

9.3 Subroutines

The second type of subprogram is the *SUBROUTINE.* Like the FUNCTION, it stands outside the MAIN program and uses LOCAL variables. The difference is that the SUBROUTINE can do more sophisticated things than the simple FUNCTION. Whereas the FUNCTION can give only single numerical answers such as the square root of a number, SUBROUTINEs can return more than a single numerical answer. For example, we can use the SUBROUTINE to sort an entire array or to carry out a curve-fitting procedure on a large data set. There is no limit to the amount of data that the SUBROUTINE can return. Therefore, it is a much more powerful tool than the FUNCTION.

As with FUNCTIONs, the use of a SUBROUTINE requires a three-step process:

- CALLing the SUBROUTINE from the MAIN program
- Passing data to the SUBROUTINE
- Setting up the SUBROUTINE to receive the passed data, process it, and return the results

The procedure for SUBROUTINEs is a little different from the FUNCTION because of the larger amount of data returned. We can use a single-valued variable with FUNCTIONs since a single numerical answer is all that it can return. But with SUBROUTINEs, we can return any amount of data, and so the single-valued variable is insufficient. We need something more flexible. Therefore, we pass data to and from the SUBROUTINE with the CALLing statement:

```
CALL subroutinename ( var1 , var2 , . . . , varN )
```

where the *subroutinename* is the name of the specific subroutine that you want to use and *var1 . . . varN* is a list of variables passing between the MAIN and the SUBROUTINE. These variables are *two-way* variables. They may pass data to the SUBROUTINE or they may receive results, depending on the context.

The corresponding statements that set up the SUBROUTINE are:

```
SUBROUTINE subroutinename( var1 , var2, . . . , varN )
                .
                .
                .
RETURN
END
```

The *subroutinename* must be identical with the one used in the CALLing statement. The name must follow the standard rules for any variable name, which usually means a maximum of six letters and numbers. Some compilers may allow you to use longer SUBROUTINE names, but you should limit these to improve the portability of your programs.

The argument list in the CALLing statement and in the SUBROUTINE statement should have the same number of variables and each must agree in type with its counterpart. They need not have the same variable names. Thus:

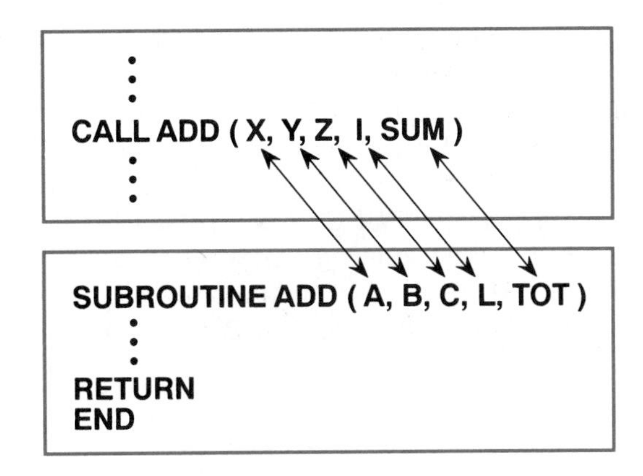

In this example the first variable in the SUBROUTINE argument list, A, receives the value of the first variable in the CALLing argument list, X. Similarly, B gets the value of Y; C gets the value of Z, and so forth. The variable sending the data and the one receiving the data must also match in type. Thus, if X is REAL, A must be REAL, and if I is an INTEGER, then L must also be an INTEGER.

An important point about the variables in the argument lists is that each variable can either send or receive data. Usually, this will be obvious from the context of the problem. In the example given, X, Y, Z, and I are probably sending data to the SUBROUTINE, whereas SUM is probably the answer returned. Anyhow, our guess is not important since each variable is a two-way variable. Thus, we may use X to send data down, but if the subprogram modifies the corresponding variable, A, then X will also change. Here is an example:

```
                 .
                 .
                 .
CALL ADD ( X , Y , Z , I , SUM )
                 .
                 .
                 .
```

```
      SUBROUTINE ADD (A , B , C , L ,TOTAL )
           DO 10  K = 1 , L
                 TOTAL = TOTAL * A / B
10         CONTINUE
           A = TOTAL / C
      RETURN
      END
```

Notice that the SUBROUTINE changes the value of A with the statement, A=TOTAL/C. Since A is equivalent to X in the MAIN program, X will also change. Thus, we use X in this example for both sending and receiving data. This can create problems if you are not careful.

The RETURN and END statements in the SUBROUTINE serve the same function as in the FUNCTION SUBPROGRAM. The RETURN statement may go anywhere in the subprogram, whereas the END statement must go at the end.

The SUBROUTINE should physically follow the last line of the MAIN program. If there are several SUBROUTINEs and FUNCTIONs, they may go in any order:

```
.MAIN.
        .
        .
        .
```

```
SUBROUTINE .TWO.
        .
        .
        .
```

```
FUNCTION .ONE.
        .
        .
        .
```

```
SUBROUTINE .ONE.
        .
        .
        .
```

The order in which you list the various subprograms is unimportant since the computer will seek the subprogram out by name when it needs them. It may help in debugging, though, if you list them in a logical order, say, the order in which you CALL them. This will reduce the amount of your work during tracing. We will discuss this in greater detail in the next chapter.

9.4 Arrays and Subprograms

All our discussion so far has focused on passing single-valued variables to subprograms. But we can pass arrays also. To do this, we need to declare the arrays in both modules. In the following example we pass the entire array, A, to the subroutine SUM. Notice that we had to declare the array twice:

```
REAL A(100)
      .
      .
      .
CALL SUM ( A, TOT )
      .
      .
      .
STOP
END
```

```
SUBROUTINE SUM ( X, TOTAL )
REAL X(100)
      .
      .
      .
TOTAL = ...
      .
      .
      .
RETURN
END
```

An important variation of the ability to pass arrays is the possibility of using a SUBROUTINE with variable size arrays. Recall that in a MAIN program we had to specify explicitly the maximum size of an array in the DECLARATION statement. FORTRAN does not permit variable size arrays. With SUBROUTINEs, though, this is no longer true.

To illustrate this point, assume that we have two arrays, A and B, with 100 and 10 elements, respectively. We wish to send A to the SUBROUTINE for processing first, followed in a second CALL with B. All we must do is to send the array A with its size of 100. Then we send B down with its size of 10. We set up the SUBROUTINE so that the local array, X, has a variable size, M. When the computer reads the value of M, either 100 or 10 in this example, it will set up the required amount of memory space for the variables:

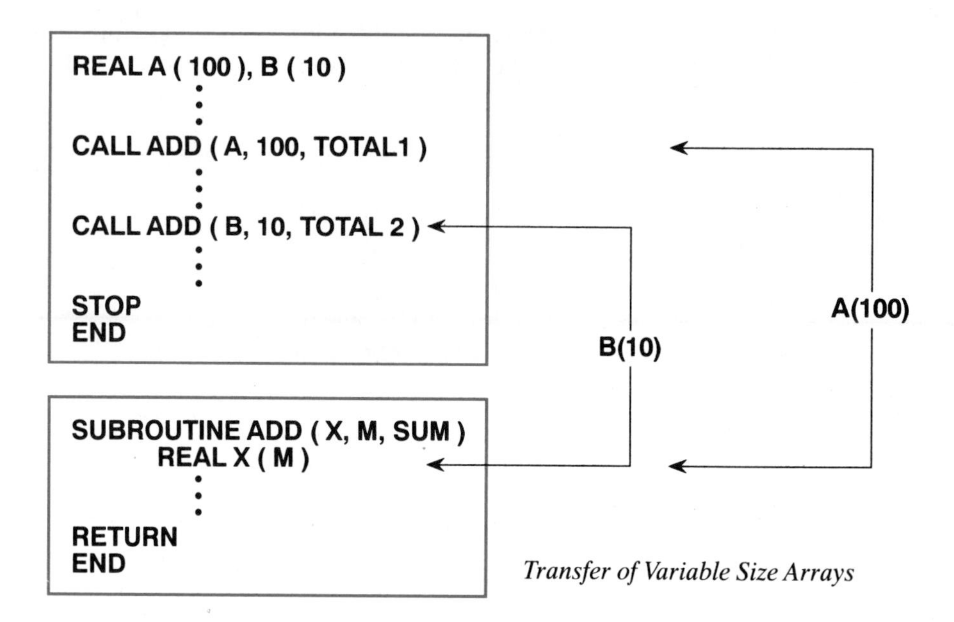

Transfer of Variable Size Arrays

Besides sending down entire arrays, it is also possible to transfer individual elements to a single-valued variable. Of course, the type of the array element must match the single-valued variable:

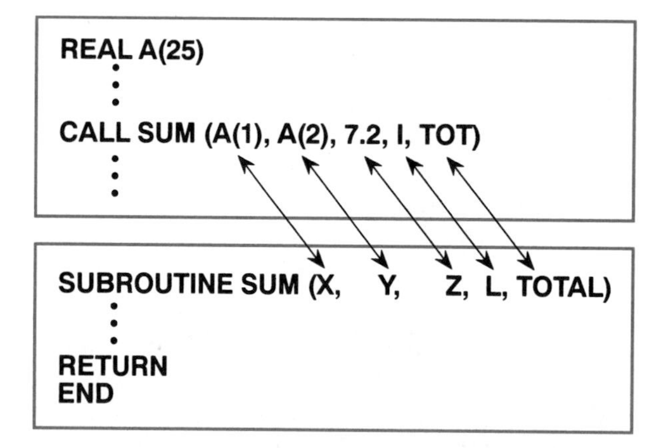

In the preceding example, A(1) and A(2) are individual elements of the properly declared array A. These elements are then assigned to the variables, X and Y, in the SUBROUTINE. Also, we have sent down a constant, 7.2, to be assigned to the variable Z.

Also, you may want to reverse the preceding process. You may transfer constants or single-valued variable to array elements in the SUBROUTINE. Again, you must be careful to declare the array in the appropriate module and make sure that the variable types match.

9.5 Creating Global Data—The COMMON BLOCK

We have stressed several times that FORTRAN uses variables that are usually local. They exist only in the modules in which you explicitly use them. Thus, the variable X in the MAIN program is different from the variable X in a subprogram. This is a very important feature since it allows you to reuse subprograms with few changes. Still, there are times when you may wish to have *global variables,* or variables that exist in several modules simultaneously. FORTRAN has a feature, called the COMMON BLOCK, that allows you to do this and to declare a list of selected variables as global.

Before showing you how to use the COMMON BLOCK, we want to show you why they exist. Suppose you have a lengthy program that makes many calls to SUBROUTINEs and that each CALL transfers many items in the argument list:

```
          .
          .
          .
CALL DUMMY(A, B, I, J, C, D, E, F, U)
          .
          .
          .
CALL DUMMY(A, B, I, J, C, D, E, F, V)
          .
          .
          .
CALL DUMMY(A, B, I, J, C, D, E, F, W)
          .
          .
          .
CALL DUMMY(A, B, I, J, C, D, E, F, X)
          .
          .
          .
```

In all four CALL statements we send down many of the same constants and variables. The only difference is in the last item in the list, U, V, W, or X. The first eight items are repetitious and can easily result in errors if you are not careful. If we can turn these eight into global variables, there will be no need to transfer them. They will already be there.

The COMMON BLOCK structure allows us to declare a list of variables to be global in several modules. The syntax of the statement is:

```
COMMON variable1, variable2, . . . variableN
```

where the statement COMMON contains a list of variables to be stored in common memory for use by each module given access to it. If a subprogram needs these variables, then the COMMON statement must be included in that module. If the subprogram will not use the common data, then you do not need the COMMON statement.

When the COMMON statement appears in the subprogram, the variables need not be the same as those in the MAIN program. Instead, the data is transferred by the location in the variable list. For instance, in our previous example we could put the eight variables in the COMMON statement and simplify the CALLing statement. We also must duplicate the COMMON statement in the appropriate subroutine (see Fig. 9.1).

The variables A (in the MAIN) and X1 (in the first SUBROUTINE) share the same memory location because of the COMMON statement. Therefore, they will always have the same value. Note that in the second SUBROUTINE, which does not have a COMMON statement, there are only local variables since there is no sharing of memory. Thus, the variable A in SUBROUTINE OUTPUT is different from A in the MAIN program.

This example shows that you can have a mixture of local and global variables. By default, variables are local. If you want global variables, you must set up a COMMON BLOCK and specify it in each subprogram that will use it. We must warn you that global variables can cause many problems if you are not careful. The COMMON BLOCK should be avoided to prevent unintended changes to variables by a faulty subprogram. Overuse of COMMON BLOCKs will greatly increase the difficulty of debugging the program containing them.

There is a second type of COMMON BLOCK called the NAMED COMMON BLOCK. We can give each COMMON BLOCK a name, which will allow us to break down all shared data into smaller packages. This will allow us to share some data with one subprogram and other data with another subprogram. The syntax of the NAMED COMMON BLOCK is:

```
COMMON /name/ variable1, variable2, ...variableN
```

We give the block a name by specifying it inside the slash marks. Suppose we wanted to share variables A, B, C, and D with one group of subroutines and variables E, F, G, and H with a second set of subroutines. Figure 9.2 shows how we would do this.

The NAMED COMMON BLOCK, /JACK/ shares the variables A, B, C, and D with the SUBROUTINE DUM variables X1, X2, X3, and X4. The second NAMED COMMON BLOCK, /JILL/ shares the remaining variables, E, F, G, and H, with the SUBROUTINE SUM variables Y1, Y2, Y3, and Y4 as illustrated earlier. Variables listed in the COMMON BLOCKs cannot be initialized with the DATA statement. Each variable must be explicitly assigned a value with either a READ statement or a direct assignment.

One requirement of the NAMED COMMON BLOCK is that it must be the same length (4, 5, or however many variables) in every module in which it appears. Even if you do not use all variables in the block, you must provide a

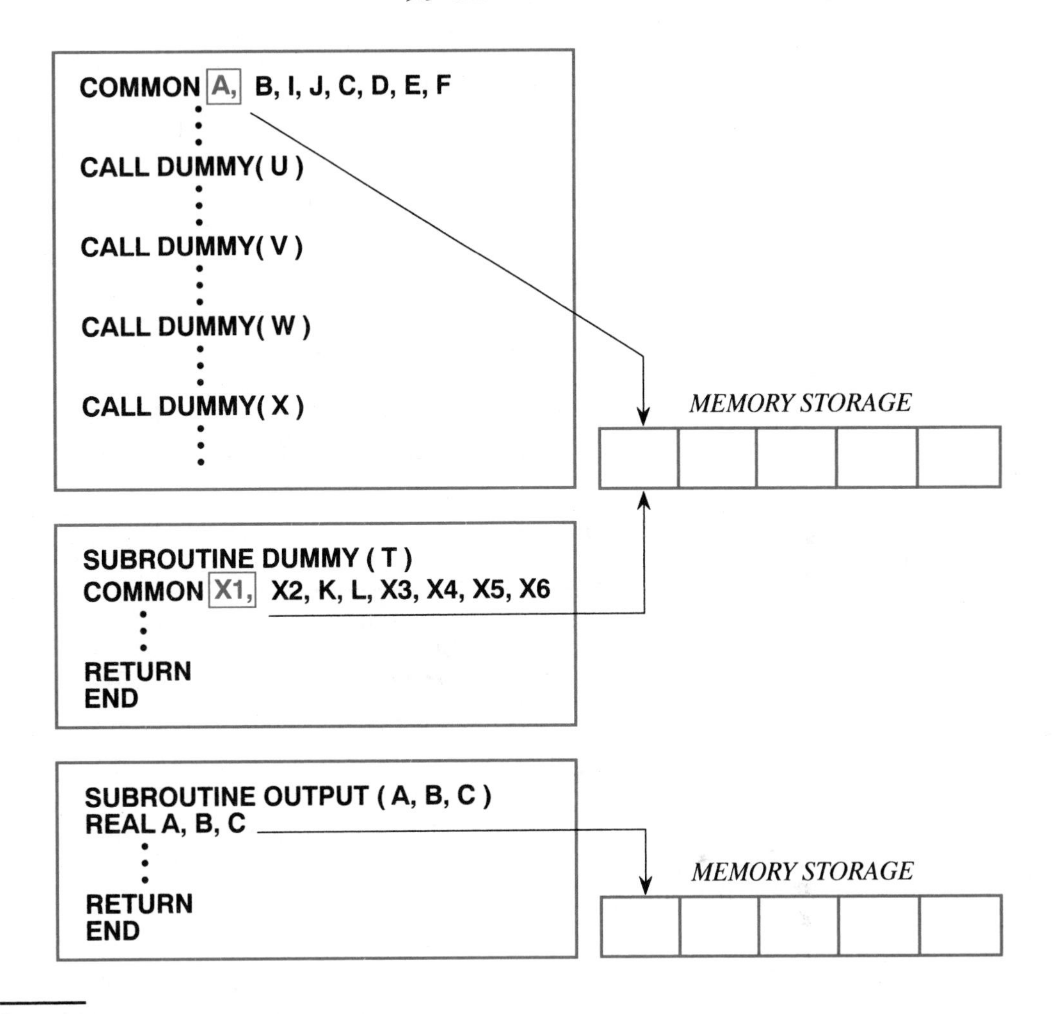

FIGURE 9.1

Use of Common Blocks for Data Transfer

variable to receive the data. For example, in SUBROUTINE SUM you may not need the variables Y2 and Y4 for any calculations. Yet, you must provide these two variables nonetheless.

Finally, you may pass CHARACTER data in a COMMON BLOCK, either NAMED or BLANK. But, if you do, you cannot pass other types of data such as REALs, INTEGERs, and so forth. You must use a separate NAMED COMMON BLOCK for these. In these blocks, though, you can mix different types, provided that there are no CHARACTERs.

In spite of its apparent usefulness, we strongly recommend that you avoid COMMON BLOCKs, if possible. Sometimes they can be used to good effect, as we will show in the next chapter on debugging. In general, though, they compromise the safety of the individual modules. We go to great lengths to set up subprograms that are independent of each other to ensure that there

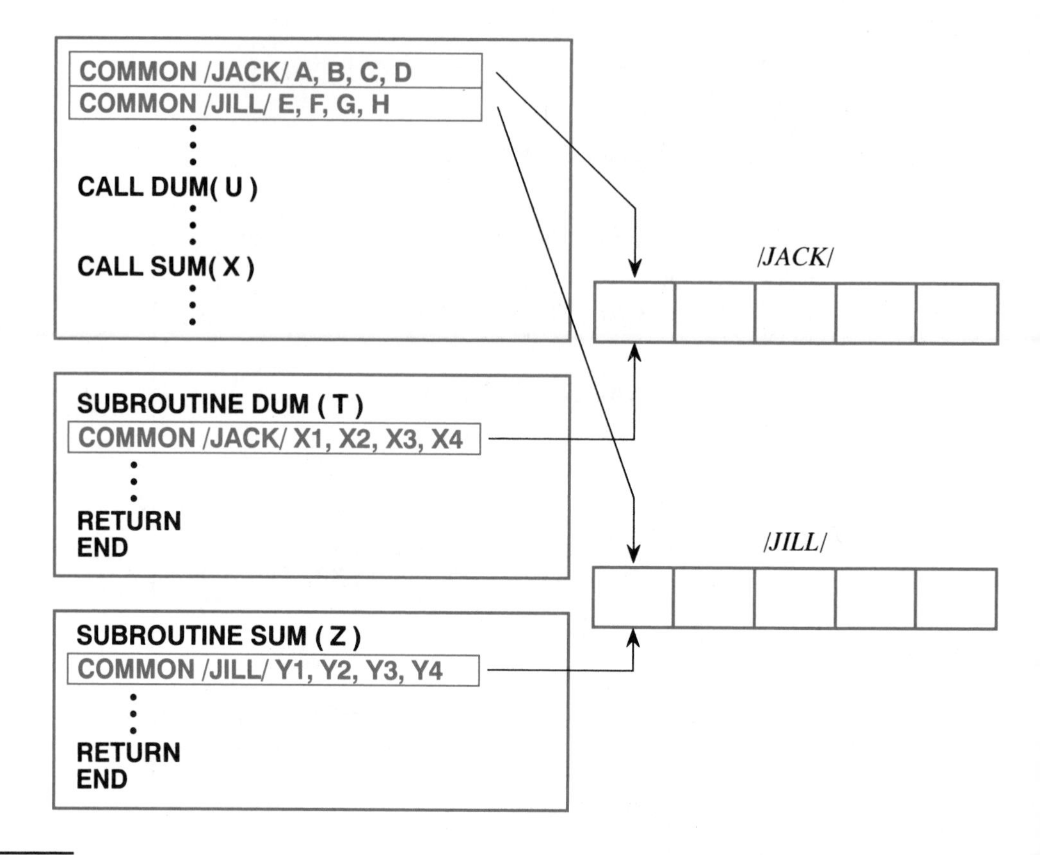

FIGURE 9.2

Use of Named Common Blocks

is no "crosstalk." COMMON BLOCKs directly attack this structure and may produce errors that are very difficult to eliminate.

9.6 Algorithms

We began this chapter by stating that there is nothing new about subprograms. They are simply a different way of packaging the same goods. Therefore, it seems appropriate here to begin the discussion by "repackaging" two of our previous algorithms as subprograms. These will include:

Dot product of two vectors
Sorting an array

The first algorithm for taking the dot product of two vectors involves a FUNCTION SUBPROGRAM since it returns only a single scalar quantity. Sorting, on the other hand, returns an array of values, and we will need a SUBROUTINE to do this.

In addition, we present three new algorithms, including:

Determine the angle between two vectors
Coordinate transformation of a vector
Numerical approximation to the derivative

Each of these algorithms is a critical component of the numerical algorithms in the later chapters of this book.

DOT PRODUCT OF TWO VECTORS

In Chap. 8 we defined the dot product as the sum of the product of the individual components of two vectors. The result of this calculation is a *scalar* quantity, which means that it only has a magnitude; it lacks a direction. Therefore, we can convert this into a FUNCTION, which we call DOT. Here is the original program segment from the DOT PRODUCT LS:

```
      REAL F1(3), F2(3)
      READ*, (F1(I), I=1,3), (F2(J), J=1,3)
      DOT=0.0
      DO 10 I=1,3
            DOT = DOT + F1(I)*F2(I)
10    CONTINUE
```

We can convert this into a FUNCTION almost directly. There are only a few changes and additions that we should make:

- Create a MAIN program to call the FUNCTION.
- Move the READ statement to the MAIN program.
- Make the arrays variable size to improve the flexibility.
- Make the LCV a variable corresponding to the size of the arrays.
- Give the name, DOT, to the FUNCTION.

Once we make these changes, the FUNCTION will be a general one that we probably can reuse without any modifications:

LS FOR DOT PRODUCT FUNCTION SUBPROGRAM

PROGRAM SEGMENT:

```
C     MAIN PROGRAM
C ------------------------------------------------------------
      REAL A(100), B(100)
            .
            .
            .
      READ*, M, (A(I), I=1,M), (B(J), J=1,M)
            .
            .
            .
      Z = DOT( A , B , M )
            .
            .
            .
      STOP
      END
C ------------------------------------------------------------
C     FUNCTION SUBPROGRAM TO CALCULATE DOT
C     PRODUCT OF TWO VECTORS
C ------------------------------------------------------------
      FUNCTION DOT ( F1 , F2 , L )
      REAL F1( L ), F2( L )
      DOT = 0.0
      DO 10 I = 1 , L
            DOT = DOT + F1( I ) * F2( I )
10    CONTINUE
      RETURN
      END
```

The MAIN program sets up the two arrays and reads in their values. Notice that the declaration statements there must specify an actual size for the arrays. Therefore, we usually set them arbitrarily high so that it is unlikely that we will exceed the reserved memory space. Then we read in M, which is the actual number of elements in the array. This is usually three for real space vectors. Once the program reads in the arrays, they go to the FUNCTION, with the value of M. The FUNCTION uses this variable to set up the arrays in the FUNCTION and to control the loop that does the dot product. Finally, the FUNCTION name, DOT, is a variable within the FUNCTION and contains the final answer.

A key point about setting up a subprogram is that you should make an effort to have the subprogram as general as possible. In this example, for instance, we could use the unmodified FUNCTION for DOTting two vectors with 3, 10, 100, or any number of elements. The critical step is to try to set up the arrays as variable size arrays. Also, you must set up any associated

DO LOOPs to run with variable limits. Finally, notice that we took the I/O statements out of the subprogram. This improves the portability of the created subprogram.

SORTING AN ARRAY

The MAX/MIN sort is one of the easiest sorts to implement, as we discussed in Chap. 8. Often we may use a sort many times within a large program. There may be many arrays to sort, for instance. To avoid having to reenter the code, we can place the sort into a SUBROUTINE, and simply issue the appropriate CALL whenever we need it. Note that we must set up the sort with a SUBROUTINE instead of a FUNCTION since a large quantity of data returns. Remember that the FUNCTION can return only a single value.

Here is the MAX/MIN SORT LS from the previous chapter:

```
      REAL A(1000)
      READ *, N
      READ *, (A( I ), I = 1,N )
      DO 10 L = 1, N-1
            BIG = A( L )
            DO 10 I = L , N
                  IF( A( I ) .GT. BIG ) THEN
                           TEMP = BIG
                           BIG = A( I )
                           A( I ) = TEMP
                  ENDIF
10    CONTINUE
      PRINT *, (A( I ), I = 1,N )
```

Fortunately, we already wrote this program segment in a general format so that we need to make only a few changes before putting it into a SUBROUTINE. These changes are:

- Make the arrays a variable size.
- Transfer the array and the size down to the SUBROUTINE with a CALL statement.
- Put all the I/O statements into the MAIN program.

Once we make these changes, the SUBROUTINE can handle any size array:

LS FOR SORTING AN ARRAY WITH A SUBROUTINE

PROGRAM SEGMENT:

```
      REAL A(1000)
      READ *, N, (A( I ), I = 1,N )
           .
           .
           .
      CALL SORT ( A , N )
           .
           .
           .
      PRINT *, (A( I ), I = 1,N )
      STOP
      END
C --------------------------------------------------------------
C     SUBROUTINE SORT: TAKES THE ARRAY, X AND SORTS
C     IT IN DESCENDING ORDER. THE SORTED ARRAY IS
C     STORED BACK INTO THE ARRAY X.
C --------------------------------------------------------------
      SUBROUTINE SORT ( X , K )
      REAL X( K )
      DO 10 L = 1, K-1
            BIG = A( L )
            DO 10 I = L , K
                  IF( A( I ) .GT. BIG ) THEN
                           TEMP = BIG
                           BIG = A(I)
                           A( I ) = TEMP
                  ENDIF
10    CONTINUE
      RETURN
      END
```

This SUBROUTINE can be used as often as you need it. For example, suppose that you have several different sized arrays for sorting. All you must do is to send down the whole array and the size of each. The SUBROUTINE is general enough so that it can handle this task:

```
REAL A(100), B(5), C(28)
     .
     .
     .
CALL SORT( A , 100 )
     .
     .
     .
CALL SORT( B , 5 )
     .
     .
     .
CALL SORT( C , 28 )
     .
     .
     .
END
```

To do this, you must set up the arrays with a variable size. Try to do this, even if the specific problem does not call for it. You may very well have use for the general SUBROUTINE at another time.

ANGLE BETWEEN TWO VECTORS

There are many occasions when we must find the orientation of two vectors with respect to each other, as in Fig. 9.3. We can find the angle, α, between the two vectors, $\mathbf{F}^1$ and $\mathbf{F}^2$, with the formula:

$$\cos(\alpha) = \frac{\mathbf{F}^1 \cdot \mathbf{F}^2}{|F^1|\,|F^2|}$$

where $\mathbf{F}^1 \cdot \mathbf{F}^2$ = DOT PRODUCT of the two vectors

$$|F^1| = \text{magnitude of } F^1 = [(F^1_X)^2 + (F^1_Y)^2 + (F^1_Z)^2]^{1/2}$$

For example, if we had two vectors, $\mathbf{F}^1 = (12, 5, 6)$ and $\mathbf{F}^2 = (3, 2, 0)$, we calculate the angle as follows:

$$\begin{aligned}
\mathbf{F}^1 \cdot \mathbf{F}^2 &= 12 \cdot 3 + 5 \cdot 2 + 6 \cdot 0 = 46 \\
|\mathbf{F}^1| &= (12^2 + 5^2 + 6^2)^{1/2} \quad = 205^{1/2} \\
|\mathbf{F}^2| &= (3^2 + 2^2 + 0^2)^{1/2} \quad = 13^{1/2} \\
\cos(\alpha) &= \frac{46}{\sqrt{205 \cdot 13}} \quad = 0.8911 \\
\alpha &= \cos^{-1}(.8911) = 26.993^\circ
\end{aligned}$$

We convert this to a FUNCTION called ANGLE, which receives the two vectors and returns the single numerical value that is α. Keep in mind that when a computer does trigonometric calculations, all answers appear in radians, not degrees! So, you may want to convert the answer before printing any results:

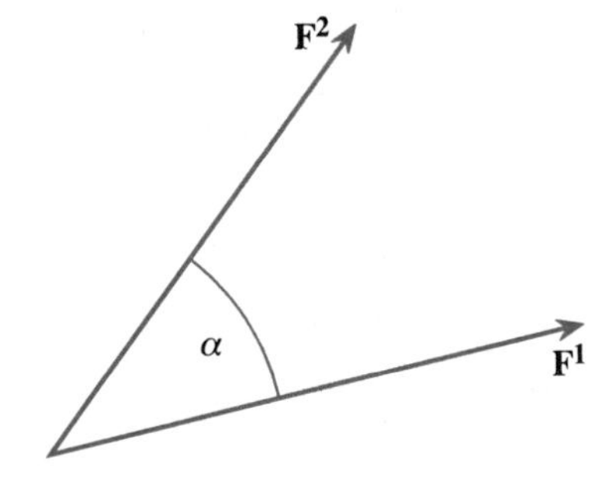

FIGURE 9.3

Definition of Angle between Two Vectors

LS FOR ANGLE BETWEEN TWO VECTORS

PROBLEM STATEMENT: Read in two vectors and determine the angle between them.

ALGORITHM DEVELOPMENT:

Step 1a: Review the problem statement. Indicate all nouns in **bold** and all verbs in *italics:*

Read in two **vectors** and *determine* the angle between them.

Formulate questions based on nouns and verbs:

Q1: How do we read in a vector?
Q2: How do we determine the angle between them?

Step 1b: Construct answers to the above questions:

A1: Use IMPLIED DO LOOP and assign components to elements of an array.
A2: Calculate the angle using the given formula.

ALGORITHM:

1. Read in vector1, assign to array1.
2. Read in vector2, assign to array2.
3. Send two vectors to FUNCTION ANGLE:
   Call FUNCTION DOT to compute DOT product.
   Compute magnitudes of each vector.
   Compute ANGLE=ARCCOS(DOTPRODUCT/(MAG1*MAG2)).
   Return ANGLE to MAIN program.
4. Print value of ANGLE.

FLOWCHART:

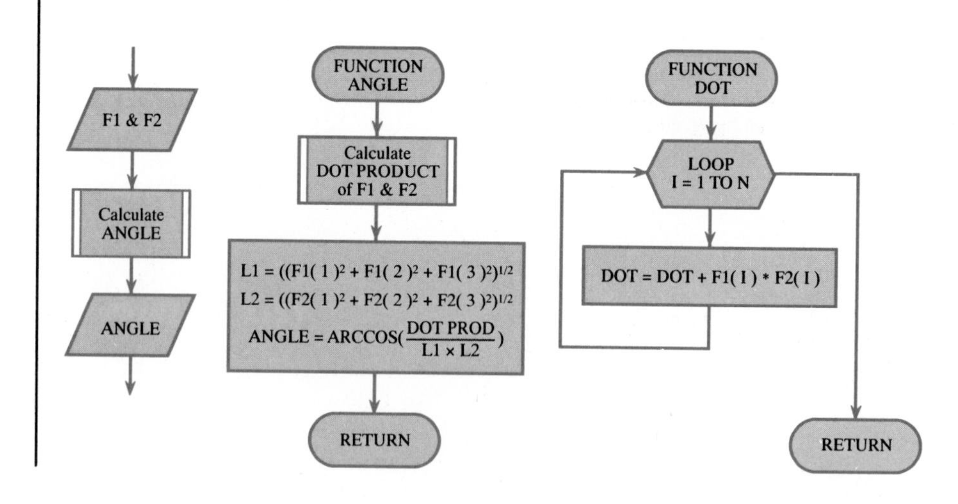

PROGRAM SEGMENT:

```
      REAL F1(3), F2(3)
      READ *, (F1(I), I=1,3), (F2(J), J=1,3)
      ANS = ANGLE ( F1, F2 )
      PRINT*, ANS*57.296
      STOP
      END
C -------------------------------------------------------------
C     FUNCTION ANGLE TO COMPUTE ANGLE BETWEEN
C     TWO VECTORS. THIS FUNCTION CALLS ANOTHER
C     FUNCTION, DOT, TO OBTAIN THE DOT PRODUCT.
C -------------------------------------------------------------
      FUNCTION ANGLE ( F1, F2 )
      REAL F1(3), F2(3), L1, L2
      L1=SQRT(F1(1)**2+F1(2)**2+F1(3)**2)
      L2=SQRT(F2(1)**2+F2(2)**2+F2(3)**2)
      ANGLE=ACOS(DOT(F1,F2,3)/(L1*L2))
      RETURN
      END
C -------------------------------------------------------------
C     FUNCTION TO COMPUTE DOT PRODUCT OF TWO
C     VECTORS OF VARIABLE SIZE
C -------------------------------------------------------------
      FUNCTION DOT (F1, F2, L)
      REAL F1(L), F2(L)
      DOT = 0.0
      DO 10 I = 1, L
          DOT =DOT + F1(I) * F2(I)
10    CONTINUE
      RETURN
      END
```

The flowchart for this algorithm contains two new symbols for a transfer to a subprogram, as shown in Fig. 9.4. In the flowcharts shown in the LS the MAIN program first transfers the arrays, F1 and F2, to the FUNCTION ANGLE. The way that we have set this up, though, requires this FUNCTION to call another FUNCTION, DOT, before it can complete its calculations. Figure 9.5 illustrates this process. Thus, the sequence of transfer is as follows:

- Main program sends F1 and F2 to FUNCTION ANGLE.
  - FUNCTION ANGLE sends F1 and F2 to FUNCTION DOT
    - FUNCTION DOT uses F1 and F2 to calculate the dot product and return answer with the variable, DOT
  - FUNCTION ANGLE then uses DOT to compute ANGLE, which is sent back to the MAIN program.
- MAIN program prints out the value of ANGLE.

FIGURE 9.4

Flowchart Symbols Used with Subprograms

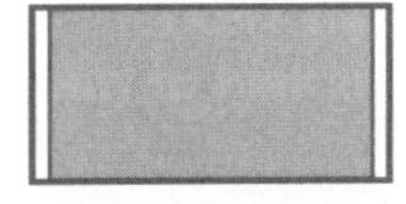

→ Indicates Transfer to a Function or a Subroutine

→ Indicates Beginning or End of a Program Segment

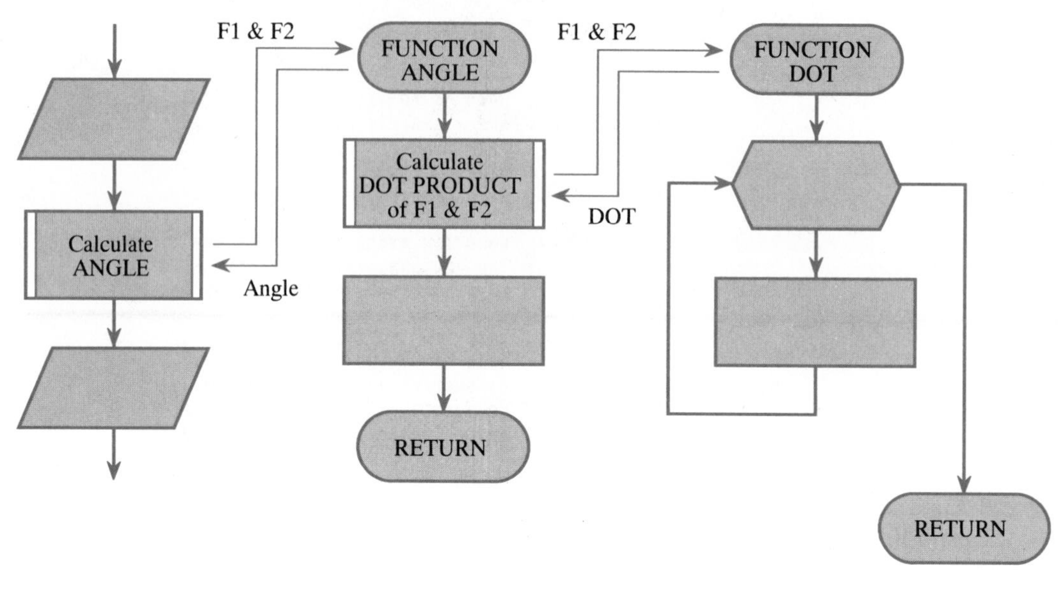

FIGURE 9.5

Transfer of Control to a Subprogram

Notice that a FUNCTION can call another FUNCTION or any other subprogram. The only restriction is that it may not call itself. We call this *recursion,* which the current version of FORTRAN does not allow. Future editions, though, will allow recursion, which we will discuss in greater detail in Chap. 11.

Before leaving this example, note that we could have incorporated the FUNCTION DOT into the FUNCTION ANGLE, thus minimizing the number of transfers. But then we could not use DOT, which we have already written and debugged. All we had to do was to send the data to the FUNCTION correctly, and we know that it will work, based on our previous efforts.

The goal of the modular approach to programming is to break down the problem into smaller, more manageable parts. This is exactly what we have done here. Yes, we could have written the entire program as a MAIN program, without any calls to subprograms. But then we would have to reinvent the algorithms. Instead, we modify the problem slightly to take advantage of preexisting modules.

TRANSFORMATION OF A VECTOR

If we have a vector, **F**, in the X-Y coordinate system, it is sometimes easier to work in a different coordinate system such as X'-Y' shown in Fig. 9.6. This is because the mathematics may be simpler in the new *transformed coordinates.* The simplest transformation is a rotation about one of the axes.

FIGURE 9.6

Rotation of a Coordinate System

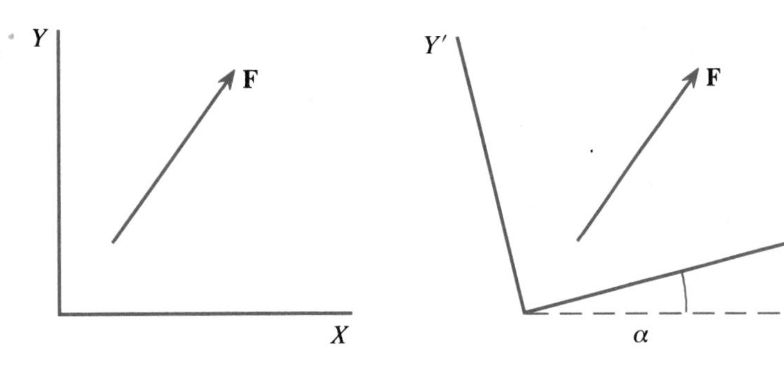

In the transformation in Fig. 9.6 we have taken a rotation about the Z axis (perpendicular to the page) through an angle α. Notice that the vector has not changed position, nor has its absolute length changed. But the ***components*** in the new coordinate system have changed. If the vector coordinates in the old X-Y system were $\mathbf{F} = (F_X, F_Y, F_Z)$, then we can calculate the components of the new vector, $\mathbf{F}' = (F'_X, F'_Y, F'_Z)$ with the equation:

$$\begin{bmatrix} F'_X \\ F'_Y \\ F'_Z \end{bmatrix} = \begin{bmatrix} \cos(\alpha) & -\sin(\alpha) & 0 \\ \sin(\alpha) & \cos(\alpha) & 0 \\ 0 & 0 & 1 \end{bmatrix} \times \begin{bmatrix} F_X \\ F_Y \\ F_Z \end{bmatrix}$$

If the rotation is 30°, for example, on a vector, $\mathbf{F} = (2, 4, -1)$, we have:

$$\begin{bmatrix} F'_X \\ F'_Y \\ F'_Z \end{bmatrix} = \begin{bmatrix} 0.866 & -0.500 & 0 \\ 0.500 & 0.866 & 0 \\ 0 & 0 & 1 \end{bmatrix} \times \begin{bmatrix} 2 \\ 4 \\ -1 \end{bmatrix}$$

Following the rules for multiplication of two matrices from the previous chapter, we get $F' = (-0.268, 5.464, -1.0)$. The Z component has not changed as you would expect, but the other two components have. To carry out a coordinate transformation, we need to multiply the original vector by a square matrix containing the trigonometric functions. This problem is somewhat different from the previous algorithm. The preceding example considered the multiplication of a L × M by a M × N matrix to produce a L × N matrix. In the transformation done here the vector is a L × 1 matrix, or a matrix with only one row. Therefore, we can simplify the algorithm for the multiplication somewhat.

There is a similar transformation for a rotation of the coordinate system around one of the other axes. For example, for a rotation about the x axis, the transformation matrix becomes

$$\begin{bmatrix} 1 & 0 & 0 \\ \cos(\alpha) & -\sin(\alpha) & 0 \\ \sin(\alpha) & \cos(\alpha) & 1 \end{bmatrix}$$

This process is discussed further in one of the end-of-chapter exercises.

In the following algorithm notice that the subroutine for the matrix multiplication has eliminated one loop from the previous algorithm. We leave it as an exercise to show that this works for multiplication of a vector by a square matrix:

LS FOR TRANSFORMING A VECTOR

PROBLEM STATEMENT: Read in a vector and determine the components in a new coordinate system rotated about the Z axis by α.

ALGORITHM DEVELOPMENT:

Step 1a: Review the problem statement. Indicate all nouns in **bold** and all verbs in *italics:*

Read in a **vector** and *determine* the **components** in a new **coordinate system** *rotated* about the Z axis by α.

Formulate questions based on nouns and verbs:

Q1: How do we read in a vector?
Q2: How do we determine the new components?

Step 1b: Construct answers to the above questions:

A1: Use IMPLIED DO LOOP and assign components to elements of an array.
A2: Calculate new components by matrix multiplication.

ALGORITHM:

1. Read in vector, assign to F(I) array.
2. Send vector and rotation angle to subroutine ROTATE:
   Form rotation matrix with given angle
   Send rotation matrix and vector to SUBROUTINE MATMUL
     MATMUL applies matrix multiplication
     Return new vector to ROTATE
   Return new vector to MAIN program.
3. Print out new vector.

Flowchart:

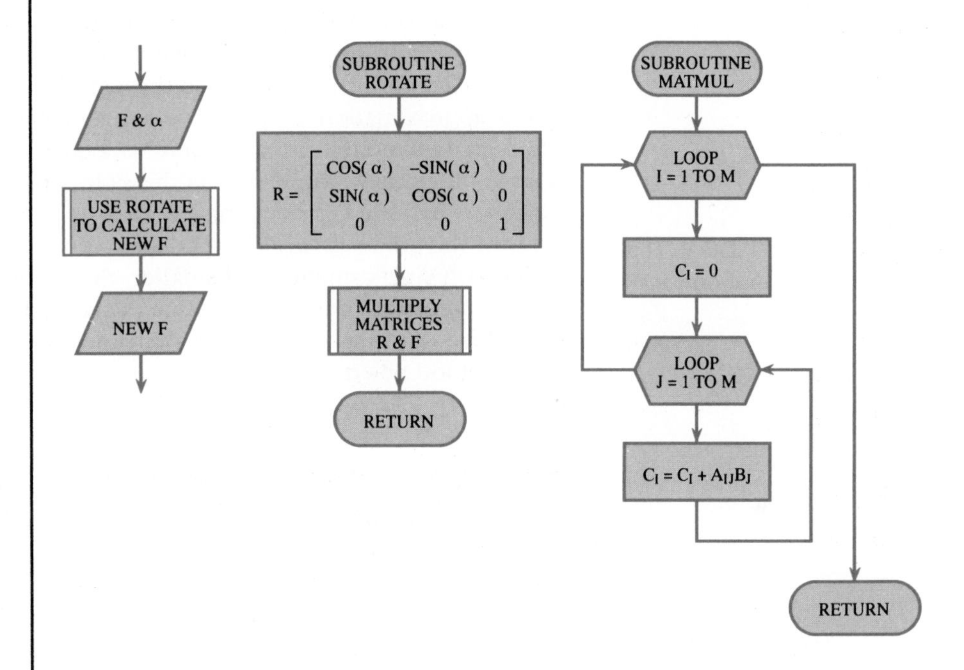

Program Segment:

```
      REAL F(3), ANG
      READ *, (F(I), I=1,3), ANG
      CALL ROTATE ( F , ANG )
      STOP
      END
C ---------------------------------------------------------------
      SUBROUTINE ROTATE ( F , A )
      REAL F(3), R(3,3), A
      R(1,1) = COS(A)
      R(1,2) = -SIN(A)
      R(2,1) = SIN(A)
      R(2,2) = COS(A)
      R(3,3) = 1
      CALL MATMUL ( F, R, F, 3, 3 )
      RETURN
      END
C ---------------------------------------------------------------
      SUBROUTINE MATMUL ( A, B, C, M, N )
      REAL A(M,N), B(N), C(M)
      DO 10 I = 1, M
            C(I) = 0.0
            DO 10 J = 1, N
                  C(I) = A(I,J)*B(J)
10    CONTINUE
      RETURN
      END
```

NUMERICAL DIFFERENTIATION

The derivative, dY/dX, at a specific point of a function, $Y = f(X)$, can be represented by the slope of the tangent line passing through that point. Graphically, we can draw a tangent line through the desired point, a, as in Fig. 9.7. A good approximation to this true tangent can be made by drawing a line through two points on the curve at equal distances from the point a. In Fig. 9.7 a straight line connects the points on the curve at $X = a - \Delta X$ and $X = a + \Delta X$. If ΔX is very small compared to a, then the slope of the line connecting the points is very close to the slope of the true tangent line at a. As the value of ΔX gets smaller and smaller, the accuracy improves.

To illustrate this, let's examine the curve $Y = X^3 - 2X + 1$ near $X = 2.3$. The slope of the true tangent at this point is 13.87000; but the approximation is very good when ΔX is only 0.1:

ΔX	True Slope	Approximation	Error
0.3	13.87000	13.96000	0.09000
0.2		13.91000	0.04000
0.1		13.88000	0.01000
0.01		13.87010	0.00010

The formula for the approximation comes from the definition, slope $= \Delta Y/\Delta X$. The numerator, ΔY, is $f(a + \Delta X) - f(a - \Delta X)$, and the denominator is $(a + \Delta X) - (a - \Delta X) = 2\Delta X$. Thus:

$$\left.\frac{dY}{dX}\right|_a \approx \frac{f(a + \Delta X) - f(a - \Delta X)}{(a + \Delta X) - (a - \Delta X)} \approx \frac{f(a + \Delta X) - f(a - \Delta X)}{2\,\Delta X}$$

We call this method of evaluating the derivative at a point as the *central difference* method since it uses small differences about a point in the center (a in this case). Two other common methods are the *forward difference* and

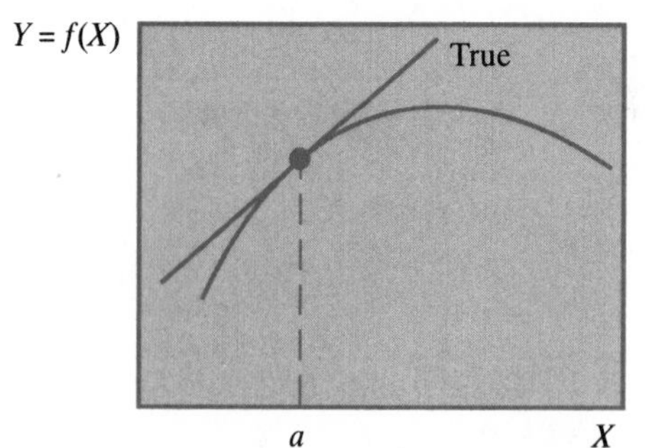

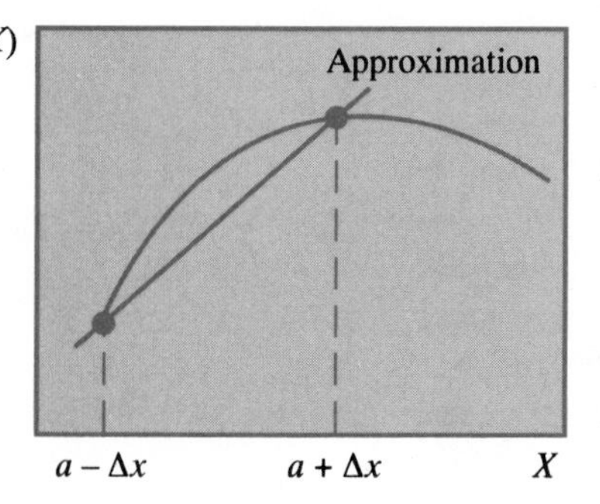

FIGURE 9.7

Graphical Definition of the Derivative and a Method for Approximating It *backward difference* methods. We will defer discussion of these methods until a later chapter.

Converting the approximation for the derivative at a point into a program is a simple matter. We only have to determine the value of the function at two different points, subtract them, and divide by twice the step size. For example, for the function given previously, $Y = f(X) = X^3 - 2X + 1$ at $X = 2.3$ and $\Delta X = 0.1$; we have:

$$\begin{aligned}
a &= 2.3 \\
\Delta X &= 0.1 \\
f(2.3 + 0.1) &= (2.4)^3 - 2(2.4) + 1 \quad = 10.02400 \\
f(2.3 - 0.1) &= (2.2)^3 - 2(2.2) + 1 \quad = 7.24800 \\
\text{approx.} &= (10.024 - 7.248)/0.2 = 13.88000
\end{aligned}$$

The best way to set this up is to use a FUNCTION to calculate the derivative and a second FUNCTION to calculate the value of $f(X)$. This will allow you to reuse the derivative FUNCTION SUBPROGRAM again without any changes. But you will need to customize the FUNCTION for $f(X)$ for each problem:

LS FOR NUMERICAL DIFFERENTIATION

PROBLEM STATEMENT: Read in a value, $X = a$ and a step size, ΔX, and determine the derivative of $Y = f(X) = X^3 - 2X + 1$ at that point.

ALGORITHM DEVELOPMENT:

Step 1a: Review the problem statement. Indicate all nouns in **bold** and all verbs in *italics:*

Read in a **value, X,** and a step size, ΔX, and *determine* the **derivative** of $Y = f(X) = X^3 - 2X + 1$ at that **point.**

Formulate questions based on nouns and verbs:

Q1: How do we use the value $X = a$ and ΔX?
Q2: How do we determine the derivative?

Step 1b: Construct answers to the above questions:

A1: Use $X = a$ and ΔX to obtain $X + \Delta X$ and $X - \Delta X$ and then use these to get $f(X + \Delta X)$ and $f(X - \Delta X)$ values.
A2: Obtain $(f(X + \Delta X) - f(X - \Delta X))$, then divide by $2\,\Delta X$.

ALGORITHM:

1. Read in a and ΔX

2. Send both to FUNCTION DERIV
   Form $a + \Delta X$ and send to FUNCTION F to get $f(a + \Delta X)$
   Form $a - \Delta X$ and send to FUNCTION F to get $f(a - \Delta X)$
   FUNCTION $F(X)$ substitutes X into $X^3 - 2X + 1$
   Return value to DERIV
   Calculate $(F(a + \Delta X) - F(a - \Delta X))/(2\Delta X)$ and return value to MAIN program.
3. Print out DERIV value.

FLOWCHART:

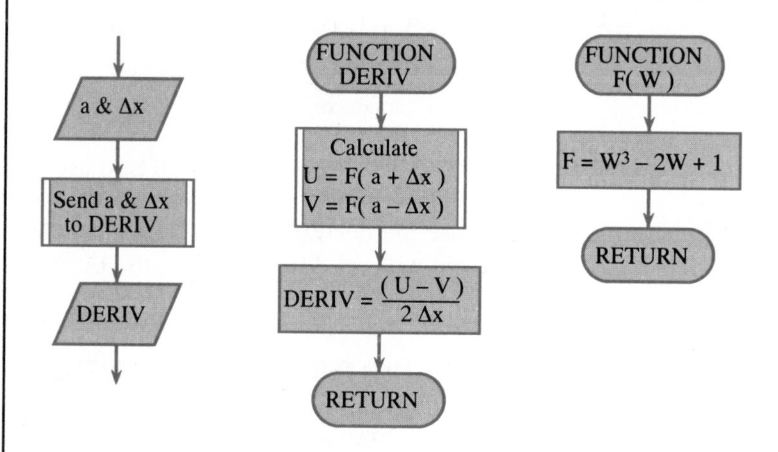

PROGRAM SEGMENT:

```
      READ*, A, DELX
      PRINT *, DERIV (A, DELX)
      STOP
      END
C ------------------------------------------------------------
      FUNCTION DERIV (A, DELX)
      DERIV = (F(A+DELX)-F(A-DELX))/(2.0*DELX)
      RETURN
      END
C ------------------------------------------------------------
      FUNCTION F ( W )
      F = W**3 - 2.0*W + 1
      RETURN
      END
```

If you want to take the derivative of another function, you only need to replace the FUNCTION F(W) with the desired function. For example, if you wanted to take the derivative of a complex function like:

$$F(X) = 2\sin(X)\frac{\exp(-X)}{\ln(X)}$$

the subprogram would become:

```
FUNCTION F(X)
F=2.0*SIN(X)*EXP(-X)/ALOG(X)
RETURN
END
```

The FUNCTION for the derivative would stay the same.

Exercises

SYNTAX ERRORS

9.1 Locate Syntax Errors. Each of the following FORTRAN segments contains syntax errors. Find and correct them:

a.
```
INTEGER I,J
   .
   .
   .
CALL MULT(I,J,K)
   .
   .
   .
END
SUBROUTINE MULT(A,B,C)
   .
   .
   .
END
```

b.
```
REAL A(10)
         .
         .
         .
CALL SUB1(A(10))
         .
         .
         .
END
SUBROUTINE SUB1(A(10))
REAL ARRAY
         .
         .
         .
END
```

c.
```
REAL A, B, C
   .
   .
   .
ANS=38.2*SUB3(A,B,C,1)-SIN(X)
   .
   .
   .
END
REAL FUNCTION SUB3(X,Y,Z,1)
REAL X, Y, Z
   .
   .
   .
END
```

d.
```
REAL FACT
I=5
PRINT *, FACT(I)
STOP
END
REAL FUNCTION FACT(I)
IF(I.LE.1) THEN
        FACT=1
ELSE
        FACT=I*FACT(I-1)
ENDIF
END
```

e.
```
CALL SUB1()
STOP
SUBROUTINE SUB1()
PRINT*, 'WORTHLESS'
RETURN
END
```

f.
```
REAL A(5), B(5)
CALL SUB4(A,B,5)
STOP
END
SUBROUTINE SUB4(A,B,I)
REAL A(I), B(I), C(I)
   .
   .
   .
RETURN
END
```

g.
```
CALL SUB(A)
STOP
END
REAL SUBROUTINE SUB(A)
PRINT*, A
RETURN
END
```
h.
```
CALL JACK(A,B)
STOP
END
JACK=A*B
PRINT*, JACK
RETURN
END
```
i.
```
COMMON Y
X=4.0
CALL SUB(X)
STOP
END
SUBROUTINE SUB(X)
COMMON X
PRINT*, X
RETURN
END
```
j.
```
REAL F, X, Y, Z
F(X)=SIN(X)*EXP(-X)**2
    .
    .
    .
PRINT*, F(X, Y, I)
STOP
END
REAL FUNCTION F(X, Y, I)
F+SIN(X)*EXP(Y)**I
RETURN
END
```
k.
```
DOUBLE PRECISION A, B
CALL SUB3(A,B)
STOP
END
SUBROUTINE SUB3(A,B)
  .
  .
  .
RETURN
END
```

1.
```
INTEGER DOT, A(3), B(3)
DATA A/1, 2, 3/, B/1, 2, 6/
C=DOT(A,B)
STOP
END
FUNCTION DOT(A,B)
INTEGER A(3), B(3)
DOT=...
RETURN
END
```

LOGIC ERRORS

9.2 Locate Logic Errors. The following FORTRAN segments contain logic errors. In some cases the errors may be errors which will escape detection by the compiler. In such a case they may show as run-time or logic errors. Find and correct all errors:

a.
```
      REAL A(10)
      DO 10 I=1,ISIZE
   10 A(I)=I
      CALL PRINT(A)
      STOP
      END
      SUBROUTINE PRINT(A)
      REAL A(10)
      PRINT 10, (A(I), I=1, ISIZE)
   10 FORMAT(1X, 10F6.2)
      RETURN
      END
```

b.
```
      REAL A, B
      DATA A,B/1.0, 2.0/
      CALL SUB1(A)
      CALL SUB2(B)
      STOP
      END
      SUBROUTINE SUB1(X)
      DATA Y/100.0/
      PRINT*, X-Y
      RETURN
      END
```

c.
```
      REAL A, B, FN
      PRINT*, FN(A,B)
      STOP
      END
      FUNCTION FN(A,B)
      ANSWER=A*B
      RETURN
      END
```

d. The following program was designed to print out the area of circles of radii between 0.1 and 1.0 in increments of $r = 0.1$. What went wrong?

```
      PI=0.0
      DO 10 R=0.1, 1.0, 0.1
10    CALL AREA(R, PI)
      STOP
      END
      SUBROUTINE SUB(R, PI)
      DATA PI/3.14159/
      PRINT*, PI*R**2
      RETURN
      END
```

e. Find the error in the following code designed to print the sum of the elements in array A and the sum of the elements in array B:

```
      REAL A(10), B(10), SUM
      PRINT*, SUM(A), SUM(B)
      STOP
      END
      REAL FUNCTION SUM(MAT)
      REAL MAT(10)
      DO 10 I=1,10
10    SUM=SUM+MAT(I)
      RETURN
      END
```

f. This program was supposed to print out sequentially the elements of an array. What went wrong?

```
      REAL A(10)
      DO 10 INDEX=1,10
10    CALL TEST(A)
      STOP
      END
      SUBROUTINE TEST(A)
      DATA I/1/
      PRINT*, A(I)
      I=I+1
      RETURN
      END
```

YOUR SYSTEM

9.3 Duplicate Names. Program segments generally must have unique names. Some compilers, though, allow a subroutine, a built-in function, or a user-defined function to share the same name. Run the following to see how your compiler responds:

```
X=0.145
PRINT*, SIN(X)
CALL SIN(X)
STOP
END
SUBROUTINE SIN(X)
PRINT*, X
RETURN
END
```

9.4 Double Precision Storage. The storage of DOUBLE PRECISION data is not standardized. On the VAX compiler, the standard default compiler option is the /D_FLOAT, which produces a different storage from the /G_FLOAT option. The latter option produces greater precision and also a greater range for the exponent. You may need some help from the system operators to find out the exact compiler options:

```
REAL*8 JACK(1), JILL(1)
JACK(1)=1437.342E21
JILL(1)=-4343.32E-23
CALL SUB(JACK,JILL)
STOP
END
SUBROUTINE(JACK, JILL)
REAL JACK(2), JILL(2)
PRINT*, JACK(1), JACK(2), JILL(1), JILL(2)
RETURN
END
```

9.5 Variable Dimensioning of Arrays. Variable dimensioning of arrays in a subprogram is really a misnomer. The storage location has been set up in the CALLing module and this cannot change, no matter what you use to dimension the array in the subprogram. Try these two examples:

```
REAL A(1)
A(1)=100
CALL SUB1(A,I)
STOP
END
SUBROUTINE SUB1(A,I)
REAL A(I)
PRINT*, A
RETURN
END
```

```
      REAL A(100)
      DO 10 1=1,100
         A(I)=I
10    CALL SUB1(A,1)
      STOP
      END
      SUBROUTINE SUB1 (A,I)
      REAL A(I)
      PRINT*, (A(J), J=1,100)
      RETURN
      END
```

TRACING

9.6 Tracing a Program. Trace through the following program segments and predict their output:

a.
```
      REAL F
      Y=2
      PRINT*, F(F(F(F(Y))))
      STOP
      END
      FUNCTION F(X)
      F=X**2
      RETURN
      END
```

b.
```
      DATA A,B/1.0,2.0/
      PRINT*, A,B,C
      CALL SUB(A,B,C)
      PRINT*, A,B,C
      STOP
      END
      SUBROUTINE SUB(X,Y,Z)
      B=X
      A=Y
      Z=48
      RETURN
      END
```

c.
```
      REAL X(5,5), Y(5,5),Z(5,5)
      DO 10 I=1,5
            DO 10 J=1,5
                  X(I,J)=I**2-J
   10 Y(I,J)=X(I,J)/2*I
      CALL MUL(5,3,X,Y,Z)
      PRINT 20, ((A(I,J),I=1,5),J=1,5)
   20 FORMAT(' ',(3(F7.1,2X),/,1X))
      STOP
      END
      SUBROUTINE MUL(I,J,A,B,C)
      REAL A(I,I),B(I,I),C(I,I)
      DO 10 K=1,J
       DO 10 L=K,I
        DO 10 M=I-J,I
   10    C(M,L)=C(K,K)+A(L,L)*B(K,K)
      RETURN
      END
```

d.
```
      INTEGER A(5,5),B(5,5),C(5,5),D(25)
      DATA ((A(I,J),I=1,5),J=1,5)/25*2/
      DATA ((B(I,J),J=1,5),I=1,5)/5,4,23*1/
      CALL DDA(A,B,C,5,5)
      DO 10 I=1,25
10          D(I)=A(I,5)-B(5,I)
      CALL SREVER(D,25)
      PRINT*,(D(J),J=1,25)
      STOP
      END
      SUBROUTINE DDA(A,B,C,I,J)
      INTEGER A(I,J),B(I,J),C(I,J)
      DO 10 K=1,I
       DO 10 L=1,J
10      C(K,L)=A(K,L)+B(K,L)
      RETURN
      END
      SUBROUTINE SREVER (D,M)
      INTEGER D(M)
      DO 10 I=1,M/2
10     CALL SWITCH(D(I),D(M-I+1))
      RETURN
      END
      SUBROUTINE SWITCH(I,J)
           TEMP=I
           I=J
           J=TEMP
      RETURN
      END
```

PROGRAMS

9.7 Creating a Personal Library of Subprograms. We have already given you several important algorithm primitives that you should be able to use in solving many important problems. To be useful, however, they should all be in the form of subprograms. Once these are created, you can then use your editor to copy them into your application program that uses them. In this exercise you are to create FUNCTIONs or SUBROUTINEs of the following algorithms:

FUNCTIONs:
- Determining if a number is odd or even
- Summation of an infinite series
- Search for MAX/MIN value in a list
- Search for MAX/MIN value in a table
- Summing elements in a list
- Summing elements in a row in a table

	Summing elements in a column in a table
	Summing elements in a table
	Dot product of two vectors
	Angle between two vectors
	Numerical approximation to the derivative
SUBROUTINEs:	Switch two numbers
	Adding two vectors
	Matrix addition
	Matrix multiplication
	Sorting an array
	Transformation of a vector

Try to make these subprograms as general as possible by using variable size arrays. You will also need to create written documentation for yourself that shows what data each subprogram needs.

9.8 Statistical Analysis. Write a FUNCTION that computes the average, the variance, and the standard deviation of an array of 100 data values. Use the following formulas:

Average: $$\overline{X} = \frac{\sum_{i=1}^{N} X_i}{N}$$

Variance: $$\sigma^2 = \frac{\sum_{i=1}^{N} (\text{Average} - X_i)^2}{(N - 1)}$$

Standard deviation: $$\sigma = (\sigma^2)^{1/2}$$

9.9 Cross Product of Two Vectors. The cross product, C, of two vectors, **A** and **B** is given by:

$$\mathbf{C} = \mathbf{A} \otimes \mathbf{B} = [(A_Y B_Z - A_Z B_Y), (A_Z B_X - A_X B_Z), (A_X B_Y - A_Y B_X)]$$

Note that **C** is a vector whose components are given in the parentheses. Write a MAIN program to read in the vectors **A** and **B** and transfers them to a SUBROUTINE CROSS, which computes the cross product.

Now modify your MAIN program and add the appropriate FUNCTION for the DOT PRODUCT to prove the identity:

$$(\mathbf{A} \otimes \mathbf{B}) \cdot (\mathbf{C} \otimes \mathbf{D}) = (\mathbf{A} \cdot \mathbf{C}) * (\mathbf{B} \cdot \mathbf{D}) * (\mathbf{B} \cdot \mathbf{C})$$

Use any arbitrary values for the vectors, **A**, **B**, **C**, and **D**. Also, try the identity for many different values of the vectors. An elegant way would be for you to use the random number generator to create the vectors and let the machine test 100 different combinations.

9.10 Determinant of a 3 × 3 Matrix. In Prob. 9.9 we presented the formula for the determinant of a 3 × 3 matrix:

$$\begin{vmatrix} a_{11} & a_{12} & a_{13} \\ a_{21} & a_{22} & a_{23} \\ a_{31} & a_{32} & a_{33} \end{vmatrix} = a_{11}(a_{22}a_{33} - a_{23}a_{32}) - a_{12}(a_{21}a_{33} - a_{23}a_{31}) + a_{13}(a_{21}a_{32} - a_{22}a_{31})$$

If you look closely at each of the terms, you may see a general formula:

$$a_{1i}(a_{2j}a_{3k} - a_{2k}a_{3j}) \qquad (\text{where } i \neq j \neq k)$$

Write a FUNCTION SUBPROGRAM that determines the DET(A) with the preceding formula.

9.11 Cramer's Method of Solving Simultaneous Equations. One of the best-known methods for solving a system of simultaneous equations uses the determinant. Assume that we have the following series of equations:

$$a_{11}X + a_{12}Y + a_{13}Z = c_1$$
$$a_{21}X + a_{22}Y + a_{23}Z = c_2$$
$$a_{31}X + a_{32}Y + a_{32}Z = c_3$$

where a_{ij} and c_i are constant and we want to solve for the variables X, Y, and Z. We can solve these equations by using Cramer's rule:

$$X = \frac{\begin{vmatrix} c_1 & a_{12} & a_{13} \\ c_2 & a_{22} & a_{23} \\ c_3 & a_{32} & a_{33} \end{vmatrix}}{\begin{vmatrix} a_{11} & a_{12} & a_{13} \\ a_{21} & a_{22} & a_{23} \\ a_{31} & a_{32} & a_{33} \end{vmatrix}} \qquad Y = \frac{\begin{vmatrix} a_{11} & c_1 & a_{13} \\ a_{21} & c_2 & a_{23} \\ a_{31} & c_3 & a_{33} \end{vmatrix}}{\begin{vmatrix} a_{11} & a_{12} & a_{13} \\ a_{21} & a_{22} & a_{23} \\ a_{31} & a_{32} & a_{33} \end{vmatrix}} \qquad Z = \frac{\begin{vmatrix} a_{11} & a_{12} & c_1 \\ a_{21} & a_{22} & c_2 \\ a_{31} & a_{32} & c_3 \end{vmatrix}}{\begin{vmatrix} a_{11} & a_{12} & a_{13} \\ a_{21} & a_{22} & a_{23} \\ a_{31} & a_{32} & a_{33} \end{vmatrix}}$$

Notice that the denominator in all three cases is the same or DET(A), where A is the matrix of the variable coefficients. The numerators consist of the DET(A′), where the new matrix A′ is formed by replacing the vector **C** in subsequent columns:

$$A'_1 = \begin{vmatrix} \boxed{c_1} & a_{12} & a_{13} \\ \boxed{c_2} & a_{22} & a_{23} \\ \boxed{c_3} & a_{32} & a_{33} \end{vmatrix} \qquad A'_2 = \begin{vmatrix} a_{11} & \boxed{c_1} & a_{13} \\ a_{21} & \boxed{c_2} & a_{23} \\ a_{31} & \boxed{c_3} & a_{33} \end{vmatrix} \qquad A'_3 = \begin{vmatrix} a_{11} & a_{12} & \boxed{c_1} \\ a_{21} & a_{22} & \boxed{c_2} \\ a_{31} & a_{32} & \boxed{c_3} \end{vmatrix}$$

Write a MAIN program that reads in the coefficient matrix, A, forms the matrix, A', and then uses the SUBROUTINE for the DET(B) to solve for the three unknowns. Be careful to see if DET(A) = 0 before you attempt to apply Cramer's rule. What is the significance of a system where DET(A) = 0? Use the following system to check your answer:

$$3X - 2Y + Z = 16$$
$$Y + 3Z = 23$$
$$7X + Y - Z = 27$$

9.12 Mesh Analysis of Resistor Network. Mesh analysis is often used to solve complex resistor networks to calculate the currents flowing in different legs. Consider the following network with a single voltage source and eight resistors arranged as shown:

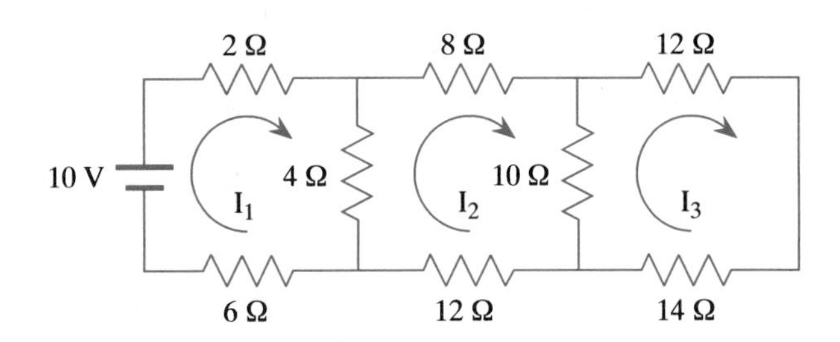

Because the voltage drop around each loop must be zero, it can be shown that the following set of simultaneous equations describe the current flow in each leg:

$$\begin{aligned} 12I_1 - \;\; 4I_2 \qquad\quad &= 10 \\ -4I_1 + 34I_2 - 10I_3 &= 0 \\ -10I_2 + 36I_3 &= 0 \end{aligned}$$

Use the program for Cramer's rule from the previous problem to solve for the current, I_i, in each of the legs.

9.13 Determinant of the Fibonnaci Series. One of the most interesting series in mathematics is that of Leonardo Fibonacci (b. 1175). This series seems to describe many different phenomena in nature and has fascinated scientists for almost 800 years. The series is a simple one:

$$1 \quad 1 \quad 2 \quad 3 \quad 5 \quad 8 \quad 13 \quad 21 \quad 34 \quad 55$$

Notice that any term in the series is simply the sum of the two previous terms, or

$$T_n = T_{n-1} + T_{n-2}$$

An interesting feature of this series is that the determinant of a matrix made from this series is always zero. For example:

$$\begin{vmatrix} 1 & 1 & 2 \\ 3 & 5 & 8 \\ 13 & 21 & 34 \end{vmatrix} = 0 \qquad \begin{vmatrix} 13 & 21 & 34 \\ 55 & 89 & 144 \\ 233 & 377 & 610 \end{vmatrix} = 0$$

This can be proven by a mathematical analysis of the sequence of terms. But it is also useful for you to try to "prove" this numerically by substituting the numbers and calculating the determinant. This does not rigorously prove the statement, but is does support the statement. Sometimes a rigorous mathematical proof does not exist and you will have to try this numerical approach.

In this problem you are to examine the first ten matrices formed in this way and then evaluate the determinant. Here is the approach that you should use:

- Generate the Fibonnaci series and store it in an array.
- Create a matrix of terms in the series, starting with the first term.
- Calculate the determinant of the matrix.
- Repeat steps 2 and 3, but now start filling the matrix with the second term for the second loop, the third term for the third loop, and so on. The two preceding matrices represent the first and seventh matrices formed.
- Repeat for ten cycles. If any of the determinants are not zero, print out an appropriate message.

9.14 Multiaxis Rotation of a Vector. In this chapter we presented a solution for calculating the components of a vector following a rotation about the Z axis. There are many times, however, when we need to apply more complex rotations. Therefore, in this problem you are to rewrite the subprogram given in the text for a rotation about *any* axis. The appropriate equations for the rotation matrix are:

Rotation about X axis: $$\begin{vmatrix} 1 & 0 & 0 \\ \cos(\alpha) & -\sin(\alpha) & 0 \\ \sin(\alpha) & \cos(\alpha) & 1 \end{vmatrix}$$

Rotation about Y axis: $$\begin{vmatrix} \cos(\alpha) & 0 & \sin(\alpha) \\ 0 & 1 & 0 \\ -\sin(\alpha) & 0 & \cos(\alpha) \end{vmatrix}$$

Rotation about Z axis: $$\begin{vmatrix} \cos(\alpha) & -\sin(\alpha) & 0 \\ \sin(\alpha) & \cos(\alpha) & 0 \\ 0 & 0 & 1 \end{vmatrix}$$

Rewrite the program and subprogram so that you may take three successive rotations about each of the axes. Use an initial vector of $(2.5, 3.0, 0.0)$ and rotate it 25° about X, $-45°$ and Y, and 90° about Z. To double check your program, take the new vector that you get from these calculations and put it back into the program. This time, though, rotate the vector in the reverse direction ($-90°$ about Z, 45° about Y, and $-25°$ about X). Did you get the original vector back?

9.15 Infinite Series. Write a program to approximate the value of the infinite series:

$$R = J_0(X) + \frac{J_0(X^2)}{2!} + \frac{J_0(X^3)}{3!} + \cdots$$

where J_0 is known as the zero-order Bessel function defined by:

$$J_0(X) = 1 - \frac{X^2}{2^2} + \frac{X^4}{2^2 4^2} - \frac{X^6}{2^2 4^2 6^2} + \cdots$$

Assume that the series for R can be terminated when the argument of the Bessel function (X^n) is smaller than a number, eps, that you read in at execution time. Also, terminate the series for $J_0(X)$ when a term in the series is less than a small number, del, that you also read in at execution time.

9.16 Maximum of a Continuous Function. The FUNCTION presented in this chapter for finding the maximum value is only useful for examining discrete data or individual data items. It cannot be used for a continuous function. In this problem we consider how to find the maximum of such a function.

The program that you write should utilize the following algorithm:

- Read in a starting point, X, a step size, DeltaX, and an allowable error, EPS.
- Start the search by calculating F(X) and F(X + DeltaX).
- If F(X) < F(X + DeltaX), then continue.
- Otherwise reduce the step size by half and repeat the comparison.
- Repeat until ABS(F(X) − F(X + DeltaX)) < EPS.

Apply this algorithm to the function $Y(X) = \sin(X)e^{-X}$. Start your search at $X = 0.7$ with a step size of 0.1.

9.17 Heart Performance. Experiments were performed to measure the performance of a dog's heart. Recordings of the aortic flow, Q (flow out of the heart), and the aortic pressure, P, were made. One cycle (one heart beat) is shown here:

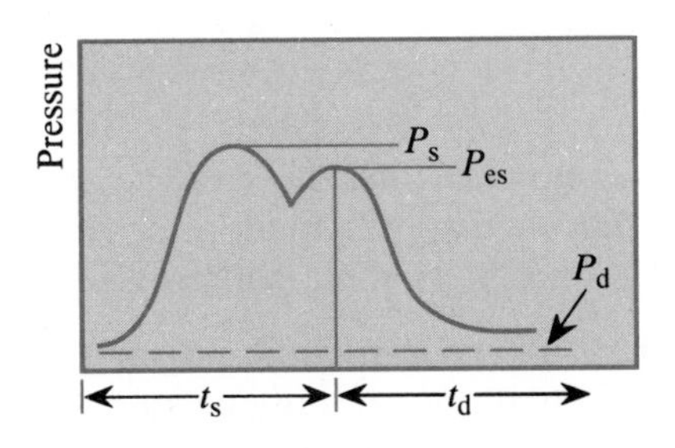

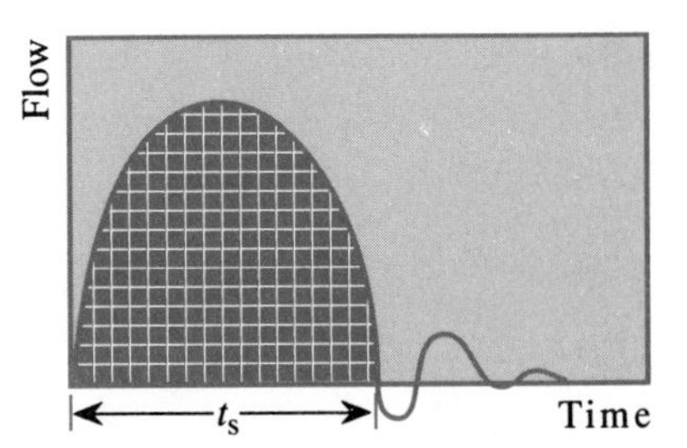

The data for the preceding curves were collected at 0.01-s intervals and are stored in data files, each consisting of 51 data points. The heart rate, HR, was 120 beats per minute. Write a MAIN program and SUBROUTINEs to do the following:

a. In the MAIN program load the data from the data files into arrays.
b. In a SUBROUTINE program determine the systolic pressure, P_s and the diastolic pressure, P_d, shown schematically in the previous diagram. Also, calculate the mean aortic pressure, *MAP*, according to the formula:

$$MAP = P_d + (P_s - P_d)/3$$

c. In a SUBROUTINE calculate the stroke volume, *SV*. This is the amount of blood ejected by the heart during a contraction. It is the shaded area shown in the preceding flow curve.
d. In a SUBROUTINE calculate the peripheral resistance, *R*, by:

$$R = MAP/(SV*HR)$$

e. In a SUBROUTINE calculate the arterial compliance, *C*, by:

$$P_d = P_{es} \exp[-t_d/(R*C)]$$

f. In the MAIN write out all calculated results for P_s, P_d, *SV*, *R*, and *C*.

9.18 Induced Voltages in a Coil. A time-varying current, like *AC*, through a wire coil will induce a current in a nearby second coil. The total voltages on each of the terminals are as follows:

$$V_1 = L_1\frac{di_1}{dt} + M\frac{di_2}{dt} \qquad V_2 = L_2\frac{di_2}{dt} + M\frac{di_1}{dt}$$

where V_1 and V_2 = voltages on inductor 1 and 2

L_1 and L_2 = inductances

i_1 and i_2 = currents through L_1 and L_2

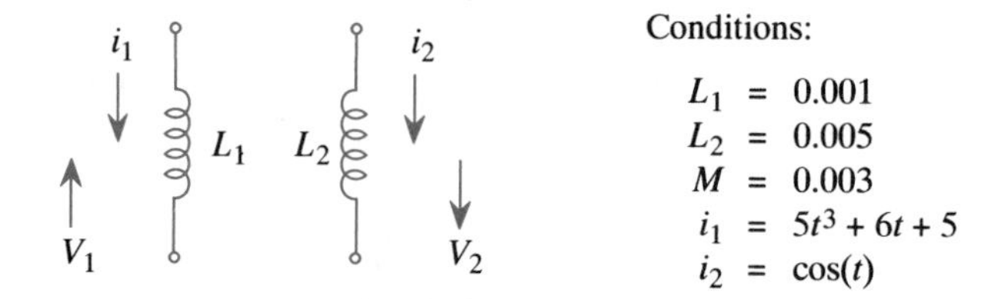

For the two coils shown here and the conditions cited, use the central difference approximation for di_1/dt and di_2/dt to calculate the voltages $V_1(t)$ and $V_2(t)$ at 0.1-s intervals for 0 to 25 s.

9.19 Three-Dimensional Heat Distribution. A solid block 10 × 10 × 10 contains a centrally located 2 × 2 × 2 heat source. The temperature of the heat source changes with time, *t*:

$$T_e(t) = T_0 e^{-\alpha t}$$

The large solid block is surrounded by water, which keeps the outer surface of the block at a fixed ambient temperature, T_a. At time $t = 0$ the heating element is turned on and the block begins to heat up. But, note, as time goes on the heating element cools off. In this exercise you are to calculate the temperature distribution in the solid block as a function of time.

Divide the block into subblocks of dimension $1 \times 1 \times 1$ and let the temperature in each subblock be represented by $T(X,Y,Z)$. At any time, t, the temperature of the subblock is equal to its temperature during the previous time period minus a quantity related to the heat flux, Q, times a material parameter, C:

$$T^{\text{new}}(X,Y,Z) = T^{\text{old}}(X,Y,Z) - C * Q$$

where the heat flux, Q, is given by:

$$\begin{aligned} Q = [&6*T(X,Y,Z) - T(X-1,Y,Z) - T(X+1,Y,Z) \\ &- T(X,Y-1,Z) - T(X,Y+1,Z) - T(X,Y,Z-1) \\ &- T(X,Y,Z+1)] \end{aligned}$$

Write a FUNCTION segment to compute Q. Then use this FUNCTION to compute the temperature distribution in the block at 10-s intervals between $t = 0$ and $t = 100$ and only for a row of elements such as $T(5,5,Z)$ passing through the center of the block. Do not print out the whole matrix and do not print it out except at 10-s intervals. (Use the following values: $A = 0.005$, $T_0 = 100.0$, $T_a = 10.0$, $C = 1.0$)

CHAPTER

10

THE ART OF DEBUGGING—PART II

10.1 A Second Look at Debugging

Since the previous chapter on debugging, we have introduced two new FORTRAN topics—arrays and subroutines. These two major tools of programming can produce unique errors that are very different from those that we discussed in Chap. 7. Therefore, these new topics merit additional discussion about the electronic battlefield that we call *debugging.*

In Chap. 7 we presented many types of errors with techniques to detect them. Those techniques usually involved the use of variable tables, compiler list-files, and tracing of programs. In this chapter we show you the types of errors that often occur when using arrays and subprograms. Also, we present some suggested techniques to help you detect these errors.

Before we begin, though, keep in mind that debugging is indeed an art. While we can present some useful methods, you should still attempt to develop some techniques on your own. Watch how other people, such as your fellow students, instructor, or computer professionals debug programs. And don't be afraid to ask others their debugging secrets. You will find that programmers are loathe to keep secrets. If they know of a good technique, they want everyone to know. So, go ahead and ask. You might find some real gems.

10.2 Debugging Programs with Arrays

In the previous chapter on debugging we presented topics in the order of ease of error detection. First came the syntax errors because they were the easiest to locate. Next came the run-time errors and logic errors, which are more difficult to locate. In Chap. 7 this ordering made sense. But for this discussion we alter the order. Since tracing of arrays plays such an important role in locating almost all nontrivial errors, we begin our discussion there. Then we talk about syntax and run-time errors.

TRACING ONE-DIMENSIONAL ARRAYS

An *array* is an efficient means to declare a group of similar variables with a single command. For example, to create the REAL array, A, with 100 elements, we use the following command:

```
REAL A(100)
```

Recall that the array is comparable to a subscripted variable in mathematics. Thus, for example, A(5) is the fifth element in the array. Since we think of each element in the array as an individual REAL variable, we must leave enough room in our tracing table for every element. The previous declaration statement defined A as a one-dimensional array with 100 elements. Thus, our trace table will consist of 100 variables, A(1), A(2), . . . , A(100).

This might seem a frightening idea. After all, who wants to trace a program with 100 variables in it? Fortunately, for most problems that use arrays, we can safely reduce the size of the array to make it more manageable for tracing. Then, after we eliminate the errors, we restore the arrays back to their original size. To show this, consider the following program to read in a list of 100 numbers and then to print them back out in reverse order:

```
*     PROGRAM TO READ IN ARRAY OF 100 NUMBERS AND
*     PRINT THEM OUT IN REVERSE ORDER
*  ______________________________________________________
      INTEGER A(100),TEMP
      DO 10 I=1,100
            PRINT *,'ENTER NUMBER ',I
            READ *,A(I)
10    CONTINUE
      DO 20 I=1,100/2
             J=100+1-I
             TEMP=A(I)
             A(I)=A(J)
             A(J)=TEMP
20    CONTINUE
      PRINT *,(A(I),I=1,100)
      STOP
      END
```

Without altering the function of the program, we can simplify our job of tracing the program by reducing the array to only ten elements:

```
*    PROGRAM TO READ IN 10 NUMBERS AND PRINT
*    THEM OUT IN REVERSE ORDER
*  ____________________________________________
     INTEGER A(10),TEMP
     DO 10 I=1,10
           PRINT *,'ENTER NUMBER ',I
           READ *,A(I)
10   CONTINUE
     DO 20 I=1,10/2
            J=10+1-I
            TEMP=A(I)
            A(I)=A(J)
            A(J)=TEMP
20   CONTINUE
     PRINT *,(A(I),I=1,10)
     STOP
     END
```

The key in knowing how to simplify the problem lies in knowing the function of the program. Performing a trace on the preceding program with an assumed input of 1, 2, . . . , 10, we have:

A(1) 1, 10

A(2) 2, 9

A(3) 3, 8

A(4) 4, 7

A(5) 5, 6

A(6) 6, 5

A(7) 7, 4

A(8) 8, 3

A(9) 9, 2

A(10) 10, 1

I 1, 2, 3, 4, 5, 6, 7, 8, 9, 10, 11, 1, 2, 3, 4, 5, 6, 1, 2, 3, 4, 5, 6, 7, 8, 9, 10, 11

J 10, 9, 8, 7, 6

TEMP 1, 2, 3, 4, 5

Output to screen:

Since the program worked for 10 elements, it is likely that it will work for 100 elements. The methodology is the same, only the number of elements has changed. While this may be true for many situations, there may be some cases where this may not work. But you are unlikely to encounter this problem anytime soon.

Just as we summarized some of our debugging tools in the form of proverbs in Chap. 7, we continue the practice here. So, our first hint for this chapter becomes Proverb 12:

> **PROVERB 12:**
> **SMALLER IS BETTER!**
> **Most programs that use large arrays are more easily debugged by using smaller arrays.**

One method that programmers often use to change array sizes easily is to use the PARAMETER statement. A PARAMETER statement provides a way to declare symbolic constants (variables whose values cannot be changed) in a nonexecutable fashion. Consequently, PARAMETER statements must come at the beginning of a program. Modifying the previous program to use a PARAMETER statement, we now have the following program:

```
*
*    PROGRAM TO READ IN 10 NUMBERS AND PRINT
*    THEM OUT IN REVERSE ORDER
*___________________________________________
     PARAMETER (NA=10)
     INTEGER A(NA),TEMP
     DO 10 I=1,NA
          PRINT *,'ENTER NUMBER ',I
          READ *,A(I)
10   CONTINUE
     DO 20 I=1,NA/2
           J=NA+1-I
           TEMP=A(I)
           A(I)=A(J)
           A(J)=TEMP
20   CONTINUE
     PRINT *,(A(I),I=1,NA)
     STOP
     END
```

By using the PARAMETER statement, we have greatly simplified the process of modifying the program for a different size array. You need only to edit a single line. This will take some tedium out of debugging. Also, it removes the possibility that you will forget to change the value of the variable somewhere within the program. This latter point is an extremely important one, especially if your program is lengthy. For example, if you are not paying close attention, it is easy to miss one or more places in the program where the array should be reduced. When you go to run such a program, the results will be erratic, as we will show later in this chapter. To avoid such problems, use the PARAMETER statement as we described, until you have finished debugging the program:

> PROVERB 13:
> **MINIMIZE THE NUMBER OF CHANGES!**
> **Use PARAMETER statements to define the size of arrays and other variables that change during debugging.**

TRACING HIGHER-ORDER ARRAYS

The tracing techniques previously illustrated can be used with any dimension array that the compiler supports. For example, a 100 × 100 array would produce 10,000 variables, which would be impossible to trace by hand. Thus, the first thing that we do with two-dimensional arrays is to reduce their size to something like 10 × 10, which we can follow more easily. Also, when we go to trace the array, we usually organize the trace as a table. Here is a program that creates a table of size 11 × 11:

- Load blank spaces into every location.
- Load the number '1' into column #6.
- Load the character '-' into row #6.
- Load the character '+' into the center of the table.

```
*
*    PROGRAM SEGMENT TO FILL IN COLUMN #6 WITH '1'
*    ROW #6 WITH '-', AND CENTER (6,6) WITH '+'
*___________________________________________________
     CHARACTER*1 PLOT (11,11)
     DO 10 I=1,11
           DO 10 J=1,11
                 PLOT(I,J)=''
10   CONTINUE
     DO 30 K=1,11
           PLOT(6,K)='-'
           PLOT(K,6)='I'
30   CONTINUE
     PLOT(6,6)='+'
     ...
```

Performing a trace on this program is no more difficult than our previous examples. The difference is that the array, PLOT, will be an 11 × 11 as the following trace table shows:

I	1, 2, , 12
J	1, . . . ,11, 12, 1, . . ,11, 12, ... , 1, . . . , 11, 12
K	1, 2, 3, 4, 5, 6, 7, 8, 9, 10, 11, 12

Plot:	1	2	3	4	5	6	7	8	9	10	11
1						1					
2						1					
3						1					
4						1					
5						1					
6	–	–	–	–	–	1+	–	–	–	–	–
7						1					
8						1					
9						1					
10						1					
11						1					

FIGURE 10.1

Illustration of Higher-Dimensional Arrays

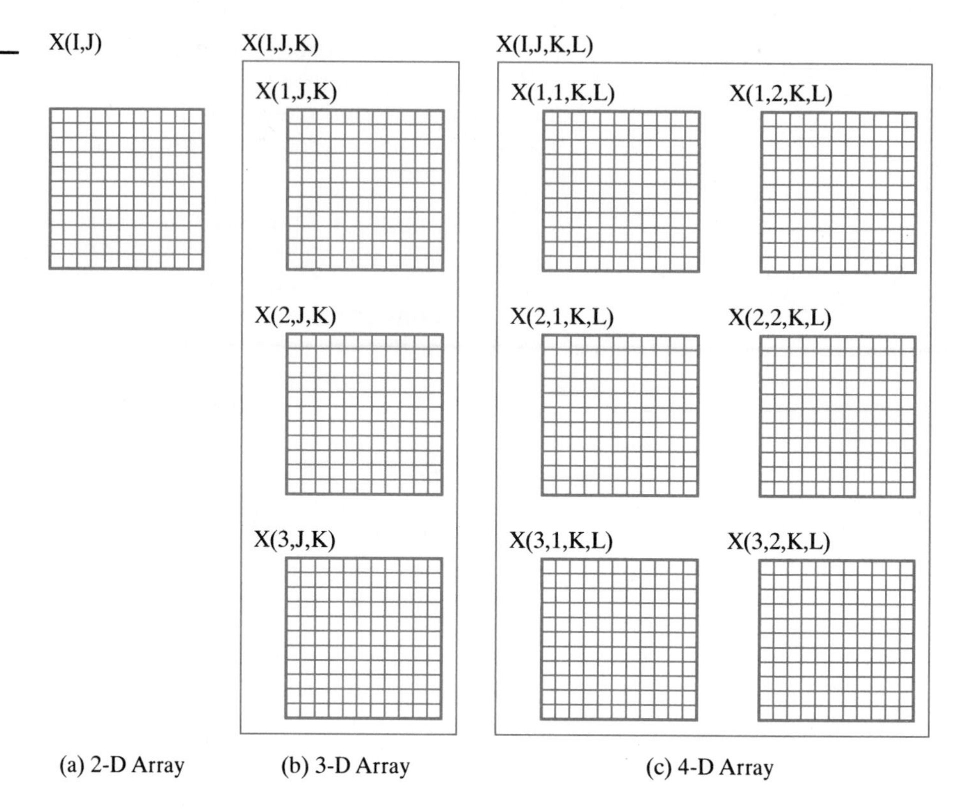

Similarly, tracing higher-order arrays can be done with lists of tables, tables of tables, and so on, as Fig. 10.1 shows.

ARRAYS AND SYNTAX ERRORS

In Chap. 7 we stated that one of the biggest problems with a compiler is that its error message may not match the actual errors. This is especially true when using arrays in your programs. Errors in the declaration of arrays typically lead to multiple errors throughout the program. For the new FORTRAN programmer, such an error can be very intimidating. It is not uncommon to write a moderately sized program (e.g., one full page of code) that generates 20 or more errors. To see this, consider the program segment of the previous example into which we have introduced a minor syntax error:

```
*
*    EXAMPLE OF PROBLEMS CREATED BY SIMPLE SYNTAX
*    ERRORS WHEN USING ARRAYS
*
     CHARACTER*1 PLOT(11,11)
     DO 10 I=1,11
           DO 10 J=1,11
                 PLOT(I,J)=''
```

```
10   CONTINUE
     DO 30 K=1,11
           PLOT(6,K)='-'
           PLOT(K,6)='l'
30   CONTINUE
     PLOT(6,6)='+'
     PRINT 35, ((PLOT(I,J),J=1,11), I=1,11)
35   FORMAT('',11A1)
     STOP
     END
```

We have added a PRINT statement to the previous program segment. Also, we have deliberately made a simple typographical error in the declaration statement for the array PLOT. We changed the letter 'O' to the number '0' in the name of the array, PLOT. Okay, this is a simple mistake, but the compiler doesn't know how to handle the problems created. Look at what happens when it tries to translate this segment:

```
%FORT-F-UNDARR, Undimensioned array or statement function definition out
      of order [  PLOT() in module ARRAY$MAIN at line 4
%FORT-F-UNDARR, Undimensioned array or statement function definition out
      of order [  PLOT() in module ARRAY$MAIN at line 8
%FORT-F-UNDARR, Undimensioned array or statement function definition out
      of order [  PLOT() in module ARRAY$MAIN at line 9
%FORT-F-UNDARR, Undimensioned array or statement function definition out
      of order [  PLOT() in module ARRAY$MAIN at line 11
%FORT-F-UNDARR, Undimensioned array or statement function definition out
      of order [  PLOT() in module ARRAY$MAIN at line 13
%FORT-F-ENDNOOBJ,DISK$USERFILES:[MACWIAKALA.BOOK]ARRAY.FOR;1 completed
      with 5 diagnostics—object deleted
```

Only one error is present, but it creates a ripple effect, which results in a total of five errors. The reason is that the mistake occurs in the statement needed to define the array. Notice that PL0T exists, but PLOT does not exist. Thus, every time we try to access the nonexistent array, we produce a compiler error. The compiler errors, though, are all the same except that they occur on different lines:

> PROVERB 14:
> **MANY FROM ONE**
> **Multiple errors in referencing a common array or function usually indicate an error in declaration.**

This type of error may occur also in long declaration statements. If you accidentally type beyond column 72, the compiler ignores the extra characters. As a result, your declaration statement may be incomplete. One way to detect this is to check your compiler list-file.

ARRAYS AND RUN-TIME ERRORS

Since arrays have indices (sometimes called subscripts), it is possible that you may try to access elements in the array that may not exist. Compilers report this as a subscript out-of-range error. This is a simple example of how it might occur:

```
REAL X(100)
    .
    .
    .
PRINT *, X(101)
```

The PRINT statement is told to print out the contents of the 101st element. But the array only has 100 elements according to the declaration. Therefore, this data point does not exist. Yet, the machine may try to get this information. A more subtle variation on this theme is:

```
REAL X(100)
    .
    .
    .
PRINT *, X(I)
```

where the variable, I, is the result of a calculation. If the value of I exceeds 100, then you will have a subscript that is out of range.

It is essential that you establish very early how your system responds to this type of error. One method of doing this is to write a small program to produce this error deliberately. An example of such a program follows:

```
*
*    PROGRAM TO TEST SYSTEM RESPONSE WHEN HANDLING
*    ARRAY INDEX "OUT OF BOUNDS"
*________________________________________________________
     REAL A(10)
     DO 10 I=1,1000
           A(I)=0
           PRINT *,I,A(I)
10   CONTINUE
     STOP
     END
```

Notice that inside the DO LOOP I takes on a value of up to 1000, but the array, A, contains only 10 elements according to the declaration statement. Therefore, when the program tries to assign a value to A(11) or to print out the value of A(11), we expect strange results. The results are strongly dependent on the compiler in use. So, we decide to run the program on two different computers. Here are the results:

OUTPUT	
Computer 'A'	Computer 'B'
1 0.000000	1 0.000000
2 0.000000	2 0.000000
3 0.000000	.
4 0.000000	.
5 0.000000	.
6 0.000000	94 0.000000
7 0.000000	95 0.000000
8 0.000000	*%SYSTEM-F-ACCVIO, access violation*
9 0.000000	*reason mask=04,virtual address=*
10 0.000000	*00000400, PC=00000418, PSL=03C00024*
0 0.000000	*%TRACE-F-TRACEBACK, symbolic stack*
1 0.000000	dump follows...
2 0.000000	
...	

Don't try to decipher the error message from machine 'B'. We show it only to indicate that something very unusual is happening. Its output like that presented here will cost you a few sleepless nights. Computer 'A' starts to modify the subscript variable, I, when it exceeds the upper limit boundary of the array. Notice that I begins to recycle back to 0 when it should be advancing to 11. Imagine the complications if you use I or A(I) for some calculations. Meanwhile computer 'B' simply allows you to exceed the array upper bound—or will it? When the subscript goes beyond the upper limit, the integrity of the data stored in those out-of-range array elements is questionable. You may find that your program works when only a few people are on the system, but will not work when the system has many users.

Our purpose in showing you these two strange results is to emphasize the point that you *must* test your compiler. If you know your compiler's poor points, you will speed debugging when a problem of this kind occurs:

> PROVERB 15:
> **KNOW YOUR COMPILER!**
> **Determine how your compiler handles out-of-range subscripts in arrays.**

We want to make one last point before leaving arrays. Be sure to check what happens when the subscript goes below the minimum subscript value, which is usually 1.

10.3 Debugging Programs Containing Subprograms

Subprograms are one of the most useful and desirable features of any programming language. They allow programmers to:

- Modularize their code to take advantage of repetitive occurrence of operations.
- Break down the problem into several smaller, more manageable steps.
- Implement user-defined primitives and structures in a consistent manner.

Also, by creating libraries of subprograms that you will use frequently, you indirectly test and debug the subprograms repetitively, resulting in high-quality code. You have probably heard the saying, "Don't reinvent the wheel." By using subprograms in the fashion just outlined, not only do you not reinvent the wheel, you take out the flat spots over time.

Now we must talk about the down side. By modularizing the code to this extent, we can find it very difficult to determine where an error originates. This poses a unique problem. For example, if your program calls a function ten times in the MAIN program and four times in a subprogram, how do you know which of the 14 calls produces the error? Also, we tend to nest subprograms (as with the example of a SUBROUTINE calling a FUNCTION). After two levels of nesting, you need a road map to keep track of what's happening.

Tracing through subprograms is a difficult task. So, we need some special tools to help us. That's what this section is all about: creating tools (more subprograms!) to help keep track of what is happening. To do this, we must use a feature of FORTRAN known as the COMMON BLOCK. With this feature, it will be much easier to locate the origin of a problem.

COMMON BLOCKs: WHEN AND WHEN NOT TO USE

COMMON BLOCKs provide another way to transfer data between program modules (SUBPROGRAM to SUBPROGRAM and SUBPROGRAM to MAIN Program). Before discussing how to use COMMON BLOCKs, we want to stress that for general programming you should not use COMMON BLOCKs. Some people feel that COMMON BLOCKs are similar to GOTO statements and that good programmers should never use them. We will not go that far. They do perform a necessary function in some special cases as we will discuss. The problem with COMMON BLOCKs is that overusing them often makes subprograms less modular. The net result is that you may need extra code to do the same thing:

> PROVERB 16:
> **USE COMMON BLOCKS SPARINGLY!**
> **COMMON BLOCKs reduce the versatility of subprograms. So, use them only when needed, such as when debugging.**

Recall from Chap. 9 that the purpose of a LABELED COMMON BLOCK (COMMON BLOCK with a name) is an additional means to transfer data between program modules. If the program assigns a value to X in the MAIN program, it's possible to pass X to a subprogram without going through the argument list. The following program illustrates the use of a COMMON BLOCK, and how a COMMON BLOCK can be used as an additional way to pass data:

```
*
*    PROGRAM TO DEMONSTRATE COMMON BLOCKS
*  ___________________________________________________
     REAL X,Y,Z
     COMMON/ONE/X
     X=1.0
     Y=2.0
     Z=3.0
     PRINT *, 'BEFORE TEST: X,Y,&Z= ', X, Y, Z
     CALL TEST(Y)
     PRINT *, 'AFTER TEST: X,Y,&Z= ', X, Y, Z
     STOP
     END
*  ___________________________________________________
     SUBROUTINE TEST(Y)
     REAL X,Y,Z
     COMMON/ONE/X
     X=X*2
     Y=Y*4
     Z=10.0
     PRINT *, 'WITHIN TEST: X,Y,&Z= ', X, Y, Z
     RETURN
     END
```

Executing the preceding program results in the following output:

```
BEFORE TEST:       X,Y,&Z = 1.0    2.0     3.0
WITHIN TEST:       X,Y,&Z = 2.0    8.0    10.0
AFTER TEST:        X,Y,&Z = 2.0    8.0     3.0
```

The variable X transfers to the subroutine TEST by way of the COMMON BLOCK, ONE. Any modifications to X's value in the subroutine will change the value of X in the MAIN program also. The variable Y transfers to the subprogram through the argument list. Also, any modifications to Y's value will propagate back to the MAIN program, too. The variable Z is not referenced in either the COMMON BLOCK or the argument list, and is therefore local. We highlight this in the output by noting that the value stored in Z within the MAIN program does not change with the execution of the subprogram TEST. The variable, X, on the other hand, is a global variable. Therefore, X in the subroutine and the MAIN program is the same variable.

Programmers typically use COMMON BLOCKs when a lot of data must pass between a system of subprograms and where it would be inconvenient to pass all that data through an argument list. It must be stressed that it is a group of subprograms working together that concerns us, not just a single program module. A typical example of this would be a library of subprograms written to use graphic terminals. These graphics routines are written by outside vendors that allow you to transfer data to the routines. Unfortunately, not all systems have graphics terminals available. So, we present here a system of subprograms to provide low-resolution plots to a terminal screen with 20 lines and 80 columns of text. Here is a .MAIN. program to do this:

```
*
*     MAIN PROGRAM TO ACCESS GRAPHICS ROUTINES
*---------------------------------------------------------------
      CHARACTER*1 STAR
      REAL PI,X,Y
      STAR='*'
      PI=3.1415927
      CALL WINDOW(-PI,PI,-1.0,1.0)
      CALL PLTCHR(STAR)
      DO 10 X=-PI,PI,2*PI/81
            Y=SIN(X)
            CALL PLOT(X,Y)
10    CONTINUE
      CALL PRINT
      STOP
      END
```

The purpose of this program is to use the SUBROUTINEs WINDOW, PLTCHR, PLOT, and PRINT to generate the sine wave of Fig. 10.2.

The process of creating a plot on the terminal screen is straightforward when using the subroutines provided in Fig. 10.3. The first subroutine, called WINDOW, sets up the *X* and *Y* axis ranges. The arguments are real, and they specify *X* minimum, *X* maximum, *Y* minimum, and *Y* maximum. The second subroutine, PLTCHR defines the character to be used for creating the plot (the

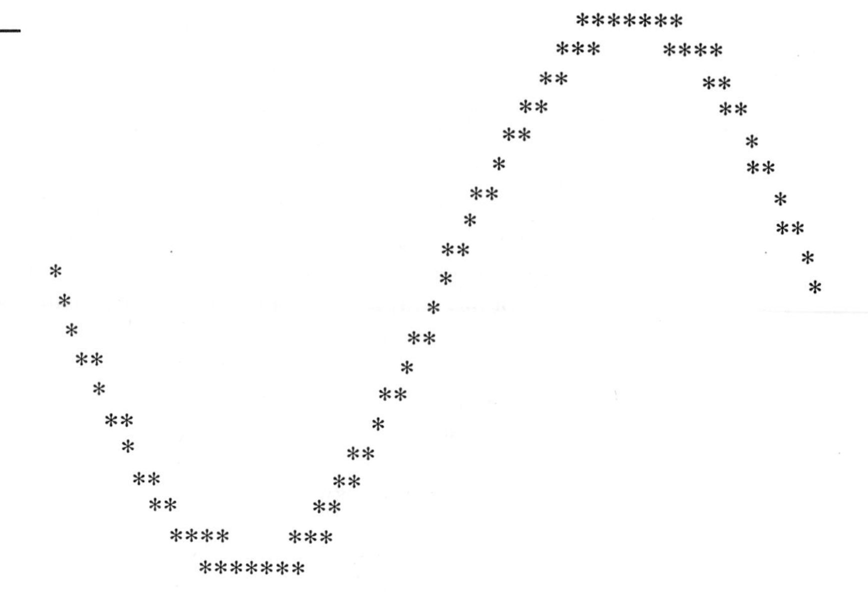

FIGURE 10.2

Use of Graphics Program to Generate a Sine Wave

plot character). The PLOT subroutine plots points using the plot character. The last subroutine, PRINT, generates the plot at the terminal screen. Each of these subroutines only passes information in their argument lists that is pertinent to the function they are performing. Note that PRINT lacks arguments:

> **PROVERB 17:**
> **COMMON BLOCKS FOR INTERNAL USE**
> **Use COMMON BLOCKs with systems of subprograms to make internal transfers between the system of subprograms.**

By examining the subroutines of Fig. 10.3, you will see that all the data necessary for the plotting process passes through a single COMMON BLOCK named GRPHCS. This data consists of the plot character in CHAR and the character array SCREEN used to store the PLOT. Also, the BLOCK contains the plotting domain and range limits stored in XMIN, XMAX, YMIN, and YMAX, respectively. Without the use of COMMON BLOCKs, you will have to declare these six variables in the MAIN program. Also, you will have to pass these six variables in the PLOT subroutine call. By using the COMMON BLOCK, you can streamline and simplify the function calls. Variables needed for internal processing are accessible from within the system of subroutines only. This eliminates the need to declare those variables in the MAIN program.

FIGURE 10.3

Subroutines for Plotting Program

```
      SUBROUTINE WINDOW(XMIN1,XMAX1,YMIN1,YMAX1)
* ________________________________
      CHARACTER*1 SCREEN(20,80),CHAR
      COMMON/GRPHCS/CHAR,SCREEN,XMIN,XMAX,YMIN,YMAX
* ________________________________
      XMIN=XMIN1
      XMAX=XMAX1
      YMIN=YMIN1
      YMAX=YMAX1
      DO 10 I=1,20
            DO 10 J=1,80
                  SCREEN(I,J)=''
10    CONTINUE
      RETURN
      END
```

```
      SUBROUTINE PLTCHR(CHAR1)
      CHARACTER*1 CHAR1
* ________________________________
      CHARACTER*1 SCREEN(20,80),CHAR
      COMMON/GRPHCS/CHAR,SCREEN,XMIN,XMAX,YMIN,YMAX
* ________________________________
      CHAR=CHAR1
      RETURN
      END
```

```
      SUBROUTINE PLOT(X,Y)
* ________________________________
      CHARACTER*1 SCREEN(20,80),CHAR
      COMMON/GRPHCS/CHAR,SCREEN,XMIN,XMAX,YMIN,YMAX
* ________________________________
      IX=(X-XMIN)/(XMAX-XMIN)*(80-1)+1+0.05
      IY=(Y-YMIN)/(YMAX-YMIN)*(20-1)+1+0.05
      SCREEN(IY,IX)=CHAR
      RETURN
      END
```

```
      SUBROUTINE PRINT
* ________________________________
      CHARACTER*1 SCREEN(20,80),CHAR
      COMMON/GRPHCS/CHAR,SCREEN,XMIN,XMAX,YMIN,YMAX
* ________________________________
      DO 20 IY=20,1,-1
             PRINT 10,(SCREEN(IY,IX),IX=1,80)
10    FORMAT('',80A1)
20    CONTINUE
      RETURN
      END
```

When using COMMON BLOCKs, you should enter the variable names in a consistent fashion. The analogy of "COMMON BLOCK like argument list" is an accurate one. It is possible to have different names for variables in different COMMON statements with the same BLOCK name. The result is to assign values by matching order. Also, it is possible to have a different number of variables within different COMMON statements. Both "features" are more of a potential trap than a potential help. We will see in a following section that tracing programs with the COMMON BLOCK can be very involved when all COMMON BLOCKs are identical. Changing variables names and number of variables present in a COMMON statement is one sure way of making code almost untraceable:

> PROVERB 18:
> **USE UNIFORM COMMON BLOCKS!**
> **When defining a COMMON BLOCK, use identical declaration and COMMON statements (name, type, and number).**

One method of doing this is by using the "insert file" feature of your editor. Most editors will have a command to allow you to insert a file at a given location. By creating a file containing the declaration and COMMON statements, you can insert the insert file in each subprogram. This eliminates the possibility of making typing mistakes. When doing this, add blank comment lines to highlight the code defining the COMMON BLOCK. This makes it easy to see which subprograms have COMMON BLOCKs and which do not:

There is one other application where COMMON BLOCKs are essential. It involves using a particular class of subprograms from a system library. We will discuss this application in a later chapter.

Before leaving the topic of COMMON BLOCK, we should mention one restriction. That restriction is that variables defined in COMMON BLOCKs cannot be used as dummy arguments for subprograms. Note in the SUBROUTINEs WINDOW and PLTCHR that additional variables had to be used to pass the necessary information. As an example, the compiler will not allow the following:

```
SUBROUTINE PLOT(XMIN, XMAX,...)
COMMON /GRPHCS/ XMIN, XMAX ...
```

In the previous example the variables XMIN and XMAX transfer with both the COMMON statement and the I/O list in the SUBROUTINE statement. This may create an ambiguity, which the compiler will not allow. Therefore, choose one method, not both.

ROUTINES FOR TRACING SUBPROGRAMS

This section presents a system of subroutines to help you trace the execution of subprograms. By using these routines, we can print out the path, or flow of control of the various subprograms. This is particularly useful when subprograms are deeply nested. To illustrate this path idea, consider the following program:

```
      PRINT *,' THIS IS OUTPUT FROM MAIN'
      CALL A
      CALL B
      STOP
      END
C ------------------------------------------------------------
      SUBROUTINE A
      PRINT *,' THIS IS OUTPUT FROM SUBPROGRAM A'
      RETURN
      END
C ------------------------------------------------------------
      SUBROUTINE B
      PRINT *,' THIS IS OUTPUT FROM SUBPROGRAM B'
      CALL C
      RETURN
      END
C ------------------------------------------------------------
      SUBROUTINE C
      PRINT *,' THIS IS OUTPUT FROM SUBPROGRAM C'
      RETURN
      END
```

If you execute this simple program, you will get the following output:

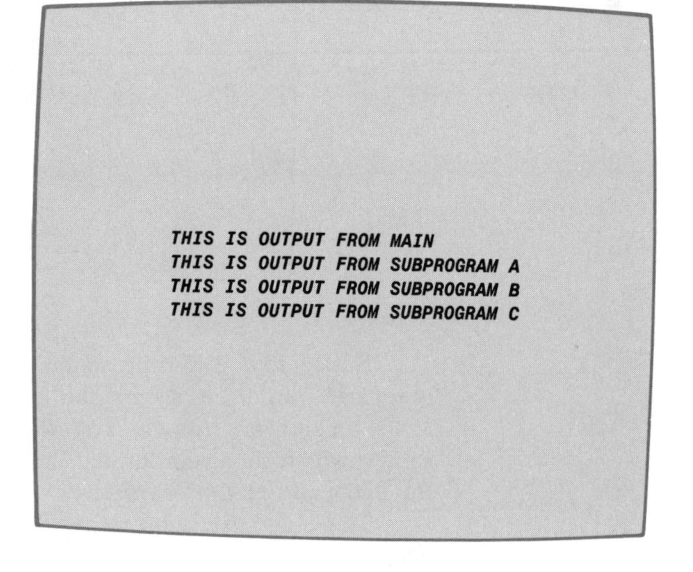

Although this output is somewhat useful, we need more interesting information like that contained in the precise path of execution:

Step 1	Step 2	Step 3	Step 4	Step 5	Step 6	Step 7
MAIN	MAIN	MAIN	MAIN			MAIN
	↓	↑	↓			↑
	A	A	B	B	B	B
				↓	↑	
				C	C	

This diagram is very useful since it shows the interrelationship between the different subroutines and would be a good starting point for finding out which subroutine is faulty. Therefore, we present here two subroutines to do this. These subroutines will print out the desired information in this format:

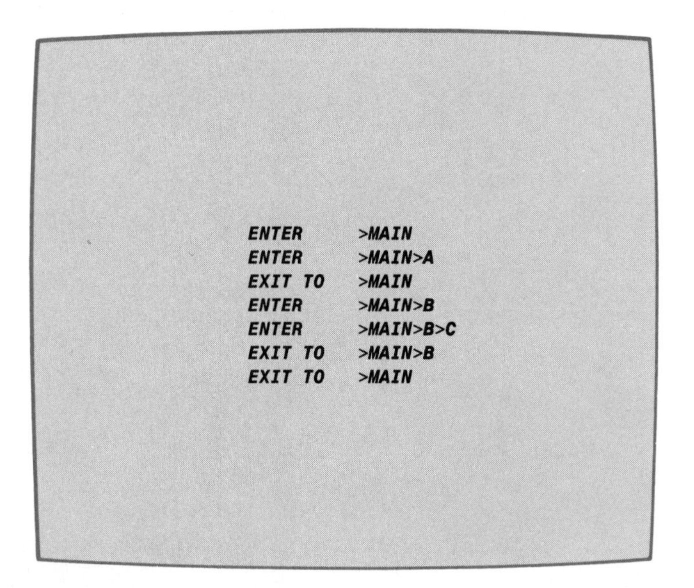

Notice that this representation contains the same information as the graphical output. It shows the transfers to the subroutine, for example, MAIN>A shows a transfer from the .MAIN. to subroutine A. Similarly, it also shows exits from a subroutine. The sole purpose of the subroutines is to print the entry and exit *path names.*

There are two subroutines that we will use. The first, INTRC indicates when control transfers to a subroutine. The second one, OUTTRC, indicates when a subroutine finishes. To streamline this system of subroutines and to minimize the amount of modification necessary to all program segments, we use a COMMON BLOCK to pass internal data.

We will not discuss here the details as to how the subroutines work, for that will obscure our purpose in introducing them. Here we will show you only how to use them:

```
      SUBROUTINE INTRC(NAME)
*     ____________________
      CHARACTER NAME*10
      CHARACTER PATH*80
      LOGICAL ONOFF
      COMMON /TRCVAR/ PATH,ONOFF
*     ____________________
      IF (.NOT.ONOFF) RETURN
      DO 10 I=80,1,-1
            IF (PATH(I:I).NE.'') GO TO 20
10    CONTINUE
20    I=I+1
      PATH(I:)='>'//NAME
      PRINT 30,PATH
30    FORMAT(' ENTERING ',A80)
      RETURN
      END
```

```
      SUBROUTINE OUTTRC
*     ____________________
      CHARACTER SPNAME*10
      CHARACTER PATH*80
      LOGICAL ONOFF
      COMMON /TRCVAR/ PATH,ONOFF
*     ____________________
      IF (.NOT.ONOFF) RETURN
      DO 10 I=80,2,-1
            IF(PATH(I:I).EQ.'>') GO TO 20
10          CONTINUE
      I=2
20    PATH=PATH(1:I-1)
      PRINT 30,PATH
30    FORMAT(' EXIT TO ',A80,/)
      RETURN
      END
```

USING THE PATH TRACING ROUTINES

Several modifications to your program are necessary before you can use the two SUBROUTINEs INTRC and OUTTRC. These include:

1. Insert the following COMMON BLOCK and assignments into the .MAIN. program and each subroutine:

```
CHARACTER NAME*10
CHARACTER PATH*80
LOGICAL ONOFF
COMMON /TRCVAR/PATH,ONOFF
```

2. In the .MAIN. program turn the trace on by inserting the line:

```
ONOFF = .TRUE.
```

   The SUBROUTINEs OUTTRC and INTRC check this logical variable. Only if it is .TRUE. will it perform the trace.
3. Initialize the character variable, PATH, to all blanks:

```
PATH = '  '
```

4. The character variable, NAME, will be the name of the routine where control resides. Since we start in the .MAIN. program, we assign 'MAIN' to this variable:

```
NAME = 'MAIN'
```

5. In each subroutine assign the appropriate names to NAME:

```
NAME = 'A'    (in subroutine A )
NAME = 'B'    (in subroutine B )
NAME = 'C'    (in subroutine C )
```

6. The last step is to call the subroutine OUTTRC from each subroutine before the RETURN statement.

Once you make these changes, several things will happen. The variable, PATH, will contain information about any change of control. Each time we transfer into or from a subroutine, PATH will change. This variable will contain the contents of the variable, NAME, which will change after every entry to a subroutine. Thus, when the program passes control to a subroutine, it prints out where it came from and where it is going. The program repeats this process until the program stops.

Let's now return to the original program and modify it according to these suggestions. We show here the original simple program and the code to add. The two subroutines INTRC and OUTTRC that you also must add are not shown.

To simplify the process of making these changes, you should create two files. One file should have the COMMON BLOCK and associated assignment statements. The second file should contain the two subprograms that need to be added. You can then use the editor to include these two files in the spots indicated earlier.

When you execute this program, the following will appear on the CRT screen:

```
Entering >MAIN
THIS IS OUTPUT FROM MAIN
Entering >MAIN>A
THIS IS OUTPUT FROM SUBPROGRAM A
Exiting to >MAIN
Entering >MAIN>B
THIS IS OUTPUT FROM SUBPROGRAM B
Entering >MAIN>B>C
THIS IS OUTPUT FROM SUBPROGRAM C
Exiting to >MAIN>B
Exiting to >MAIN
```

Notice that the computer automatically traces the program without any further help from you. This trace shows the entry to and exit from each of the modules. Thus, from the above, we see that the transfer sequence is:

$$\text{MAIN} \rightarrow \text{A} \rightarrow \text{MAIN} \rightarrow \text{B} \rightarrow \text{C} \rightarrow \text{B} \rightarrow \text{MAIN}$$

You will find that the ability to trace control transfers between different modules is an extremely important debugging tool.

```
          CHARACTER NAME*10
          CHARACTER PATH*80
          LOGICAL ONOFF
          COMMON/TRCVAR/PATH,ONOFF
          ONOFF=.TRUE.
          PATH=''
          NAME='MAIN'
          CALL INTRC(NAME)
    PRINT *,' THIS IS OUTPUT FROM MAIN'
    CALL A
    CALL B
    STOP
    END
C ----------------------------------------------------------
    SUBROUTINE A
          CHARACTER NAME*10
          CHARACTER PATH*80
          LOGICAL ONOFF
          COMMON /TRCVAR/ PATH,ONOFF
          NAME = 'A'
          CALL INTRC(NAME)
    PRINT *,'THIS IS OUTPUT FROM SUBPROGRAM A'
          CALL OUTTRC
    RETURN
    END
C ----------------------------------------------------------
    SUBROUTINE B
          CHARACTER SPNAME*10
          CHARACTER PATH*80
          LOGICAL ONOFF
          COMMON /TRCVAR/ PATH,ONOFF
          NAME = 'B'
          CALL INTRC(NAME)
    PRINT *,'THIS IS OUTPUT FROM SUBPROGRAM B'
    CALL C
          CALL OUTTRC
    RETURN
    END
```

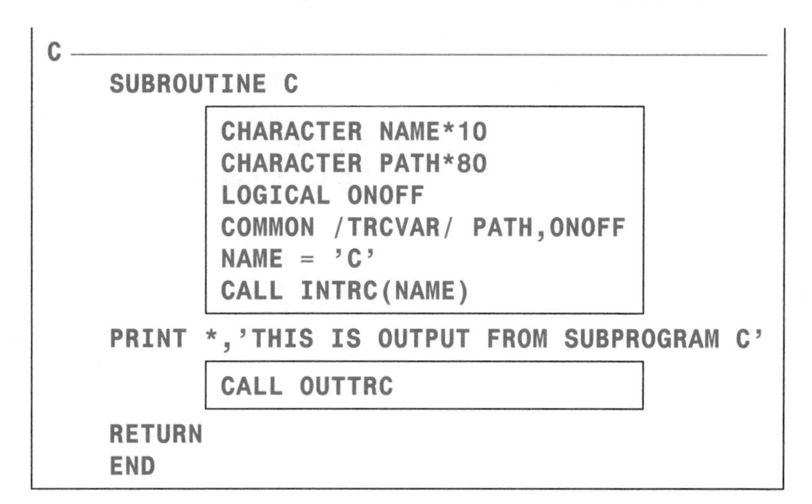
```
C ----------------------------------------------------------
      SUBROUTINE C
             CHARACTER NAME*10
             CHARACTER PATH*80
             LOGICAL ONOFF
             COMMON /TRCVAR/ PATH,ONOFF
             NAME = 'C'
             CALL INTRC(NAME)
      PRINT *,'THIS IS OUTPUT FROM SUBPROGRAM C'
             CALL OUTTRC
      RETURN
      END
```

By having the subprograms print the control path, we can put output from each subprogram into context. For example, by making multiple calls to a subprogram, we make it easier to determine which call generated the error.

The process of adding COMMON BLOCK statements to each program module may seem cumbersome or clumsy because the statements create many additional lines of code. To get around this problem, some compilers have an INCLUDE instruction that allows you to reference a file that contains information repeated throughout the program. Here is a simple example:

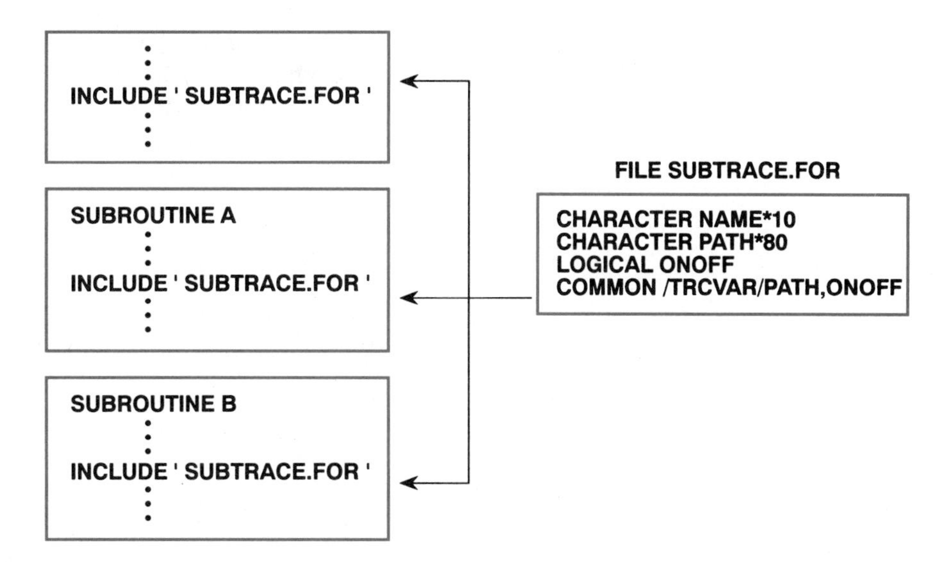

The compiler will copy the file indicated in the INCLUDE statement whenever it sees the statement. This may save you much repetitious typing. If your compiler does not support this extension, you must enter the repeated lines wherever needed. As with any nonstandard extension, these shortcuts will make your code less portable. This will prevent you from using it on a different kind of computer without modifications. The biggest problem with the INCLUDE statement is that other users will have no idea what code is in the INCLUDE file unless you are careful to document it.

10.4 Run-Time Errors in Subprograms

Subprograms are functionally no different from the simple stand-alone program segment. Therefore, you would expect that the sources of errors are the same. Accordingly, the information that we presented in Chap. 7 on locating run-time errors are equally valid for subprograms.

There is one source of run-time errors, though, which are unique to subprograms. This occurs in the transfer of data to and from the subprograms. Because subprograms must pass data with the same number, order, and type, run-time problems may arise if you violate this rule. The only way to find out what will happen if you should mistype a variable, or leave one out of an argument list, is to test your system.

> PROVERB 19:
> **CHECK YOUR ARGUMENTS!**
> **Determine how your compiler handles errors in passing arguments to subprograms.**

10.5 Tracing Programs with Subprograms

When tracing with subprograms, you need two additional steps. First, the data in the argument list must be transferred to the subprogram at the beginning of the trace. The trace then continues for the subprogram as a separate program. After completing the trace, data returns to the calling program again through the argument list.

The only additional information you need to do this is the names of the variables passing from the calling program. These variable names can be stored in the trace table. Consider the following program that uses both a FUNCTION SUBPROGRAM and a SUBROUTINE. You must execute three separate traces, one each for the .MAIN., the FUNCTION, and the SUBROUTINE:

```
*
*    PROGRAM TO DEMONSTRATE TRACING OF SUBPROGRAMS
*  ______________________________________________________________
     REAL A,B,C,D,MAXI
     READ *, A, B, C, D
     PRINT *, 'The maximum value entered was ',
 1             MAXI(A,B,C,D)
     CALL SWAP(A,B)
     CALL SWAP(C,D)
     PRINT *, 'A,B,C, and D have values ', A, B, C, D
     STOP
     END
*  ______________________________________________________________
     REAL FUNCTION MAXI(X1,X2,X3,X4)
     REAL X1,X2,X3,X4
     MAXI=X1
     IF (X2.GT.MAXI) MAXI=X2
     IF (X3.GT.MAXI) MAXI=X3
     IF (X4.GT.MAXI) MAXI=X4
     RETURN
     END
*  ______________________________________________________________
     SUBROUTINE SWAP(X,Y)
     REAL X,Y,TEMP
     TEMP=X
     X=Y
     Y=TEMP
     RETURN
     END
```

Assume that we have 2, 1, 4, 3 as the initial input. The trace table will then consist of three separate columns:

.MAIN.	FUNCTION .MAXI.	SUBROUTINE .SWAP.	
A 2, 2, 1	MAXI 2, 4	X↔A 2, 1	↔C 4, 3
B 1, 1, 2	X1↔A 2	Y↔B 1, 2	↔D 3, 4
C 4, 4, 3	X2↔B 1	TEMP 2	4
D 3, 3, 4	X3↔C 4		
MAXI 4	X4↔D 3		

We have introduced a new notation in tracing the FUNCTION and the SUBROUTINE. Since these variables are local to the subprogram, we show these variables with the corresponding variables from the calling routine. Thus, in the FUNCTION .MAXI. the computer associates the local variable, X1, with the variable A from the .MAIN.. Similarly, the program calls the SUBROUTINE .SWAP. twice. Thus, we show the local variable, X, with A for the first transfer and with C for the second transfer. Follow this example closely to make sure that you understand how we traced this simple program. To help you, we show here the trace documentation:

MAIN TRACE
READ values for A,B,C,D
Call function MAXI(A,B,C,D)

MAXI TRACE
Transfer variable names and values to dummy variables
Assign X1 to MAXI
X2>MAXI, 1>2 false
X3>MAXI, 4>2 true, MAXI=4
X4>MAXI, 3>4 false
Return MAXI to MAIN
Return argument values to MAIN
A=2,B=1,C=4,D=3

MAIN TRACE
Print MAXI value
Call subroutine SWAP(A,B)

SWAP TRACE
Transfer variable names and values to dummy variables
Assign temp value of X
Assign X value of Y
Assign Y value of TEMP
Return argument values to MAIN
A=1,B=2

MAIN TRACE
Call SWAP(C,D)

SWAP TRACE
Transfer variable names and values to dummy variables
Assign temp value of X
Assign X value of Y
Assign Y value of TEMP
Return argument values to MAIN
C=3,D=4

MAIN TRACE
Print A,B,C,D

END of TRACE

Output:

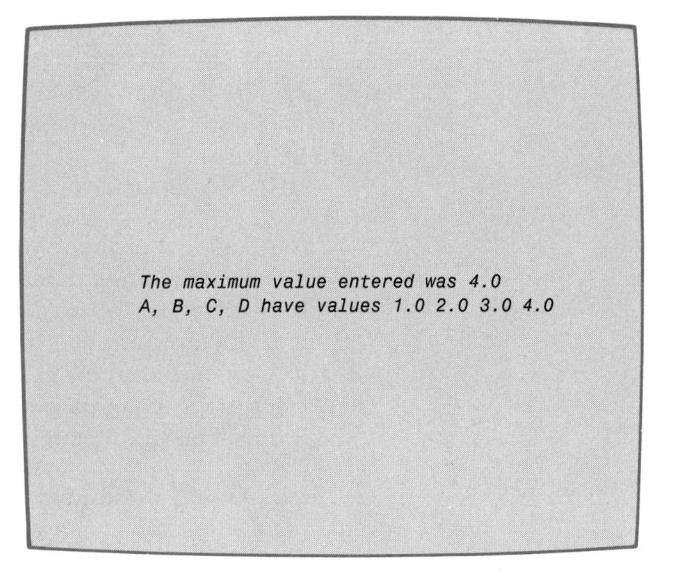

A few points from the preceding trace should be noted. First, each program module requires its own variable table. Second, when tracing a function, you must consider that the FUNCTION name is also a variable. Finally, you must be careful that when the dummy variable in the subprogram changes, its corresponding variable in the calling routine also changes.

Our advice, then, for tracing programs with subroutines is:

1. Create a variable table for each program module.
2. When tracing a FUNCTION, also include its name as a variable.
3. Show dummy arguments and the names of the corresponding calling variables together.

10.6 Summary

In this chapter we presented the topic of debugging programs with array and subprograms. In doing so, we presented two additional language primitives that you may find useful in debugging:

1. PARAMETER statement
2. LABELED COMMON BLOCKs

By using PARAMETER statements, you can adjust the size of arrays with the fewest program modifications. The PARAMETER statement provides a nonexecutable means to declare symbolic constants. These constants can be used in declaration statements.

We used COMMON BLOCKs to streamline argument lists when using a system of subroutines. Internal data between the subprograms can be passed using the COMMON BLOCK. The only arguments pertinent to the function of the subprogram pass through its argument list. This reduces the amount of information required to use the subroutines, since internal arguments do not need to be declared.

We also developed several user primitives that you might find useful:

1. Generating a printer plot that can:
   a. Specify plotting domain and range (WINDOW)
   b. Specify PLOT character (PLTCHR)
   c. Plot points at a specified X,Y (PLOT)
   d. Generate the plot at the terminal screen (PRINT)
2. Print subprogram path when entering a subprogram (INTRC)
3. Print subprogram path when exiting a subprogram (OUTTRC)

These LSs are useful in two important areas. The PLOT routines allow you to examine visually the results of your calculations for a quick check on their accuracy. Thus, if something is wrong with the computations, you will know immediately. The second set of primitives, the subroutines, OUTTRC and INTRC, will be useful for tracing complex systems of subroutines and isolating the source of any error. Once you locate the faulty subroutine, the methods discussed in Chap. 7 can be used to correct the error.

As we did in Chap. 7, we summarize here our hints on debugging in proverbs. We introduced eight additional proverbs in this chapter, bringing the total to 19. Briefly, these are:

PROVERB 12	Most programs that use large arrays are more easily debugged by using smaller arrays.
PROVERB 13	Use PARAMETER statements to define the size of arrays and other variables that change during debugging.
PROVERB 14	Multiple errors in referencing a common array or function usually indicate an error in declaration.
PROVERB 15	Determine how your system handles out-of-range subscripts in arrays.
PROVERB 16	COMMON BLOCKs reduce the versatility of subprograms. So, use them only when needed, such as when debugging.

PROVERB 17 Use COMMON BLOCKs with systems of subprograms to make internal transfers between the system of subprograms.

PROVERB 18 When defining a COMMON BLOCK, use identical declaration and COMMON statements (same name, type, and number).

PROVERB 19 Determine how your compiler handles errors in passing arguments to subprograms.

We stated at the beginning of Chap. 7 that debugging is somewhat of an art. Therefore, you will undoubtedly develop your own technique as you gain more experience. The advice that we have given you here is only to get you started. We are not smart enough to anticipate every error that students are likely to make. Thus, we make no claim to being all-inclusive. Yet, we think that you will find these 19 suggestions very useful.

Exercises

YOUR SYSTEM

10.1 Check Out Your Compiler. Write small FORTRAN program segments to see how your compiler reacts to the following common errors:

a. How does your system respond to an index out-of-range? Use the following program to determine:

```
      REAL A(10)
      DO 10 I=1,1000
            A(I)=0.0
10    CONTINUE
```

b. Modify the previous program to see what happens when the array index becomes negative.

c. How does your compiler respond to using REAL numbers for an array index? Use the following program to find out:

```
REAL A(10),I
PRINT *, 'ENTER I'
READ *, I
A(I) = 0.0
PRINT *, A(I)
```

Will the system respond differently if the REAL index is a whole number (1.0) or a fractional index (1.5)?

d. How does your compiler respond to the following subprogram errors?
   i. mismatched argument types
   ii. missing variables in the argument list
   iii. excess variables in the argument list

Use the following program to find out. (*Note:* Depending on how your machine traps the following errors, you may have to break this up into several smaller programs).

```
REAL A,B
INTEGER I,J
A=1.234
B=2.468
I=4
J=5
PRINT*,' THIS CALL SHOULD WORK CORRECTLY'
CALL TEST(A,I)
PRINT*,' BOTH ARGUMENTS ARE REAL'
CALL TEST(A,B)
PRINT*,' BOTH ARGUMENTS ARE INTEGER'
CALL TEST(I,J)
PRINT*,' ARGUMENT TYPES ARE SWITCHED'
CALL TEST(I,A)
PRINT*,' NO ARGUMENTS PASSED'
CALL TEST
PRINT*,' ONLY 1 ARGUMENT PASSED'
CALL TEST(A)
PRINT*,' PASSED TOO MANY ARGUMENTS'
CALL TEST(A,I,J)
STOP
END

SUBROUTINE TEST(A,I)
REAL A,I
PRINT*, 'REAL NUMBER WAS ', A
PRINT*, ' INTEGER NUMBER WAS ', I
RETURN
END
```

TRACING

10.2 Document a trace through the plotting routine of Fig. 10.1.

10.3 Using our method to represent higher-dimensional arrays, draw a schematic of a five-dimensional and a six-dimensional array.

10.4 Trace through the following program and predict its output:

```
INTEGER INT(5,5), NEW(5,5)
READ*, ((INT(I,J),I=1,5), J=1,5)
DO 10 I=1,5
      DO 10 J=1,5
            NEW(I,J) = INT(J,I)
```

```
10  CONTINUE
    DO 20 I=2,4
          DO 20 J=2,4
                NEW(I,J)=INT(J,I)-INT(I-1,J-1)
20  CONTINUE
    PRINT*, ((NEW(I,J),I=1,5), J=1,5
    STOP
    END
```

Input: $12, 5, 6, -1, 0, 3, 1, 0, -7, 9, 11, 10, 12, 7, 2, 3, 1, -3, 5, 6, 2, 0, -1, -4, 4$

10.5 Trace through the following program and predict its output:

```
    INTEGER A(20,20), B(20,20), C(20,20)
    READ*, M,N,((A(I,J),I=1,M), J=1,N)
    CALL DDA(A,B,C,M,N)
    CALL SREVER(A,M,N)
    PRINT*,((C(I,J),I=1,M),J=1,N)
    STOP
    END
    SUBROUTINE SREVER(B,M,N)
    INTEGER B(M,N)
    DO 10 I=1,M/2
          DO 10 J=1,N/2
                T=B(I,J)
                B(I,J)=B(M-I+1, N-J+1)
                B(M-I+1,N-J+1)=T
10  CONTINUE
    RETURN
    END
    SUBROUTINE DDA(A,B,C,M,N)
    INTEGER A(M,N),B(M,N),C(M,N)
    DO 10 I=1,M
          DO 10 J=1,N
                C(M,N)=A(M,N)+B(N,M)
10  CONTINUE
    CALL SREVER(C,M,N)
    RETURN
    END
```

Input: $4, 4, 1, 2, 3, 4, 5, 6, 7, 8, 7, 6, 5, 4, 3, 2, 1, 0$

10.6 The following program is a simple method to *smooth* data. The eight nearest neighbors of a data point are scanned. If the point differs radically from the majority of its neighbors, it is replaced by the average value. Trace the program and predict its output:

```
      INTEGER A(4,4), COUNT, SUM
      READ*,((A(I,J),I=1,4),J=1,4)
      DO 10 I=2,3
            DO 10 J=2,3
                  COUNT=0
                  SUM=0
                  DO 20 M=I-1,I+1
                        DO 20 N=J-1,J+1
                              SUM=SUM+A(M,N)
                              IF(ABS(A(I,J)-A(M,N))
     1                           .GT.4) COUNT=COUNT+1
20                CONTINUE
            IF(COUNT.GT.5) A(I,J)=(SUM-A(I,J))/8
10    CONTINUE
      PRINT*,((A(I,J),I=1,4),J=1,4)
      STOP
      END
```

Input: 1, 2, 3, 4, 6, 9, 8, 1, 1, 1, 1, 3, 4, 4, 4, 4

10.7 Produce a trace documentation for the following program. Check it by adding the subroutines INTRC and OUTTRC and running it on your computer. (Assume as input: 1, 2, 3, −1, −2, −3, 7, 2, 9).

```
      INTEGER X(3,3), TOT
      READ*, ((X(I,J),I=1,3),J=1,3)
      DO 10 I=1,3
            DO 10 J=1,3
                  IF(MOD(X(I,J),X(I+1,J+1)).NE.0)
                  THEN
                         CALL A(X,TOT)
                  ELSE
                         CALL B(X,TOT)
                  ENDIF
10    CONTINUE
      PRINT*,((X(I,J),I=1,3),J=1,3), TOT
      STOP
      END
      SUBROUTINE A(X,TOT)
      INTEGER X(3,3),TOT
      DO 10 I=1,3
            DO 10 J=1,3
                  X(I,J)=X(I,J)-FUNK(X(I,I),X(I,J))
                  TOT=TOT+X(I,J)
```

```
10  CONTINUE
    RETURN
    END
    SUBROUTINE B(X,TOT)
    INTEGER X(3,3),TOT
    DO 10 I=1,3
          DO 10 J=1,3
                X(I,J)=X(I,J)+FUNK(X(J,J),X(I,J))
10  CONTINUE
    RETURN
    END
    FUNCTION FUNK(I,J)
    FUNK=I*MOD(I,J)
    RETURN
    END
```

DEBUGGING

10.8 Summing Columns in a Table. The following program is supposed to add up all elements in each column of a table. Find the errors and eliminate them:

```
   REAL X(100,100)
   READ *, M,N,((X(I,J),I=1,M),J=1,N)
   DO 10 I=1,M
         CALL SUM (X,I,M,N,TOT)
         PRINT*, 'FOR COLUMN = ', I, 'SUM = ',TOT
10 CONTINUE
   STOP
   END

   SUBROUTINE SUM(Z,I,M,N,TOT)
   REAL Z(M,N)
   DO 10 I=1,M
         TOT=TOT+Z(I,I)
10 CONTINUE
   RETURN
   END
```

PLOTTING PROGRAMS

10.9 Half a Plot. Modify the plotting subroutines given here to use PARAMETER statements for the declaration of the SCREEN array. Make sure that portions of the code that depend on the upper limits of the array subscripts use the new parameters. Test your program by plotting on only half the screen.

10.10 Hard Plot. Write a second plotting subroutine, PRINT2, which generates an output file of your graph. The output file should be called PLOT.OUT.

10.11 Trap Points. Modify the PLOT subroutine to *trap* points that are out of subscript range for the array SCREEN. Do not report any error messages; simply do not plot those points. Use your modified program to plot the tangent function from 0 to 2π. Be careful to skip those points where the tangent function is not defined. For example, TAN(90°) = ∞ and TAN(180°) = $-\infty$. Also, as you approach these points, called *singularities,* the TAN values may go off scale. So, be careful!

10.12 Grid Lines. For the plotting subroutines, write an additional subroutine named GRID to initialize the SCREEN array by adding grid lines. (*Note:* This new subroutine must be called *after* you call the WINDOW SUBROUTINE.)

10.13 Combination Graph. Modify the PRINT SUBROUTINE to print the Y values of the plot in the last 10 columns of the graph, and X values at four locations determined by the user. You must modify other subroutines as well to do this task. Use the G format for printing the numerical results.

10.14 Dual Graphs. Modify the MAIN plotting program to plot a sine and cosine curve on the same plot. Plot the sine curve using the "S" symbol and the cosine curve using the "C" symbol. Plot with a sufficiently small step to ensure that no horizontal gaps are present between points.

10.15 Parametric Equations. The following *parametric equations* (*X* and *Y* are a function of another variable) generate a curve known as a *bowditch* curve. Plot this curve with the subroutines given in the text.

$$X(t) = \cos(t) \qquad \text{range: } 0 \le t \le 6\pi$$
$$Y(t) = \sin(2t/3)$$

CHAPTER

11

THE NEW FORTRAN STANDARD

11.1 FORTRAN—A Dynamic Language

In the early days of FORTRAN, language revisions and improvements appeared almost yearly. The original version, FORTRAN I, first appeared in 1957, but it lasted only 1 year. FORTRAN II was born in 1958 and lasted about 4 years, until FORTRAN IV came along in 1962. There was a FORTRAN III, but it never gained wide acceptance and was quickly displaced by version IV. By this time, though, there were so many different versions of the language in use that a great deal of confusion ensued. Therefore, the American Standards Association (now the American National Standards Institute or ANSI) set up a task force to design a standard which all "FORTRANs" should follow. The result was FORTRAN '66. It was very similar to the then widely used FORTRAN IV, and programmers often confuse the two versions.

During the late 1960s and early 1970s major changes were taking place in the development of structured programming languages. Many new languages appeared incorporating these ideas and great progress was made in the area of programming efficiency. As a result, ANSI once again revised FORTRAN in 1978. We now refer to the new standard that resulted as ANSI FORTRAN 77. This has been the language that we have presented so far and is the one that almost all commercial suppliers of FORTRAN compilers use.

The science of programming has matured significantly since 1978, and it is about time to revise the standard again. Since FORTRAN is the oldest of the languages, it possesses many obsolete features. In addition, it lacks some advanced niceties of other modern languages. The new standard attempts to address both concerns with their revision. No features of FORTRAN 77 will be eliminated, but the committee is discouraging their use by placing them on a "deprecated" list for possible future elimination. Programmers are being warned that these obsolete language elements may be eliminated in the next standard. More importantly, the new standard introduces an impressive list of new features. Among these are:

- Streamlining of array processing
- Additional data types, including user-defined data
- Recursive subprogram calls
- Additional control structures
- Many new BUILT-IN FUNCTIONS

We will not attempt to give you an exhaustive introduction. A long discussion would obscure the important features and you would not get much out of this chapter. Also, some of the new language elements do not belong in a course at this level. You will learn them as you gain more experience. We prefer, instead, to select those features of the new language that we think will be of interest to you, and therefore we present an overview, with only a short and very selective list of topics.

At the time of this writing there are still no compilers that follow this standard. Thus, you cannot use any of the new features yet. But when the new compilers arrive, the amount of documentation that comes with them may be bewildering. Our purpose here is to condense some of this information and to contrast the important elements of the new language with the old. When the new compilers arrive, this chapter will be a good place to start your background reading.

11.2 Overview

We present the changes and additions to FORTRAN in the same order that we discussed them in the previous chapters. Where possible, we will try to compare and contrast the old versus the new. Briefly, here is the order of presentation:

- Program format
- Variables
    - Allowed names
    - Precision and magnitude
    - User-defined data types
- Control structures
- Array manipulation
- Subprograms
    - Recursive calls
    - New intrinsic FUNCTIONS

The discussion of each listing will be brief. Since compilers using the standard do not yet exist, there is no sense in trying to give many examples.

Also, the examples that we present may not be the exact form that will be used in new compilers. So, be careful when you try to implement these ideas.

11.3 Program Format

At the beginning of Chap. 4 we introduced the standard form of the FORTRAN program line. It contains 80 columns divided into three sections. Columns 1 to 6 contain statement labels and continuation indications; columns 7 to 72 contain the program itself; and the last eight columns (73 to 80) are for comments that the compiler ignores.

The new standard does not bind you to this rigid structure. You may begin a FORTRAN line anywhere you wish. Since the sixth column is no longer reserved for the continuation indicator, we need a way to show that a line is continued on the next line. This is done with the ampersand, &, as follows:

```
X = 1.23456 * X * Y + Z * Y      &
    -2.56784 * V + SIN(X)        &
    *COS(2*X)
```

The ampersand at the end of the line will reduce the common error in FORTRAN 77 of putting the continuation mark in the wrong column.

Another effect of eliminating special columns for special functions is that the method of putting in *comments* must change also. With FORTRAN 77, we do this with a C or * in column 1. Instead, we now do this with the exclamation point anywhere in the line:

```
!    THIS IS THE USUAL AREA FOR COMMENTS IN A PROGRAM
! -----------------------------------------------------
            REAL X(100), Y(100)       ! ANOTHER COMMENT
```

The final feature that will be of interest is that you can put more than one statement on the same line, if you separate them with a semicolon:

FORTRAN 77	New Standard
X = 12.4 Y = 24.5 Z = SQRT(X*Y)	X=12.4; Y=24.5; Z=SQRT(X*Y)

While this structure may appear to be very useful, don't abuse it. Too many statements on a line make it difficult to follow the logic. In cases where you are initializing many variables this structure makes sense.

11.4 Variables

Some rules for naming variables have been relaxed. Specifically, when giving a variable a name, you may now:

- Use up to 31 characters including A to Z, 0 to 9, and the underscore (_).
- Define any precision or magnitude you wish.
- Create your own data type.

As an example of the first point, the following are valid variables:

```
DERIVATIVE_OF_THE_FUNCTION
GROSS
DERIVATIVE_SQUARED_2
```

FORTRAN 77 limits a variable name to six characters and no underscore. The underscore is useful for simulating spaces in names to make them more English-like. We warn you not to abuse these relaxed rules on variable names. Repeating long variable names throughout a program increases the amount of typing and greatly increases your chances of making an error. You probably should use this only for special occasions.

The ability to specify the accuracy and magnitude of your data is an important feature of the new language. Recall that REAL numbers in FORTRAN 77 had only 6 to 7 digits of accuracy and DOUBLE PRECISION numbers had 13 to 14 digits. You can still use these, but you can also specify accuracy at any level that you want. Here's how:

```
REAL (PRECISION = num1 , EXPONENT_RANGE = num2 )
      var1, var2, ...
```

where the statement **PRECISION = *num1*** specifies the desired precision and the statement **EXPONENT_RANGE = *num2*** specifies the magnitude of the number. For example:

```
REAL (PRECISION = 20, EXPONENT_RANGE = 120) X
```

states that the variable X must have 20 digits of accuracy and a range of 10^{-120} to 10^{+120}. The computer will use this information to determine automatically how much memory to use. If you want the conventional accuracy for X, drop the statements inside the parentheses.

The third new feature of variables is the ability to create your own data type. With the current standard, there are six intrinsic data types—REAL, INTEGER, DOUBLE PRECISION, COMPLEX, CHARACTER, and LOGICALs. The new standard still has these, but also allows DERIVED data types such as matrices, lists, or mixed information types. Suppose you were running an experiment and you wanted a data type that contained sample identification, sample weight, and chemical formulas. With the DERIVED data type, you can create a SAMPLE data type, or anything else that you choose to call it:

```
TYPE SAMPLE
      CHARACTER*25 IDENTIFICATION, FORMULA
      REAL (PRECISION = 10) WEIGHT
END TYPE SAMPLE
```

This example combines two conventional data types, REAL and CHARACTER, to create the new data type, SAMPLE. You can manipulate this new data type as you would any conventional type, or you can even create your own operators. The general structure of the DERIVED data types is:

```
TYPE name
                .
                .
                .
      desired structure
                .
                .
                .
END TYPE name
```

where *name* is any unique name that you supply to identify how to set up the desired type. Inside the body of the TYPE-ENDTYPE block are the statements that specify how to define the TYPE.

Besides this definition, you must also declare any variables to use the newly defined structure. For example, if we have samples X1, X2, and X3 that use this new structure, your program may look like this:

```
TYPE SAMPLE
      CHARACTER*25 IDENTIFICATION, FORMULA
      REAL (PRECISION = 10) WEIGHT
END TYPE SAMPLE
REAL A, B, C
INTEGER ICOUNT, ISUM
TYPE(SAMPLE) X1, X2, X3
```

The DERIVED data type goes before declaration statements. Once you have defined SAMPLES, you may then declare X1, X2, and X3 as having the attributes associated with the SAMPLE data types. You put this declaration with all other declaration statements as shown in the preceding example.

11.5 Control Structures

There are several important changes and additions in the control structures. These include, among others:

- Simpler logical comparisons
- NAMED CONTROL structures
- SELECT-CASE structure
- Simplified form of the DO LOOP

The change in logical comparisons is very simple and quite logical. Now instead of using mnemonics such as .EQ., .NE., .GT., .GE., .LT. or .LE., you may substitute conventional mathematical symbols:

FORTRAN 77	New Standard
.EQ.	=
.NE.	<>
.GT.	>
.GE.	>=
.LT.	<
.LE.	<=

You may use either type of logical operator. For example, the following are equivalent:

```
IF (X .NE. 0.0) PRINT *, 'X IS NONZERO'
                         or
IF (X < > 0.0) PRINT*, 'X IS NONZERO'
```

The basic control structures in FORTRAN 77 are simple to use, but it is sometimes difficult to follow the logic when you nest these structures. As an example, if you nest two BLOCK-IFs within each other, it is sometimes difficult to see where one block ends and the other begins:

```
IF (       ) THEN
    IF(           ) THEN
          .
          .
          .
    ELSE
          .
          .
          .
    ENDIF
```

```
ELSE
     IF(          ) THEN
           .
           .
           .
     ELSE
           .
           .
           .
     ENDIF
ENDIF
```

Of course, the computer has no problem following the logic, but we humans are not so fortunate. The new standard has a device that may make these kinds of problems obsolete. This is the NAMED CONTROL structure. You have the option of adding names to the BLOCK structures:

```
[name:]   IF(        ) THEN
                .
                .
                .
          ELSE  [name]
                .
                .
                .
          ENDIF [name]
```

What we have done is to add the *name* at any of the three key points in the block. This will help you identify to which block a given ELSE or ENDIF statement belongs:

```
JACK:      IF (      )THEN
JILL:                IF (      ) THEN
                           .
                           .
                           .
                     ELSE  JILL
                           .
                           .
                           .
                     ENDIF JILL
           ELSE JACK
```

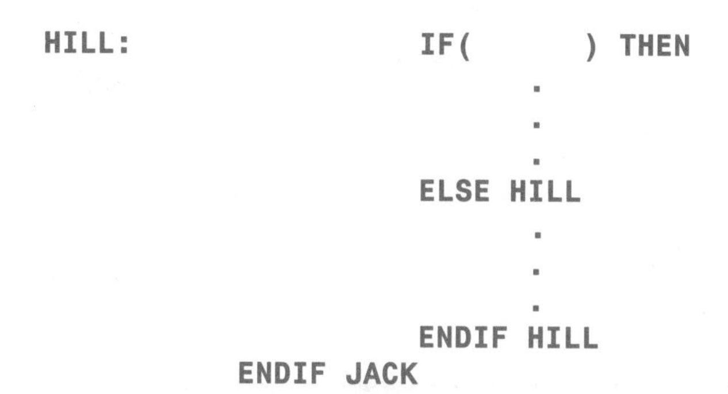

```
HILL:                    IF(      ) THEN
                              .
                              .
                              .
                         ELSE HILL
                              .
                              .
                              .
                         ENDIF HILL
               ENDIF JACK
```

The BLOCK-IF structure operates the same way that it did in FORTRAN 77. The only difference is the ability to add a name to improve readability. The *name* is optional at any of the three points in the structure. So, you might use it only with the IF statement, with the ELSE statement, or with the END. But it is likely that if you use the name at all, you will want to use it at all three locations. You may use this type of device also with DO LOOPs. Here, though, it is less useful since DO LOOPs pose far fewer tracing problems than BLOCK-IFs.

In the previous chapter on the "Common Threads of Programming," we described a control structure, the SELECT CASE, which FORTRAN 77 does not support. Fortunately, the new standard will add this very useful structure and combine it with the NAMED CONTROL feature described earlier. The SELECT CASE is like the IF-THEN-ELSE structure in the sense that it performs a test before a branch occurs. With the SELECT-CASE structure, though, there can be more than two alternatives. The BLOCK-IF is limited to only two possible outcomes. If you want to choose among three outcomes, you must use two BLOCK-IFs. In general, if you have N outcomes, you would need (N − 1) BLOCK-IFs. The SELECT CASE is a much more efficient way to do the same thing.

Schematically, the SELECT CASE behaves as shown in Fig. 11.1.

In this structure only a single expression is evaluated. The value of this expression (numerical or character data) determines the block of instructions for execution. Think of this as a switch where the switch can have many different positions, and once you set the switch, control will go in that direction. The syntax of the SELECT CASE is shown on the following page.

FIGURE 11.1

The Select Case Structure

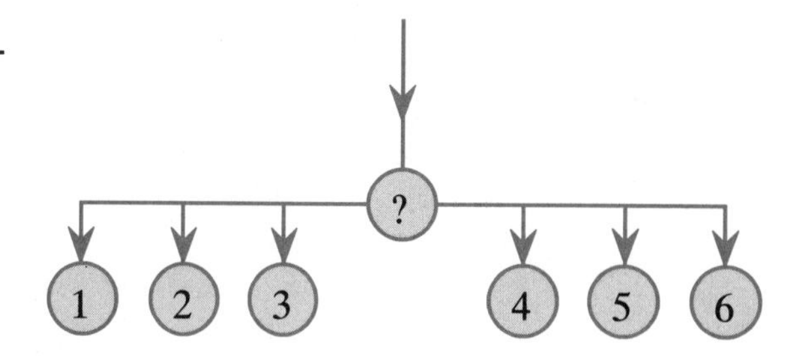

```
[name:]     SELECT CASE (expression )
            CASE (value1 ) [ name ]
                 instruction 1
                 instruction 2
            CASE (value2 ) [ name ]
                 instruction 1
                 instruction 2
                 .
                 .
                 .
            END SELECT [ name ]
```

If the expression evaluates to *value1,* the intructions under CASE(*value1*) execute. Note that the expression must evaluate to one of the values in the CASE statements. This may be a serious problem since you may not consider all possible values of the expression. One way to handle this is with the CASE DEFAULT statement that takes care of any CASE not specifically listed:

```
READ *, NUM
SELECT CASE (NUM)
CASE (-1)
      PRINT *, 'NUMBER = -1'
CASE ( 1)
     PRINT *, 'NUMBER = +1'
CASE DEFAULT
     PRINT *, 'NEITHER +1 NOR -1 ENTERED'
END SELECT
```

In this example you will type in an integer for the variable, NUM. If NUM = −1, the CASE corresponding to (−1) executes and the computer prints the statement "NUMBER = −1." Similarly, if NUM = +1, the second CASE is selected and the message "NUMBER = +1" appears. If you enter any other number, the CASE DEFAULT takes over and the message "NEITHER +1 NOR −1 ENTERED" appears.

There are three new forms of the DO LOOP that appear in the new standard. The first eliminates the need for a statement label to show the end of the loop:

FORTRAN 77	New Standard
DO *sl* LCV = *start, stop [,step]*	DO LCV = *start, stop [,step]*
⋮	⋮
sl CONTINUE	END DO

This form of the DO LOOP is already available on several FORTRAN 77 compilers and eliminates the need for the statement label and CONTINUE statement marking the end of the loop. One advantage to this form of the DO LOOP is that you can eliminate the possibility of accidentally using the same statement label more than once.

As with the other new forms of control structures, you may name the LOOP if you wish:

```
[ name: ]        DO  LCV = start , stop [, step ]
                          .
                          .
                          .
                 END DO [ name ]
```

The name is optional in either location, as indicated by the brackets. Otherwise, the LOOP executes exactly like the FORTRAN 77 DO LOOP.

Another form of the DO LOOP that you might find useful is the DO TIMES LOOP with the following syntax:

```
DO (       ) TIMES
      .
      .
      .
END DO
```

As you might guess, this loop executes the number of times indicated inside the parentheses, which may contain an INTEGER constant, variable, or expression. In this form of the DO LOOP there is no LCV that you can use inside the LOOP. Therefore, you may need a combination of both kinds of loops. Use the DO ( ) TIMES LOOP when you don't need the LCV and the DO LCV=... structure when you do need the LCV.

11.6 Simplified Array Manipulations

The new FORTRAN makes giant strides in our ability to manipulate arrays. Of greatest interest is the extremely simple way to do some mathematics on arrays. For example, if we wanted to add the arrays, A and B, together element by element, FORTRAN 77 would need something like this:

```
      REAL A(100,100), B(100,100), C(100,100)
            .
            .
            .
      DO 10 I=1,100
            DO 10 J=1,100
                  C(I,J) = A(I,J) + B(I,J)
10    CONTINUE
```

The new FORTRAN will do the same thing with:

```
REAL A(100,100), B(100,100), C(100,100)
      .
      .
      .
C = A + B
```

The statement C = A + B implies that C(I, J) = A(I, J) + B(I, J) for all values of I and J. All the DO LOOPs needed in FORTRAN 77 are no longer necessary. You can perform any other type of calculations using the same technique. To set all values to an initial value, say, 2.0, all we must do is:

```
B = 2.0          Old Way:    DO 10 J=1,100
                                    DO 10 I=1,100
                                     B(I,J) = 2.0
                          10    CONTINUE
```

Here's how you increment each element of the array X by 1.0:

```
X = X + 1.0      Old Way:  DO 20 I=1,100
                            DO 20 J=1,100
                             X(I,J) = X(I,J) + 1.0
                        20 CONTINUE
```

We also can assign values to just a few elements of an array by using a *subarray*. A subarray is a small section of a larger array. For a large one-dimensional array, A, the subarray is indicated by A(lower:upper), where lower or upper equals the element where the subarray begins or ends, respectively. For example, if A has 100 elements, then A(3:13) represents a smaller array starting from element 3 through element 13. We can use this in assignment statement as follows:

```
A(3:13) = 1.0      Assigns 1.0 to elements 3 to 13
A(14:100) = 0.0    Assigns 0.0 to elements 14 to 100
```

In a similar way, you can have arrays on both sides of the assignment statements:

```
REAL A(100,100), B(100)
      .
      .
      .
A(24 , 10:20) = B( 90:100 )
```

and can assign the B(90), B(91), ..., B(100) into A(24, 10), A(24, 11), ..., A(24, 20), respectively. These examples show how powerful this simple structure is and the dramatic reduction in program complexity that is possible.

Another useful structure that can be used with arrays is the WHERE statement with the syntax:

```
WHERE ( logical condition )
                .
                .
                .
     array assignment statements
                .
                .
                .
END WHERE
```

The WHERE statement is like an IF statement. The difference is that the IF statement will execute only once but the WHERE structure will execute the instructions for every element of the array. For example, if you wanted to assign only nonzero terms from the array X into array Y, you would use this structure:

New Way:
```
      REAL X(100,100), Y(100,100)
      WHERE ( X < > 0.0)
           Y = X
      END WHERE
```

Old Way:
```
      REAL X(100,100), Y(100,100)
      DO 10 I=1,100
            DO 10 J=1,100
     1              IF(X(I,J) . NE. 0.0)
     1                Y(I,J) = X(I,J)
   10 CONTINUE
```

Array processing with the new standard will be greatly simplified as these examples show. In addition, there are still several things not shown here that you will find useful. There are now more concise ways to declare groups of arrays, and also ways to create arrays whose size is determined only after completing some calculations. Recall that FORTRAN 77 requires that you fix the size of an array before execution. You may find these features to be useful in more complex programs than we develop here.

11.7 Subprograms

There are two main items of interest in the new standard related to subprograms. The first item is the RECURSIVE SUBPROGRAM. Recall that *recursion* occurs when a subprogram calls itself either directly or indirectly through other subprograms. We sometimes call this indirect method daisy-chaining. Up to now, FORTRAN has forbidden RECURSIVE CALLS. But in the new standard recursion will be allowed if you define the subprogram as follows:

```
RECURSIVE FUNCTION name ( argument list )
                    .
                    .
                    .

                    .
                    .
                    .
END
```

In this form the FUNCTION can CALL itself, but only through an intermediate FUNCTION or SUBROUTINE. If the FUNCTION calls itself directly, we must provide an additional clue to the FUNCTION statement that will allow the data to be returned:

```
RECURSIVE FUNCTION name ( list ) RESULT ( list )
                    .
                    .
                    .

                    .
                    .
                    .
END
```

This form contains the new statement, RESULT (*list*). The variables inside this argument list are the ones that contain the result of the recursive calculation. Usually, it is the FUNCTION name that returns the answer, but in a RECURSIVE call this method will not work. That is why we need the RESULT statement.

An example of a RECURSIVE CALL is the search process when we want to know the position of an item in a list. We start the process by sending the array, the value for matching, and the initial guess for the position, usually 1:

```
              .
              .
              .
K = SEARCH (A, 1, 6.5)
              .
              .
              .
```

```
RECURSIVE FUNCTION SEARCH (A, I, VALUE) RESULT (J)
          .
          .
          .
IF (A(I) = VALUE) THEN
        J = I
ELSE
        I = I + 1
        J = SEARCH (A, I, VALUE)
ENDIF
END
```

The FUNCTION checks to see if A(I) is the value sought. If it is not, then I increments and the FUNCTION calls itself, but now it looks at the I + 1 element of A. When the FUNCTION finds the value, it assigns the value of the position to the variable J, which is in the argument list for the RETURN statement.

Many problems in programming are best solved by recursion. Of course, if not properly done, recursion can generate problems of its own. But if you are careful, you will find that RECURSIVE subprograms will be a very useful addition to the FORTRAN language.

Our final topic from the new version of FORTRAN is the addition of many new INTRINSIC FUNCTIONs. Some of these already exist on different compilers, but they are not universal. Different compilers may implement the FUNCTIONs in different ways, making it difficult to transport programs from one machine to another. The standard will require that these FUNCTIONs follow the form set in the standard.

The new FUNCTIONs that may be of interest to you are:

DOTPRODUCT (A, B)	Takes the dot product of two vectors A and B
MATMUL (A, B)	Multiplies matrix A by matrix B. Able to multiply two-dimensional and two-dimensional; two-dimensional and one-dimensional; or one-dimensional and two-dimensional matrices and vectors
MAXVAL (A)	Returns maximum value of array A
MINVAL (A)	Returns minimum value of array A
PRODUCT (A)	Determines product of all elements of matrix A
SUM (A)	Determines sum of all elements of array A
TRANSPOSE (A)	Determines transpose of matrix A

There are many other FUNCTIONs that you may want to use, but these are the ones that you will probably use more frequently. Also, these FUNCTIONs are similar to those we have developed here as algorithm primitives. Therefore, they should be familiar to you.

11.8 Final Comments

Before leaving this section, we want to emphasize that many changes are coming to FORTRAN that will greatly streamline and enhance the language. This brief survey of the new standard cannot begin to discuss all these changes. We suggest that you consult the book cited here for more detailed information. This chapter should give you an idea of what to expect when the new compilers arrive. Hopefully, we have also aroused an interest that will lead you to see whether your compiler already supports these features. You may be surprised to find that much of the information presented in this chapter is already available.

Further Reading

1. M. Metcalf and J. Reid, *FORTRAN 8X Explained,* 1st ed., Clarendon Press, Oxford, 1987.

CHAPTER 12

WHY NUMERICAL METHODS?

12.1 Overview

This section of the textbook deals with using the computer to solve engineering problems. From that simple statement come two major questions that should help you focus on the purpose of the next few chapters: (1) What are engineering problems? (2) Why use a computer?

Engineers are problem solvers and the types of problems that they deal with generally fall into two categories. Those that you can solve analytically and those that you cannot solve analytically. Analytical solutions are those for which we can write a mathematical expression. Of course, we need to know the underlying physical principles and must have the necessary mathematical tools to solve these types of problems. For example, the forces in the members of a statically determinant truss can be calculated analytically. Based on the geometry and loading on the truss, we can generate a system of equations that come from the knowledge that all forces and moments sum to zero. Solving that system of equations results in determining the unknown forces in each truss member.

There are many problems in engineering and science for which we are not so fortunate to have precise analytical solutions. As an example, consider the problem of computing the area under the curve generated by the equation $\sin(X)/X$ for X ranging from 1 to 10. This problem has no known analytical solution, but a value does exist. As you can see from Fig. 12.1, the area under the curve is finite and you might be able to approximate it by many different means. However, you cannot write down a mathematical formula that you can evaluate.

Analytical methods are often very difficult to obtain, and so we must often resort to numerical approximations. We use numerical methods to "get answers" when analytical solutions are not possible by using what is usually a calculation-intensive process. For example, we find the area under a curve by

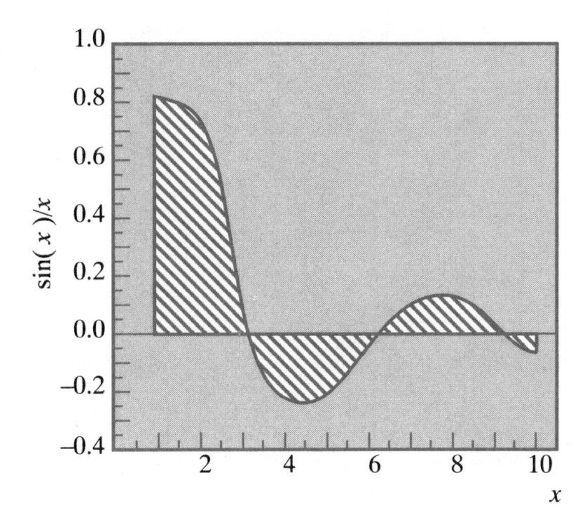

Figure 12.1

Area under $\sin(x)/x$ Curve with x Ranging between 1 and 10

dividing the domain into an infinite number of sections, calculating the area of each section, and then summing them together. Of course, breaking the curve into an infinite number of sections is impossible, and so we may have to make some compromises to get an answer. A fundamental issue in numerical analysis is that if we must make compromises, we do not sacrifice accuracy. To show this, consider the following program to calculate the area under the curve shown in Fig. 12.1:

```
C    PROGRAM TO APPROXIMATE THE AREA UNDER THE CURVE
C    GIVEN BY Y=SIN(X)/X FOR X BETWEEN 1.0 AND 10.0.
C    THE APPROXIMATION IS BASED ON THE IDEA OF
C    BREAKING THE CURVE INTO A FINITE NUMBER OF
C    RECTANGLES, DETERMINING THE AREA OF EACH
C    RECTANGLE AND ADDING THEIR AREAS.
C ----------------------------------------------------
        AREA=0.0
        XMIN=1.0
        XMAX=10.0
        N=9
        DELTAX=(XMAX-XMIN)/N
        DO 10 X=XMIN,XMAX-DELTAX,DELTAX
              Y=SIN(X)/X
              AREA=AREA+Y*DELTAX
    10  CONTINUE
        PRINT*, 'AREA = ', AREA
        STOP
        END
```

We base this program on the idea illustrated in Fig. 12.2, where we break up the curve into small rectangles. The base of each rectangle is the same length,

FIGURE 12.2

Approximating the Area under the Curve $\sin(x)/(x)$ for x Ranging from 1 to 10

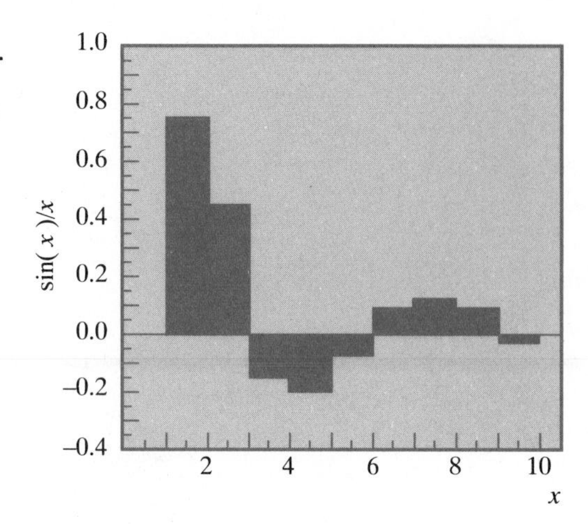

DELTAX = 1.0, if we divide the curve into nine rectangles. The height of each rectangle is the value of $\sin(X)/X$ at that point on the curve. To approximate the area under the full curve, we add up the areas of the rectangles.

We expect that the accuracy will not be very good if we use only ten rectangles. We also expect the accuracy to improve as we make more rectangles. To see whether this is true, we ran the preceding program with finer and finer divisions and obtained the area estimate each time. Here are the results:

N	Estimated Area
9	1.12481
49	1.56159
89	1.60514
129	1.62166
169	1.63035
209	1.63571
249	1.64140
289	1.64198

We have also plotted the results for you in Fig. 12.3. As you can see from these tabulated results and from the graph in Figure 12.3, the area under the curve is approximately 1.644. Increasing the number of subdivisions past 249 did not change the approximated answers very much.

After finishing a calculation of the sort shown previously, a legitimate question is—If there is no known analytical solution to the problem just solved, how do we know the answer is correct? We can only answer this question indirectly by showing that our process works for a function where we *do* know the precise answer. We can then infer that since our program works for

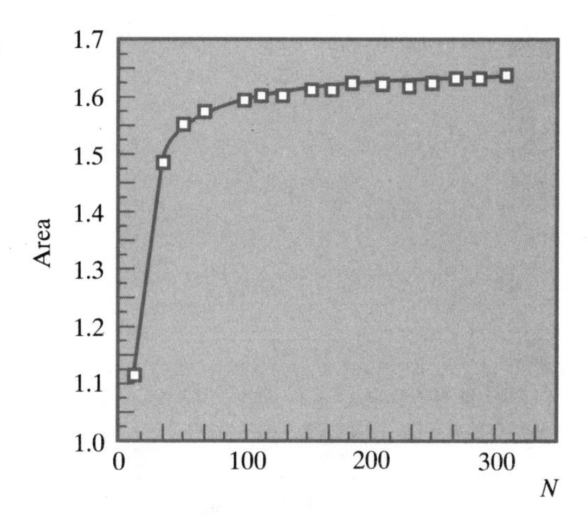

Figure 12.3

Estimated Area under $\sin(x)/x$ Ranging from 1 to 10 with Different Number of Rectangular Areas

the known case, then it probably works for the unknown case. We call this process *validation*.

The process of validating a program usually involves solving a problem with a known analytical solution. To verify that our previous program is working correctly, we replace $\sin(X)/X$ with the function, Y = X, over the range 0 to 10. We know that the area under this curve is exactly 50.0. Now, when we run our program with this function, the results are encouraging:

N	Estimated Area
9	44.4444
49	48.9796
89	49.4382
129	49.6124
169	49.7042
209	49.7607

These results show that the method we are using works correctly. The area that we calculated is within 0.5 percent of the true value. If we had carried these calculations out to an even higher number of rectangles, our answer would improve even further. But there is a limit. As we will see in the next sections, small errors are inherent in computer calculations. As we increase the number of calculations, these errors will grow and destroy any benefit gained from a larger number of rectangles. Thus, there is an optimum region for calculating the best estimate. We will discuss this in greater detail later.

Numerical methods are typically calculation-intensive. The process of obtaining the results themselves usually requires a great deal of calculation. Furthermore, validating the program also requires much calculation. Because of the massive number of calculations involved, and the extra work to validate a numerical method, you have no choice but to implement them on a computer.

The example just provided was typical. As we present various numerical methods, we will discuss topics such as accuracy, validation, and numerical parameters.

12.2 Numerical and Analytical Solutions

Earlier we stated that because numerical methods are calculation-intensive, we must perform them on the computer. Using the computer for less intensive analytical solutions also has advantages that we should describe.

An engineer may need to use a set of calculations to do his or her job. For example, an engineer that designs pressure vessels (tanks) would most likely use a set of calculations present in the ASME Pressure Vessel Code. A code is a list of specifications showing how something should be designed, inspected, and tested. Such calculations are usually not calculation-intensive. However, putting them on the computer has many advantages.

First, by automating the calculation process, you can make changes in design or analysis more quickly. Automation also makes it possible to look at many designs to determine which is *optimal*. Second, by putting such calculations on the computer, we can ensure their accuracy. Whenever you try to do these calculations manually, there is a chance of making a mistake every time you make the calculation. By creating a program, you only enter the equations once. During the validation process you have an opportunity to find any errors and correct them. After you finish with this stage, you can be confident in any future calculations. Because of these advantages, we often automate common calculations. For established design codes, commercial vendors can supply prepackaged programs.

12.3 General Overview of Numerical Methods

In the previous section we illustrated the use and need for numerical methods. All too often practical engineering problems cannot be solved by conventional analytical methods. As an example, we showed you how to calculate the area under a curve. When solving this type of problem, you need to consider many things. Among these are:

Accuracy
Efficiency
Stability
Programming simplicity and versatility
Computer storage requirements
Previous experience

Accuracy deals with how closely the numerical result matches the true value. Different methods may have different degrees of accuracy because of the number of calculations and the method of solving the problem. As we stated earlier, every calculation on a computer has an error associated with it because of the way that computers store numbers. Also, mathematical functions that run on the computer have a small inaccuracy built into them. Thus, intensive numerical calculations can have appreciable errors because of these *truncation errors*. We will discuss this source of error in more detail in the next chapter.

Efficiency deals with the number of steps or calculations needed to arrive at a solution. If a method provides accurate results but requires a million calculations to do so, this may not be an acceptable method. In the example that we presented in the previous section the method that we used was inefficient. While many better techniques are available to perform this type of calculation, the method that we used was at least understandable. Thus, sometimes we may wish to trade off accuracy for the sake of efficiency.

Stability is an important issue in numerical methods and tells us whether there is a possibility that the method may not return an answer at all. This is especially a concern with techniques that use successive approximations based on previously calculated answers. Depending on the characteristics of the problem, it is possible that the method diverges instead of converges. Converging on an answer means that each successive iteration brings you closer to the answer. Diverging means that at some point something has happened to make the solution process move away from the solution. Typically, this condition will result in a run-time error. An application in which this consideration comes into play is finding the root (zero) of an equation.

Programming *simplicity* and *versatility* are concerned with how easily we can program and modify the program. A method might provide very accurate results in an efficient manner. But, if the programming language you are using does not provide an easy way to implement it, your method most likely will not be used.

Computer storage requirements address the question of how much memory the computer needs to implement a procedure. If you are using a personal computer with limited memory, memory-intensive methods may not be able to run on your machine. If you are using a mini- or mainframe computer with virtually unlimited memory available to you, then this concern will not be important. Sometimes even these large machines are insufficient, and only the latest supercomputer will be able to handle your program. The point here is that at times you may have an elegant, efficient, and accurate program to solve a problem. But, if it cannot fit into your computer and return an answer in a reasonable amount of time, then you must find another approach to a solution.

Finally, after having used several methods, you may find that you have certain favorite methods that work for your type of problems. You will learn that experience with various techniques is invaluable when it comes to choosing the correct one.

12.4 Program Validation

The process of ensuring that your program is working correctly is known as *program validation.* Every program should undergo a validation process. The most desirable method of validation is to solve a problem with a known analytical solution, and then compare the numerical results to the analytical one. Even this approach may have its limitations, though.

Consider writing a program that simulates the flight of a solid fuel rocket. As the rocket engine burns, the rocket loses mass. Also, the thrust of the engine is a function of time. At lift-off the engine may provide maximum thrust, then level off to some predetermined value for a length of time, after which the engine burns out. How would we validate such a program? One step in the validation process might be to test the program with zero engine thrust, no air resistance, but with an initial launch velocity. This would then be simple motion of a projectile, which has a known analytical solution. If the program worked well for this, you would then have greater confidence in your flight simulation algorithm. Still you would not know whether the engine thrust simulation, or whether the air resistance calculations are correct. To verify these portions of the algorithm, you would need to test independently these sections of the program. For example, we can test just the portion of code that is responsible for calculating the thrust versus time function and compare it with laboratory measurements. Finally, knowing the terminal velocity of the rocket (from experimental data) might help to validate the air resistance calculations. It should be clear that validation can be an intensive operation requiring an intensive investment of effort.

When performing a validation, you should look for key parameters that change the directions of a calculation. For example, consider the solution to the quadratic equation:

$$aX^2 + bX + c = 0$$

This equation has three possible sets of roots, which depend on the value of the discriminant:

$(b^2 - 4ac) > 0$	two real roots
$(b^2 - 4ac) = 0$	one real root
$(b^2 - 4ac) < 0$	two imaginary roots

Once you write a program to solve this equation, you can then use the preceding information to validate your program. For example, you can enter data that will give each of these three results. If your program does not give the correct answer for each, then your program is faulty. We previously described the idea of a *magic bullet,* which we use to check out the different branches of your algorithm. You must do these types of validation procedures since even the most conscientious programmer might overlook some special cases. These missed special cases will eventually surface with repeated use of the program and might escape detection.

It is for such reasons that you should create a subroutine library of your user LSs. By generating a subroutine code and validating it, you can build programs on a sound foundation. By not "reinventing the wheel," you can make your life easier and fine tune your programs. If you find an error in one of your routines, fixing the routine will ensure that future programs do not have the same problem. Recompiling old programs that have used that routine will implement the new fix. This process also reduces validation time because you do not have to validate the subroutines that you are using.

12.5 Documentation

It is essential that you provide documentation within your code. The documentation within the program has no effect on the execution of your program or the accuracy of your answer. The primary benefits of documentation are the improved ease of debugging and improved readability. Well-thought out documentation will provide feedback while you are constructing the program and when you are trying to debug it. For example, you can use it to describe the function of a small segment of the program. Without these comments, it is unlikely that you will remember what that section of the program is doing.

You can document your programs with the comment statement. The kinds of information that you should include in your documentation in the program itself are:

- Name of the program or subroutine
- Date of creation
- Description of the code's purpose. This should include what information to enter and what answers the program returns
- Description of how you validated the code
- Date of the last modification, with a description of the modification
- A list describing all variables used in the program
- Comments describing major sections of the program code. Some of these may be embedded in the program itself
- Use of the indentation method to highlight structured commands

By including such documentation, you will have better success when modifying existing programs. Without such documentation, the process of fixing a bug in the code you have written in the past becomes a chore. The first step is usually trying to figure out how you were attacking the problem on the day when you originally wrote the program. This involves remembering or deducing what each variable was used for, and what techniques were being implemented. By using comment statements to record such information, you shorten the reacclimation process.

When using comments to delineate main sections of the program, indicate the beginning and end of the section. One method for doing this is to use the following format:

```
C
C   Description indicating END of preceding section
C   ==============================================
C   Description indicating BEGINNING of next section
C
```

When you wish to indicate the purpose of single statements, consider this suggested format:

```
C
C   ... Description of following line ...
C
```

Both of these comment statement formats have the quality that they *do not* look like FORTRAN statements. This makes them easy to identify with only a glance. Some compilers have extensions that allow you to place comment lines directly on the command line. For example, some compilers will recognize a special symbol to indicate a comment on the same line as an instruction:

```
RADCL=B**2-4*A*C /* Define term under the radical
```

The following program to solve the quadratic formula is a good example of careful documentation. Study it carefully to see how we provide all the necessary information:

```
C  QUAD.FOR  BY M. Cwiakala  Date: February 4, 1990
C
C  PURPOSE:  This program reads in the coefficients
C  for a quadratic equation (AX**2+BX+C) from the
C  keyboard, and returns to the terminal screen
C  the corresponding roots. Cases handled by this
C  program include real, repeated, and complex roots.
C  In the event that a linear equation is entered
C  (A=0), a single root is returned.
C
C  VALIDATION: This code was validated by entering
C  coefficients of equations with known roots. In
C  particular (X-2)(X+3)=X*X+X-6 having roots 2
C  & -3, (X-2)(X-2) = X*X-4X+4 having repeated
C  root 2, X*X+4 having complex roots 2i and -2i,
C  and finally the linear case, 3X-4 having the
C  single root 4./3.
C
```

```
C  REVISION HISTORY: Date ______  Description ______
C
C  VARIABLES: A    =COEF OF X SQUARED TERM
C             B    =COEF OF X TERM
C             C    =CONSTANT TERM
C             X1   =FIRST ROOT
C             X2   =SECOND ROOT
C             RADCL=RADICAL TERM OF QUADRATIC FORMULA
C             RL   =REAL PART OF A COMPLEX ROOT
C             IM   =IMAGINARY PART OF A COMPLEX ROOT
C ========================================
C Declare variables and read in coef.
C ----------------------------------------------------
  REAL A,B,C,X1,X2,RADCL,RL,IM
  PRINT *,' For the equation A*X**2+B*X+C=0'
  PRINT *,' Enter values for A,B, and C'
  READ *,A,B,C
C
C Variables declared and coef read in
C ========================================
C Define Radical term and select appropriate
C calculations. Print results.
C
  RADCL=B**2-4*A*C
  IF (A.EQ.0.0) THEN
               X1=-C/B
               PRINT *,'Equation enter was linear'
               PRINT *,'The root is ',X1
        ELSE IF (RADCL.LT.0.0) THEN
               RL=-B/(2*A)
               IM=SQRT(-RADCL)
               PRINT *,'Solution is complex'
               PRINT *,RL,U + IU,IM
               PRINT *,RL,U - IU,IM
        ELSE IF (RADCL.EQ.0.0) THEN
               X1=-B/(2*A)
               PRINT *,'Repeated roots are present'
               PRINT *,'Root is ',X1
        ELSE IF (RADCL.GT.0.0) THEN
               X1=(-B+SQRT(RADCL))/(2*A)
               X2=(-B-SQRT(RADCL))/(2*A)
               PRINT *,'Two real roots are present'
               PRINT *,'X1 = ',X1,' and X2 =',X2
        ENDIF
   STOP
```

Most of this documentation is straightforward. The only portion of it that may provide difficulty is the list of variables used. If your compiler has the list-file options, then that problem is most likely solved. Most compiler listings generate a list of variables used. That list can be used to create your index of variables. Even better, if you have an advanced editor that allows you to edit two files simultaneously, you can copy the list of variables from the listing file to your program directly.

12.6 Where Do We Go from Here?

The remaining chapters of this text will deal predominantly with specific numerical methods. The topics to be covered include:

- Simple data analysis: Topics to be covered include statistical analysis on data and methods of smoothing data with noise
- Curve fitting: How to fit straight lines, exponential functions, and polynomial functions to data
- Roots of equations: Finding the solution to a single equation with a single unknown
- Systems of equations: Finding the solutions to a system of equations
- Derivatives and integrals: Presenting various methods to perform numerically differentiation and integration.

EXERCISES

All the problems in this section (and in all remaining chapters) require that you provide complete documentation (which includes validations).

12.1 Using the LS for MAX/MIN search in Chap. 9, create two functions that return the maximum value stored in an array. One function will be for real arrays, the second for integer arrays. Arguments to be passed must include the array name and size. How will you perform the validation process for these functions?

12.2 Using the LS for MAX/MIN search in Chap. 9, create two functions that return the minimum value stored in an array. One function will be for real arrays, the second for integer arrays. Arguments to be passed must include the array name and size. How will you perform the validation process for these functions?

12.3 Using the LS for MAX/MIN sort in Chap. 9, create three subroutines that sort values stored in a one-dimensional array. One subroutine will be for real arrays, a second for integer arrays, and a third for string arrays. Arguments to be passed must include the array name and size.

The argument used to pass the array name will be used also to return the result. How will you perform the validation process for these subroutines?

12.4 Using the LS for summing elements in an array in Chap. 9, create two functions that return the sum of elements stored in an array. One function will be for real arrays, the second for integer arrays. Arguments to be passed must include the array name and size. How will you perform the validation process for these functions?

12.5 Using the LS for *vector addition* in Chap. 9, create a subroutine that returns the resultant vector. What type and number of arguments do you need for this subroutine? Why can't a function be used for this LS? How will you perform the validation process for this subroutine?

12.6 Using the LS for *vector dot-product* in Chap. 9, create a function that returns the scaler product of two vectors. How will you perform the validation process for this function?

12.7 Using the LS for MAX/MIN to search a table in Chap. 9, create two functions that return the maximum value stored in a two-dimensional array. One function will be for REAL arrays, the second for INTEGER arrays. Arguments to be passed must include the array name and size. How will you perform the validation process for these functions?

12.8 Using the LS for MAX/MIN to search a table in Chap. 9, create two functions that return the minimum value stored in a two-dimensional array. One function will be for real arrays, the second for integer arrays. Arguments to be passed must include the array name and size. How will you perform the validation process for these functions?

12.9 Using the LS for *matrix addition* in Chap. 9, create a subroutine that returns the resultant matrix. What type and number of arguments are needed for this subroutine? Why can't a function be used for this LS? How will you perform the validation process for this subroutine?

12.10 Using the LS for *matrix multiplication* in Chap. 9, create a subroutine that returns the resultant vector. What type and number of arguments are needed for this subroutine? Why can't a function be used for this LS? How will you perform the validation process for this subroutine?

CODE CALCULATION PROBLEMS

Whenever you use code in a design, calculation sheets must be provided showing the code section, year, and addenda (an addenda is a yearly revision to the code.) The calculation sheet also gives a description of the calculations, the equations used, values of design variables, and the computed value. The calculation sheet should be understandable by all and able to be checked manually. Thus, you need to show the equations being used. The reason for the involved output is that in code work a number of individuals may inspect the work. One key individual, known as the authorized inspector (typically someone from an insurance company), checks calculations and other documentation. Before a job can go into production, certain documentation must be approved by the authorized inspector. Whether calculations are performed on a computer or by hand, the necessary documentation must be provided to the inspector. With this background, consider the following problems:

12.11 The strength of a piece of metal typically depends on two parameters: the type of metal used and the temperature range at which it operates. The ASME Pressure Vessel Code, Section VIII, Division 1 has a table listing the allowable stress values (a measure of strength) as a function of temperature for many materials. Write a program to request a material specification and lower and upper limits for the temperature range of operation. At the end values of the temperature range the program should report the allowable stress. Finally, indicate the minimum value that should be used in all design calculations.

a. Read material specification (e.g., SA-240-316, a common type of stainless steel).
b. Open the data file (e.g., SA-240-316.DAT). If the program cannot open it, report an error and end the program (see ERR option for an open statement). Format of a data file should look like this:

```
'ASME CODE 1986, Addenda 1989, figure ...'      <-Source of data (string)
'SA-240-316 (PLATE)'                            <-Type of material (string)
14                                              <-Number of data points
-20  100  200  300  400  500 ...                <-Temperature values
18.8 18.8 16.2 14.6 13.4 12.5 ...               <-Allowable stress
```

c. Read in lower and upper temperature limits.
d. For the lower temperature limit, look up allowable stress. If the temperature value is between values specified, perform linear interpolation.
e. For the upper temperature limit, look up allowable stress. If the temperature value is between values specified, perform linear interpolation.
f. Determine minimum allowable stress from values calculated in steps 4 and 5.
h. Output must include: the source of data (string value in data file), material specification (second string value), the lower limit and upper limit of the temperature range, allowable stress values at each temperature value, interpolation calculations (if necessary), and finally, the allowable stress value to be used for following design calculations.

   How will you validate this program? (*Hint*: Consider making a data file named TEST.DAT that has allowable stress values equal to the temperature.)

12.12 ASME Pressure Vessel Code calculations for shell thickness (wall thickness in a straight portion of the pressure vessel) based on internal pressure and internal radius is performed using the following equations:

From ASME Code, Section VIII, Division 1, Section UG-27(c)1. Courtesy of ASME.

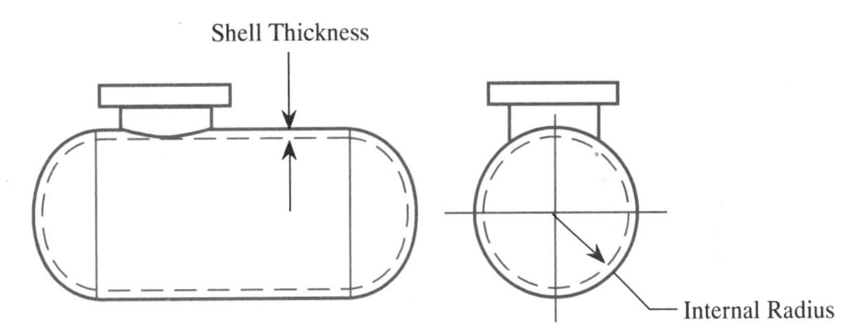

$$t = PR/(SE - 0.6P)$$

where t = required thickness in inches

P = internal pressure in psi

R = inside radius of shell in inches

S = allowable stress in psi

E = joint efficiency factor (For welded seams: Value ranges from 0.4 to 1.0.)

Write a program to read in the specified parameters and generate a calculation sheet.

12.13 ASME Pressure Vessel Code calculations for shell thickness (wall thickness in a straight portion of the pressure vessel) based on internal pressure and external radius is performed using the following equations:

From ASME Code, Section VIII, Division 1, Appendix 1.1, (a)(1). Courtesy of ASME.

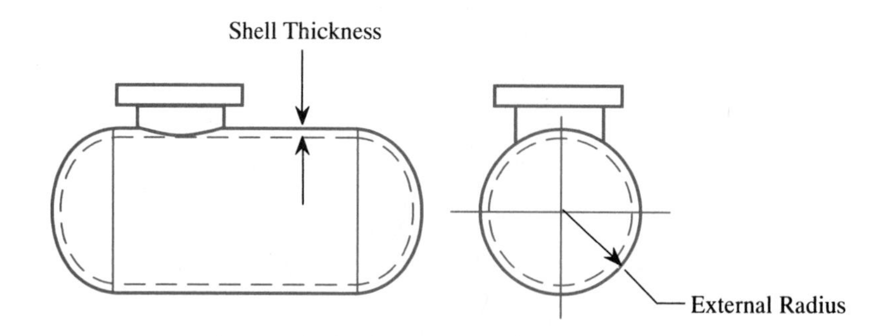

$$t = PR/(SE + 0.4P)$$

where t = required thickness in inches
P = internal pressure in psi
R = outside radius of shell in inches
S = allowable stress in psi
E = joint efficiency factor (For welded seams: Value ranges from 0.4 to 1.0.)

Write a program to read in the specified parameters and generate a calculation sheet.

CHAPTER

13

ERROR ANALYSIS AND DATA REDUCTION

13.1 Data—An Imperfect Medium

As much as engineers hate to admit it, all data has errors associated with it. We are not referring to gross errors such as obviously incorrect calculations, but to small errors that occur in every experiment, no matter how well the experiment was thought out. For example, if you ask several of your friends to weigh a sample carefully in a chemistry lab, you probably will get many different values. Some of your friends may not be very careful; others may be careful, but the instrument they are using is inaccurate. Still others may be careful and use an accurate instrument, but they could still introduce a personal bias such as always reading the scale to the next highest digit.

As a simple experiment, ask several of your friends to read a thermometer like the one in Fig. 13.1. Is the temperature 17 or 18°F? Or, is it something in between? The correct way to read the thermometer is to use the top of the mercury meniscus, but some people will use the lowest point. Also, we should ask about the accuracy of the thermometer. Can the standard thermometer give an accurate reading at this temperature? There are many potential sources of error in any data reading operation and you must understand their origin before you can begin to correct them.

Errors occur in many different ways, some of which we will discuss in this chapter. Among the types of error to look out for are:

Experimental and systematic errors
Computational errors
Random errors

FIGURE 13.1

Errors in Reading a Thermometer

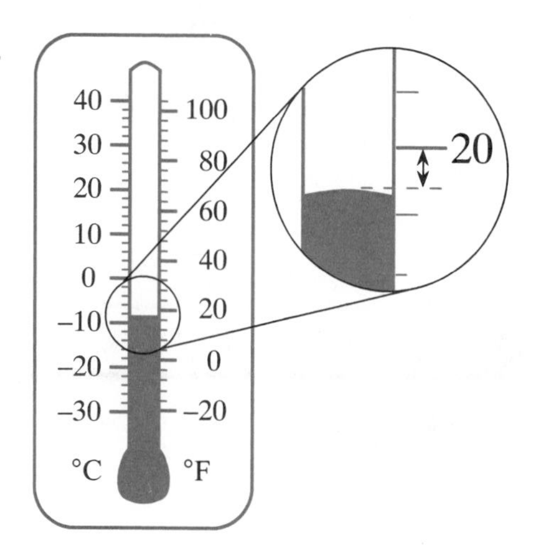

Some of these errors, especially the first two, can be minimized by careful experimental technique, but the third type will always occur. Therefore, we must resort to mathematical methods to correct for these errors.

To illustrate the difference between systematic and random errors, we present in Fig. 13.2 and 13.3 a series of figures. Assume that we have weighed the same sample many times. It is often convenient to plot the data as a histogram in which the horizontal axis represents different measured values, such as weight. The vertical axis then represents the number of times that each weight occurred.

The average value of the weight is what we really want as shown in the figure. Note that it would be a mistake to use any single value for the weight since the reproducibility of the data is not too good. While the average value is our ultimate goal, we should also be concerned with the spread of the data for this will affect our confidence in the results. Also, note that this histogram tells us nothing about how precise the instrument itself is.

When systematic errors are present, the measured value will differ from the true value. Random errors influence the spread of the data, or the ability to reproduce the measured value as shown in Fig. 13.3.

Notice that the systematic errors result in an average value that is significantly different from the "true" value. For an unknown sample, it is impos-

FIGURE 13.2

Distribution of Values from Repeated Weighings

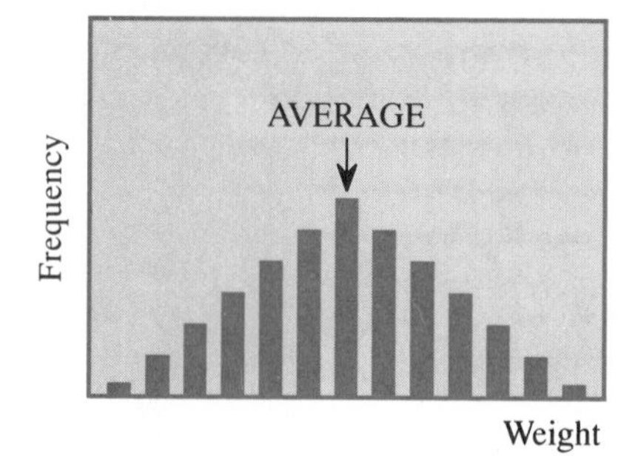

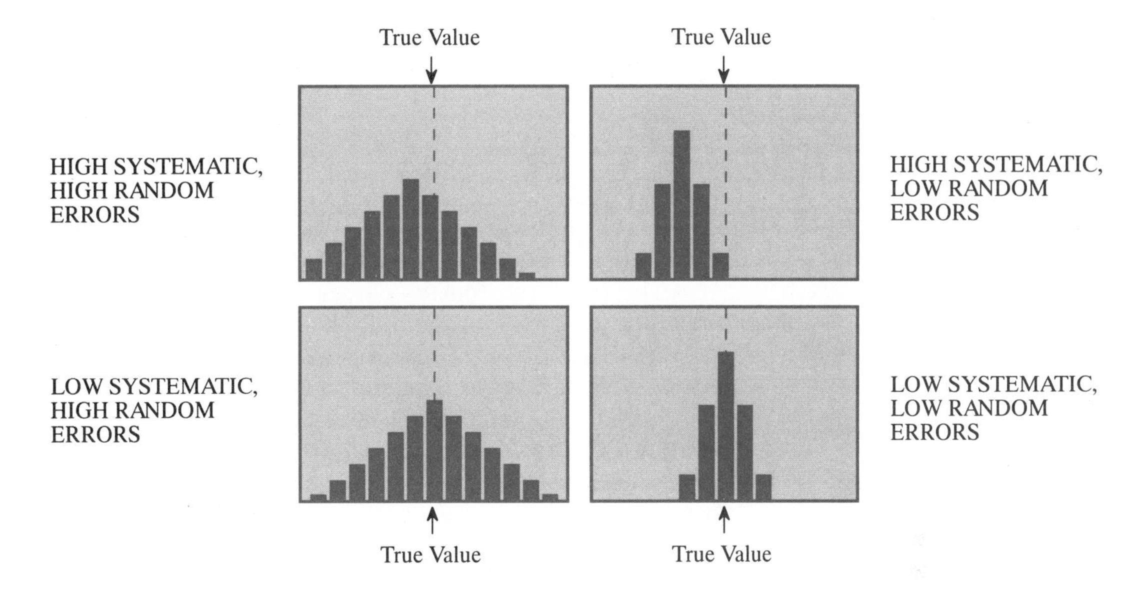

FIGURE 13.3

Illustration of Differences between Systematic and Random Errors

sible to measure the magnitude of the systematic error by just examining the histogram. You can usually do this, though, with a *calibration standard* whose value you know accurately.

The random errors, on the other hand, affect the spread of the data around the average. If the random errors are large, you must make many measurements before you have a good average value. Obviously, the best situation is when both the random and the systematic errors are low.

Because all experimental data has errors associated with it, we must always concern ourselves with the quality of our data and the associated degree of uncertainty. For example, if we can determine that the errors are small, then we feel confident with the data. But, if we determine the errors to be high, then our confidence drops. Thus, the determination of the magnitude of the errors plays a major role in experimental science.

In this chapter we describe the different types of errors and methods for determining the degree of uncertainty. We will also show you a few useful methods for smoothing data to reduce random errors.

13.2 Experimental and Systematic Errors

The first type of error that we introduce is the experimental error. This can arise from one of several sources:

- Carelessness in collecting data
- Errors in calculations

- Parallax errors in reading an analog scale
- Bias in reading a scale

The first two sources are obvious, as are their remedies. If any data point is obviously incorrect, you should repeat the measurement since human error is the most likely problem. Do not just throw out data that appears to be incorrect because this will bias unjustly the results. Repeat the measurements at the suspicious points and report *all* the data.

The third and fourth sources of error are more difficult to deal with when you are working with analog measuring devices. Analog devices are those that provide a *continuous* range of possible values. A mercury thermometer is such an instrument since the mercury expands in a continuous way and does not jump between discrete values. A digital device, on the other hand, does not move in continuous steps. Instead, it jumps between values. Look at your watch and you will see that the hand moves only every second, for example. With such a device, you cannot interpolate between seconds because of the discrete jumps. When you measure the time with a conventional watch, the time is either 4:30:54 or 4:30:55. It cannot be 4:30:54.327467, for example.

PARALLAX

There are two problems with analog devices with which you must be concerned. The first is the problem of *parallax*. Parallax is a phenomenon that you have undoubtedly experienced in which the value measured depends on the angle from which you view the scale. To illustrate this, consider a simple voltmeter, Fig. 13.4, that has a needle approximately 2 mm away from the scale.

Notice that if you view the scale from an angle perpendicular to the scale, you will get the correct reading. But, if you view it from a slightly differ-

FIGURE 13.4

Illustration of Errors Due to Parallax

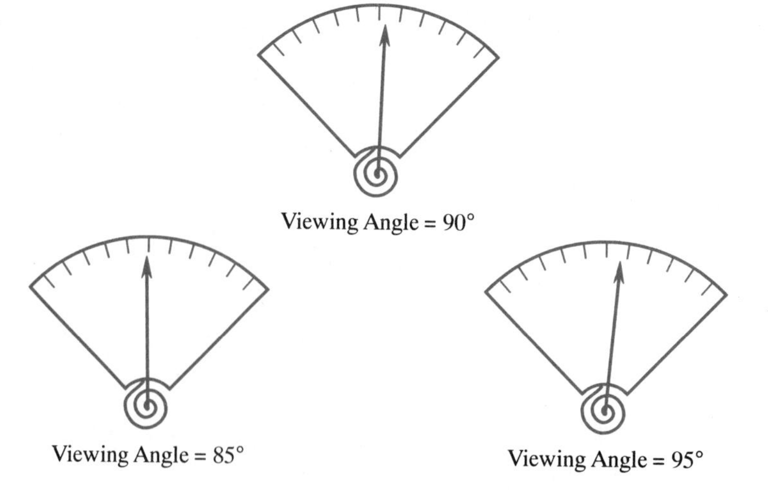

ent angle, the needle will appear to move. This is the error due to parallax. The same kinds of errors occur when you use any analog measuring instrument such as a ruler, a scale, or a meter. One way to minimize this error is to try to view the device always at 90° or place the scale as close as possible to the object. With a meter, of course, this latter option is not possible since the needle is a fixed distance from the scale. The only other option is to use a digital device such as a digital voltmeter, a digital scale, and so forth. These will give you much better accuracy.

BIAS

The second source of experimental error is a bias. An example is someone who always rounds data to the next lower scale, even though it is very close to the higher scale. In the example of the thermometer of Fig. 13.1 someone may read the temperature as 16°F. Sometimes bias may also come in when you "know" what the answer is supposed to be. As a result, you may inadvertently read the data as higher or lower to match your expectations. The cure for bias errors is the same as that for carelessness. You should attempt to take your data carefully, and report every data point, even if it falls outside the expected range.

SYSTEMATIC ERRORS

Systematic errors are the next problem that we consider. Systematic errors can be due to at least four major sources. These are:

- Improperly zeroed equipment
- Improperly calibrated instrument
- Nonlinear equipment
- Background noise

Note that three of these causes are related to the equipment itself. Thus, you can minimize the problem to some extent by using carefully calibrated and high accuracy equipment. For example, to measure a sample whose thickness is 0.356 mm, you would not use a meter stick that typically has a spacing of 1.0 mm between divisions. It is obvious that you need something with higher resolution than a simple ruler. A better choice would be a micrometer that typically has a resolution of 0.0025 mm.

Most electronic and mechanical measuring devices require a zeroing operation before you use it. The purpose of the zeroing is to make sure that the device reports a value of zero when it is not measuring anything. For example, a voltmeter should read 0 V when nothing is connected to it, or a scale should read 0 g when it is not loaded. It is especially important to check the zero after changing ranges. Thus, your meter may be zeroed at the 10 V range, but when you change to the 10 mV range, the zero is most likely incorrect. Even mechanical systems, such as balances and micrometers, need to be zeroed.

Improper calibration of equipment also leads to faulty data. Calibration consists of measuring a "known" value, sometimes called a *standard,* with the equipment. If the measured value is incorrect, the equipment may be adjusted until it is correct. An important point concerning calibration is that you should be careful to test the calibration at several different points over a wide range.

The best example of improperly calibrated equipment is a thermometer that measures 0°C in ice water but 100.5°C in boiling water, as shown in Fig. 13.5. The thermometer is correct at the ice point, but 0.5°C too high at the higher temperature. Since mercury thermometers have no means of adjusting the calibration, we must make corrections to our data.

Sometimes you may use equipment that gives a nonlinear response. A nonlinear system is one where the response is not proportional to the signal to be measured. The simplest nonlinear measuring system is the spring scale shown in Fig. 13.6. When the load is small, the scale responds linearly. For example, when we double a small load, the scale response doubles. But at some point a nonlinear response begins. A double load may then result in a fourfold increase in the response. At higher loads the response may be even more dramatic.

Generally, you should avoid highly nonlinear systems if possible. But, if you have no choice, you should carefully calibrate the device and use the resulting curve for correcting the data. This may often require some curve-fitting techniques, which we will describe in later chapters.

The final source of error in this category is background noise. Background error can come from either the equipment itself or the sample or system. You cannot eliminate these easily. Fortunately, it is usually an easy matter to subtract out the background as Fig. 13.7 shows. The background in this ex-

FIGURE 13.5

Illustration of Errors Due to Faulty Calibration

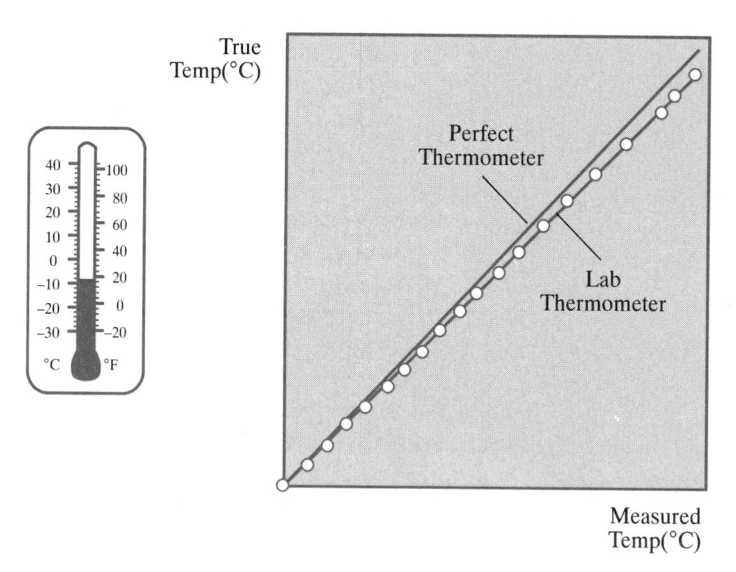

Figure 13.6

Example of Nonlinear Measuring System

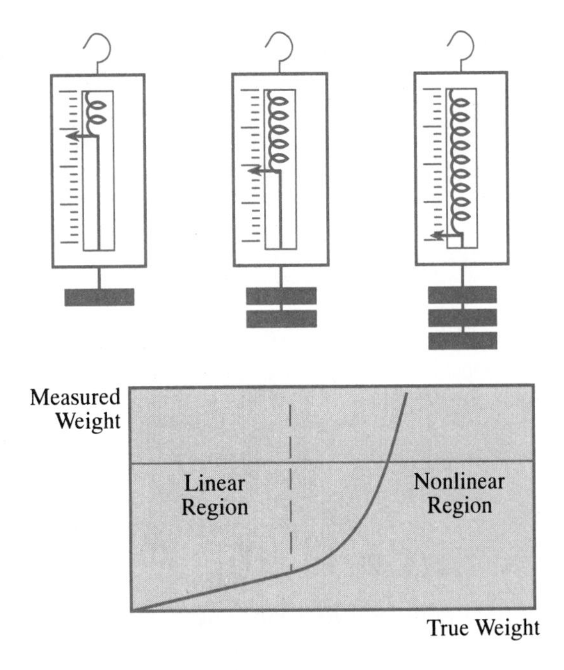

Figure 13.7

Background Subtraction

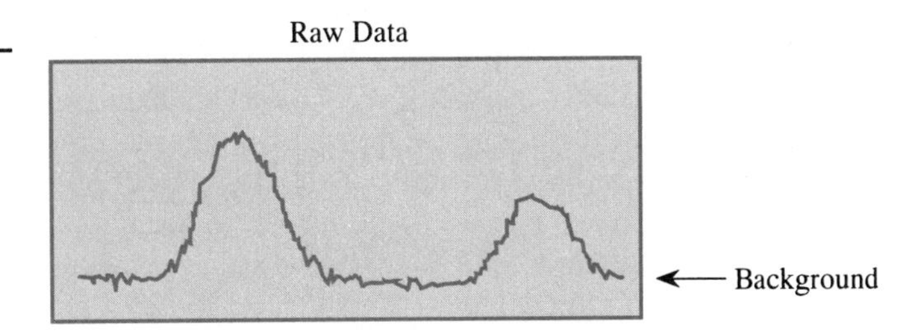

Background Removed

ample is almost constant. So, we can correct the data by subtracting the background from each data point:

$$Y_i = Y_i - \text{background}$$

Sometimes the background level can be determined from an independent measurement with a dummy sample. A dummy sample is one that gives no response of its own. An example might be a straight wire in an electrical circuit that shorts out the elements to be tested. Other times you can determine the background from the data itself. Note in Fig. 13.7 that if you go far away from the peaks, the only signal is the background. This is probably the most common technique for measuring the background.

13.3 Computational Errors

Computational errors are due to the manner in which we treat numbers and computations based on them. Some of the most common sources of errors in this category are:

- Limited accuracy of the raw data
- Round-off errors
- Accumulation errors during multiple calculations
- Truncation errors in series approximations

All data collected in a laboratory has limited accuracy that strongly depends on the quantity measured. For example, we can easily measure time to an accuracy of 1 μs, and with special equipment, we can push this to an accuracy of 1 ps (10^{-12} s). But quantities like temperature are difficult to measure to an accuracy of better than 0.01°. Therefore, if you see a time measurement reported as 1.234567 s, it is likely that this level of accuracy is justified. Conversely, if you see a temperature measurement of 192.23523°C, you should be suspicious. Do not fall into the trap of reporting or using an accuracy higher than your equipment can attain. As we will see shortly, the number of ***significant digits*** in the raw data will have an important impact on the estimated error in the calculated results. So, to get a realistic estimate of the final error, you should not use more significant digits than the raw data justifies.

When you report a measured quantity, the value should have as many digits of accuracy as the equipment allows plus one additional digit that you estimate. These accurate and one estimated digits make up the ***significant digits*** of the value. The number of digits that you include in the reported value should follow these generally accepted rules:

- The last digit is estimated. It may be written with a ( ) to highlight this fact. Thus, 1.2345(6) would show six significant digits, five of which come directly from the equipment. The last digit is estimated, usually by interpolation.
- We call the last digit the *least significant digit* and the leftmost digit the *most significant digit.*

ROUND-OFF ERRORS

There are many occasions when you will round off data. Sometimes this is intentional; at other times it is inadvertent. For example, when you enter a number into a program, the computer will automatically round off the number to match the amount of memory reserved. Thus, if you type in a number with 28 digits of accuracy and try to store it into a REAL variable, the computer will automatically round the number to 7 or 8 digits. Therefore, we need to examine how rounding-off produces errors.

When the computer rounds a number, it drops all insignificant digits. In the previous example the computer treats the last 20 to 21 digits like a fractional number. If this fraction is greater than 0.5, the computer increases the least significant digit by 1. But if the fraction is less than 0.5, the least significant digit does not change. A third situation also can occur. What if the fraction exactly equals 0.5? The adopted convention is that you round the number up only if the least significant digit is an odd number. Thus, half the time the number rounds up and half the time the number rounds down. Otherwise you will introduce an artificial bias to the rounding process. Here are some examples of numbers rounded to seven significant digits:

0.012345678901	→ 0.01234568	(*fraction = 0.8901, round up*)
32.71401267	→ 32.71401	(*fraction = 0.267, round down*)
12345678901234.1234	→ 12345680000000.0	(*fraction = 0.8901, round up and fill out with zeros*)
65234.05501	→ 65234.06	(*fraction = 0.5, round up since least significant digit is odd*)
65234.04501	→ 65234.04	(*fraction = 0.5, round down since least significant digit is even*)

Note in these examples that we do not count leading or trailing zeros when determining the number of significant digits.

The reason that we must be concerned with significant digits and round-off errors is that these errors tend to "creep" in when we manipulate the data. If we add two numbers together, for example, we should not expect the result to have a greater accuracy than either of the two numbers. Thus, if we add two numbers with four significant digits each, the result also should have four significant digits:

$$\begin{array}{r} 12.3(4) \\ +\ 13.6(9) \\ \hline 26.0(3) \end{array}$$

In this situation, where the two numbers are of approximately the same size, the result seems reasonable. But, what if the magnitude of two numbers is very different and we attempt to add them? By the same rule, we must round off the result to the same number of significant digits as the numbers being added:

$$\begin{array}{rl} & 123(4). \\ + & 0.00123(4) \\ \hline & 123(4). \end{array}$$

Adding a small number to a very large number often results in the smaller number being lost in the rounding process. Clearly, this is a process that you want to avoid.

A loss of significant digits may occur when we subtract two numbers of approximately the same magnitude and the same number of significant digits:

$$\begin{array}{r} 123(4). \\ -\ 117(3). \\ \hline 6(1). \end{array}$$

This is an important conclusion that arises from a survey of similar problems. The number of significant digits after a calculation is always the same or fewer than the number of significant digits of the least accurate number involved in the calculation. In other words, the calculated answer can never be better than the input data.

ACCUMULATION ERRORS

The preceding examples for round-off errors occur in every calculation done on a computer. If we do the calculation only once, the relative error is usually not substantial. But, if we perform the calculation repeatedly, the error grows with each step. Before giving you an example of an accumulation error, we must define the terms *error* and *relative error*.

$$ERROR = (TRUE\ VALUE - ESTIMATED\ VALUE)$$

$$RELATIVE\ ERROR = \frac{ERROR}{TRUE\ VALUE}$$

According to these definitions, we express the error in measurable units such as seconds, meters, grams, and so forth, but we express the relative error in dimensionless units or percentages.

The relative error is usually more revealing in estimating accuracy of a calculation. As an example, a constant error of 1 g in weighing is very significant when weighing a 2.4-g sample, but insignificant when weighing a 14,300-g sample. Therefore, all our discussions will use the relative error wherever possible.

We can show the problem with error accumulations with a simple example of adding 1/3 to itself repeatedly. If we add 1/3 three times, the true answer of course, is 1. But if we do this on a computer, 1/3 is not stored precisely because of the finite size of the computer memory. Therefore, a very small error occurs. If we then add 1/3 on the computer three times, the relative error increases by a small amount. As we increase the number of times the program adds 1/3 to itself, the relative error increases. Here are some actual computer results that show how these many small errors can accumulate into very large errors:

Number of Additions	True Value	Calculated Value	Relative Error (%)
3333	111	111.0003	0.00027
33333	1111	1110.986	0.00126
333333	11111	11109.75	0.01125
3333333	111111	111036.4	0.06714
33333333	1111111	1086658.	2.20077

As this example shows, the relative errors due to accumulation are not large until the number of calculations is of the order of several million. Yet, there are times when you will run a program with this many calculations. So, you must always be wary of errors of this sort. This will become an important topic in higher-level numerical methods courses that many of you will undoubtedly take during your careers.

TRUNCATION ERRORS

The final source of computational error is *truncation errors.* These errors are due to the computer's inability to compute various mathematical functions to an infinite degree of precision. Good examples are the trigonometric functions, such as the sine of an angle, which requires the use of an infinite series approximation to obtain a value:

$$\sin(X) = \frac{X^1}{1!} - \frac{X^3}{3!} + \frac{X^5}{5!} - \frac{X^7}{7!} + \cdots$$

As an example, if $X = 0.7$, then each term in the series is:

$$\begin{aligned}\sin(0.7) &= \frac{0.7^1}{1!} - \frac{0.7^3}{3!} + \frac{0.7^5}{5!} - \frac{0.7^7}{7!} \\ &= 0.7 - 0.05717 + 0.001401 - 0.00001634 \\ &= 0.6442\end{aligned}$$

Note that when $X = 0.7$ the answer is correct to four decimal places after only three terms in the series:

Term		Value	Series Total	Change (%)
1	+	0.70000000	0.70000000	–
2	–	0.05716667	0.64283333	−8.17
3	+	0.00140058	0.64423391	+0.22
4	–	0.00001634	0.64421757	−0.0025

Additional terms in the expansion improve the accuracy, but the series approximation never gives the precise answer because we must truncate the series at some point. Any function that uses the series approximation will behave this way. These include the most common mathematical functions such as trigonometric functions, the square root, the logarithm, the exponent, or any transcendental constant like e (2.71828...) or π (3.14159...).

13.4 Random Errors

Systematic errors control the *accuracy* of a measurement as we showed in the histograms at the beginning of this chapter. Thus, if the systematic errors are small, or if you can mathematically correct for them, then you will obtain an accurate estimate of the "true" value. The *precision* of the experiment, on the other hand, is related to random errors. The precision of a measurement is directly related to the uncertainty in the measurement.

Random errors are the statistical fluctuations that occur during a measurement. Whenever you measure the length of a sample, for example, you will obtain slightly different values each time. If these values are too close to each other, then the random errors are small. But, if the values are not too close, then random errors are large. Thus, random errors are related to the reproducibility of a measurement. In Fig. 13.8 we show a sample of data that

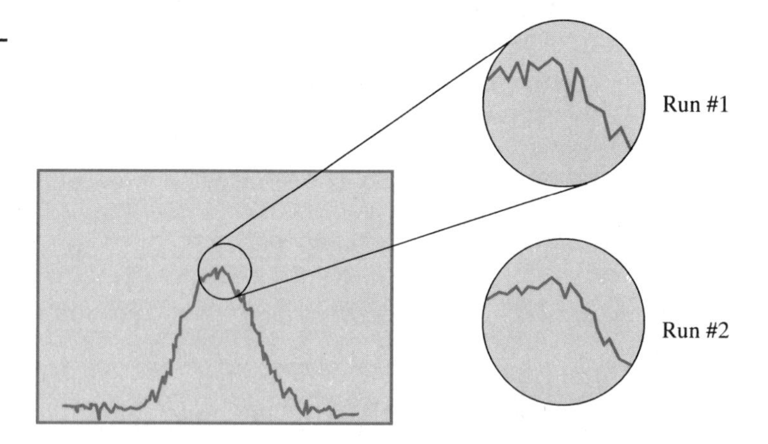

FIGURE 13.8

Illustration of Random Errors

we collected twice. Notice that individual data points from the two runs do not match, but the general shape of the two curves match very well.

Small fluctuations in a repeated measurement are a common occurrence. For example, when weighing a sample, air currents near the weighing pan can produce a small mechanical vibration that slightly changes the weight read off the scale. Another example would be small voltage fluctuations in an electrical circuit. Thus, you can always reduce random errors by using more precise equipment or modifying the experimental technique. Thus, you might want to use a scale that reduces air currents by enclosing the weighing pan in an enclosure.

Even with the best equipment, though, there will be random errors. Fortunately, we can do something about random errors after finishing the data collection. For the data in the preceding curve, Fig. 13.9 shows how the noise in the data can be reduced by using a mathematical filter. We will discuss some of these methods at the end of this chapter. While these methods do a good job of reducing the noise in the data, we warn you that smoothing should not substitute for a good experimental technique.

13.5 Statistical Analysis of Data

If we conduct an experiment to measure a quantity, some error sources cited earlier prevent us from obtaining an exact measure. Instead, we obtain an estimate. By repeating the experiment, we obtain a second estimate, which usually is different from the first value. As we obtain more data of this type, we begin to see a distinct frequency distribution. A frequency distribution indicates how often a particular value has occurred. To demonstrate, here is a

Figure 13.9

Reduction of Random Errors by Filtering

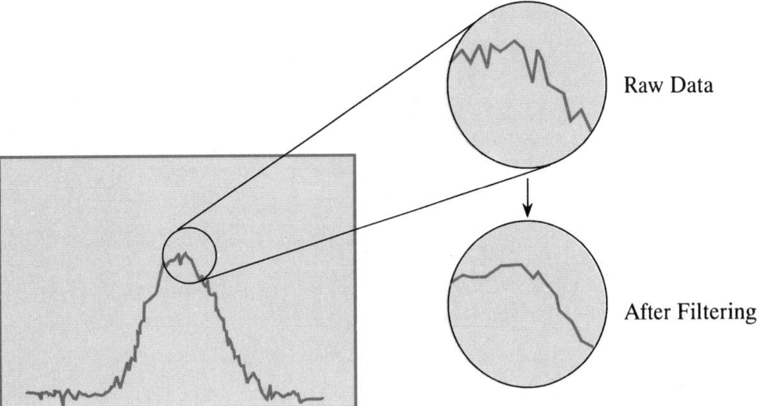

table of consecutive temperature readings (in °C) taken from a mercury thermometer of the type shown previously:

19.6	20.3	19.9	20.0	20.0	19.7	19.9	20.0	19.6	19.8	19.9	20.0
20.3	19.9	20.1	20.1	20.0	20.1	20.0	20.3	19.9	20.0	19.8	20.2
20.0	20.0	19.9	20.3	20.1	19.9	20.0	19.8	20.1	20.1	19.8	20.1
19.9	20.0	19.6	20.3	20.1	20.0	19.9	20.3	20.0	19.8	20.2	20.1
20.2	20.3	20.0	20.1	20.1	20.0	19.8	20.1	20.0	20.3	20.1	20.1
20.1	19.9	20.1	20.1	20.1	20.3	20.0	19.8	20.1	20.0	20.1	20.1
20.0	20.0	20.1	20.0	20.1	20.0	19.6	19.8	20.0	19.9	20.1	20.0

We often arrange data of this type as a histogram such as that in Fig. 13.10a. Notice that this data yields an average value of approximately 20°C but that there is a large spread in the data. In Fig. 13.10b we show how the distribution looks after taking twice as many data points. The spread has improved and our initial guess of an average temperature of 20°C now looks more reasonable. Still, there is room for improvement. If we collect a very large number of data points as in Fig. 13.10c, the frequency distribution now is very symmetric and well behaved.

When using a finite number of data points to obtain a distribution, we call this a *sample,* which we believe represents the *entire population*. In this example the population would be an infinite number of temperature readings, which, of course, would be an impossible task. Therefore, we must rely on a small sample of temperature readings to obtain a representative distribution. When the sample size is too small, as in our first example, the possible errors are very large. But as the sample size increases, the errors begin to decrease.

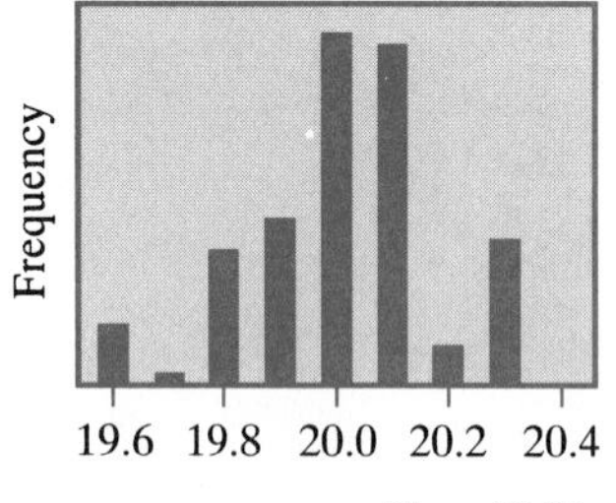

(a) Small Sample

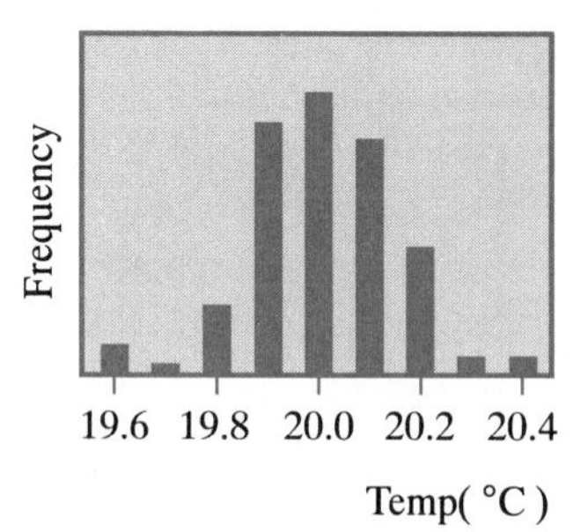

(b) Medium Sample

Frequency
19.6 19.8 20.0 20.2 20.4
Temp(°C)

(c) Very Large Sample

Figure 13.10

Frequency Distribution for (a) Small, (b) Medium, and (c) Very Large Sample Size

We use the previous histograms to represent the frequency distribution with a limited sample size. The continuous curve in Fig. 13.10c shows what we assume the frequency distribution would look like for the entire population. Notice that as the sample size increases, the agreement between the continuous curve and the histogram improves.

Once we assemble the data into a histogram, we are ready to analyze it for key pieces of information. Among the things that we usually look for are:

- The *mean,* which is the average value
- The *median,* which is the value separating the top 50 percent of the sample from the bottom 50 percent
- The *deviation,* which is the difference between the actual values and the average value

There are many other parameters that you can extract from the data, but these three are the customary ones. Therefore, our discussions will focus only on these three.

THE MEAN

We define the mean value (also called the average value), $\overline{X}$, by

$$\overline{X} = \frac{X_1 + X_2 + X_3 + \cdots + X_N}{N} = \frac{1}{N}\sum_{i=1}^{N} X_i$$

For the preceding temperature readings, we can calculate the mean value;

$$\overline{T} = \frac{19.6 + 20.3 + \cdots + 20.0}{84} = \frac{1}{84}\sum_{i=1}^{84} T_i = 20.0$$

To be consistent with our discussion on significant digits, we have rounded the temperature down to 20.0 from 20.0131 because the input data has only three significant digits. If you examine the histogram representing this data, you will see that the mean value falls in the middle of the distribution.

THE STANDARD DEVIATION

The *deviation* from the mean, d_i is simply the difference between any data point, X_i, and the mean. We define this by:

$$d_i = X_i - \overline{X}$$

where we use the mean value from the previous equation. If you examine this equation closely, you will realize that the deviation is nothing more than the apparent error for each measurement. For the temperature readings used in the calculation of the mean, we obtain:

$$d_1 = 19.6 - 20.0 = -0.4$$
$$d_2 = 20.3 - 20.0 = +0.3$$
$$d_3 = 19.9 - 20.0 = -0.1$$
$$\vdots$$

Notice that if we try to calculate an average deviation, the value probably will be zero because positive and negative values will cancel. Yet, an "average" measure of the error would be very desirable since it tells us how good the data is in a quantitative way. Therefore, we need a different way to obtain the measure of the scatter of the data. One way to do this is with the *standard deviation,* σ, defined by:

$$\sigma^2 = \frac{d_1^2 + d_2^2 + \cdots + d_3^2}{N - 1} = \frac{1}{N - 1} \sum_{i=1}^{N} d_i^2$$

which for the preceding data becomes:

$$\sigma^2 = \frac{(-0.4)^2 + (+0.3)^2 + \cdots + (0.0)^2}{83} = 0.0280$$

which gives a standard deviation of $\sigma = (\sigma^2)^{1/2} = (0.0280)^{1/2} = 0.167$. Notice that this definition of the average error does not have any positive and negative values canceling each other. All deviations make a contribution. We call the term σ^2 the *variance,* but we rarely use it in discussions of error. The standard deviation, or the square root of the variance, is the term that is most widely used to tell us about the spread of our data.

The definition of the standard deviation differs slightly for small samples and for entire populations. When you use a small sample, the square of the deviations should be divided by $(N - 1)$ instead of (N) used in the previous equation. Thus:

$$\sigma^2 = \frac{1}{N} \sum_{i=1}^{N} d_i^2 \qquad \text{(for large samples)}$$

$$\sigma^2 = \frac{1}{(N - 1)} \sum_{i=1}^{N} d_i^2 \qquad \text{(for small samples)}$$

We can illustrate the physical significance of the standard deviation with the aid of Fig. 13.11. For the "bell-shaped" data distribution, there is a 68 percent probability that any measurement is within 1σ of the mean and a 95 percent chance that it is within 2σ of the mean. This rule is strictly valid, though, only for the specific distribution shown, which we call the Gaussian, normal, or bell-shaped curve. It should be clear by now that the standard deviation is a measure of the spread in the data. When σ is small, most of the data points are nearly identical and the calculated value for the mean is precise. But, if σ is large, the precision is low and you need to take many measurements to have confidence in the calculated mean.

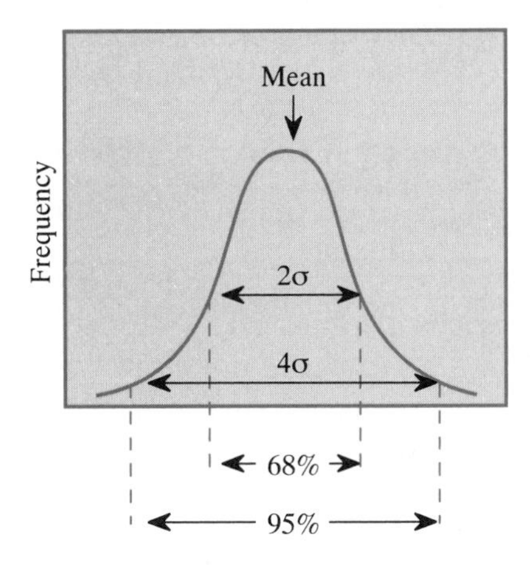

FIGURE 13.11

Significance of the Standard Deviation

Now let's write a program that computes the mean and standard deviation of a sample of data points. This is not a difficult task once you understand how the equations work. This process should consist of the following:

- Enter an unknown number of data points.
- Compute the *mean*.
- Compute the *standard deviation* after the mean has been calculated.

Note that you cannot compute σ until you have first determined the *mean*. Also, you should try to set up the program to operate with a variable number of data points so that it will be as general as possible. Therefore, in the following algorithm we establish a variable sized array in two separate subprograms to compute the mean and the standard deviation:

LS FOR STATISTICAL DESCRIPTION OF DATA

PROBLEM STATEMENT: Read in a set of data and calculate the mean and standard deviation.

ALGORITHM DEVELOPMENT:

Step 1a: Review the problem statement. Indicate all nouns in **bold** and all verbs in *italics:*

Read in a **set of data** and *calculate* the **mean** and **standard deviation.**

Formulate questions based on nouns and verbs:

Q1: What is a set of data?
Q2: How do we read the data in?

Q3: How do we calculate the mean?
Q4: How do we calculate the standard deviation?

Step 1b: Construct answers to the above questions:

***A1:** A set of data is a large number of values that we can store in an array.
***A2:** We can use an IMPLIED DO LOOP to read in the set of data.
***A3:** To calculate the mean, we add up all values in the set of data and divide by the number of items.
***A4:** To calculate the standard deviation, we use a LOOP to subtract the mean from each data value, square the result, and add it to a running total. After finishing with all data points, we take the square root of the total. This is the standard deviation.

ALGORITHM:

MAIN:

1. Read in N, which is the number of data points.
2. Use N to control an IMPLIED DO LOOP to read in the data points and assign them to the array, X.
3. CALL the subroutine XBAR to calculate the MEAN:
   Send down N and X
   Get back the value, MEAN.
4. CALL the subroutine DEV to calculate the standard deviation:
   Send down N, X, and MEAN
   Get back the value, STDDEV.
5. Print out the results, MEAN and STDDEV.

SUBROUTINE
MEAN

1. Use the value, N, transferred from the MAIN to set up an adjustable array. Also, N will control the loop to compute the sum.
2. Use a LOOP to compute the sum of the data.
3. Set MEAN to the value of the sum/N.
4. Return to the MAIN.

SUBROUTINE
DEV

1. Use the value, N, transferred from the MAIN to set up an adjustable array. Also, N will control the loop to compute the standard deviation.
2. Use a LOOP to compute the sum of (data point $-$ mean)^2.

3. Set STDDEV equal to the value of the sum/(N − 1).
4. Return to the MAIN.

FLOWCHART:

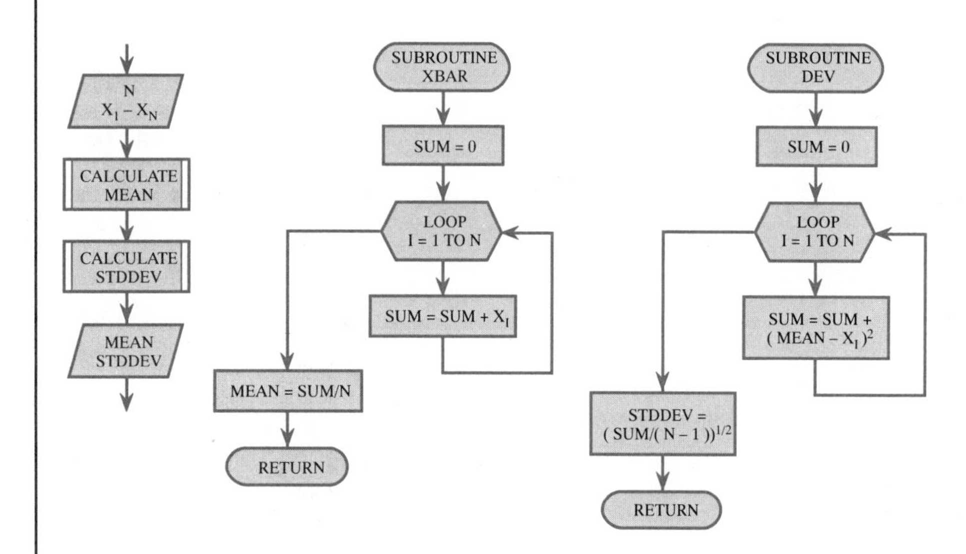

PROGRAM:

```
C      PROGRAM TO CALCULATE MEAN AND STANDARD
C      DEVIATION OF A SET OF C NUMBERS. IT READS
C      IN THE NUMBER OF POINTS AND SENDS THEM TO
C      SUB XBAR TO DETERMINE THE MEAN AND THEN TO
C      SUB DEV ALONG WITH THE MEAN TO DETERMINE
C      THE STANDARD DEVIATION.
C             N      - NUMBER OF DATA POINTS
C             X      - ARRAY OF DATA POINTS
C             MEAN   - VALUE OF THE MEAN
C             STDDEV - STANDARD DEVIATION
C -----------------------------------------------
       REAL X(1000), MEAN
       PRINT *, 'ENTER NUMBER OF POINTS YOU WISH
                 TO ENTER:'
       READ*, N
       PRINT *, 'ENTER ALL DATA POINTS:'
       READ *, (X(I), I=1,N)
       CALL XBAR (N, X, MEAN)
       CALL DEV (N, X, MEAN, STDDEV)
       PRINT*, 'MEAN=', MEAN, ' STANDARD
     1          DEVIATION=', STDDEV
       END
```

```
C ------------------------------------------------------------
C     SUBROUTINE XBAR TO CALCULATE THE MEAN OF AN
C     ARRAY OF NUMBERS STORED IN X.
C           N        - NUMBER OF POINTS
C           SUM      - RUNNING TOTAL OF SUM OF
                       ARRAY ELEMENTS
C           MEAN     - MEAN OF ARRAY VALUES
C ------------------------------------------------------------
      SUBROUTINE XBAR (N, X, MEAN)
      REAL X(N), MEAN
      SUM=0.0
      DO 10 I=1,N
            SUM=SUM + X(I)
10    CONTINUE
      MEAN=SUM/N
      RETURN
      END
C ------------------------------------------------------------
C     SUBROUTINE DEV TO CALCULATE THE STANDARD
C     DEVIATION OF AN ARRAY OF NUMBERS STORED
C     IN X.
C           N      - NUMBER OF POINTS
C           SUM    - RUNNING TOTAL OF SUM OF
                     (ARRAY ELEMENTS-MEAN)
                     SQUARED
C           STDDEV - STANDARD DEVIATION =
C                    SQRT(SUM/(N-1))
C ------------------------------------------------------------
      SUBROUTINE DEV (N, X, MEAN, STDDEV)
      REAL X(N), MEAN
      SUM=0.0
      DO 10 I=1,N
            SUM=SUM + (X(I)-MEAN)**2
10    CONTINUE
      STDDEV=SQRT(SUM/(N-1))
      RETURN
      END
```

13.6 Data Reduction Procedures

At the beginning of this chapter we introduced several examples of removal of systematic and random errors from sets of data. In this section we explore these ideas in more detail. Specifically, we present algorithms for the following:

- Subtracting background noise
- Correcting for calibration errors

- Simple three- and five-point smoothing to remove random noise
- The weighted average method for removing random noise

Before beginning this section, we must warn you that not all scientists and engineers agree about the appropriateness of "massaging the data." They argue that correcting the data may introduce artifacts of its own, and the results that you obtain may be misleading. Therefore, you must apply some common sense when using these techniques. For example, you should compare the raw data to the smoothed profile to see if there is good correspondence between the two. Also, if you apply statistical methods to the data, you should use the raw, not the smoothed data. In spite of the objections of some, these techniques are useful for experimentalists. But, again, we warn you that these methods are not a substitute for poor experimental technique. They should be used in conjunction with careful experimental design, calibration, and execution.

SUBTRACTING BACKGROUND NOISE

If you have a set of data that contains a significant amount of background noise, you may be able to subtract it from the data set. The key to success, however, is the ability to identify what is noise and what is real. One way to do this, for example, is to run an experiment with a dummy sample that produces no real data of its own. The resulting spectrum is the noise in the system. The second way, and the one usually used, is to examine the data itself for clues about the noise level.

Look at the curve (Fig. 13.12) that we showed previously. In this figure we choose a *constant* background level as indicated. Once we know the magnitude of this background level, we can simply subtract it from each data point. The question is, How do we determine this value? There are two ways. The first is to scan all the data for the lowest value and then assume that this is the background. The second way would be to average the values near the minimum value, thus accounting for small random errors. Choosing a

Figure 13.12

Constant Background Level

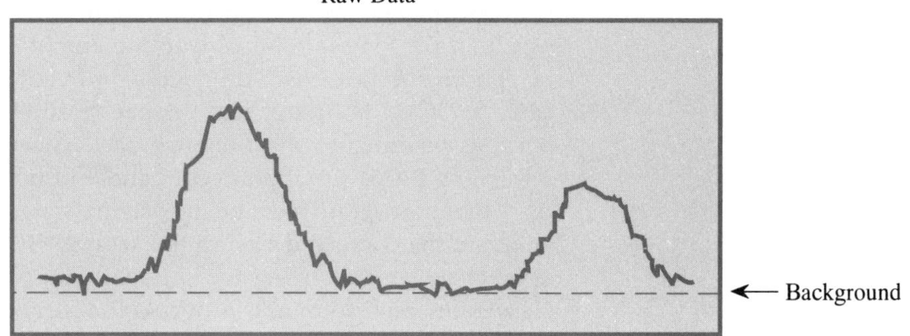

FIGURE 13.13

Determination of Background Level

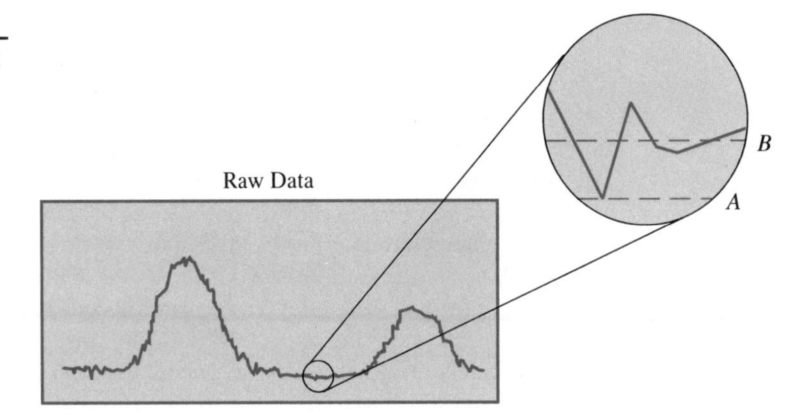

single data point is dangerous with experimental data since it *always* underestimates the background noise. Figure 13.13 illustrates the two different determinations of the constant background noise. Line A is the value based on the single lowest point in the curve, whereas curve B is the value based on the average of several points. Notice that curve B is more realistic in describing the true background level.

The only problem with choosing line B as the background is that when we subtract this value from all the data points, a few of them may become a negative value. This is clearly unrealistic. Therefore, we may need to take care of this potential artifact. For example, we set any corrected value to zero if the subtraction of the background brings it below zero. We can do this with the BUILT-IN FUNCTION, MAX, which will return the larger of the two numbers, (X-background) or zero.

The procedure that we use is the following:

- Find the smallest value in the data set.
- Average over several data points on either side of the lowest value. This is then the average background level.
- Subtract this background from every data point.

For convenience, we average over five points near the lowest value. We must be careful if the lowest point is near one end of the spectrum since some of these neighbors may not exist. Although we can adjust our program for these special cases, we must point out that the background determination may not be valid because we did not collect enough data. Therefore, our program will print out a warning if this situation occurs and will not calculate the background.

In our program we take the original data, stored in the array XOLD(I), and generate the corrected data in the array XNEW(I). In keeping with our philosophy of never destroying the original data, we keep both sets of data intact. Here is the algorithm and program for correcting for a constant background:

LS FOR SUBTRACTING A CONSTANT BACKGROUND

PROBLEM STATEMENT: Read in a set of data and correct the data for the constant background level.

ALGORITHM DEVELOPMENT:

Step 1a: Review the problem statement. Indicate all nouns in **bold** and all verbs in *italics:*

Read in a **set of data** and *correct* the **data** for the constant **background level.**

Formulate questions based on nouns and verbs:

Q1: How do we read in a data set?
Q2: What is the background level?
Q3: How do we correct for the background?

Step 1b: Answer questions:

***A1:** Use an IMPLIED DO LOOP to read in the data into an array, XOLD(I).

***A2:** The background level is the average of the five data points near the lowest point in the data set. We can find the lowest point by using the MAX/MIN SEARCH LS. Then compute the average of that point and the two neighbors on either side.

***A3:** Once we determine the background, we correct the data set by subtracting the background from each data point.

ALGORITHM:

MAIN:

1. Read in N, which is the number of data points.
2. Use N to read in the data set, XOLD(1) to XOLD(N).
3. CALL subroutine BACK to determine the background; data to be returned is XNEW, which is the corrected data.
4. Print out XNEW(1) to XNEW(N).

SUBROUTINE
MEAN

1. Use N, transferred from the MAIN, to set up two adjustable arrays, XOLD for the original data, XNEW for the corrected data.
2. Use MAX/MIN SEARCH LS for finding the position and the value of the lowest data point:
   Use a LOOP to scan every data point:
      Check to see if the data point is smaller than SMALL, if it is, then SMALL receives the value of XOLD(I).

3. Check to see if the position of the smallest data point is near the ends;
   If it is: Print a message.
   If not: Compute the average of five points near the smallest value.
4. Use a LOOP:
   Subtract BACKGROUND from each data point and store in XNEW.
5. Return to MAIN.

FLOWCHART:

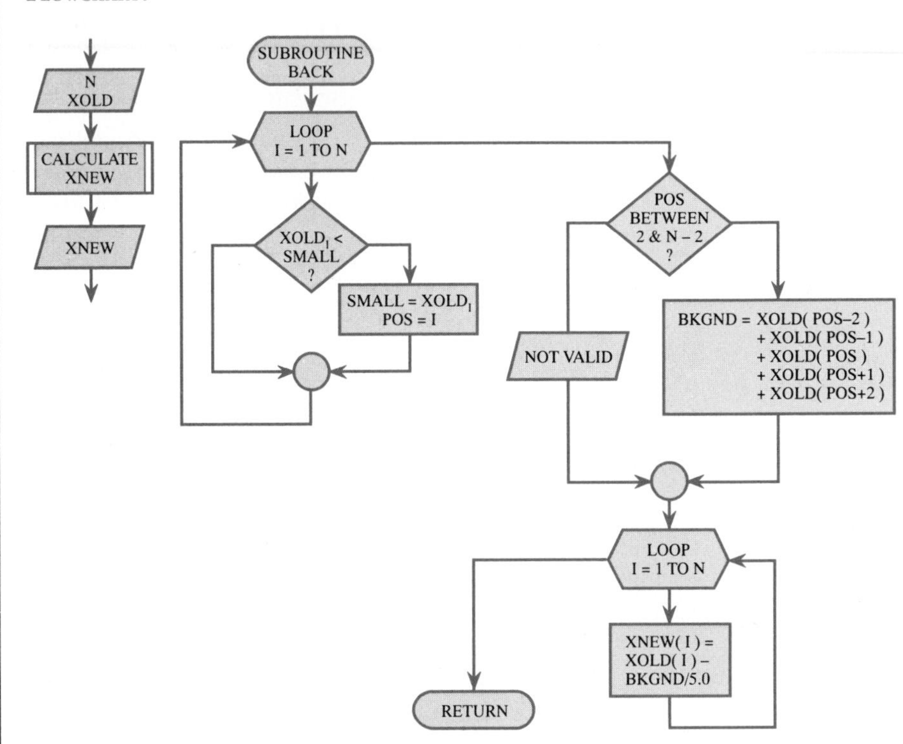

PROGRAM:

```
C      PROGRAM TO DETERMINE AND REMOVE A CONSTANT
C      BACKGROUND. USES A SUBROUTINE BACK TO DO
C      THIS
C            XOLD - ARRAY OF ORIGINAL DATA POINTS
C            XNEW - ARRAY OF CORRECTED DATA POINTS
C            N    - NUMBER OF POINTS
C ------------------------------------------------------
       REAL XOLD(1000), XNEW(100)
       PRINT *, 'ENTER NUMBER OF POINTS TO BE
     1           ANALYZED'
       READ *, N
       PRINT *, 'ENTER DATA POINTS:
```

```
      READ *, (XOLD(I), I=1,N)
      CALL BACK (N, XOLD, XNEW)
      PRINT *,' CORRECTED DATA POINTS FOLLOW:'
      PRINT *, (XNEW(I), I=1,N)
      STOP
      END
C ---------------------------------------------------------
C     SUBROUTINE BACK TO DETERMINE THE BACKGROUND
C     NOISE IN A SET OF DATA. READS IN AN ARRAY
C     XOLD CONTAINING THE ORIGINAL DATA. THEN
C     SCANS THE DATA TO FIND THE POSITION OF THE
C     SMALLEST VALUE. THE BACKGROUND IS THEN 1/5
C     OF THE SUM OF THE SMALLEST VALUE AND ITS 4
C     NEAREST NEIGHBORS. AFTER THE BACKGROUND IS
C     DETERMINED, EACH DATA POINT HAS THE
C     BACKGROUND SUBTRACTED FROM IT. THE NEW
C     VALUES ARE STORED IN XNEW.
C           N     - NUMBER OF DATA POINTS
C           XOLD  - ARRAY OF ORIGINAL DATA POINTS
C           XNEW  - ARRAY OF CORRECTED DATA
C                   POINTS
C           BKGND - AVERAGE BACKGROUND LEVEL
C ---------------------------------------------------------
      SUBROUTINE BACK ( N, NOLD,XNEW)
      REAL XOLD(N), XNEW(N)
      INTEGER POS
      DO 10 I=1,N
            IF(XOLD(I).LE.SMALL) THEN
                  SMALL=XOLD(I)
                  POS=I
            ENDIF
10    CONTINUE
      IF(POS.GE.3.AND.POS.LE.N-3) THEN
         BKGND=(XOLD(POS-2)+XOLD(POS-1)+XOLD(POS)
     1         +XOLD(POS+1)+XOLD(POS+2))/5.0
      ELSE
            PRINT*, ' POSSIBLE ERROR'
            RETURN
      ENDIF
      DO 20 I=1,N
            XNEW(I)=MAX( (XOLD(I)-BKGND) ,0.0)
20    CONTINUE
      RETURN
      END
```

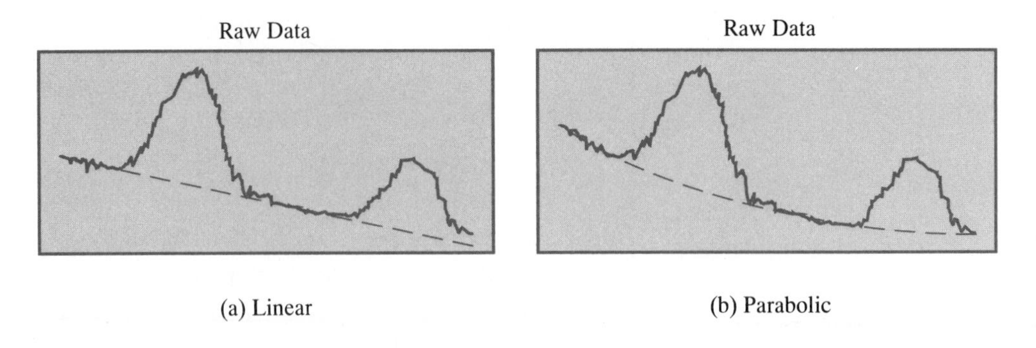

FIGURE 13.14

Common Forms of Background Noise

There are many occasions, though, when the background level is not a constant as in the previous example. Figure 13.14 shows two possibilities that you are likely to encounter.

Sometimes the background is either linear or parabolic. If it is, then you can fit it to these functions using the methods discussed in the next chapter. Once you obtain a fit, the subtraction proceeds in a way that is identical to the previous procedure, or:

```
XNEW(I) = XOLD(I) - BKGND            (constant background)
XNEW(I) = XOLD(I) - (M*X + B)        (linear background)
XNEW(I) = XOLD(I) - (C - D*X**2)     (parabolic background)
```

In these equations you must determine the constants M, B, C, and D from the least squares fit, which we discuss in the next chapter. But no matter what the form of the background is, once you know the mathematical form, you can subtract it from the raw data. Therefore, you will need to modify the preceding program for your situation. There is no automatic way to determine the noise and remove it. Instead, you must examine each case individually and decide how to approach the problem. What we have provided earlier is merely an example of the approach you should use.

CORRECTING FOR ZERO ERRORS

As we described at the beginning of this chapter, one important source of error is an instrument that has a faulty zero position. A good example is a voltmeter that always reads 0.5 V too high. If the zero is off, all the data will be off by the error in the zero. Consider the set of voltage and current readings in Fig. 13.15, taken from a simple resistor circuit. Because the current-voltage (*I-V*) curve does not pass through the origin, we know immediately that the zero of the voltage meter probably is incorrect. After all, if there is no applied voltage, there can be no current. So, all the voltage values are incorrect by an error that we sometimes call the *offset* shown in the figure. This would not

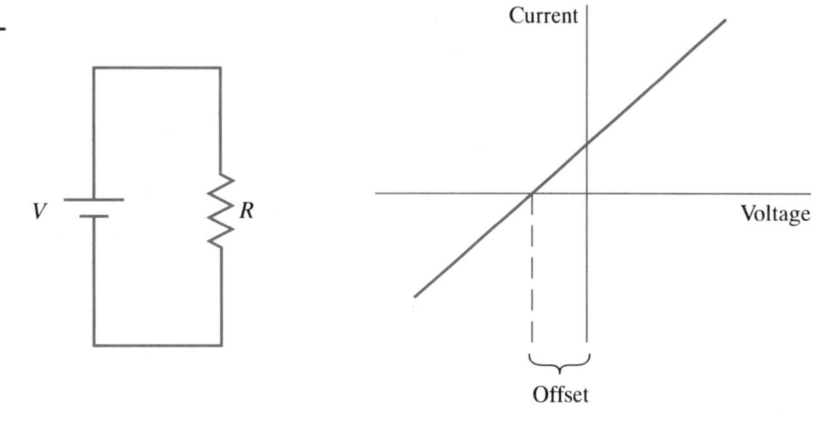

FIGURE 13.15

Current/Voltage Characteristics of a Resistive Circuit

present a serious problem in this simple experiment since we could simply shift all the voltage readings by the offset and obtain the corrected voltages:

$$V_i^{\text{corrected}} = V_i^{\text{raw}} + \text{offset}$$

The code for doing this would look like this:

```
      SUBROUTINE ZERO(N, XOLD, XNEW, OFFSET)
      REAL XOLD(N), XNEW(N), OFFSET
      DO 10 1=1,N
            XNEW(I)=XOLD(I)-OFFSET
   10 CONTINUE
      RETURN
      END
```

Notice that you must determine the value of the offset and supply it to the program for analysis. Since every experiment will have a different offset, there is no general way that the program can determine the amount of the offset. Fortunately, this is experimentally a simple thing to measure. So, we do not include a discussion here. Of course, you could avoid this problem by resetting the zero position frequently. In those situations where it is not possible to set a zero, then you may need to correct the data by this simple mathematical operation. Finally, if you need to correct for the offset error, you should do this correction first before doing any other processing.

CORRECTING CALIBRATION ERRORS

In Fig. 13.5 we illustrated how an improperly calibrated thermometer can lead to experimental errors. In that example we found that the calibration was correct at one temperature but incorrect at the higher temperature. Usually, there is a well-defined mathematical relationship between true and measured values, as shown in Fig. 13.16.

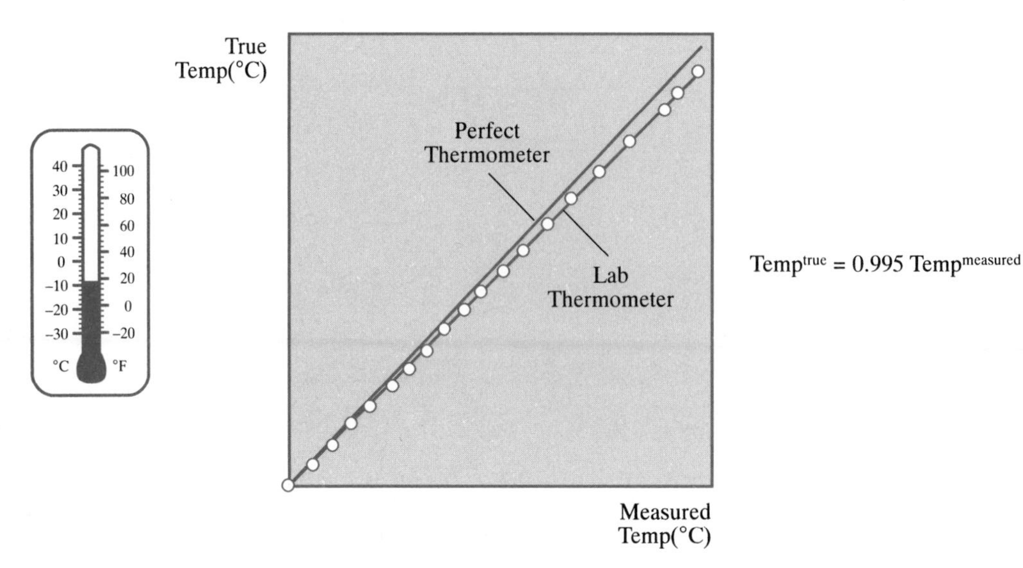

FIGURE 13.16

Calibration Relationship between Measured and True Values

Notice that the simple offset approach that we used for adjusting a faulty zero does not work here. If we simply subtract 0.5°C from every temperature reading, we would be correct at 100°C, but incorrect at every other temperature. So, we need to make a different type of correction. Based on the curve shown in Fig. 13.16, the true temperature can be determined from the measured temperature by:

$$T^{true} = 0.995T^{measured}$$

where the constant is the measured slope of the curve (100 − 0)/(100.5 − 0). Notice that the program cannot conduct the calibration test for you. It is something that you must do yourself. Once you obtain the information about the calibration errors, you can use the following code to correct it:

```
      SUBROUTINE CALIB(N, XOLD, XNEW, C1)
      REAL XOLD(N), XNEW(N)
      DO 10 1=1,N
            XNEW(I)=XOLD(I) * C1
   10 CONTINUE
      RETURN
      END
```

The constant, C1, in the subroutine must come from your calibration measurements. You must pass this data from the MAIN program at execution time. Also, you may need to apply simultaneously an offset correction. Do the offset correction first, then do the calibration correction.

This example is a simple one where the equation for the correction is linear. Sometimes the correction is more complicated and you may need to spend a great deal of time correcting the data. You can avoid this, of course, by carefully adjusting the calibration before you begin experiment. Only if the calibration cannot be adjusted, will you need to correct the data as we did for the thermometer.

THREE-POINT SMOOTHING

The previous three sections focused on removing or correcting systematic errors. When you encounter systematic errors, every experiment will be different. Therefore, you must tailor your corrections to suit the type of error. For example, you may have a parabolic background instead of a constant background. This will require you to change the program to account for the parabolic noise. Therefore, the previous examples are models only and you should be careful when using them.

The tools that we will show you in this and the following sections are more general than those in the previous three sections. You will find these tools useful for minimizing random errors in most situations without modification of the code that we provide.

The first of the methods is *three-point smoothing*. In this method we only replace a data point by the average of itself and its two nearest neighbors. In the *five-point smoothing* method, though, we average over the four nearest neighbors:

```
X(I) = (X(I - 1) + X(I) + X(I + 1))/3.0
                                  (three-point smoothing)
X(I) = (X(I - 2) + X(I - 1) + X(I) + X(I + 1)
          + X(I + 2))/5.0
                                  (five-point smoothing)
```

These formulas present a special problem when we get close to the end of the spectrum. For example, if we try to smooth the first data point, X(1), the previous formula becomes:

```
X(1) = (X(0) + X(1) + X(2))/3.0
```

Notice that the point X(0) does not exist! In a similar way, we have the same problem at the other end of the spectrum, X(FINAL):

```
X(FINAL) = (X(FINAL - 1) + X(FINAL) +
            X(FINAL + 1))/3.0
```

Here, X(FINAL + 1) does not exist. So, how will we handle these special cases? The easiest way, Fig. 13.17, is to make an assumption about the needed data point. Since the point X(0) does not exist, it seems reasonable to assume

FIGURE 13.17

Creation of a Fictitious Point

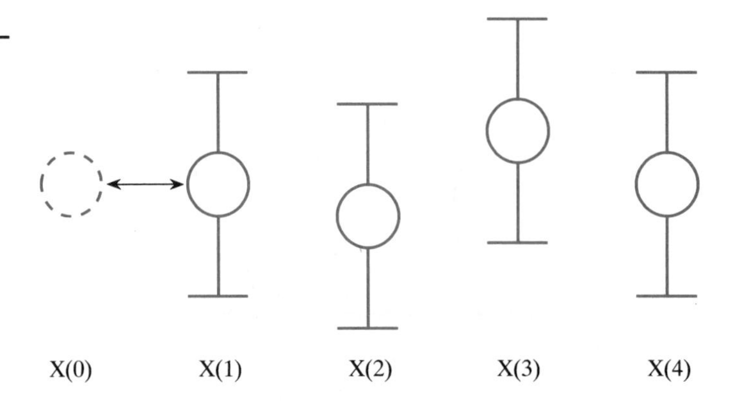

that the value of X(0), if it does exist, will be approximately equal to X(1). Similarly, we assume that X(FINAL + 1) is the same value as X(FINAL). That is exactly what we do in the equations for the first and last data points:

$$\begin{aligned} X^{new}(1) &= (X^{old}(0) + X^{old}(1) + X^{old}(2))/3.0 \\ &= (X^{old}(1) + X^{old}(1) + X^{old}(2))/3.0 \\ &= (2 * X^{old}(1) + X^{old}(2))/3.0 \end{aligned}$$

and for X(FINAL):

$$\begin{aligned} X^{new}(FINAL) &= (X^{old}(FINAL - 1) + X^{old}(FINAL) + X^{old}(FINAL + 1))/3.0 \\ &= (X^{old}(FINAL - 1) + X^{old}(FINAL) + X^{old}(FINAL))/3.0 \\ &= (X^{old}(FINAL - 1) + 2 * X^{old}(FINAL))/3.0 \end{aligned}$$

All remaining data points are analyzed with the first formula:

$$X^{new}(I) = (X^{old}(I - 1) + X^{old}(I) + X^{old}(I + 1))/3.0$$

The algorithm will have the following major components:

1. Smooth the first and last points first and store in XNEW.
2. LOOP: over data points 2 to FINAL − 1
   XNEW(I) = average of point and two neighbors.
3. Return the smoothed data points stored in XNEW to the MAIN program.

and the subroutine will look like this:

```
      SUBROUTINE SMOOTH (N, XNEW, XOLD)
      REAL XOLD(N), XNEW(N)
      XNEW(1) = (2.0 * XOLD(1) + XOLD(2))/3.0
      XNEW(N) = (XOLD(N-1) + 2.0 * XOLD(N))/3.0
      DO 10 I=2, N-1
           XNEW(I) = (XOLD(I-1) + XOLD(I) +
     1          XOLD(I+1))/3.0
10    CONTINUE
      RETURN
      END
```

Once again, we have set up the arrays with a variable size so that we can use the subroutine without modification on any size data set. Also, note that we have used two arrays, XOLD and XNEW, so that we can keep the original data set for later comparison.

Converting the algorithm to five-point smoothing is an easy process, and we leave it up to you to do it. Before doing that, keep in mind that special cases occur at each end of the spectrum. With three-point smoothing, there were only two special cases. But with five-point smoothing, there are four special cases, X(1), X(2), X(FINAL − 1), and X(FINAL). Therefore, your algorithm must add these special cases to those that we already discussed.

You must be careful that the peak you are trying to smooth must consist of many more data points than the interval over which you smooth. For example, if you smooth with three points, you must be very careful that there are no real peaks smaller than about five to ten times this smoothing inverval. Thus, if a peak has less than 15 data points, you will artificially reduce the peak height. However, if the peak has more than 30 data points, there will be no problems.

WEIGHTED AVERAGE METHOD

A more sophisticated smoothing process uses the *weighted average* method and requires advanced mathematical tools to derive. But the results are simple to understand and apply and require no advanced knowledge of mathematics. These methods are based on the idea that some points are more important than others. For example, if we wanted to give the point X(I) twice as much weight in the three-point smoothing, we would write:

$$X^{new}(I) = (X(I-1) + 2*X(I) + X(I+1))/4.0$$

where the 2*X(I) shows that the point is twice as important as the other two. In general, the weighted average, X_W is given by:

$$X_W(I) = (W(1)*X(I-1) + W(2)*X(I) + W(3)*X(I+1))/(W(1) + W(2) + W(3))$$

where W(I) = weight factor. Each data point is multiplied by its weight. We then divide that value, plus the corresponding values for the two neighbors, by the sum of the weights. In the three-point, equal weight smoothing, the total weight is three and the preceding equation reduces to our earlier formula when W(1) = W(2) = W(3) = 1.0.

This equation is useful because we can give any weight distribution that we like. The only question then is, What distribution is best? The equal distribution (W(1) = W(2) = W(3) = 1.0) is easy to apply but does not always give satisfactory results. So, we should explore other distributions.

The question of which distribution to use for optimum results requires some calculus to solve. We will not do that here. Instead, we will just list the results. If you are interested in the details, you should consult the references listed at the end of the chapter. Here is the best choice of the weighting function for a seven-point smoothing operation:

W(1) = 5.0
W(2) = −30.0
W(3) = 75.0
W(4) = 131.0
W(5) = 75.0
W(6) = −30.0
W(7) = 5.0

Therefore, the equation for a weighted seven-point smoothing becomes:

$$X^{new}(I) = (5.0*X^{old}(I - 3) - 30.0*X^{old}(I - 2) + 75.0*X^{old}(I - 1) + 131.0*X^{old}(I) + 75.0*X^{old}(I + 1) - 30.0*X^{old}(I + 2) + 5.0*X^{old}(I + 3))/231.0$$

We leave the conversion of this equation into a subroutine as an exercise. Note in this situation, though, that there are now six special cases, X(1), X(2), X(3), X(FINAL − 2), X(FINAL − 1), and X(FINAL). The best approach for these values is to assume their values and rewrite the equations.

Further Reading

1. P. R. Bevington, *Data Reduction and Error Analysis for the Physical Sciences,* McGraw-Hill, New York, 1969.
2. A. Savitzky and M. J. E. Golay, Smoothing and Differentiation of Data by Simplified Least Squares Procedures, *Analytical Chemistry,* Vol. 36, No. 8, July 1964, p. 1627.

Exercises

13.1 The following data represents the results of a survey of heights (in cm) of a small sample of freshmen and sophomore men. Use the program given in the text to compute the average height and the standard deviation:

Freshmen:	191	184	180	162	174	163	190	181	177	174	163	160	177	190	157	161	163	180
	174	183	182	164	170	165	182	179	163	157	196	183	177	155	170	166	181	179
	190	184	149	178	164	159	180	181	184	174	169	168	168	173	183	182	170	166
Sophomore:	190	182	159	166	173	167	184	147	178	167	187	165	190	174	191	180	168	156
	190	184	169	170	171	181	160	173	160	181	158	170	169	181	198	180	173	168
	174	164	191	167	156	170	172	174	168	160	165	174	169	188	160	173	170	175

In a hypothetical basketball game between these two, whom would you bet on?

13.2 Write a computer program to simulate the toss of a coin. You will find the random number generator helpful to solve this problem. The random number generator is a BUILT-IN FUNCTION that is usually called like this:

Y = RND(2.5)

where the number inside the parentheses is a *seed,* which has no function other than to start the calculation. The value that the function returns is a fraction between 0.0 and 1.0. Therefore, to simulate a coin toss, you might have any fraction less than 0.5 to be counted as heads. Any returned fraction larger than 0.5 is tails. Run the random number generator 500 times and keep track of the number of heads and tails. Are they equal? Then run it 1000 times and see if the fraction approaches 0.5 more closely.

13.3 Write a computer program using the random number generator to simulate the rolling of a pair of dice. The only possible values are whole numbers from 2 to 12. Now roll the dice 1000 times and keep track of the results. Then calculate the mean and standard deviation of the distribution. Also, find out what fraction of the time the numbers 7 and 11 were rolled.

13.4 The following voltage readings were taken with an analog voltmeter:

9.20	9.15	9.32	9.16	9.22	9.30	9.18	9.25	9.20	9.41
9.32	9.16	9.08	9.16	9.24	9.20	9.15	9.19	9.24	9.34
9.25	9.21	9.26	9.24	9.21	9.19	9.19	9.24	9.14	9.23

Since the data was taken late at night, there is a good chance that many of these readings are incorrect. Therefore, to help the poor student out, the instructor has allowed the student to eliminate all readings which are more than 1 standard deviation away from the mean. Write a program to:

a. Calculate the mean and standard deviation of the preceding data.
b. Remove any data more than 1 standard deviation away from the mean.
c. Recalculate the mean and standard deviation.
d. Report the first set of data with its mean and standard deviation along with the second set of data.

13.5 Often we find ourselves wondering whether a sample is large enough to draw accurate conclusions about a population. The question arises: Does our sample truly represent the population? Consider the following data:

05	95	42	39	51	05	66	27	33	24	33	72
83	72	75	62	81	77	91	87	81	50	70	99

Compute the mean and standard deviation. Now take the first ten values and repeat the calculations. Do the statistics change significantly? Often we can improve our statistics by taking a sample which is not selected randomly. Trimmed data is such an example. Trim the previous data by removing the two greatest and two smallest values, and repeat the calculations for the mean and the standard deviation.

13.6 Write a program to show that the sum of the deviations is zero, even though the standard deviation is not. Use the following data:

1 2 2 3 3 3 4 4 4 4 5 5 5 5 5 6 6 6 6 7 7 7 8 8 9

13.7 The following program takes a function f(*X*) and adds to it some random quantity between 0.0 and 1.0 times the square root of f(*X*):

```
      REAL D(1000)
      DATA ISEED/1234981/
      PRINT*,'ENTER THE NUMBER OF DATA POINTS'
      READ*, N
      PRINT*, 'ENTER THE LIMITS OF THE FUNCTION'
      READ*, A,B
      PRINT*, 'ENTER THE PERCENTAGE ERROR'
      READ*, PERCENT
      F2SUM=0.0
      DX=(B-A)/(N-1)
      X=A
      DO 10 I=1,N
           D(I)=F(X)
           F2SUM=D(I)**2
           X=X+DX
   10 CONTINUE
      FMEAN=F2SUM/N
      WEIGHT=S*SQRT(F2SUM)*PERCENT
      OPEN(10,FILE='RANDOM.DAT')
      WRITE(10,*) (D(I)+WEIGHT*RAND(ISEED),I=1,N)
      STOP
      END
```

Enter this program on your system, making any necessary changes to account for your compiler peculiarities. In particular, pay attention to how the random number generator (RAND above) is used. Now write a FUNCTION SUBPROGRAM F(X) corresponding to $f(X) = Ae^{-X}\sin(X)$. Then write a program to read the data that the preceding program segments create, and compute the mean and standard deviation of the data.

13.8 As we pointed out in the previous chapter, it is very important for you to validate your programs with a convenient set of data. A common way to generate this set of data is to use the random number generator as in the previous example. In this problem you will simulate the generation of noise into a two-dimensional image like a television picture. You might then use this "noisy" data set to test your algorithms to remove the noise.

Write a program that reads data from the data file, IMAGE.DAT, into a two-dimensional array. Your program should then add to *random* elements, a *random* noise value. Note that there are two random processes that your program must simulate: the selection of the random elements to contaminate and the selection of the contamination level. Write the contaminated array to the file IMAGE.OUT.

13.9 We saw already the technique for three-point smoothing of a data set where the data points are evenly spaced. Actually, this is a special case of a more general smoothing technique:

$$Y(I) = Y(I-1) + \frac{Y(I) - Y(I-1)}{X(I+1) - X(I-1)}(X(I) - X(I-1))$$

With this formula, we can smooth any set of data, whether or not the data are evenly spaced. Write a program which:

a. Reads in a data set up to 1000 elements

b. Smooths the data with the preceding formula

Take extra care near the ends of the spectrum since some of the data points may not exist.

13.10 In this chapter you learned several smoothing techniques that you may want to try out. Unfortunately, massive listings of data are never easy to read, and it is difficult to see the results of your calculations. To help you, the following program prints a very rough graph of the (X,Y) data pairs to the CRT screen. Enter the subroutine in your system and make any necessary changes to get it to run with your compiler.

```
SUBROUTINE GRAPH(X,Y,POINTS)
INTEGER DIM,WIDTH,HEIGHT,POINTS
REAL X(POINTS),Y(POINTS),SCALE
PARAMETER(WIDTH=70,HEIGHT=20,SCALE=1.1)
CHARACTER*1,CARRAY(0:HEIGHT,0:WIDTH)
INTEGER ZLINE
```

```
      XMIN=X(1)
      XMAX=X(1)
      YMIN=Y(1)
      YMAX=Y(1)
      DO I=2,POINTS
           XMIN=AMIN1(XMIN,X(I))
           XMAX=AMAX1(XMAX,X(I))
           YMIN=AMIN1(YMIN,Y(I))
           YMAX=AMAX1(YMAX,Y(I))
1     CONTINUE
      DO 2 I=0,HEIGHT
             DO 2 J=0,WIDTH
2     CARRAY(I,J)=' '
      XLEN=XMAX-XMIN
      YLEN=SCALE*(YMAX-YMIN)
      XFACT=WIDTH/XLEN
      YFACT=HEIGHT/YLEN
      YMIN=YMIN-0.05*YLEN
      YMAX=YMAX+0.05*YLEN
      IF(XMAX*XMIN.LE.0.0) THEN
            ZLINE=-XMIN*XFACT+0.5
            DO 20 I=1,HEIGHT
20          CARRAY(I,ZLINE)='|'
      ENDIF
      IF(YMAX*YMIN.LE.0.0) THEN
            ZLINE=-YMIN*YFACT+0.5
            DO 25 I=1,WIDTH
25          CARRAY(ZLINE,I)='-'
      ENDIF
      DO 150 I=1,POINTS
            II=XFACT*(X(I)-XMIN)+0.5
            JJ=YFACT*(Y(I)-YMIN)+0.5
            CARRAY(JJ,II)='*'
150   CONTINUE
      DO 200 I=HEIGHT,1,-1
200   PRINT*, (CARRAY(I,J),J=1,WIDTH)
      PRINT210,XMIN,XMAX,YMIN,YMAX
210   FORMAT('XMIN:',F8.3,'XMAX: ',F8.3,
      'YMIN: ',F8.3,'YMAX: ',F8.3)
    1 RETURN
      END
```

13.11 The following voltage-current data has been collected from a certain electronic circuit by taking measurements every 10 μs. Smooth the data and compute the power consumed at each interval (P = voltage × current). Use the technique for smoothing where each data element is examined and replaced by the average of its two nearest neighbors only if it is "significantly different" from them. Choose a sensible criterion for what is significantly different and support your choice:

Time:	0	10	20	30	40	50	60	70	80	90	100
Volts:	5.5	6.3	6.4	7.1	7.9	8.1	8.4	9.0	9.6	10.0	11.0
Current:	1.1	1.8	2.7	4.7	7.5	12.3	20.1	34.1	56.6	93.0	151.4

13.12 A solid steel sphere 10 cm in diameter is embedded with 10 thermocouples located 1 cm from the surface. The sphere is heated to a temperature of 500°C and then quenched in a large turbulent oil bath at 30°C. Since the quenching process is initially very rapid but then slows, the temperature readings are taken at ten nonconstant time intervals given by:

$$t_i = \frac{1}{(i - 11)} \qquad 1 \leq i \leq 10$$

At each of these intervals the following temperature readings were taken:

Time:	0.000	0.100	0.111	0.125	0.143	0.167	0.200	0.250	0.333	0.500	1.00
Temperature:	200	180	145	140	120	102	63	59	45	34	22

Smooth the data, keeping in mind that the data was collected at nonequal time intervals.

13.13 In certain cases you can identify data points that are clearly in error. In such cases the data is usually rejected completely in favor of some approximate data point. In other cases we have some degree of faith in the data point and are unsure if we should replace it with an interpolated data point. One way to smooth data taking this into account is to replace every data point with a weighted average of itself and an interpolated value. The more faith we have in the data, the more heavily we weight the data point itself. The more we suspect that noise is present, the more we weight the interpolated value. Hence, our smoothing formula might look like this:

$$\text{Y(I)} = \text{WY(I)} + (1 - \text{W})\text{F(X(I))} \qquad \text{W} \leq 1$$

where (X_i, Y_i) are the data pairs and F(X(I)) is the interpolated of the data at X(I) taken from any two pairs of data points except (X_i, Y_i). In the simplest case the interpolated value at X_i would involve linear interpolation of the data points at (X_{i-1},Y_{i-1}) and (X_{i+1},Y_{i+1}). Write a program that interpolates the data of Prob. 13.11 using this method of curve smoothing. Select different values of w and comment on the results.

13.14 Many of the techniques that we discussed to smooth one-dimensional arrays can be extended to smooth higher-order arrays. Suppose, for instance, that an image is digitalized to produce an INTEGER array whose elements represent a grey scale with possible values from 0 to 255. Generate the image yourself (10 × 10 is sufficient). Then add to random elements a random amount of noise. Now develop and implement a smoothing technique that will replace each element of the array with some average of the surrounding elements if that element is significantly different from that average.

13.15 After many hours in the laboratory collecting data, two engineering students found that the pressure gauge that they were using was improperly calibrated. With some help from the laboratory assistant, they were able to find the following relationship between the data that they had collected, $P^{meas}(I)$, versus the true pressure, $P^{true}(I)$:

$$P^{meas}(I) = 1.024P^{true}(I) - 0.00165(P^{true}(I))^2 + 0.0074$$

In addition to the calibration problem, they also discovered that the needle in the gauge is slightly bent, thus increasing all their readings by 0.14. Write a program to correct the measured pressure values taking into consideration the new calibration curve and the bias error.

CHAPTER 14

CURVE FITTING

14.1 Introduction

Fitting a curve to experimental data is a mainstay of undergraduate laboratories. The usual procedure is for you to collect data of a property Y versus a parameter X, plot the data, and then determine the mathematical relationship between X and Y. If there were no experimental errors, the job would be simple. For example, in an ideal laboratory, if your instructor asked you to measure the length of a bar as a function of temperature, you might produce data like that shown in Fig. 14.1. The straight-line relationship between temperature and length would be obvious, and you would immediately write the equation:

$$L = mT + \mathrm{b}$$

where m = slope of the curve
b = intercept at $T = 0$

In a real experiment, though, you are more likely to produce data like that illustrated in Fig. 14.2.

In this situation the linear relationship is still obvious, but the scatter in the data presents a dilemma. How will we draw the straight line through the data to give us the "best fit"? Until now we probably would resort to taking a ruler and simply drawing a line that approximately passes through the center of the data band. For many experiments, this is not very precise nor scientific, but it is usually good enough.

We can remove the guesswork from finding the best fit to the data by a process known as *linear regression analysis.* There is a line that will pass through or near the data points that will minimize the total error of the points. We will see shortly that there is a simple equation that can give us the value of the slope and the intercept for this "best" fit. All that we are doing in this sec-

FIGURE 14.1

Straight-Line Fit of Data with No Errors

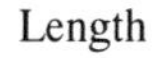

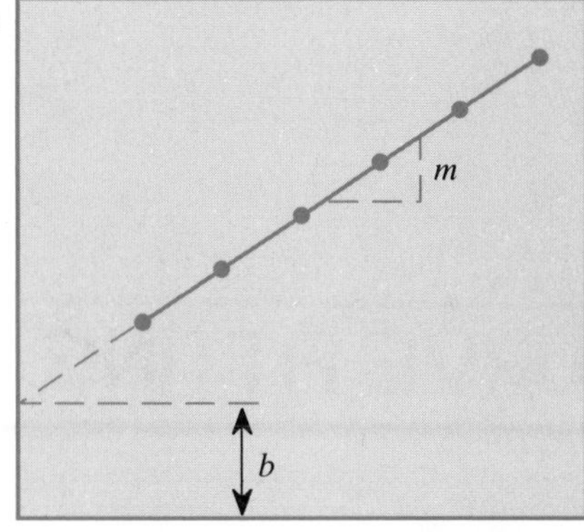

FIGURE 14.2

Straight-Line Fit of Data with Random Errors

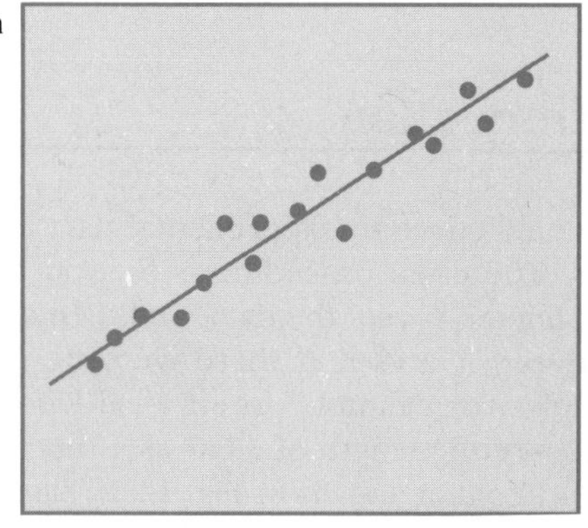

tion is being a little more accurate in drawing the line through the center of the data band. By doing it this way, we are removing any personal bias introduced by hand drawing the line. For example, in most undergraduate labs you know that the line representing the data should pass through a specific point, say, the origin of the graph. Yet, the results may not show that. It is very tempting, therefore, when drawing the line by hand, as in Fig. 14.3a, to force it to pass through the "known" point. But, if you do a linear regression on the data, the true trend in the data will be discovered.

This may not seem like a big problem since these two curves appear to be very similar. But now let's extrapolate these two curves to very high values of X and see which one predicts more accurately the value of Y (see Fig. 14.4).

The small box in the corner of this figure is the region where we collected that data for the first two graphs of the straight-line fit. Notice that when we extrapolate these lines to higher X values, even small errors in the slope can lead to large errors in estimating the value of Y. Our point in showing you this graph is that personal bias in fitting data can lead to significant errors. The linear regression method does a much better job of finding the

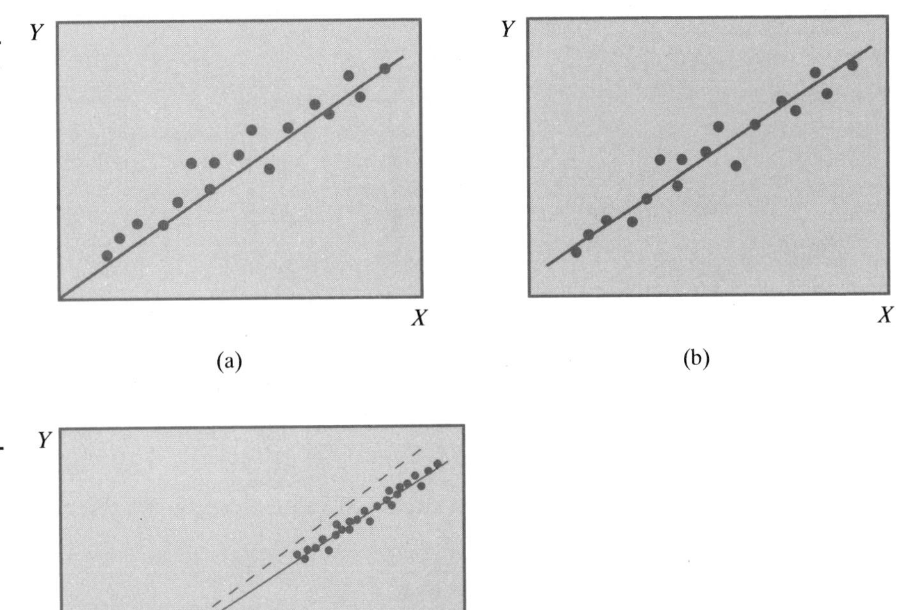

Figure 14.3

Comparison of (a) Fit Forced to Pass through a "Known" Point with (b) Fit Obtained with the Linear Regression Method

Y

X

– – Force fit by hand

— Linear regression

Figure 14.4

Extrapolation of Force Fit and Linear Regression Curves

best fit than the simple hand drawing method. Of course, the data may not fit a straight line at all. But that is another matter.

In this chapter we explore the linear regression method for curve fitting. We will not go into the mathematics of the derivation of the method, only the applications. The method is useful not only for straight-line fits such as the previous ones, but also for many other situations that you are likely to encounter. We will examine several of these situations.

14.2 The Least Squares (LSQ) Method

In Chapter 13 we defined the idea of the deviation:

$$d_i = \overline{Y} - Y_i$$

where $\overline{Y}$ is an expected value and Y_i is the actual value. In the examples of the previous chapter the expected value is the mean value, and so we defined the equations in those terms. In the present case the expected value is more meaningful, so we will use the expected value instead. To give you an idea of the physical significance of the deviations, examine Fig. 14.5. Here, we define the deviations of the individual data points with respect to a line passing near the point.

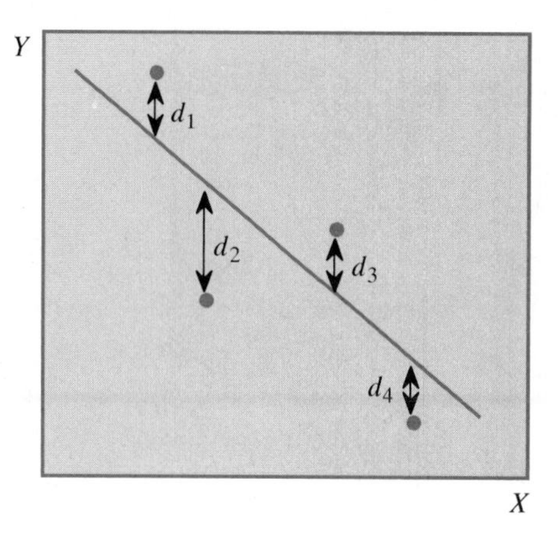

FIGURE 14.5

Definition of the Deviation, d_i

The expected values come from the equation of the line passing through the points, or:

$$\overline{Y} = mX_i + b$$

The process of least squares fitting involves minimizing the total value of:

$$\sum_i d_i^2$$

To do this, we need to use calculus, which we will not discuss. From this analysis comes the results of *m* and *b* for the best fit. These variables are:

$$m = \frac{n\sum X_i Y_i - \sum X_i \sum Y_i}{n\sum X_i^2 - \left(\sum X_i\right)^2}$$

and

$$b = \frac{\sum Y_i - m\sum X_i}{n}$$

We sum over i = 1 to n, where n is the number of data points and X_i and Y_i are the individual data points.

The best way to illustrate how these formulas work is with an example. Assume that we have the following set of data to which we wish to know the best linear fit:

X	Y	
2.00	6.30	$\sum X_i = 30.0$
4.00	3.70	$\left(\sum X_i\right)^2 = 900.0$
6.00	0.90	$\left(\sum X_i^2\right) = 220.0$
8.00	−1.80	$\sum X_i Y_i = -27.6$
10.00	−4.60	$\sum Y_i = 4.50$

The summations that we need appear next to the data. You should try to reproduce these results by hand to ensure that you understand how we obtain them. With these results, we can then calculate the best fit:

$$m = \frac{n\sum X_i Y_i - \sum X_i \sum Y_i}{n\sum X_i^2 - \left(\sum X_i\right)^2} = \frac{5(-27.6) - 30(4.50)}{5(220) - 900} = -1.37$$

$$b = \frac{\sum Y_i - m\sum X_i}{n} = \frac{4.50 + 1.37(30)}{5} = 9.12$$

Just to make sure that these calculated values fairly represent the original data, here is a comparison of that data with the calculated values from the best fit:

X	Y (original)	Y (fit)	Error
2.00	6.30	6.38	+0.08
4.00	3.70	3.64	−0.06
6.00	0.90	0.90	0.00
8.00	−1.80	−1.84	−0.04
10.00	−4.60	−4.58	+0.02

The match between the original data set and the calculated values obtained from the least squares fit is a good one. But scanning data such as this does not tell us how good the match is. For example, it is *always* possible to find the values of m and b by the least squares method, even if the data does not fit a straight line, as illustrated in Fig. 14.6.

FIGURE 14.6

Illustration that the Best Linear Fit May Not Accurately Describe the Data

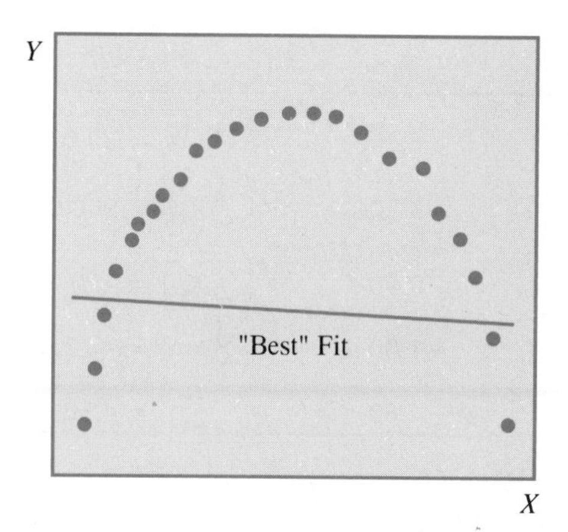

Clearly, the data does not fit the straight line that the least squares method predicts is the best fit. Yet, the least squares method did find a fit. Therefore, after we obtain a fit by using these formulas, we must ask if the fit is a valid one. Ideally, we should have a numerical value that indicates the degree of validity of the fit. After all, we should expect a continuous range of degrees of fit, ranging from a perfect fit to a nonexistent fit, and points between.

The magic number that we seek to show the goodness of fit is the *correlation coefficient, r*, defined by:

$$r = \frac{n\sum X_iY_i - \left(\sum Y_i\right)\left(\sum X_i\right)}{\left[\left(n\sum Y_i^2 - \left(\sum Y_i\right)^2\right)\left(n\sum X_i^2 - \left(\sum X_i\right)^2\right)\right]^{1/2}}$$

For the preceding example, the correlation coefficient becomes:

$$r = \frac{5(-27.6) - (4.5)(30)}{[(5(78.59) - (4.5)^2)(5(220) - (30)^2)]^{1/2}} = -0.99993$$

The magnitude of r will vary between -1.0 and 1.0. The negative sign is unimportant. All that you should focus on is the magnitude of the number itself. A perfect correlation has a magnitude of 1.0. A curve that has no correlation has an r value of 0.0. As the magnitude of r decreases from 1 to zero, the goodness of fit deteriorates. Figure 14.7 shows what the scatter of data about the least squares line might look like for different values of r.

Although the value of r may be high, you must still be careful to compare the calculated results and the experimental results. This is usually best accomplished by visually examining the two. So, before you use the data from the least squares fit, check the results to see if they make any sense.

To convert the preceding ideas into a program, we must focus on the summations. There are a total of seven summations that we must carry out. These are:

Figure 14.7

Relationship between Correlation and the "Goodness" of the Fit

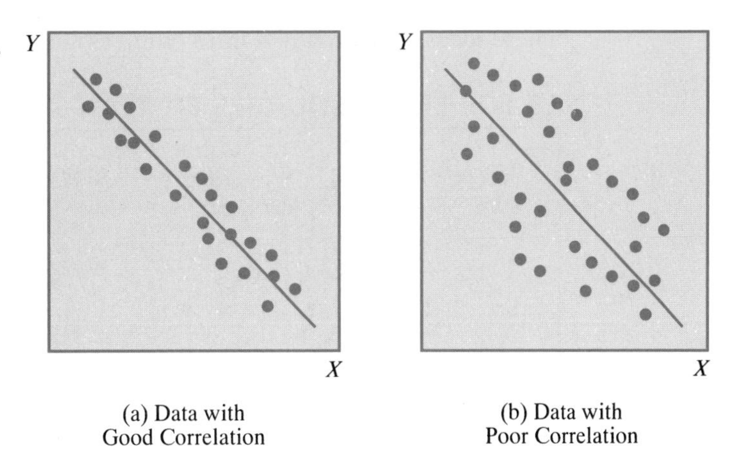

(a) Data with Good Correlation

(b) Data with Poor Correlation

$$\sum X_i = \text{sum of all } X \text{ values}$$

$$\left(\sum X_i\right)^2 = \text{sum of all } X \text{ values squared}$$

$$\sum X_i^2 = \text{sum of all } X \text{ values, individually squared}$$

$$\sum Y_i = \text{sum of all } Y \text{ values}$$

$$\left(\sum Y_i\right)^2 = \text{sum of all } Y \text{ values squared}$$

$$\sum Y_i^2 = \text{sum of all } Y \text{ values, individually squared}$$

$$\sum X_i Y_i = \text{sum of the product of all } X,Y \text{ pairs}$$

Thus, all we need to perform these calculations is the values of all the (X,Y) data points and the number of points to be fit. Sometimes students confuse the terms $(\Sigma X_i)^2$ and ΣX_i^2. The difference is that in the first case we add all the X values first and then square the total. But in the second case we square each X values first and then add. We do a similar process with the Y values to obtain (ΣY_i^2) and ΣY_i^2. Therefore, our algorithm will consist of the following steps:

- Read in all (X,Y) pairs.
- Compute the seven summations.
- Substitute into the equations for m, b, and r.
- Print out the values for m, b, and r.
- Print out a comparison of the original values and the calculated values from the least squares fit.

These ideas are contained in the following program listing:

LS FOR LEAST SQUARES FIT TO A STRAIGHT LINE

PROBLEM STATEMENT: Read in a set of data points, (X,Y), and compute the least squares fit by using the following equations:

$$m = \frac{n \sum X_i Y_i - \sum X_i \sum Y_i}{n \sum X_i^2 - \left(\sum X_i\right)^2}$$

$$b = \frac{\sum Y_i - m \sum X_i}{n}$$

$$r = \frac{n \sum X_i Y_i - \left(\sum Y_i\right)\left(\sum X_i\right)}{\left\{\left(n \sum Y_i^2 - \left(\sum Y_i\right)^2\right)\left(n \sum X_i^2 - \left(\sum X_i\right)^2\right)\right\}^{1/2}}$$

ALGORITHM DEVELOPMENT:

Step 1a: Review the problem statement. Indicate all nouns in **bold** and all verbs in *italics:*

Read in a **set of data points,** (X,Y), and *compute* the **least squares fit** by using the following **equations:**

Formulate questions based on the nouns and verbs:

Q1: How do we read in a set of data pairs?
Q2: How do we compute the least squares fit?

Step 1b: Answer the questions:

***A1:** We first read in the value of N, which is the number of points. Then we use IMPLIED DO LOOPs to read in an array of X data points, and then an array of Y data points.

***A2:** Use a LOOP:

Compute the sum of all X values
Compute the sum of all Y values
Compute the sum of all X^2 values
Compute the sum of all Y^2 values
Compute the sum of all XY products

End LOOP.

Compute additional terms $(\Sigma X)^2$ and $(\Sigma Y)^2$ by squaring terms obtained inside the LOOP.

ALGORITHM:

1. Read in N and X and Y arrays.
2. LOOP:
   SUMX=SUMX+X(I)
   SUMY=SUMY+Y(I)
   SUMX2=SUMX2+X(I)**2
   SUMY2=SUMY2+Y(I)**2
   SUMXY=SUMXY+X(I)*Y(I)
3. Use assignment statements to get $(\text{SUMX})^2$ and $(\text{SUMY})^2$.
4. Use these seven values to obtain m, b, and r from equations given.
5. Print out m, b, and r; original (X,Y) pairs and calculated (X,Y) pairs.

FLOWCHART:

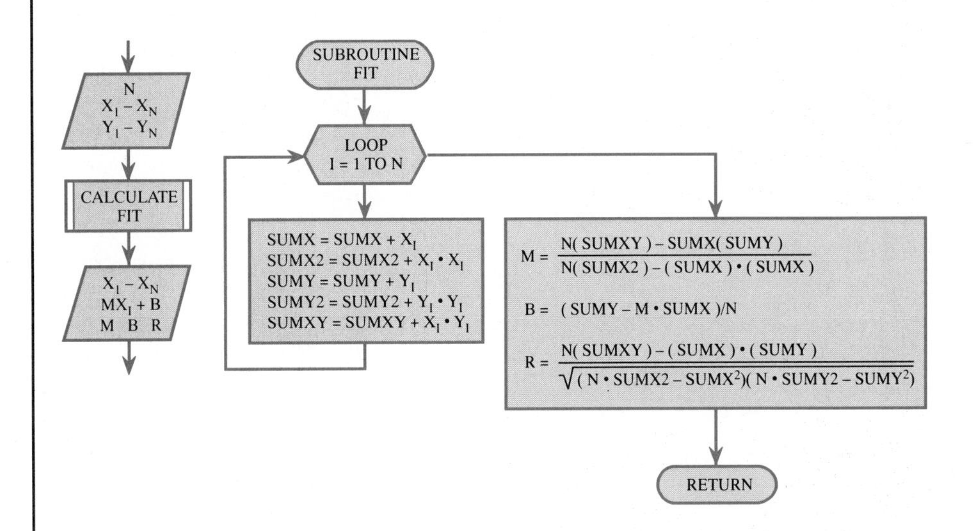

PROGRAM:

```
C     THIS PROGRAM COMPUTES THE LEAST SQUARES FIT
C     TO A SET OF DATA POINTS. IT REPORTS M AND B
C     (SLOPE & INTERCEPT) OF THE BEST FIT. IT
C     THEN DETERMINES THE CORRELATION
C     COEFFICIENT.
C           X, Y -ARRAYS CONTAINING THE DATA SET
C           SUMX - SUM OF X ELEMENTS
C           SUMX2 - SUM OF X ELEMENTS SQUARED
C           SUMY  - SUM OF Y ELEMENTS
C           SUMY2 - SUM OF Y ELEMENTS SQUARED
C           SUMXY - SUM OF PRODUCTS OF X AND Y
C ----------------------------------------------
```

```
      REAL X(1000), Y(1000), M
      PRINT *, 'ENTER NUMBER OF DATA POINTS: '
      READ *, N
      PRINT *, 'ENTER X,Y DATA PAIRS: '
      READ *, (X(I), Y(I), I=1,N)
      CALL LSQ( N, X, Y, M, B, R)
C ----------------------------------------------------------
C
C- THIS SECTION PRINTS OUT THE ACTUAL VALUES AND
C   THE COMPUTED VALUES FOR COMPARISON
C
C ----------------------------------------------------------
      PRINT 25
 25   FORMAT('X',10X,'Y(MEAS)',10X,'Y(FIT)')
      PRINT 30, (X(I), Y(I), M*X(I)+B, I=1,N)
 30   FORMAT( F5.3, 5X, F5.3, 5X, F5.3)
      PRINT 35, M, B, R
 35   FORMAT('SLOPE = ',F7.2,'INTERCEPT=',F7.2,
     1     CORRELATION COEFFICIENT = ', F6.4)
      STOP
      END
C ----------------------------------------------------------
C     THIS SUBROUTINE COMPUTES THE LEAST SQUARES
C     FIT OF A SET OF DATA POINTS (X,Y), STORED
C     IN ADJUSTABLE ARRAYS OF SIZE N READ IN FROM
C     THE CALLING ROUTINE. RETURNS THE VALUES, M,
C     B (SLOPE AND INTERCEPT) AND R, THE
C     CORRELATION COEFFICIENT.
C ----------------------------------------------------------
      SUBROUTINE LSQ (N, X, Y, M, B, R)
      REAL X(N), Y(N), M
      DO 10 I=1,N
            SUMX=SUMX+X(I)
            SUMX2=SUMX2+X(I)**2
            SUMY=SUMY+Y(I)
            SUMY2=SUMY2+Y(I)**2
            SUMXY=SUMXY+X(I)*Y(I)
 10   CONTINUE
      M=(N*SUMXY-SUMX*SUMY)/(N*SUMX2-SUMX**2)
      B=(SUMY-M*SUMX)/N
      R=(N*SUMXY-SUMX*SUMY)/SQRT((N*SUMX2
     1      -SUMX**2)*(N*SUMY2-SUMY**2))
      RETURN
      END
```

LINEARIZING THE FITTING FUNCTION

So far we have only talked about fitting data to a straight-line function. This class of functions is an important one, and many physical systems can be described by it. But there are many other types of functions that you will encounter, as shown in Fig. 14.8. Among the most common functions are the exponential and logarithmic functions.

The following equation gives the form of the exponential curve:

$$Y = ae^X + b$$

and the logarithmic curve is of the form:

$$Y = a \log(X) + b$$

These types of functions occur frequently in many engineering disciplines and you will undoubtedly encounter them often. Fortunately, we can fit data to these functions by the least squares method of the previous section, even though the functions are not linear. Note that if we rewrite the exponential equation, we can obtain a straight-line function. For example, if we let $Z = e^X$, then the equation for the exponential curve becomes:

$$Y = aZ + b$$

which is the same equation that we used previously for the least square fit. We can do this graphically also by plotting X versus Y on a linear plot, or plotting e^X versus Y, as in Fig. 14.9. We call this process *linearizing the fitting function*. By plotting e^X against Y, we are creating a linear fitting function, provided that the data follows this relationship. In a similar way, you can plot Y versus $\log(X)$ if that curve is more appropriate.

The table on the following page is an example of fitting data which follows to an exponential curve, which will give us the best fit for a and b:

FIGURE 14.8

Common Relationships Found in Engineering and Science

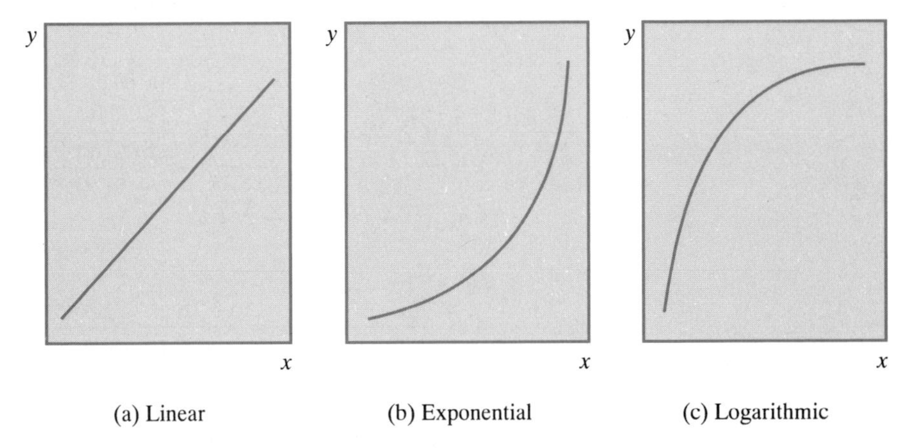

FIGURE 14.9

Linearizing the Fitting Function

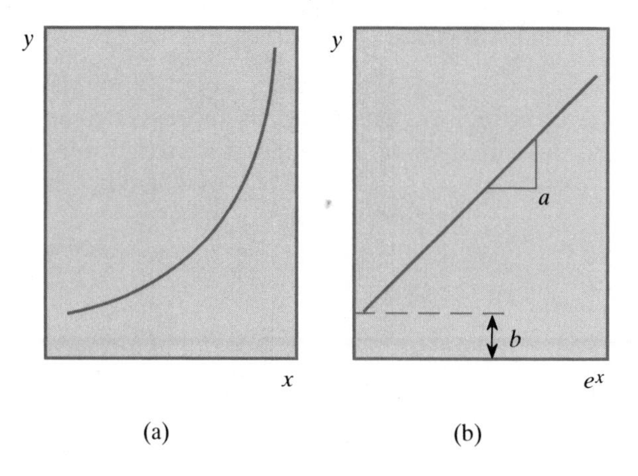

X	e^X	Y	
0.200	1.221	1.39	$\sum X_i = 9.479$
0.400	1.492	2.05	$\left(\sum X_i\right)^2 = 89.851$
0.600	1.822	2.75	$\left(\sum X_i^2\right) = 19.379$
0.800	2.226	3.75	$\sum X_i Y_i = 31.269$
1.000	2.718	4.84	$\sum Y_i = 14.780$

$$a = \frac{\mathrm{n}\sum X_i Y_i - \sum X_i \sum Y_i}{\mathrm{n}\sum X_i^2 - \left(\sum X_i\right)^2} = \frac{5(31.269) - 9.479(14.78)}{5(19.379) - (9.479)^2} = 2.306$$

$$b = \frac{\sum Y_i - \mathrm{a}\sum X_i}{\mathrm{n}} = \frac{14.78 - 2.306(9.479)}{5} = -1.416$$

As in the previous case, you should always compare the calculated results to the original data to make sure that the fit is reasonable:

X	Y (original)	Y (fit)	Error
0.200	1.39	1.40	+0.01
0.400	2.05	2.02	−0.03
0.600	2.75	2.78	+0.03
0.800	3.75	3.72	−0.03
1.000	4.84	4.85	+0.01

Notice in these calculations that we used the data from the e^X column to generate the various summations. The only time that we use the actual X(I) data is when we create the e^X column. For these data, the correlation coefficient is 0.99985. This value, along with a visual comparison of the original data and the calculated ones, gives us a strong indication that we have found a good fit. You should trace through these calculations to be sure that you understand how we derived the results.

We can use the subroutine that we previously generated for the least squares fit to analyze the exponential fit. In the following MAIN program we take the X values and create a new array XNEW containing the values of exp(X(I)). The array XNEW will then be sent down to the subroutine for the least squares fit. Here is the program:

```
C    THIS PROGRAM COMPUTES THE LEAST SQUARES FIT TO AN
C    EXPONENTIAL FUNCTION FOR A SET OF DATA POINTS. IT
C    REPORTS M AND B (SLOPE & INTERCEPT) OF THE BEST
C    FIT. IT THEN DETERMINES THE CORRELATION
C    COEFFICIENT.
C         X, Y  - ARRAYS CONTAINING THE DATA SET
C         XNEW - ARRAY OF EXP(X(I)) POINTS FOR FITTING
C ---------------------------------------------------------
     REAL X(1000), XNEW(1000), Y(1000), M
     PRINT *, 'ENTER THE NUMBER OF DATA POINTS: '
     READ *, N
     PRINT *, 'ENTER THE DATA PAIRS: '
     READ *, (X(I), Y(I), I=1,N)
     DO 10 I = 1,N
           XNEW(I) = EXP(X(I))
10   CONTINUE
     CALL LSQ ( N, XNEW, Y, M, B, R )
C
C---THIS SECTION PRINTS OUT THE ACTUAL VALUES AND THE
C        COMPUTED VALUES FOR COMPARISON
C
     PRINT 25
```

```
25  FORMAT( ' X', 10X, ' Y(MEAS) ', 10X, 'Y(FIT)')
    PRINT 30, (X(I), Y(I), M*XNEW(I)+B, I=1,N)
30  FORMAT( F5.3, 5X, F5.3, 5X, F5.3)
    PRINT 35, M, B, R
35  FORMAT( 'SLOPE = ', F7.2, ' INTERCEPT= ', F7.2,
  1     CORRELATION COEFFICIENT = ', F6.4)
    STOP
    END
```

The SUBROUTINE LSQ remains unchanged. We have only modified the X array sent for the analysis. In other cases you may want to take $\log(X)$ or $1 - \exp(-X)$ or whatever function seems appropriate. This method is a very powerful tool that you will find useful in a many other applications.

14.3 Interpolation

Experimentalists perform curve fitting on data that has experimental scatter in it. There are many occasions, however, when we know precisely the value of a function, but only at discrete values. For example, tables often list the value of a complex mathematical function such as this one for a function known as the error function, $\text{erf}(Z)$:

Z:	1.800	1.900	2.000	2.100	2.200	2.300	2.400	2.500
$\text{erf}(Z)$:	0.9641	0.9713	0.9772	0.9821	0.9861	0.9893	0.9918	0.9938

As long as you want the error function value at one of the listed values of Z, there is no problem. But what do you do if you need the value between two values of Z, for example, $Z = 2.225$? Unless you repeat the calculations used to generate the table with the new value of Z, you will be forced to ***interpolate.*** The process of interpolation, Fig. 14.10, involves estimating a value for the desired point by drawing a straight line through the two nearest points.

Assume that we know the values of a function at points X_1 and X_2. We can then estimate the value at the intermediate point, X, by drawing a straight line between the two points. Notice that we do not obtain the exact value of the function at this point. Instead, we have only an estimate since the function itself is not a straight line. In many cases, though, this estimate is very good. This is especially true if the two points, X_1 and X_2 are close.

FIGURE 14.10

Linear Interpolation

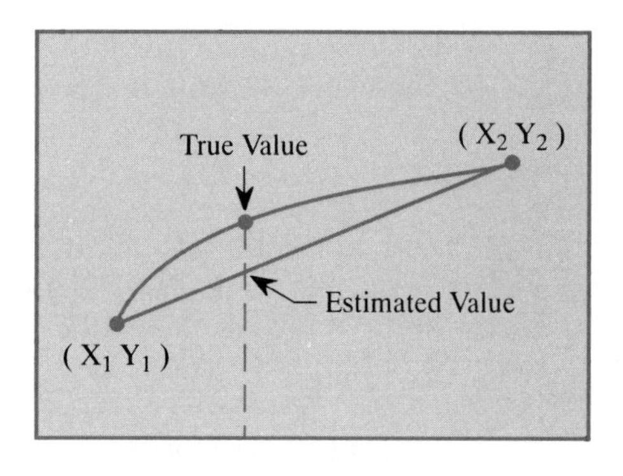

FIGURE 14.11

Similar Triangles Used for Linear Interpolation

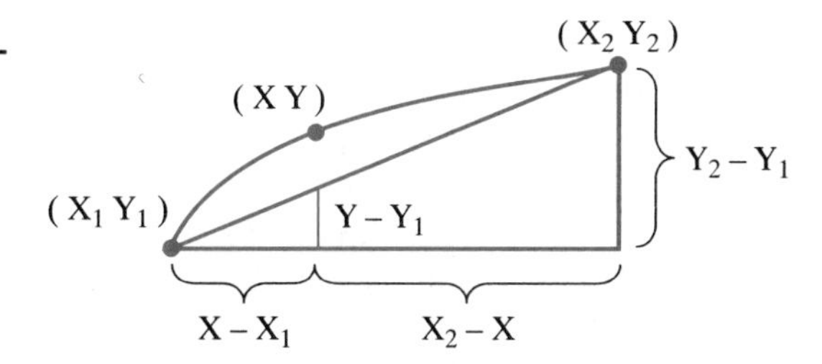

To obtain the estimated value, we must reexamine the preceding curve more closely as shown in Fig. 14.11. Due to the two similar triangles, we can write:

$$\frac{X - X_1}{X_2 - X_1} = \frac{Y - Y_1}{Y_2 - Y_1}$$

Since it is the value of Y that we are looking for, we can rearrange the equation as follows:

$$Y = Y_1 + (Y_2 - Y_1)\frac{(X - X_1)}{(X_2 - X_1)}$$

Since we know all values in the equation, it is a simple matter to calculate the estimated value, Y. For the preceding example for the error function at $X = 2.225$:

$X_1 = 2.200$

$X_2 = 2.300$

$Y_1 = 0.9861$ $\qquad Y = 0.9861 + (0.9893 - 0.9861)\dfrac{(2.225 - 2.200)}{(2.300 - 2.200)}$

$Y_2 = 0.9893$

For these values, the estimate for Y is 0.9869, which is very close to the correct value of 0.98695. Converting this process into a program is very easy. All we must do is to create a FUNCTION SUBPROGRAM that executes the previous formula:

```
C   FUNCTION SUBPROGRAM TO COMPUTE THE INTERPOLATED
C   VALUE, Y, AT SOME POSITION, X, BETWEEN TWO KNOWN
C   DATA POINTS, X1 & X2. THE VALUES OF Y AT THE
C   POINTS X1 and X2 MUST BE KNOWN.
C ------------------------------------------------------
    REAL FUNCTION INTER(X1, X2, X, Y1, Y2)
    INTER = Y1+(Y2-Y1)*(X-X1)/(X2-X1)
    RETURN
    END
```

We can call this from a MAIN program, such as the following:

```
REAL INTER
PRINT*, ' ENTER X1, X2, Y1, Y2 VALUES'
READ*, X1, X2, Y1, Y2
Y = INTER (X1, X2, X, Y1, Y2)
      .
      .
      .
PRINT *, 'INTERPOLATED VALUE = ', Y
STOP
END
```

For functions that are smooth, the preceding linear interpolation will give good results. Sometimes you will need to use more advanced methods. You may have noticed that the linear interpolation forces the function to fit a straight line, although only over a very small interval. If the function is more complex, you may need to fit the function to a parabola or higher-order function.

Exercises

14.1 Write a program to interpolate between the data points for $X = 0.35$, $X = 0.8$, and $X = 0.85$, given the following data:

X:	0.1	0.3	0.5	0.7	0.9
sin(X):	0.0998	0.2955	0.4794	0.6002	0.7833

Compare your answers with the values obtained from the intrinsic function.

14.2 Write a program which computes the correlation coefficient of the following data, then smooth the data (using any smoothing algorithm you like) and compute again the correlation coefficient. Is the correlation coefficient significantly better, about the same, or perhaps worse?

X:	0.0	1.0	4.0	9.0	16.0	25.0	36.0	49.0	64.0	81.0
Y:	10.0	12.0	22.0	28.0	45.0	64.0	88.0	100.0	140.0	172.0

14.3 Consider the following data, which was collected three times using different techniques:

t:	1.00	1.10	1.20	1.30	1.40	1.50	1.60	1.70	1.80	1.90
Set 1:	1.69	1.85	2.07	2.32	2.72	3.12	3.65	4.20	5.15	6.02
Set 2:	1.65	1.92	2.14	2.33	2.71	2.95	3.58	4.06	5.29	6.33
Set 3:	1.64	1.72	1.97	2.18	2.69	3.11	3.79	4.47	5.33	6.09

It is known that the phenomena being observed obeys the law:

$$Y(t) = e*(a*t**2)$$

Using this information, transform the data into data which shows a linear relationship, and determine which method of measurement is most accurate. Use the correlation coefficient.

14.4 Consider the following data:

X:	1.00	1.10	1.20	1.30	1.40	1.50	1.60	1.70	1.80	1.90	2.00
Y:	2.50	2.80	3.24	3.85	4.68	5.80	7.25	9.12	11.47	14.41	18.00

It has been shown that this data must obey the law:

$$Y = C_1 * X ** e + C_0$$

where C_1, C_0, and e must still be determined. It is also known that e is an integer between 1 and 7. Using the initial and final data values (which are known to be accurate), compute the coefficients C_0 and C_1. Now using the correlation coefficient, predict the exponent *e*. (*Hint*: You will have to linearize the data and compare the correlation coefficients for the trials involving each suspected exponent.)

14.5 Employing the model of the following equation and linear regression methods discussed, fit the following data for viscosity (V) as a function of temperature (T) to the equation:

T	40	50	60	70	80
V	1.6	1.3	1.1	0.95	0.85

Read the data from a file. Determine the best fit and use the following relationship to predict the viscosity at any temperature which is entered by the user.

$V = V_0 * T ** M$

14.6 An experiment was conducted in which measurements of current (I) flow through a wire was recorded at various voltages (V). This data is shown here:

V	0.6	1.6	1.9	2.5	3.0	3.5	4.2	4.5	5.0	5.1
I	2.0	1.3	2.6	2.5	2.7	4.7	5.1	6.4	7.8	6.1

After plotting the data, it was not conclusive if the relationship were linear or exponential. Write a program to perform a linear and exponential fit on the data. Determine the correlation coefficient for each plot. Finally, indicate which function fits best and what values were obtained for the coefficients in the fitting equations. Possible fitting functions are:

$I = A_0 + A_1 V$ (linear)

$I = C_1 * E ** (C * V)$ (exponential)

14.7 The following data was obtained from stress-strain measurement of gold at 68° F:

Stress (10**6 psi)	1	2	3	4	5	6	7	8	9
Strain (10**−1%)	0.18	0.32	0.55	0.75	0.84	1.09	1.15	1.40	1.61

a. Plot the data using the plotting routines provided in Chapter 10.
b. Use linear regression to determine the strain when the stress is 4.6×10^6 psi.
c. Using the three-point smoothing method, smooth the data and repeat (a) and (b). Compare the results.

14.8 An experiment was performed to measure the dependence of the viscosity (V) of a polymeric material at various temperatures (T). The resulting data is:

T (°F)	40	60	80	100	120	140
V	282	252	228	210	192	180

Which curve fits the data better, a straight line or an exponential equation? Support your conclusions using the correlation coefficient.

14.9 The concentration of a certain chemical during a reaction is known to obey the exponential law. The concentration versus time measurements during an experiment are as follows:

Time (min)	1.8	16.3	37.0	62.1	79.5	109.7	150.1
Concentration	0.83	0.71	0.56	0.45	0.36	0.28	0.17

Extrapolate values for time equal to 5, 50, and 200 min. How good do you think these values are?

14.10 We can determine how fast one chemical species diffuses in another species by the diffusion equation:

$$D = D_0 \exp(-Q/kT)$$

where R = 1.387 Kcal/mol/deg

T = the temperature in Kelvin

D = diffusion distance in micrometers

Q = an activation energy

Given the following diffusion data, compute the value for Q:

T (°K):	300	450	600	750	900	1050
D (μm):	1.09	1.20	1.31	1.44	1.57	1.72

14.11 This problem involves a two-dimensional interpolation of data. The data that follows are part of a steam table and list the specific volume that steam would occupy for a given pressure and temperature. For example, at 300°C and 10 atm pressure the steam would have a specific volume of 45.0:

	Temperature (°C)				
Pressure (atm)	200	250	300	350	400
1	392.6	404.5	482.1	512.0	541.8
5	78.16	80.59	90.25	102.26	108.24
10	38.85	40.09	45.00	48.03	51.04
20			22.36	23.91	25.43
40			11.04	11.84	12.62

Write a program that will accept temperature and pressure as input and print out the volume interpolated from the preceding table. Generally, this problem calls for three interpolations, and the order may cause different answers. Use the following algorithm:

a. Find the two rows on either side of the input pressure.
b. Interpolate between these two rows for the input temperature. This will yield two volume values, one for each of the tabular pressure entries.
c. Using the two pressure values, interpolate a third time for the temperature. This will yield the desired volume.

CHAPTER

15

ROOTS OF EQUATIONS

15.1 Introduction

Many equations that you encounter in engineering and science can be solved analytically. An analytical solution is one that is a simple, closed-form mathematical solution. The solution to the quadratic equation is the best-known example. But there are many times when an analytical solution to an equation is not possible. An example would be to find the value of X for which the following equation becomes zero:

$$f(X) = e^X[\sin^X(X) - \ln(1/X)] \tag{15.1}$$

We call the value of X where the function value goes to zero the *root*. We can express this by:

$$f(X_r) = 0 \tag{15.2}$$

For the Eq. (15.1), a root exists at $X = 0.981713248$. You can verify this by substituting this value of X into the equation and seeing that the value of $f(X)$ is indeed very close to zero.

To solve Eq. (15.1), we have to resort to one of the numerical techniques described in this chapter. While an analytical solution may exist, it is not easy to derive. Therefore, the numerical approach is an alternate method to obtain the desired result.

As Fig. 15.1 shows, there are four possible types of roots for a general function. Roots exist at any point where the function $f(X)$ crosses the X axis. Thus, the possible situations are:

1. A single root, where the function crosses the axis only once
2. Multiple roots, where the function crosses the axis more than once
3. Repeated roots, where the function is tangent to the axis
4. No roots, where the function does not cross the axis at all.

FIGURE 15.1

Examples of Roots of Equations: (a) Single Root, (b) Multiple Roots, (c) Repeated Roots, (d) No Roots

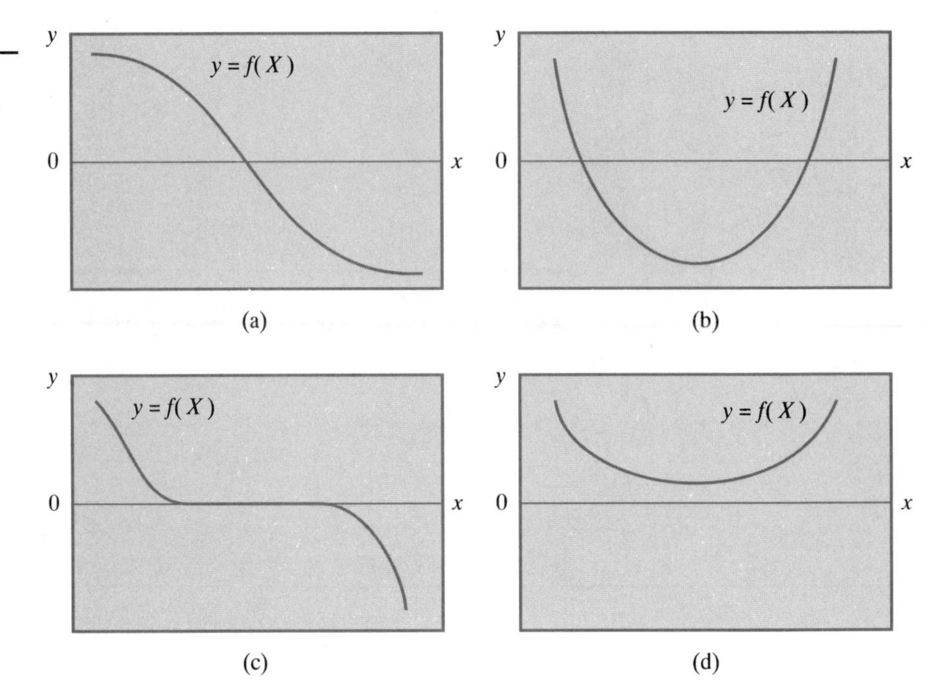

This chapter presents four methods for obtaining the roots of algebraic equations. The first and simplest of the methods is the *search* method. This is a "brute force" approach in which we try many different values of X to find a first approximation for the value of the root. The second method is the *bisection* method where we choose an ever-decreasing range of X values over which we want to search for the root. Both methods are examples of a *bracketing process*. In a bracketing process we search over a small range of X values for the root. In the search method the size of the bracket remains fixed, whereas in the bisection method the bracket size constantly decreases.

The other methods that we describe are the method of *successive substitution* and the *Newton-Raphson* technique. Unlike the bracketing methods that require two data points on either side of the root, these last two techniques use only a single data point to locate the root. These single point methods find the root much faster than the bracketing methods. But there are many equations where these methods do not work well. In these cases you may have to fall back on the bracketing methods. Consequently, there is no one *best* root-solving method. Instead, we may need to use a combination of several methods to find a root. Therefore, we must present all four methods and describe the conditions under which each works.

15.2 The Search Method

All root-solving algorithms require an estimate, or guess, about the value of the root. Another way of stating this is by saying that none of the methods is *self-starting*. Fortunately, providing an initial guess is a simple matter since engineering- and science-related problems usually provide some insight into an approximate range of values for the solution. For example, the value for the coefficient of friction is between 0 and 1.0; the maximum strength of a steel I-beam is 100,000 psi, for example. Simple common sense often is an important first step in locating roots.

The search method is a brute force method. The approach is one of evaluating the function at many points over a wide range of values that seem reasonable. For example, in Eq. (15.1) we started a search at $X = 0.1$ up to $X = 1.0$. We chose this range by examining the functions involved. Since the logarithm of a negative number does not exist, the lower limit on X is zero. Also, note that the logarithmic term in the equation becomes negative when X exceeds 1.0. Therefore, a root, if it exists, must lie between 0 and 1.

The method consists of the following:

- Establish a bracket, $(X_{i+1} - X_i)$, which is a constant size.
- Determine the value of the function at each end of the bracket.
- If the sign of the function changes within the region inside the bracket, then a root exists within that region.
- If the sign of the function does not change inside the bracket, then move the bracket to the next interval and repeat the preceding process.

We illustrate this process for the function, $f(X) = X^4 - 3X^3 + 6$, shown schematically in Fig. 15.2. We know that one of the roots lies in the range between $X = 1$ and $X = 2$. We break this range into a number of convenient intervals. For the purposes of this illustration, the length of each interval will be 0.1. Thus, we set up brackets over the range (1.0 to 1.1), (1.1 to 1.2), and so forth. Here is a listing of the different intervals and the values of the function at each end of the interval:

Figure 15.2

The Search Method

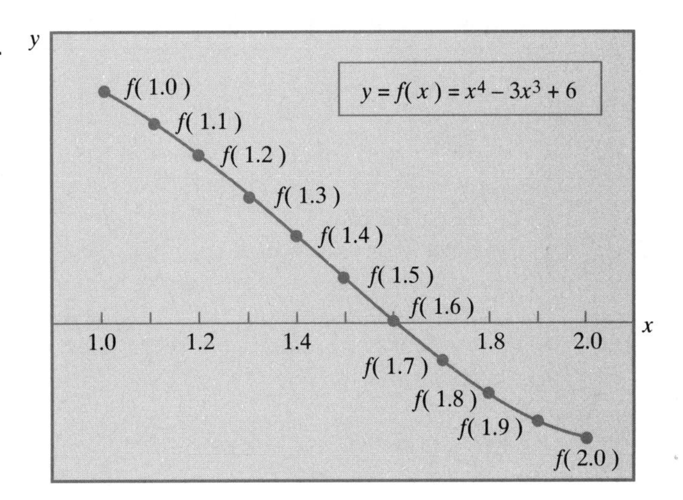

Interval	$f(X_i)$	$f(X_{i+1})$	Product
1.0–1.1	4.000	3.471	+
1.1–1.2	3.471	2.890	+
1.2–1.3	2.890	2.265	+
1.3–1.4	2.265	1.610	+
1.4–1.5	1.610	0.938	+
1.5–1.6	0.938	0.266	+
1.6–1.7	0.266	−0.387	−
1.7–1.8	−0.387	−0.998	+
1.8–1.9	−0.998	−1.545	+
1.9–2.0	−1.545	−2.000	+

In this method we check to see if a root occurs in the selected interval. We do this by determining the sign of the product, $f(X_i)*f(X_{i+1})$, where the two X values are the values at the ends of the intervals. For example, for the interval (1.5 to 1.6), $X_i = 1.5$ and $X_{i+1} = 1.6$. If there is a root inside this interval, there will be a sign change. Thus, if there is a root in the interval, the product will always be negative or zero, as we show in the previous calculations. Note in this example data set that the product of the two functions is always positive, except over the interval, (1.6 to 1.7). Thus, $f(1.6) = +0.266$ and $f(1.7) = -0.387$ and their product is a negative value. Notice that the magnitude of the product is unimportant—only the sign is. If you compare this with the graph of Fig. 15.2, you will see that this is the range where the curve crosses the axis. Shown below is the algorithm and program element to carry out the search. Note once again that the method requires initial guesses to the roots which you must supply.

LS FOR SEARCH METHOD FOR FINDING ROOTS

ALGORITHM:

1. For a given function f, values of Xmin, Xmax, and number of division:
2. Calculate δX.
3. LOOP: For values of X from $X_{MIN} + \delta X$ to X_{MAX}, in steps of δX:
   Multiply $f(X)$ and $f(X - \delta X)$.
   If value of $f(X)$ and $f(X - \delta X)$ is less than or equal to zero, then a root has been encountered (sign change).
4. Print values of $X - \delta X$ and X and the function at these values.

PROGRAM SEGMENT:

```
      SUBROUTINE SRCH ( XMIN, XMAX, N )
      DELTAX = (XMAX-XMIN)/N
      DO 10  X = XMIN+DELTAX, XMAX, DELTAX
            IF (F(X-DELTAX)*F(X) .LE. 0.0) THEN
                    PRINT *,' ROOT IS BETWEEN ',
                             X-DELTAX, X
                    PRINT *,'    F(', X-DELTAX,')
     1                       = ', F(X-DELTAX)
                    PRINT *,'    F(', X,        ')
     1                       = ', F(X)
            ENDIF
   10 CONTINUE
      RETURN
      END
```

The SUBROUTINE SRCH reports all occurrences of roots within the range of consideration. Often we employ the search method to obtain coarse approximations for roots, and then use other methods to refine the values. One advantage of this method is that if a root is present in the range searched, the method will locate it. This is true, provided the step size, δX, is small enough to ensure that the program does not skip over a root. Mathematicians call this quality *convergence*.

A disadvantage with this method is that it is slow, and it will have problems locating repeated roots illustrated in Fig. 15.1c. It should be noted, though, that most methods have problems with repeated roots. To show how this subroutine works, let's use it to find the root(s) of $f(X) = X^2 - 4$ in the range -5 to 5. For convenience, we divide the range into 100 sections. Thus, the value of XMIN, XMAX, and N are -5.0, 5.0, and 100, respectively. Here is a program that calls the SUBROUTINE SRCH to find the roots. Also shown is the FUNCTION SUBPROGRAM that defines $f(X)$:

```
      REAL XMIN,XMAX
      INTEGER N
      PRINT *, ' ENTER Xmin, Xmax, and Number of
     1           divisions '
      READ *, XMIN , XMAX , N
      CALL SRCH (XMIN , XMAX , N )
      STOP
      END
C ----------------------------------------------------------
C     FUNCTION SUBPROGRAM TO EVALUATE F(X) = X**2 - 4
C ----------------------------------------------------------
      REAL FUNCTION F(X)
      F = X**2 - 4
      RETURN
      END
```

When we run this program, this is the dialogue printed on to the screen:

BOLD = COMPUTER RESPONSE *ITALIC = USER INPUT*

```
ENTER Xmin, Xmax, and Number of divisions
        -5   5   100
Root is between -2.00000 -1.90000
        F( -2.00000)= 1.144409E-05
        F( -1.90000)= -.389989
Root is between  2.00000  2.10000
        F( 2.00000)= -9.059906E-06
        F( 2.10000)= .409990
```

If you solve the equation by hand, you will find that it has two roots at exactly $X = 2$ and $X = -2$. But note that the program cannot give the precise answer. It can only give the values of the interval surrounding the root. The program does not provide the precise root. In this example the program reports that it found one root between -1.9 and -2.0 and a second root between 2.0 and 2.1. Thus, the program found the approximate position of both roots.

15.3 The Bisection Method

The search method of the previous section is very limited since the size of the interval over which it searches for the root is fixed. Consequently, you can never find a root to an accuracy smaller than this interval size. In the bisection method we allow the computer to change the interval size so that it can "zero in" on a solution. The bisection method starts with a large interval over which the function changes sign. The user usually enters these values by hand. An alternate method would be to use the search method to make the first approximation. Once the program does this, the computer divides the interval in half and determines in which half of the interval the root lies. The computer re-

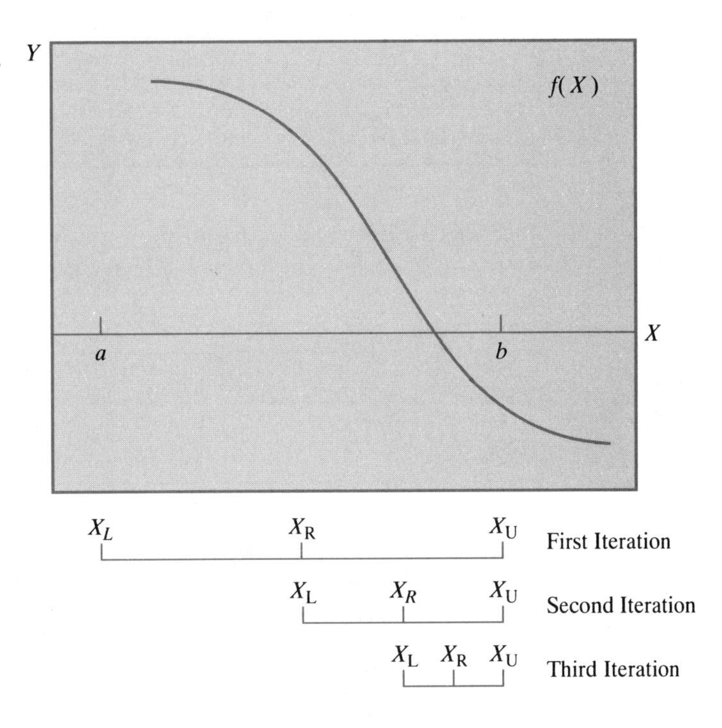

FIGURE 15.3
The Bisection Method

peats this process until it finds a root of the desired precision. The illustration shown in Fig. 15.3 should help you to understand this process.

We start the process by examining the interval (a, b) that we know contains a root. We then divide the interval into two halves, $(a, (a + b)/2)$ and $((a + b)/2, b)$, and find in which half the root lies. Recall from the previous section that we can do this by checking the sign of the product of the functions. Once we find which section contains the root, we repeat the process, but this time only over the half interval containing the root. To show this, let's return to our previous example of $f(X) = X^4 - 3X^3 + 6$, which we saw in our previous example has a root between 1.6 and 1.7. Thus, we start the search over the interval (1.6 to 1.7):

	Interval	$f(X_i)$	$f(X_{i+1})$	Product
(Left half interval)	1.60–1.65	0.266	−0.064	−
(Right half interval)	1.65–1.70	−0.064	−0.387	+

This data tells us that the left half, (1.60 to 1.65), contains the root. Therefore, we take this region and divide it into two halves, (1.60 to 1.625) and (1.625 to 1.65) and repeat the search:

	Interval	$f(X_i)$	$f(X_{i+1})$	Product
(Left half interval)	1.600–1.625	0.266	0.099	+
(Right half interval)	1.625–1.650	0.099	−0.064	−

Since the product is negative in the right half of the interval, the root lies between 1.625 and 1.650. Once again, we divide this region in half and repeat:

	Interval	$f(X_i)$	$f(X_{i+1})$	Product
(Left half interval)	1.6250–1.6375	0.099	0.018	+
(Right half interval)	1.6375–1.6500	0.018	−0.064	−

It may not be obvious that we are approaching the root, but if you look closely, you will see that the $f(X)$ values are getting smaller and smaller. Since the value of $f(X)$ goes to zero when $X = X_R$, we are getting very close after only three iterations. If we continue this process, we will get an even better approximation. To demonstrate this, we carry out this process for 12 iterations. For the purposes of this comparison, we assume that the root lies in the center of the interval that we are examining, or:

$$X_R = \frac{X_L + X_U}{2} \tag{15.3}$$

where X_L = lower X value at the left side of the interval

X_U = upper X value at the right side of the interval

Figure 15.3 schematically shows all these values. The following data summarizes this rapid approach to the root:

Iteration	Interval	X_R	$F(X_R)$
1	1.60000–1.70000	1.65000	−0.06437
2	1.60000–1.65000	1.62500	0.09985
3	1.62500–1.65000	1.63750	0.01753
4	1.63750–1.65000	1.64375	−0.02347
5	1.63750–1.64375	1.64063	−0.00298
6	1.63750–1.64063	1.63906	0.00727
7	1.63906–1.64063	1.63984	0.00214
8	1.63984–1.64063	1.64025	−0.00052
9	1.63984–1.64025	1.64005	0.00081
10	1.64005–1.64025	1.64015	0.00015
11	1.64015–1.64025	1.64020	−0.00019
12	1.64015–1.64020	1.64018	−0.00002

There are two key considerations when performing such an analysis:

- How do we determine the interval for the next iteration?
- How do we stop the process?

We can answer the first question by noting that one end of the interval moves closer to the position of the root. Sometimes it is the left point that moves closer. Other times it is the right point that moves. The decision about which one moves hinges on which side the root is on. Once we figure out which half of the interval contains the root, we reassign values to X_L and X_U for the next iteration. Thus, in the previous example the first iteration has $X_L = 1.6$ and $X_U = 1.7$. When we find that the root is in the left half, we do not need to reassign X_L. We only need to reassign $X_U = X_R$. There are other occasions, such as the third iteration, where we need to reassign the lower bound of the interval.

There are two possible methods that we can use to stop the iteration process. The first is based on the magnitude of the interval and the second is based on the value of the function at the assumed root. If we keep dividing the interval in half, it will quickly become so small that the error between the assumed X_R and the true root will be insignificant. The more common method to terminate the root-solving process is to check the value of the function. If its absolute value evaluated at the assumed root is below some preset limit, the process stops. In the following LS for the bisection method we use the value of the function itself to stop the root-solving process.

The steps, then, for performing the bisection method are as follows:

- Assign values to X_L and X_U that define the interval for the search.
- Divide this interval into two regions, $(X_L - X_R)$ and $(X_R - X_U)$.
- Determine the values of $f(X_L)$, $f(X_R)$, and $f(X_U)$.
- Check to see which half contains the root by checking product $f(X_L)*f(X_R)$:
  If < 0, then the root is in the left half and replace X_U with X_R.
  Otherwise the root is in the other half, and replace X_L by X_R.
- Repeat this process until the value of $f(X_R)$ is below an acceptable limit.

Before attempting to follow the algorithm and the following program, we suggest that you try to perform these steps by hand to make sure that you understand the process:

LS FOR BISECTION METHOD OF ROOT SOLVING

ALGORITHM:

1. Check to see if end points A and B are valid.
2. Assign A to XL, B to XU, and (XL + XU)/2 to XR.

3. LOOP: (While ABS(f(XR)) > eps)
   If f(XR)*f(XL) < 0, then update the end point XU, otherwise update the end point XL.
   Define the new midpoint, XR = (XL + XU)/2.
4. Return the final value of XR.

FLOWCHART:

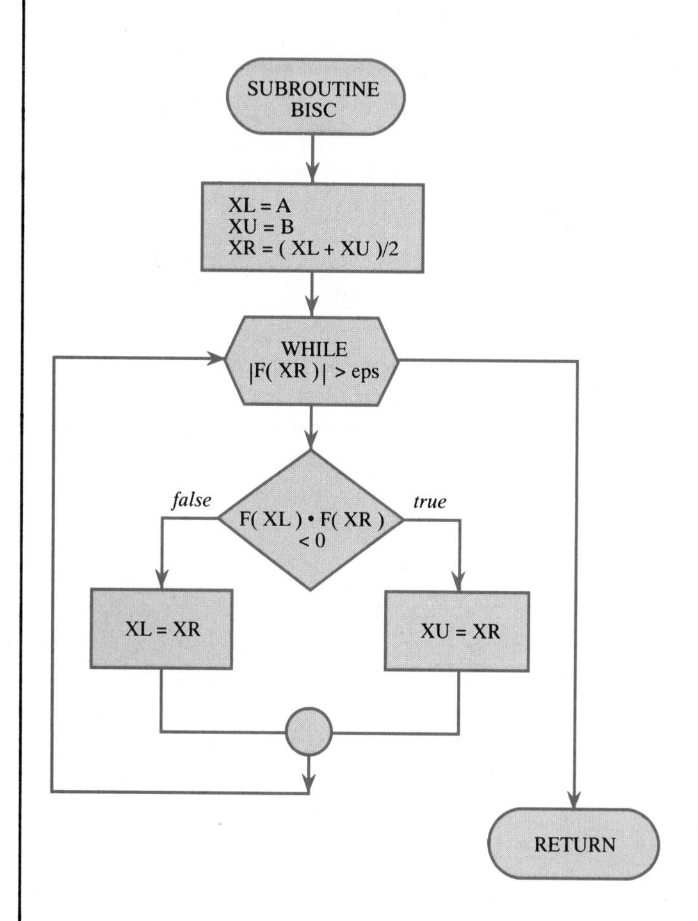

PROGRAM SEGMENT:

```
      SUBROUTINE BISC(A,B,XR,EPS)
      XL=A
      XU=B
      XR=(XL+XU)/2.0
      IF (F(XL)*F(XU) .GT. 0) STOP 'ERROR BISC:
   1     Invalid endpoints'
```

```
DO WHILE (ABS(F(XR)) .GT. EPS)
     IF(F(XL)*F(XR) .LT. 0) THEN
             XU=XR
     ELSE
             XL=XR
     ENDIF
     XR=(XL+XU)/2.0
END DO
RETURN
END
```

To demonstrate how to use the bisection method, we use the preceding subroutine to solve the equation (F(X) = 3.2 X**3 − 5.7 X**2 + 4.25 in the range −10 to 10. Then we refine those value(s) to four-place accuracy using the bisection method. For convenience, we divide the range into 50 divisions. Here are the results from the search method:

BOLD = COMPUTER OUTPUT *ITALIC = USER INPUT*

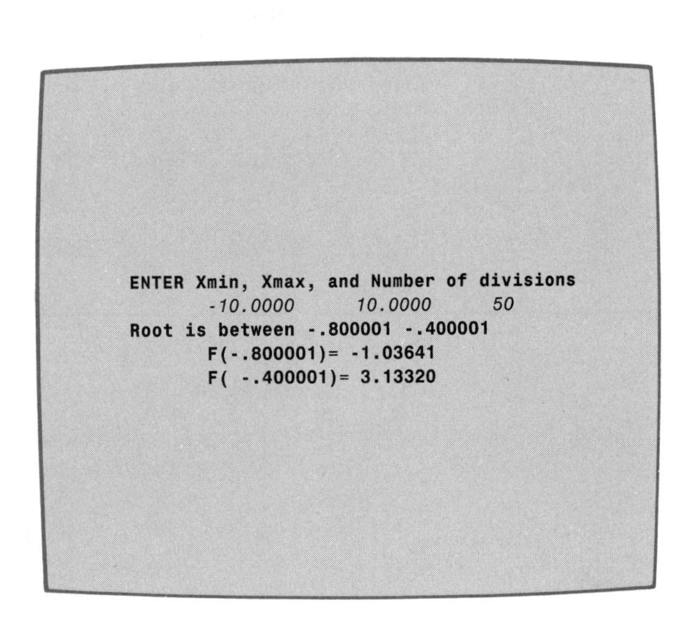

```
ENTER Xmin, Xmax, and Number of divisions
        -10.0000       10.0000       50
Root is between -.800001 -.400001
        F(-.800001)= -1.03641
        F( -.400001)= 3.13320
```

From these results we found only one root in the range between −10 to 10. We made a second run with 500 intervals, but we found no additional roots. Therefore, we feel confident that there is only a single root in this region. The next step is to use the range from these results as the input for the bisection method:

```
C ------------------------------------------------------------
C   MAIN PROGRAM TO CALL THE BISECTION SUBROUTINE.
C   STARTING VALUES MUST BE DETERMINED FROM ANOTHER
C   METHOD AND ARE INPUT AT EXECUTION TIME.
C ------------------------------------------------------------
    REAL XL , XU , ROOT , EPS
    PRINT *,'ENTER LOWER & UPPER VALUES FOR ROOTS AND
 1           DESIRED ERROR'
    READ *, XL , XU , EPS
    CALL BISC(XL , XU , ROOT , EPS)
    PRINT *, 'The root is at ', ROOT
    PRINT *, 'F( ', ROOT,')= ',F(ROOT)
    STOP
    END
C ------------------------------------------------------------
C   FUNCTION SUBPROGRAM FOR THE EQUATION TO BE SOLVED
C ------------------------------------------------------------
    REAL FUNCTION F(X)
    F = 3.2*X**3 - 5.7*X**2 + 4.25
    RETURN
    END
```

After execution the following results appear on the CRT screen:

```
ENTER LOWER & UPPER VALUES FOR ROOTS AND DESIRED ERROR
      -0.8       -0.4       0.0001
The root is at -.727585
      F(-.727585)= -3.814697E-06
```

Although it is not apparent from the preceding dialogue, the bisection method is much faster at locating the root than the search method. However, it requires a guess to establish the range to begin the search. This method looks for the position of the root by looking for a sign change. Therefore, it has the same limitation that the search method has in locating repeated roots.

15.4 The Successive Substitution Method

The previous methods are known as bracketing methods since they create a bracket around the suspected root. Given a range of X values in which a root does exist, these methods will always find a root. Consequently, we need two end points to establish the search range. The next two methods that we present are nonbracketing methods. They only require a single data point to find the root. But under certain circumstances that we will discuss later they do not always find a root. Therefore, you must exercise great care when using these single point methods.

Successive substitution is an easy method to set up. For many problems it will not converge, but since the method takes so little time to execute on a computer, it is worth considering. The method is based on rewriting the equation $f(X_R) = 0$ in the form:

$$X = g(X) \tag{15.4}$$

This is an iterative method, which requires an initial guess for the root value. We then use this initial guess to calculate the next value of the root. Thus, the process looks like this:

First iteration:	X_R guess
Second iteration:	$X_R^{new} = g(X_R^{old})$
Third iteration:	$X_R^{new} = g(X_R^{old})$
Fourth iteration:	$X_R^{new} = g(X_R^{old})$

We will find an approximation to the root when the absolute value of $X - g(X)$ is below some acceptable limit. To illustrate how this works, let's use this approach to find the friction factors, f, for a pipe. This value is a parameter used in other equations to determine the amount of pressure loss that occurs as fluid flows through the pipe. The friction factor depends on two variables: Reynolds number (R_e) and relative roughness (e/d). Reasonable values for R_e and (e/d) are 5×10^5 and 0.003, respectively. A typical range of values for f is from 0.008 to 0.08. The equation relating the friction factors and Reynolds number to the relative roughness is:

$$\frac{1}{f^{0.5}} = -2.0 \log\left(\frac{e/d}{3.7} + \frac{2.51}{R_e f^{0.5}}\right) \tag{15.5}$$

To use the successive substitution method, we must know the function, $g(X)$. We can calculate this by rearranging the equation into the form $f = g(f)$:

$$f = 1 \bigg/ \left[-2.0 \log\left(\frac{e/d}{3.7} + \frac{2.51}{R_e f^{0.5}} \right) \right]^2 \tag{15.6}$$

We now equate $g(f)$ to f:

$$g(f) = 1 \bigg/ \left[-2.0 \log\left(\frac{e/d}{3.7} + \frac{2.51}{R_e f^{0.5}} \right) \right]^2 \tag{15.7}$$

We perform the successive substitution by substituting our old guess at the solution into this equation and using the resulting $g(f)$ value as the new guess. To illustrate, let's make an initial guess of $f = 0.08$ and follow through the first three iterations:

Iteration	f^{old}	f^{new}	$f^{old} - f^{new}$
1	0.08000000	0.02617778	0.05382222
2	0.02617778	0.02617224	0.00000554
3	0.02617224	0.02617227	0.00000003

The first value for f^{old} is an initial guess, which we base on our previous knowledge about similar problems. We then calculate the first value of f^{new} by substituting f^{old} into Eq. (15.7). We then check the difference between f^{old} and f^{new} to see if we should stop. If the difference between the values is small, there is no sense in continuing the process. But here the first iteration has produced a large change in the estimate to the root, so we need to repeat the process at least one more time. Therefore, the f^{new} value for the first iteration becomes the f^{old} value for the second iteration. By the third iteration, the difference between the values is extremely small, which shows that we are very close to the true value for the root. Notice how rapidly the method converges to a solution. We only needed three iterations, although our initial guess was off by more than 300 percent!

Here then is the LS for the successive substitution method for root solving:

LS FOR SUCCESSIVE SUBSTITUTION METHOD FOR ROOT SOLVING

ALGORITHM:

1. For a given starting guess X:
2. LOOP: (while ABS(G(X) − X ≥ eps))
   Evaluate G(X)
   Update X = G(X).

3. Return the latest value of X.

PROGRAM:

```
      SUBROUTINE SUCCSB(X,EPS)
      DO WHILE (ABS(G(X)-X) .GE. EPS)
           X=G(X)
      END DO
      RETURN
      END
```

For the example that we just gave, here is the program that uses the preceding subroutine to solve for the friction within a pipe:

```
C ----------------------------------------------------------------
C    PROGRAM TO DEMONSTRATE THE USE OF THE SUCCESSIVE
C    SUBSTITUTION METHOD FOR ROOT SOLVING. IT UTILIZES
C    A FUNCTION SUBPROGRAM TO CALCULATE G(X).
C ----------------------------------------------------------------
     COMMON /FF/ ED , RE
     PRINT *,'ENTER Reynolds-number and Relative
 1            Roughness:'
     READ *, RE , ED
     PRINT*, 'ENTER initial Guess for friction factor
 1            and limit:'
     READ *, F , EPS
     CALL SUCCSB (F , EPS)
     PRINT *, 'Friction factor is ' , F
     STOP
     END
C ----------------------------------------------------------------
C    THE FUNCTION SUBPROGRAM TO CALCULATE G(X). NOTICE
C    THAT THIS WILL CHANGE FOR EACH EQUATION TO BE
C    SOLVED.
C ----------------------------------------------------------------
     REAL FUNCTION G(F)
     COMMON /FF/ ED, RE
     G=1.0/(-2.0*ALOG10((ED)/3.7 + 2.51/
 1     (RE*F**0.5)))**2
     RETURN
     END
```

Note that we used a COMMON BLOCK in this program. This is because the function $g(X)$ includes other parameters that we may want to change in the future. By using the COMMON BLOCK, we can use the SUBROUTINE SUCCSB without modification. Another approach might have been to rewrite SUCCSB and G to accept the additional variables ED and RE. If SUCCSB had been a subroutine from the system library, access to the FORTRAN code used to create SUCCSB might not be available. Here the only alternative would be to use a COMMON BLOCK.

When we run the program with an initial guess of 0.08, and EPS of 0.00001, the results are:

```
ENTER Reynolds-number and Relative Roughness:
      5.0E5       0.003
ENTER initial Guess for friction factor and limit:
      0.08        0.00001
Friction factor is 2.644210E-02
```

For this problem the successive substitution method converged upon a solution. We leave it to you to verify that the value returned is indeed a valid root. It turns out that for practical ranges of (e/d) and R_e, the method is suitable for calculating the friction factor. It should be noted that there are many problems in which the method never converges on a root. This is a problem that the method shares with all other single point methods. But because of the method's simplicity in programming, it is a good first process to try.

15.5 Newton-Raphson Method

The final root-solving method that we discuss is the Newton-Raphson method. Like the method of the previous section, Newton-Raphson is also a single point method, in which we require only a single initial guess to start the process. This method is one of the most rapid and versatile of the root solvers. However, like the successive substitution method, convergence (finding a root) is not always guaranteed.

The approach that the Newton-Raphson method uses is to attempt to find the root by fitting the curve to a straight line near the initial guess. We determine the root for this straight line, and then use this for the next iteration. Figure 15.4 illustrates this process.

Figure 15.4

Illustration of the Newton-Raphson Method

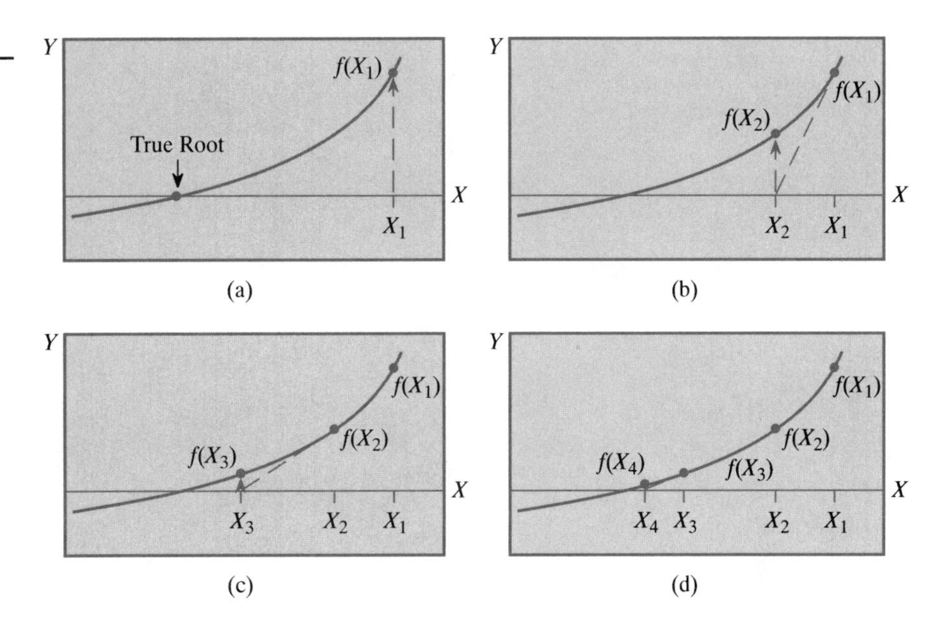

We start with an initial guess for the root such as X_1 as in Fig. 15.4a. Then we construct a tangent line that passes through the curve at $f(X_1)$ with a slope of $f'(X_1)$, Fig. 15.4b. This is the tangent line in the figure passing through the point $(X_1, f(X_1))$. Notice that this constructed line intersects the X axis at point X_2. This will be our second guess for the root. Once we find X_2, we draw a new tangent line, which now passes through the point $(X_2, f(X_2))$ and with a slope $f'(X_2)$. We repeat this process, Figs. 15.4c and 15.4d, until we are sufficiently close to the root. Usually, we use a condition based on the value of f at the suspected root, X_{root}. Examine the curves in Fig. 15.4 carefully and follow the process for finding the root before reading on.

There are conditions when the Newton-Raphson method cannot find a root. This typically occurs when the slope is close to zero at the point under examination. This causes the next guess at the root to be very far from the current location. Figure 15.5a illustrates this. Note that if we follow the procedure outlined that the subsequent guesses for the roots are actually farther away from the real root, rather than closer to it. For example, we show in the figure a slope of approximately zero at our initial guess of X_0. This then suggests that the next guess, X_1, is at an unreasonable distance from the real root.

There is another situation, shown in Fig. 15.5b, where the method cannot find the root. This is where the presence of a local minimum can cause the search to go in the wrong direction. A similar thing will happen if the local minimum is between the initial guess and the actual root.

The process of deriving the Newton-Raphson method is:

- Given the point $(X_n, f(X_n))$ and the slope $f'(X_n)$, derive an equation for a line using the point-slope equation (which follows).

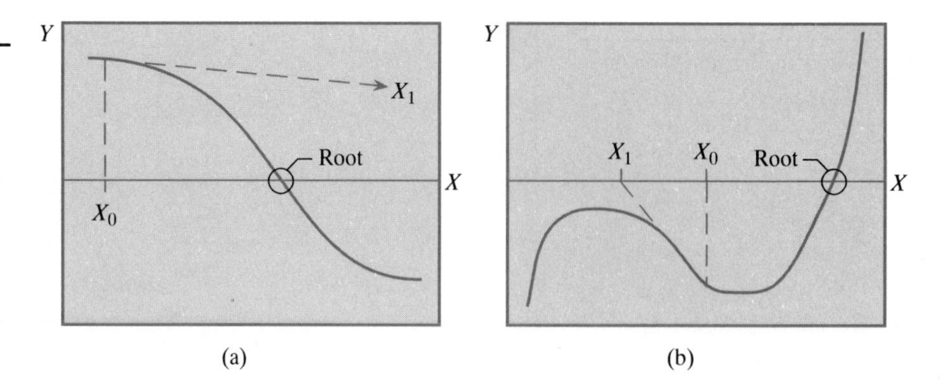

FIGURE 15.5

Cases Where the Newton-Raphson Method Does Not Converge on a Solution

- Solve for the root of the linear equation just developed. The resulting value is the next guess for the root.

The equation for the slope at the point under investigation is:

$$m = (Y - Y_0)/(X - X_0) \tag{15.8}$$

where m = slope at point X_0, Y_0
X_0, Y_0 = point that the line must pass through
X = independent variable
Y = dependent variable

Rewriting, we obtain

$$Y = m(X - X_0) + Y_0 \tag{15.9}$$

We can find the root by setting $Y = 0$:

$$X_{\text{root}} = X_0 - \frac{Y_0}{m} \tag{15.10}$$

For the specific example that we have set up, we make the following substitutions:

$$X_0 \to X_n$$
$$Y_0 \to f(X_n)$$
$$m \to f'(X_n)$$
$$X_{n+1} \to X_{\text{root}}$$

When making these substitutions, we obtain the final equation:

$$X_{n+1} = X_n - \frac{f(X_n)}{f'(X_n)} \tag{15.11}$$

Before showing you the program to carry out this method, we should review the meaning of the individual terms in the last equation. The most convenient way is as follows:

$$\begin{matrix}\text{New guess} \\ \text{for root}\end{matrix} = \begin{matrix}\text{previous guess} \\ \text{for root}\end{matrix} - \frac{\text{(value of function at previous root)}}{\text{(slope of curve at previous root)}}$$

Note that this method requires the evaluations of the derivative to locate the root. Recall, we presented the LS for the derivative in the chapter for subroutines. Therefore, everything that we need to implement the Newton-Raphson method is available. The algorithm and the subroutine for this method follows:

LS FOR NEWTON-RAPHSON METHOD FOR ROOT SOLVING

ALGORITHM:

1. Given the initial guess X, the function F and its derivative DFDX, and EPS.
2. LOOP: (while abs(F(X)) is greater than or equal to EPS):
   If false, then the root was found and returned.
   If true, then implement Eq. (15.11).

FLOWCHART:

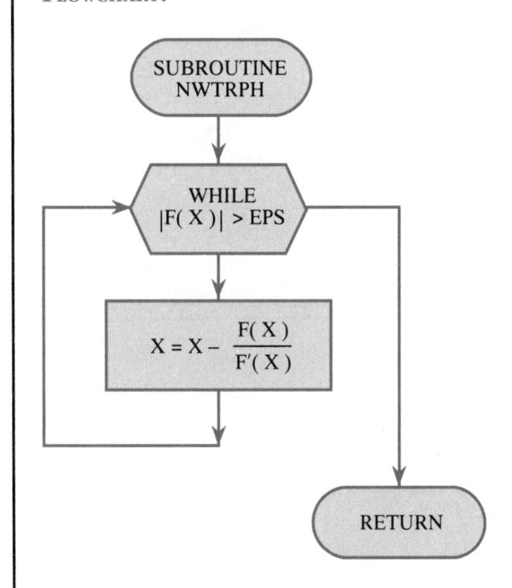

PROGRAM:

```
SUBROUTINE NWTRPH(X, EPS)
DO WHILE ( ABS(F(X)) .GE. EPS)
      X = X - F(X)/DFDX(X)
END DO
RETURN
END
```

A simple example of this method is the problem of finding the square root of a number, Z. This may seem like a trivial example, but it illustrates the procedure for a large class of problems. To solve for the square root of a number, we write the equation:

$$f(X) = X^2 - Z \tag{15.12}$$

If X is the square root of Z, then notice that $f(X)$ will be zero. Thus, we can solve for the square root by finding the root of this equation. Notice in the program that follows that we must make an initial guess for the square root of the number. Of course, we could use the search method to do that for us. But here we enter a number manually. Note also that we must supply a value for EPS, which is the allowable error in the root before the process terminates.

In the following example program we set EPS to 1.0E-5 and Z to 123.45:

```
C ------------------------------------------------------------
C    PROGRAM TO DEMONSTRATE THE USE OF THE
C    NEWTON-RAPHSON METHOD TO SOLVE AN EQUATION. IN
C    THIS EXAMPLE, WE ARE SOLVING THE EQUATION
C    X**2-Z=0.
C ------------------------------------------------------------
     PRINT *,'ENTER initial guess for the square root
 1           of 123.45'
     READ *, X
     CALL NWTRPH(X, 1E-5)
     PRINT *, 'The estimated square root of 123.45
 1             is ',X
     ERROR = SQRT(123.45) - X
     PRINT *, 'The error with the system value is ',
               ERROR
     STOP
     END
C ------------------------------------------------------------
C    FUNCTION SUBPROGRAM TO EVALUATE THE DESIRED
     FUNCTION
C ------------------------------------------------------------
     REAL FUNCTION F(X)
     F=X**2-123.45
     RETURN
     END
C ------------------------------------------------------------
C    FUNCTION SUBPROGRAM TO EVALUATE DERIVATIVE
C ------------------------------------------------------------
     REAL FUNCTION DFDX(X)
     DFDX=2*X
     RETURN
     END
```

Here is what appears on the CRT screen when we run this program, along with the NWTRPH SUBROUTINE:

BOLD = COMPUTER OUTPUT *ITALIC = USER INPUT*

```
ENTER initial guess for the square root of 123.45
      10.0
The estimated square root of 123.45 is 11.1108
The error with the system value is .000000
```

The Newton-Raphson method is versatile and quick to converge; however, divergence is possible. A drawback of the method is that it requires the evaluation of the derivative. In the preceding example we have written an explicit FUNCTION subprogram for the derivative since it is an easy function to evaluate. In many cases, however, it may be easier for you to use the numerical derivative. We presented an LS for this purpose in Chap. 9.

15.6 Validation

The validation process for finding a root is straightforward. Simply substitute the answer into the equation to be solved and see if it returns a value close to zero. Most of the methods that we presented so far have used this approach to determine when to stop. So, we consider the methods to be self-validating. One point, though, should be considered. The method used to solve the root is validated by the process described, but the function to be solved may not be valid. Consider the function that we used for the successive substitution problem:

$$g(f) = 1 \Big/ \left[-2.0 \log\left(\frac{e/d}{3.7} + \frac{2.51}{R_e f^{0.5}}\right)\right]^2$$

A common mistake is to enter the preceding line as follows:

```
G = 1.0/(-2.0*ALOG((E/D)/3.7 + 2.51/(RE*F**0.5)))**2
```

The error present is that ALOG(X) on most compilers is the natural log(log base e), not the intended log base 10. The method will find a root, but it is not the solution required. The function required is ALOG10(X), which we used in the example. To detect such problems, it is a good idea to check manually the root once. This will validate the function's definition.

15.7 Summary

In this chapter we have presented four methods for solving for the root of an algebraic equation. The root of an equation is the value of the independent variable that gives the function a value of zero. They were the search method, the bisection method, the successive substitution method, and the Newton-Raphson method.

The search method divides an interval to be searched into many points. We evaluate each point, and at locations where the sign of the function changes we report those points. This method is simple, and will locate a root if the step size is small enough to ensure that we do not inadvertently skip over the root.

The bisection method requires an interval in which a sign change is already present. This ensures that a root is present within the interval being searched. The bisection method divides the interval into two equal parts. By looking at the sign of the function at the midpoint, you can reduce the range in which the root exists by 1/2. the process repeats until the root is within a practical limit. This method is much faster than the search method, but because it requires an interval with a root present, we often use the search method to obtain such information.

The successive substitution method requires only a single point to start its iteration. By rewriting the equation $f(X) = 0$ into the form $X = g(X)$, we make successive approximations for the value of the root. This method does not always converge. For convergence, the slope of $g(X)$ must be small near the root. The process also must start near the root. Its key advantage is simplicity. Because of this, we commonly use the successive substitution method as a first attempt to solve a problem.

The last method discussed was the Newton-Raphson method. This method uses the derivative of the function to obtain better estimates of the root. Based on an initial guess, the program projects a tangent line from this point on the curve. This line passes through the point and is tangent to the curve at the guess. We assume that the location where the line crosses the X axis is the next approximation for the root. The process repeats until we obtain a satisfactory answer for the root. This method does have conditions where it will not converge. But, when it does locate a root, it usually converges quickly. This method is the only method presented that can locate repeated roots. For repeated roots, though, the convergence is slower.

Exercises

15.1 Solve the example used in the successive substitution section by the Newton-Raphson method.

15.2 Using Newton-Raphson's method, determine how many positive roots can be evaluated for the function:

$$f(X) = \tan(X) - X$$

Use eps equal to 0.0001. What is the limiting factor?

15.3 Using the successive substitution method, determine the root for the equation:

$$f(X) = 20 - 50 \times 10^{-6}(80 + 10R^2)^2$$

Use 0.0001 for eps.

15.4 The log base Y of a number X can be determined by using the definition of logarithm:

$$X = Y^{[\log_Y(X)]}$$

or, rearranging to make it convenient to find the root, we have:

$$0 = Y^{[\log_Y(X)]} - X$$

Write a program that will return the value of $\log_Y(X)$ for any value of X (positive) and Y (positive) to four-decimal places of accuracy. Use any method suitable.

15.5 Using the search method, determine how many real roots exist for the following polynomial:

$$X^4 - 10X^3 + 35X^2 - 50X + 24 = 0$$

Then use the bisection method to obtain four-decimal place accuracy.

15.6 For turbulent pipe flow, the friction factor for a smooth pipe can be approximated by the equation:

$$\frac{1}{f^{(1/2)}} = 2\log(R_e f^{(1/2)}) - 0.8$$

f is positive and below 0.1. Solve this equation for f by any of the methods presented for R_e equal to 10^5 and 10^6. Three-decimal place accuracy is required.

15.7 Find all roots for the equation over the range $X = 0$ to 5:

$$f(X) = \sin(X) - 2\cos(2X)$$

Four-place decimal accuracy is required.

15.8 Solve the equation:

$$\cos(q) = 0.6\cos(q/2)$$

using any of the methods stated. Four-place accuracy is required.

15.9 A four-bar linkage is shown in the following figure:

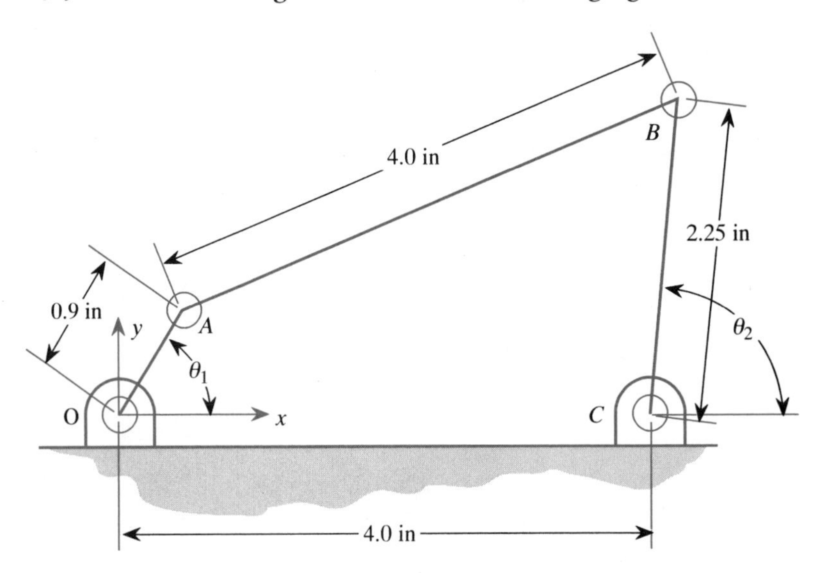

As the input crank *O-A* rotates (changing θ_1), the output link *B-C* rocks. To calculate the corresponding output angle θ_2, we can use the following equation:

$$4.0 = [(0.9\cos(\theta_1) - (4 + 2.25\cos(\theta_2)))]^2 + (0.9\sin(\theta_1) - (2.25\sin(\theta_2)))^2]^{1/2}$$

This equation states that the distance between points *A* and *B* is fixed by the coupler link. For values of θ_1 equal to 0, 45, and 90 through 315°, determine the corresponding output link position. Use the Newton-Raphson method. Results should be accurate to four decimal places in degrees.

To approximate the derivative, use either the following function or the subroutine given in Chap. 10:

```
REAL FUNCTION DFDX(X)
DFDX=(F(X+0.001)-F(X-0.001))/0.002
RETURN
END
```

15.10 A 1 kg steel mass is supported by a spring over a magnet. The force which the magnet exerts on the steel is inversely proportional to the square of the distance between the two, $F_m = C/(X^*X)$. At equilibrium the spring force $F_s = k\,dx$ is exactly balanced by the force of the magnet and the force of gravity. Hence:

$$C/(X^*X) + mg = k\,dx$$

When the mass is removed from the spring, its free end is at a height of 10 cm over the magnet surface. Let $c = 0.1$, $k = 95$, and $g = 9.8$. Write a program to compute the equilibrium position of the mass.

15.11 A beam simply supported at both ends is subject to a concentrated load, P, at the point "a" measured from the left end. Beam theory tells us that the deflection of the beam at some point $a < X < L$ is given by the formula:

$$EIY = \frac{P(L - a)X^{**}3}{6L} + \frac{P(X - a)^{**}3}{6} + \frac{Pa(X - L)^{**}3}{6L} + \frac{xP(L - A)^{**}3}{6L} - \frac{XP(L - A)L}{6}$$

where E is Young's modulus, I is the area moment of inertia of the cross section, Y is the deflection, and L is the length of the beam. Let $P = 10{,}000$, $a = 4.0$, $L = 10$, $E = 30{,}000$, and $I = 0.02$. Find the location $X \geqslant a$, where the beam deflection is 0.4.

15.12 Use the bisection method to determine the root for the following equation in the interval $0 < X < \pi/2$.

$$f(X) = \frac{\tan(X)}{\tanh(X) - 1}$$

15.13 Sectioning methods are guaranteed to converge if an appropriate interval is chosen. On the other hand, these methods converge much slower than the dangerous but fast Newton-Raphson method. One way to take advantage of both these methods while avoiding their pitfalls is to use a sectioning method to close-in on the root and then apply the Newton-Raphson method to finish off the solution. For the objective function:

$$f(X) = 0.1\cos(X) + (5X)^{**}3$$

write a program that uses sectioning to get a "rough" initial guess for the root, then apply the Newton-Raphson method to fine tune the solution.

15.14 Of course, there are problems which baffle even the most powerful algorithm. Consider, for example, the function $f(X) = \sin(1/X)$. This function has an infinite number of roots in the interval $0 < X < 1$. Sectioning methods will always find roots in some subinterval with a sign change, but the accuracy may suffer if we don't choose our termination condition carefully. The Newton-Raphson method may work, but the root we get will probably not be the root closest to our initial guess. Experiment with the bisection method and the Newton-Raphson method using this function and see how many roots you can locate.

15.15 Finding multiple roots is always a difficult problem. Sometimes we know enough about the function—for instance, it might be periodic—so that we can focus our attention on specific intervals. A more general but less fool-proof method is to find first any root, r, in an interval, then divide the objective function by $(X - r)$ and continue searching for other roots. Write a program which uses the bisection method to find multiple roots. Test your problem on the function $f(X) =$ ((X − 10)**3)*(X + 1)**2 whose roots are obvious by inspection.

15.16 A repeated root occurs when a function touches the X axis but does not cross. Clearly, such a root cannot be found by the bisection technique and even the Newton-Raphson method converges slower than expected. Suppose you have located a root using this method, how can you tell if it is a higher-order root? (*Hint*: See the previous problem.) Write a program which uses the Newton-Raphson method to find a root. Before the program terminates it should determine the order of the root.

15.17 There is a technique known as the *false positioning method* which is something like a cross between the bisection method and the Newton-Raphson method. In this method an interval XL to XU is chosen. In the bisection method the estimate of the root is the average of the interval limits. If an additional iteration is necessary, the original interval is cut in half before proceeding. In the false position method, however, the estimate for the root takes into account the magnitudes of F(XL) and F(XU). For example, if F(XL) is much closer to zero than F(XU), it is likely that the root is closer to XL than to XU. Based on this method, the estimated root, XR is:

$$XR = XU - \frac{F(XU)\,(XL - XU)}{F(XL) - F(XU)}$$

Write a program to implement the preceding method. How will you validate this program?

15.18 The cost required for a company to produce certain items, as well as the income derived from the units' sales, or revenue, is projected by the following formulas:

Cost = −0.012X**2 + 28X + 5000

Revenue = 0.048X**2 + 5X

An analysis is to be made to determine the number of units that must be manufactured so that the production costs balance the returned income. The point at which this happens is called the breakeven point. Write a program to determine the breakeven point using the bisection method.

CHAPTER

16

SOLVING SYSTEMS OF LINEAR EQUATIONS

16.1 Systems of Equations

Many problems in engineering involve *systems of equations,* which you may also know as *simultaneous equations.* The general form of such systems is:

$$
\begin{array}{l}
a_{11}X_1 + a_{12}X_2 + a_{13}X_3 + \cdots + a_{1n}X_n = b_1 \\
a_{21}X_1 + a_{22}X_2 + a_{23}X_3 + \cdots + a_{2n}X_n = b_2 \\
\vdots \\
a_{n1}X_1 + a_{n2}X_2 + a_{n3}X_3 + \cdots + a_{nn}X_n = b_n
\end{array}
\tag{16.1}
$$

For such systems, X_1 through X_n represent the unknowns for which we must solve. The a_{ij}'s represent coefficients of the unknowns, and the b_i's are constants. To illustrate how often these systems of equations may arise, we reexamine the simple problem for an end-loaded, hinged beam, reproduced in Fig. 16.1.

As in Chap. 4 when we first encountered this problem, we create force and torque balance equations describing the equilibrium of the system. However, we arrange the resulting equations in a way to be consistent with our notation for systems of equations given in Eq. (16.1):

$$
\begin{array}{lll}
T(X\sin(\theta)) & & = W_1L + W_2(L/2) \\
T(\cos(\theta)) & - H & = 0 \\
T(\sin(\theta)) & + V & = W_1 + W_2
\end{array}
\tag{16.2}
$$

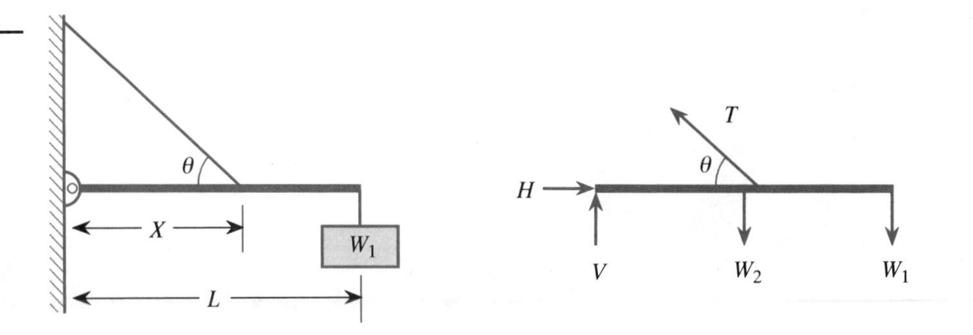

FIGURE 16.1

Force Diagram of End Loaded, Hinged Beam

As before, the variables for which we need solutions are T, V, and H. All other quantities are constants. Thus, in the notation of Eq. (16.1):

$$
\begin{aligned}
a_{11} &= X\sin(\theta) & a_{12} &= 0 & a_{13} &= 0 \\
a_{21} &= \cos(\theta) & a_{22} &= -1 & a_{23} &= 0 \\
a_{31} &= \sin(\theta) & a_{32} &= 0 & a_{33} &= 1
\end{aligned}
\tag{16.3}
$$

Similarly, we can write the constants, b_i, as:

$$
\begin{aligned}
b_1 &= W_1 L + W_2(L/2) \\
b_2 &= 0 \\
b_3 &= W_1 + W_2
\end{aligned}
\tag{16.4}
$$

We will find this notation to be very useful in the following sections. This subscripted notation allows us to manipulate more easily large quantities of data and will permit us to find the solutions for the unknowns in an efficient manner.

In this simple example so many coefficients are equal to zero that we can solve these three equations with a minimum of effort as we demonstrated in Chap. 4. In other problems, though, most of the values of a_{ij} and b_i will not be zero, and it is a more challenging job to find the solutions. We review several methods in this chapter for solving such systems. Some of these you already know. Others will be new to you. But in all cases we present a complete description of the method, the development of the algorithm, and the computer program. At the end we also present more advanced techniques that are necessary for very large systems (20 or more variables). These large systems pose special problems that cause the simpler methods to be prohibitively tedious.

We begin our discussion by developing Cramer's rule for the special case of two equations with two unknowns. This is an important place to start since it gives us the opportunity to introduce the concept of *matrices*. We will find these invaluable for further discussion of more complex systems. Matrices are the mathematical equivalent of the two-dimensional array that we have used so often. Matrices are so important for the solution of systems of equations that we will spend a considerable amount of time discussing them. Once we do that, we will then present Cramer's rule for larger systems. We typically use Cramer's rule for small systems of equations (usually up to 6 equations with 6 unknowns). For larger systems (up to 20 equations) we will need to introduce a second technique, the gaussian elimination method. The chapter concludes with a discussion of very large systems of equations.

16.2 System of Two Linear Equations with Two Unknowns

Consider the following system of equations:

$$\begin{aligned} a_{11}X_1 + a_{12}X_2 &= b_1 \\ a_{21}X_1 + a_{22}X_2 &= b_2 \end{aligned} \tag{16.5}$$

We can solve for the unknowns, X_1 and X_2, by taking the first equation and solving for one of the unknowns in terms of the other. We then substitute that equation into the second equation resulting in a single equation with a single unknown. By following this procedure, we can simply derive equations for X_1 and X_2 in terms of constants:

$$\begin{aligned} X_1 &= \frac{b_1 a_{22} - b_2 a_{12}}{a_{11}a_{22} - a_{21}a_{12}} \\ X_2 &= \frac{a_{11}b_2 - a_{21}b_1}{a_{11}a_{22} - a_{21}a_{12}} \end{aligned} \tag{16.6}$$

Let's apply these equations to the following simple case:

$$\begin{aligned} 5X_1 - 2X_2 &= 26 \\ X_1 + 4X_2 &= -8 \end{aligned} \tag{16.7}$$

Now substituting the appropriate values of a_{ij} and b_i into the equations given by Eqs. (16.6), we obtain the two unknowns, X_1 and X_2:

$$X_1 = \frac{b_1 a_{22} - b_2 a_{12}}{a_{11}a_{22} - a_{21}a_{12}} = \frac{(26)(4) - (-8)(-2)}{(5)(4) - (-2)(1)} = 4$$

$$X_2 = \frac{a_{11}b_2 - a_{21}b_1}{a_{11}a_{22} - a_{21}a_{12}} = \frac{(5)(-8) - (26)(1)}{(5)(4) - (-2)(1)} = -3$$

You can verify that these values of X_1 and X_2 are valid by simply substituting them back into Eq. (16.7).

The principal problem with this method of solving systems of equations is that it is very limited. While the equations for solving for two unknowns are simple, the solutions for larger systems, say, five unknowns, are extremely complex. Therefore, this method is impractical for systems containing more than three variables. In addition, the method is not a general one since the equations change every time we introduce an additional unknown. Consequently, we need to develop a more general approach to solving systems of equations. In the next section we will introduce such a method with the aid of matrices. This approach is applicable to systems containing up to six unknowns. For larger systems, we need even more powerful techniques.

16.3 Cramer's Method for Two Equations with Two Unknowns

It is useful for us to rewrite the system of equations for two unknowns, Eq. (16.5), as a matrix equation:

$$\mathbf{AX} = \mathbf{B} \tag{16.8}$$

where the matrix **A** is a matrix of the coefficients, **X** represents the unknowns, and **B** represents the constants:

$$\mathbf{A} = \begin{bmatrix} a_{11} & a_{12} \\ a_{21} & a_{22} \end{bmatrix}$$

$$\mathbf{X} = \begin{bmatrix} X_1 \\ X_2 \end{bmatrix}$$

$$\mathbf{B} = \begin{bmatrix} b_1 \\ b_2 \end{bmatrix}$$

Instead of writing the solutions in the form shown in Eq. (16.6), we write them in a form that utilizes *determinants*. A determinant is a special combination of the coefficients making up the matrix. For example, if we have a 2 × 2 matrix, **A**, then the determinant of **A**, or det[**A**], is:

$$\det[\mathbf{A}] = \det\begin{bmatrix} a_{11} & a_{12} \\ a_{21} & a_{22} \end{bmatrix} = \begin{vmatrix} a_{11} & a_{12} \\ a_{21} & a_{22} \end{vmatrix} = a_{11}a_{22} - a_{21}a_{12} \tag{16.9}$$

A common convention for the determinant is the set of vertical lines around the matrix. So, for example, the symbol $|\mathbf{A}|$ indicates the determinant of the matrix **A**. You may recall that the determinant of a 2 × 2 matrix is the difference of the products of the two diagonals:

$$\det[\mathbf{A}] = \begin{vmatrix} a_{11} & a_{12} \\ a_{21} & a_{22} \end{vmatrix} = (a_{11}a_{22} - a_{12}a_{21})$$

Notice that to form the determinant, we multiply the diagonal elements a_{11} and a_{22}, and then subtract the product of the other two diagonal elements, a_{12} and a_{21}. As a simple example, if we have the 2 × 2 matrix shown here, we can easily calculate the determinant:

$$\det\begin{vmatrix} 5 & -2 \\ 1 & 4 \end{vmatrix} = (5)(4) - (-2)(1) = 22$$

Let's now return to the solution of a system of linear equations and see what relationship determinants have to these solutions. If you reexamine the solution, Eq. (16.6), for two unknowns, you will note that the denominator in both equations is the same and is equal to $|\mathbf{A}|$. Also, if you examine the numerators of Eq. (16.6), you will note that these also look like determinants. With these thoughts in mind, the matrix form of the solutions to a linear system of equations with two unknowns becomes:

Algebraic Form	Matrix Form
$X_1 = \dfrac{b_1 a_{22} - b_2 a_{12}}{a_{11}a_{22} - a_{21}a_{12}}$	$X_1 = \dfrac{\begin{vmatrix} b_1 & a_{12} \\ b_2 & a_{22} \end{vmatrix}}{\begin{vmatrix} a_{11} & a_{12} \\ a_{21} & a_{22} \end{vmatrix}}$
$X_2 = \dfrac{a_{11} b_2 - a_{21} b_1}{a_{11}a_{22} - a_{21}a_{12}}$	$X_2 = \dfrac{\begin{vmatrix} a_{11} & b_1 \\ a_{21} & b_2 \end{vmatrix}}{\begin{vmatrix} a_{11} & a_{12} \\ a_{21} & a_{22} \end{vmatrix}}$

(16.10)

We form the numerator in each matrix equation by substituting the b values for one column in the $\mathbf{A}$ matrix. When we solve for X_1, we remove the *first* column of $\mathbf{A}$ and substitute the b_i values. Similarly, when we solve for X_2, we replace the *second* column of $\mathbf{A}$.

$$X_1 = \frac{\begin{vmatrix} \boxed{\begin{matrix} a_{11} \\ a_{21} \end{matrix}} & \begin{matrix} a_{12} \\ a_{22} \end{matrix} \end{vmatrix}}{\begin{vmatrix} a_{11} & a_{12} \\ a_{21} & a_{22} \end{vmatrix}} \quad \leftarrow \boxed{\begin{matrix} b_1 \\ b_2 \end{matrix}}$$

$$X_2 = \frac{\begin{vmatrix} \begin{matrix} a_{11} \\ a_{21} \end{matrix} & \boxed{\begin{matrix} a_{12} \\ a_{22} \end{matrix}} \end{vmatrix}}{\begin{vmatrix} a_{11} & a_{12} \\ a_{21} & a_{22} \end{vmatrix}} \quad \leftarrow \boxed{\begin{matrix} b_1 \\ b_2 \end{matrix}}$$

If you expand each determinant, you will find that these equations are equal to those previously given. These definitions assume that det $\mathbf{A}$ is not equal to zero. We call Eq. (16.10) Cramer's rule for the special case of a system of two equations with two unknowns. To demonstrate how this works, we return to the example presented in Eq. (16.7) with:

$$\begin{array}{lll} a_{11} = 5 & a_{12} = -2 & b_1 = 26 \\ a_{21} = 1 & a_{22} = 4 & b_2 = -8 \end{array}$$

By substituting these values into Eq. (16.10), you will obtain the values, $X_1 = 4$ and $X_2 = -3$ just as we obtained with the direct method of Sec. 16.2. The advantage of Cramer's method, as we will see shortly, is that we can readily scale it upward to solve more complex systems. The direct method, on the other hand, cannot be scaled, which severely limits its use.

To implement Cramer's rule for two unknowns, we need two subprograms. The first, a subroutine, will form the matrices that make up the numerator of Eq. (16.10). Notice that all we must do is to create a new matrix, NUM, in which we have replaced the appropriate column by the b_i values. In the following algorithm the calling module will transfer the coefficient array, a_{ij}, and the constant array, b_i, in addition to an integer that indicates the column for switching. For example, if the integer is 2, then the subroutine substitutes b_i into the *second* column of NUM as illustrated in the preceding diagram. But, if the constant is 1, then the subroutine substitutes b_i into the *first* column of NUM. The process for doing this is a simple one as the following algorithm and program segment show:

LS FOR PREPARING MATRICES FOR CRAMER'S RULE

ALGORITHM:

1. LOOP: for J = 1 to 2
   LOOP: for I = 1 to 2
     Is J = N (desired column for switching)?
       Yes: assign A(I,J) to NUM(I,J)
       No: assign B(I) to NUM(I,J).
2. Return the NUM matrix.

PROGRAM SEGMENT:

```
C     SUBROUTINE TO PREPARE NUMERATOR FOR
C     CRAMER'S RULE. VARIABLE LISTING:
C       N = COLUMN TO BE SWITCHED;
C       NUM = ARRAY RETURNED
C       A = ORIGINAL MATRIX OF COEFFICIENTS
C       B = 1-D ARRAY OF EQUATION CONSTANTS
C ----------------------------------------------
      SUBROUTINE CRAMER (N, A, B, NUM)
      REAL A(2,2), B(2), NUM(2,2)
      DO 10 I=1,2
        DO 10 J=1,2
          IF ( J .EQ. N ) THEN
            NUM(I,J) = A(I,J)
          ELSE
            NUM(I,J) = B(I)
          ENDIF
   10 CONTINUE
      RETURN
      END
```

The second subprogram that we need is one that calculates the determinant. For the 2 × 2 matrix of the example, this is a simple matter since we can use the equation det[**A**] = $a_{11}a_{22} - a_{12}a_{21}$ in an assignment statement. But this subroutine only works for the determinant of a 2 × 2 matrix. Higher-order matrices such as 3 × 3, 4 × 4, and so on, require different methods, which we will discuss in the next section. The algorithm and program for computing the determinant of a 2 × 2 matrix is very simple as the following LS demonstrates:

LS FOR CALCULATING 2 × 2 DETERMINANT

ALGORITHM:

1. Use an assignment statement: Det = a11 a22 − a12 a21

PROGRAM SEGMENT:

```
REAL FUNCTION DET2X2 ( A )
REAL A ( 2,2 )
DET2X2 = A( 1,1 ) * A( 2,2 ) - A( 1,2 ) * A( 2,1 )
RETURN
END
```

Once we create these subprograms, we can set up a simple MAIN program to initiate the calculations:

```
C
C    MAIN PROGRAM TO DETERMINE 2 UNKNOWNS WITH
C    2 EQUATIONS USING CRAMER'S RULE
C -----------------------------------------------
     REAL A(2,2), B(2), NUM(2,2)
     PRINT *, ' ENTER VARIABLE COEFFICIENTS:'
     READ *, ((A(I,J), I=1,2), J=1,2)
     PRINT *, 'ENTER EQUATION CONSTANTS:'
     READ *, (B(I), I=1,2)
     CALL CRAMER (1, A, B, NUM)
     X1 = DET2X2(NUM)/DET2X2(A)
     CALL CRAMER (2, A, B, NUM)
     X2 = DET2X2(NUM)/DET2X2(A)
     PRINT *, ' X1=', X1
     PRINT *, ' X2=', X2
     STOP
     END
```

The A array corresponds to the coefficient matrix, a_{ij}, and the B array corresponds to the equation constants, b_i. The first CALL statement creates the numerator in the equation for X_1. Notice that the value of N is 1 here, and so the subroutine will place the B values into the first column of NUM. The assignment statement following this CALL statement computes the value of X_1 by taking the determinants of NUM and A. The second CALL statement does the same thing for X_2. The only difference is that the constant in the CALLing statement is 2, indicating that the B values should be placed in column 2 of the array, NUM.

When we run this program, this is what the dialogue will look like:

BOLD = COMPUTER RESPONSE *ITALIC = USER RESPONSE*

```
ENTER VARIABLE COEFFICIENTS:
     5.0   1.0   -2.0   4.0
ENTER EQUATION CONSTANTS:
     26.0  -8.0
X1 =  4.0
X2 = -3.0
```

The values that we have input are those corresponding to the linear system given in Eq. (16.7). Of course, the results that the computer returns are the same as those that we have obtained by the direct method of Sec. 16.2.

One of the advantages of Cramer's method is that it can be extended very easily to higher-order systems of equations. We commonly use it for solving systems containing up to six equations and six unknowns. One stumbling block to doing this, however, is the problem of how to determine the determinants of higher-order matrices. So, before we talk about the formulation of Cramer's rule for the general case, we must first find a way to calculate the determinant of matrices larger than 2 × 2.

16.4 General Formulation of the Determinant

There are several techniques for calculating the determinant of an $n \times n$ square matrix. The method that we show you here is one of the simplest and easiest to understand. Also, some of the steps in this process are the same as those required for more advanced methods for solving linear systems of equations. Therefore, this discussion serves two purposes.

The method that we present to calculate the determinant manipulates the terms of a matrix so that it is in the form of an *upper triangular matrix:*

$$\begin{bmatrix} 4 & 3 & 2 & 8 \\ 0 & 5 & 6 & 2 \\ 0 & 0 & 1 & 3 \\ 0 & 0 & 0 & 2 \end{bmatrix}$$

An upper triangular matrix is one that has all elements below the diagonal equal to zero. This arrangement of the matrix is very useful because the determinant of such a matrix is simply the product of the diagonal terms:

$$\begin{vmatrix} 4 & 3 & 2 & 8 \\ 0 & 5 & 6 & 2 \\ 0 & 0 & 1 & 3 \\ 0 & 0 & 0 & 2 \end{vmatrix} = (4)\,(5)\,(1)\,(2) = 40$$

It should be apparent that a matrix in this triangular form is the easiest one to work with since the determinant is now so easy to calculate. However, you will only rarely find a matrix in this form. Fortunately, this is not a serious problem since there are two basic operations that permit us to manipulate almost any matrix into the desired form. The operations are:

1. The determinant changes sign if we switch two rows (or columns) in a matrix.
2. The value of the determinant does not change if we multiply a row (or column) by a constant and add it to another row (or column).

To see how this works, consider the following 3×3 matrix, in which we modify each row to produce the triangular form:

First Step

$$\begin{bmatrix} 4 & 3 & 1 \\ 2 & 2 & 3 \\ 4 & 2 & 6 \end{bmatrix} \leftarrow \begin{matrix} \text{Work on row 2} \\ \text{Add } [\text{ROW}_2 - 0.5(\text{ROW}_1)] \end{matrix} \rightarrow \begin{bmatrix} 4 & 3 & 1 \\ 0 & 0.5 & 2.5 \\ 4 & 2 & 6 \end{bmatrix}$$

Second Step

$$\begin{bmatrix} 4 & 3 & 1 \\ 0 & 0.5 & 2.5 \\ 4 & 2 & 6 \end{bmatrix} \leftarrow \begin{matrix} \text{Work on row 3} \\ \text{Add } [\text{ROW}_3 - \text{ROW}_1] \end{matrix} \rightarrow \begin{bmatrix} 4 & 3 & 1 \\ 0 & 0.5 & 2.5 \\ 0 & -1 & 5 \end{bmatrix}$$

Third Step

$$\begin{bmatrix} 4 & 3 & 1 \\ 0 & 0.5 & 2.5 \\ 0 & -1 & 5 \end{bmatrix} \leftarrow \begin{matrix}\text{Work on row 3} \\ \text{Add } [\text{ROW}_3 + 2(\text{ROW}_2)]\end{matrix} \rightarrow \begin{bmatrix} 4 & 3 & 1 \\ 0 & 0.5 & 2.5 \\ 0 & 0 & 10 \end{bmatrix}$$

Once we get the matrix into the desired triangular form, it is easy to compute the determinant. All we must do is to multiply together all the diagonal terms:

$$\det[\mathbf{A}] = a_{11}a_{22}a_{33} = (4)\,(0.5)\,(10) = 20$$

Notice that to get the matrix into the triangular form, we had to do the following general steps:

- Eliminate the first variable from rows 2 to N.
- Eliminate the second variable from rows 3 to N.

⋮

- Eliminate the $(N - 1)$th variable from row N.

The procedure to remove the mth variable from the jth row is:

- Calculate the constant, $c = -(a_{jm})/(a_{mm})$.
- Add $c * a_{mi}$ to a_{ji} for $i = m$ to N.

The following are the algorithm and program segment to carry out this process:

LS FOR CREATING AN UPPER TRIANGULAR MATRIX

ALGORITHM:

1. LOOP: for each variable, X_m, m = 1 to N:
2. LOOP: for each row j = m + 1 to N
3. Calculate c = −a(j, m)/a(m, m)
4. LOOP: for each row position, i = m to n

PROGRAM:

```
      SUBROUTINE TRIANG ( N, A )
      REAL A( N, N)
      DO 10 M = 1, N
        DO 10 J = M + 1, N
          C = -A(J, M )/A( M, M )
            DO 10 I = M, N
               A( J, I ) = A( J, I ) + C *
     1                     A( M, I )
   10 CONTINUE
      RETURN
      END
```

Here is a simple MAIN program that makes use of this subroutine to create an upper triangular matrix and then to calculate the determinant:

```
C    MAIN PROGRAM TO DEMONSTRATE THE USE OF THE
C    SUBROUTINE TRIANG TO CREATE AN UPPER TRIAGONAL
C    MATRIX. ONCE THIS IS DONE, THE DETERMINANT IS
C    CALCULATED IN A FUNCTION SUBPROGRAM.
C -----------------------------------------------------
     REAL A(20,20)
     PRINT *, 'ENTER SIZE OF SQUARE MATRIX: '
     READ *, N
     PRINT *,' ENTER COEFFICIENT MATRIX BY ROWS: '
     READ *, (A(I,J), J=1,N), I=1,N)
     CALL TRIANG( N, A )
     PRINT *, 'DETERMINANT =', DET( N, A )
     STOP
     END
C -----------------------------------------------------
C    FUNCTION SUBPROGRAM TO COMPUTE THE DETERMINANT OF
C    AN UPPER TRIAGONAL MATRIX BY MULTIPLYING ALL THE
C    DIAGONAL TERMS.
C -----------------------------------------------------
     FUNCTION DET( N, A )
     REAL A( N, N )
     DET = 1.0
     DO 10 I = 1, N
           DET = DET * A( I, I )
10   CONTINUE
     RETURN
     END
```

There are several special cases where the preceding algorithm will not work. Of greatest concern is when one of the diagonal terms becomes zero during the many operations. For example, consider the following matrix:

$$\begin{bmatrix} 2 & 1 & 3 & 8 \\ 3 & 0 & 7 & 2 \\ 1 & 4 & 1 & 5 \\ 5 & 1 & 4 & 2 \end{bmatrix}$$

The coefficient, $a_{22} = 0$, will cause a divide-by-zero error when we go to eliminate X_2 from the third row. The way to avoid this problem is to switch rows to move the zero out of the diagonal position. Recall that if we switch two rows (or columns), only the sign of the determinant changes. Thus, we should move the second row of the previous matrix to any of the other rows, for example, the third row:

$$\begin{bmatrix} 2 & 1 & 3 & 8 \\ 1 & 4 & 1 & 5 \\ 3 & 0 & 7 & 2 \\ 5 & 1 & 4 & 2 \end{bmatrix}$$

We will address this problem in more detail in Sec. 16.6 when we discuss the gaussian method for solving systems of equations.

16.5 Generalized Cramer's Rule

The extension of Cramer's rule to systems larger than 2 × 2 is straightforward. Recall the process for computing the roots in a system of two equations with two unknowns:

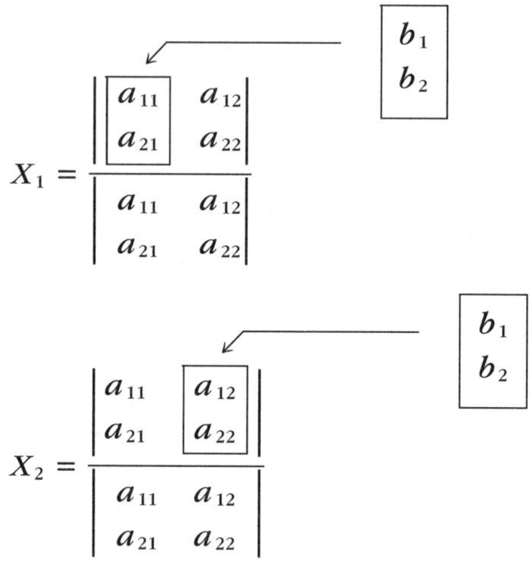

$$X_1 = \frac{\begin{vmatrix} \boxed{\begin{matrix} a_{11} \\ a_{21} \end{matrix}} & \begin{matrix} a_{12} \\ a_{22} \end{matrix} \end{vmatrix}}{\begin{vmatrix} a_{11} & a_{12} \\ a_{21} & a_{22} \end{vmatrix}} \qquad \leftarrow \boxed{\begin{matrix} b_1 \\ b_2 \end{matrix}}$$

$$X_2 = \frac{\begin{vmatrix} \begin{matrix} a_{11} \\ a_{21} \end{matrix} & \boxed{\begin{matrix} a_{12} \\ a_{22} \end{matrix}} \end{vmatrix}}{\begin{vmatrix} a_{11} & a_{12} \\ a_{21} & a_{22} \end{vmatrix}} \qquad \leftarrow \boxed{\begin{matrix} b_1 \\ b_2 \end{matrix}}$$

To extend this process to higher-order matrices, we need only to repeat the same process as before. To demonstrate this, let's examine a system of four equations with four unknowns. For each of the unknowns, we set up a denominator that is the determinant of the 4 × 4 coefficient matrix, **A**. The numerator will also consist of a determinant, but we replace the appropriate column of the matrix with the equations constants, b_i:

$$X_1 = \frac{\begin{vmatrix} b_1 & a_{12} & a_{13} & a_{14} \\ b_2 & a_{22} & a_{23} & a_{24} \\ b_3 & a_{32} & a_{33} & a_{34} \\ b_4 & a_{42} & a_{43} & a_{44} \end{vmatrix}}{\begin{vmatrix} a_{11} & a_{12} & a_{13} & a_{14} \\ a_{21} & a_{22} & a_{23} & a_{24} \\ a_{31} & a_{32} & a_{33} & a_{34} \\ a_{41} & a_{42} & a_{43} & a_{44} \end{vmatrix}} \qquad X_2 = \frac{\begin{vmatrix} a_{11} & b_1 & a_{13} & a_{14} \\ a_{21} & b_2 & a_{23} & a_{24} \\ a_{31} & b_3 & a_{33} & a_{34} \\ a_{41} & b_4 & a_{43} & a_{44} \end{vmatrix}}{\begin{vmatrix} a_{11} & a_{12} & a_{13} & a_{14} \\ a_{21} & a_{22} & a_{23} & a_{24} \\ a_{31} & a_{32} & a_{33} & a_{34} \\ a_{41} & a_{42} & a_{43} & a_{44} \end{vmatrix}}$$

$$X_3 = \frac{\begin{vmatrix} a_{11} & a_{12} & b_1 & a_{14} \\ a_{21} & a_{22} & b_2 & a_{24} \\ a_{31} & a_{32} & b_3 & a_{34} \\ a_{41} & a_{42} & b_4 & a_{44} \end{vmatrix}}{\begin{vmatrix} a_{11} & a_{12} & a_{13} & a_{14} \\ a_{21} & a_{22} & a_{23} & a_{24} \\ a_{31} & a_{32} & a_{33} & a_{34} \\ a_{41} & a_{42} & a_{43} & a_{44} \end{vmatrix}} \qquad X_4 = \frac{\begin{vmatrix} a_{11} & a_{12} & a_{13} & b_1 \\ a_{21} & a_{22} & a_{23} & b_2 \\ a_{31} & a_{32} & a_{33} & b_3 \\ a_{41} & a_{42} & a_{43} & b_4 \end{vmatrix}}{\begin{vmatrix} a_{11} & a_{12} & a_{13} & a_{14} \\ a_{21} & a_{22} & a_{23} & a_{24} \\ a_{31} & a_{32} & a_{33} & a_{34} \\ a_{41} & a_{42} & a_{43} & a_{44} \end{vmatrix}} \tag{16.11}$$

Cramer's rule should now be clear. To find an unknown variable, X_n, you must do the following:

- Form a denominator by replacing the nth column of the **A** matrix with the vector, b_i.
- Calculate the determinant of this matrix and divide by the determinant of the unmodified **A** matrix.

We can accomplish the first step with SUBROUTINE CRAMER, if we modify it for the general case. This is a simple process as shown here:

LS FOR PREPARING MATRICES FOR CRAMER'S RULE (general case)

PROGRAM:

```
C      SUBROUTINE TO PREPARE NUMERATOR FOR
C      CRAMER'S RULE. VARIABLE LISTING:
C        N = COLUMN TO BE SWITCHED;
C        M = SIZE OF MATRICES;
C        NUM = ARRAY RETURNED;
C        A = ORIGINAL MATRIX OF COEFFICIENTS;
C        B = 1-D ARRAY OF EQUATION CONSTANTS;
C ------------------------------------------------
```

```
      SUBROUTINE CRAMER (N, M, A, B, NUM)
      REAL A( M,M ), B( M ), NUM( M,M )
      DO 10 I=1,M
        DO 10 J=1,M
          IF ( J .EQ. N ) THEN
            NUM( I,J ) = A( I,J )
          ELSE
            NUM( I,J ) = B( I )
          ENDIF
   10 CONTINUE
      RETURN
      END
```

We use this subroutine to construct the matrix in the numerator. Then we use SUBROUTINE TRIANG to create the triangular matrix for the determinant calculation. Thus, the general procedure is:

1. Calculate the DENOM = det[**A**]:
2. To calculate X_n:
   Use SUBROUTINE CRAMER to create a matrix, **NUM**, which consists of the matrix, **A**, except for column n. This column consists of the constants, *b***i**.
   Convert the matrix, **NUM**, into an upper triangular matrix.
   Calculate the unknown, X_n, by det[**NUM**]/det[**A**].

You must be careful in carrying out this process since the calculation of det[**A**] in step 1 alters the array. One way to avoid this problem is to copy **A** first into the new array **A2** so that we can calculate the denominator, det[**A2**], instead. If we did not do this, then the triangularized matrix, **NUM**, would be incorrect. The following program carries out this process and stores the values in the array, X(I):

```
C    MAIN PROGRAM TO DEMONSTRATE CRAMER'S RULE TO
C    SOLVE SYSTEM OF N UNKNOWNS.
C ------------------------------------------------------------
     REAL A(10,10),A2(10,10),NUM(10,10),B(10),X(10)
     PRINT *, 'ENTER SIZE OF MATRICES: '          ]
     READ *, M                                    |
     PRINT *, 'ENTER COEFFICIENT                  |
    1 MATRIX BY ROWS: '                           } Input section
     READ *, ((A(I,J), J=1,M), I=1,M)             |
     PRINT *, 'ENTER CONSTANT VECTOR: '           ]
```

```
      READ *, (B(I), I=1,M)
      DO 10 I=1,M                                     }
            DO 10 J=1,M                               }Copy A into A2
                  A2(I,J)=A(I,J)                      }
10    CONTINUE
      CALL TRIANG ( M, A2)                            }Calculate
      DENOM = DET( M, A2 )                            } det[A2]
      DO 20 I=1,M
         CALL CRAMER ( I, M, A, B, NUM )              }Calculate
         CALL TRIANG ( M, NUM )                       } unknowns
         X( I ) = DET( NUM )/DENOM                    } X1 to Xn
20    CONTINUE
      PRINT *, 'VALUE OF UNKNOWNS: '                  }Output section
      PRINT *, (X(I), I=1,M)                          }
      STOP
      END
```

All necessary FUNCTIONs and SUBROUTINEs have been given to you already, so we do not repeat them here. To demonstrate how this program works, we use it to solve the system:

$$
\begin{aligned}
3A + 4B - 8C - 2D &= -48 \\
A - 2B \phantom{{}- 8C} + D &= 13 \\
4A + 7B + 2C \phantom{{}-2D} &= -5 \\
- 3B + 4C \phantom{{}-2D} &= 25
\end{aligned}
\tag{16.12}
$$

When we run the program, the dialogue looks like this:

```
ENTER SIZE OF MATRICES:
  4

ENTER COEFFICIENT MATRIX BY ROWS:
  3.0     4.0    -8.0    -2.0
  1.0    -2.0     0.0     1.0
  4.0     7.0     2.0     0.0
  0.0    -3.0     4.0     0.0

ENTER CONSTANT VECTOR:
-48.0    13.0    -5.0    25.0

VALUE OF UNKNOWNS:
  2.0    -3.0     4.0     5.0
```

Notice when we run this program, we must explicitly type in a value when the coefficient is zero. You cannot simply omit a value when you should type in a value of zero. If you substitute the values that the program returns in Eq. (16.12), you should be able to verify that these numbers are correct.

We have set up the previous program for a maximum size of ten unknowns. This is because Cramer's rule is too slow for systems larger than this. There are more efficient methods, as we will soon see, and so Cramer's method is rarely used for systems larger than 6×6.

16.6 Gaussian Elimination

We do not recommend the use of Cramer's rule for large systems because of the enormous number of calculations involved. For example, a system with 20 unknowns would require approximately 10^6 years, even on a computer able to do 10^6 multiplication operations per second. A careful analysis of that method shows that the number of calculations is $(n + 1)!$ where n is the number of equations in the system. Consequently, we typically limit the use of Cramer's rule to small systems with six equations or less.

A better method for solving a linear system is the gaussian elimination method. We have already explored a critical component of this method—namely, the process of converting a matrix into an upper triangular matrix. This is the first step in calculating the unknown quantities in the system. Briefly, the process of gaussian elimination involves:

1. Creating a single matrix of size $n \times (n + 1)$ that contains the coefficient matrix plus the constant vector.
2. Converting this *augmented* matrix into an upper triangular matrix.
3. Solving the last equation, which is now an equation of one unknown.
4. Working backward and substituting the results from one equation into the preceding one, until you solve all the equations.

To demonstrate this more clearly, consider the system of equations:

$$\begin{aligned} 2X_1 + 3X_2 + 4X_3 &= 29 \\ X_1 + X_2 + X_3 &= 9 \\ 2X_1 + X_2 - X_3 &= 3 \end{aligned} \tag{16.13}$$

Our first job is to create an upper triangular array. To do this, we multiply the first equation by $-1/2$ and add it to the second equation, and then multiply the first equation by -1 and add it to the third equation:

$$\begin{aligned} 2X_1 + 3X_2 + 4X_3 &= 29 \\ -\left(\tfrac{1}{2}\right)X_2 - X_3 &= -\tfrac{11}{2} \\ -2X_2 - 5X_3 &= -26 \end{aligned} \tag{16.14}$$

Next, multiply the second equation by $-(-2)/(-\frac{1}{2})$, which is -4, and add it to the last equation:

$$\begin{aligned} 2X_1 + 3X_2 + 4X_3 &= 29 \\ -(\tfrac{1}{2})X_2 - X_3 &= -\tfrac{11}{2} \\ -X_3 &= -4 \end{aligned} \tag{16.15}$$

Notice that the last equation is now easy to solve for $X_3 = 4$. We then substitute this back into the second equation, collect terms, and solve for X_2:

$$\begin{aligned} 2X_1 + 3X_2 + 4X_3 &= 29 \\ + X_2 + \quad &= 3 \\ + X_3 &= 4 \end{aligned} \tag{16.16}$$

Finally, we substitute $X_2 = 3$ and $X_3 = 4$ into the first equation, collect terms, and solve for X_1:

$$\begin{aligned} X_1 + \quad + \quad &= 2 \\ + X_2 + \quad &= 3 \\ + X_3 &= 4 \end{aligned} \tag{16.17}$$

Now let's examine this process in more detail. Notice that in Eqs. (16.14) and (16.15) we created an upper triangular matrix. We term this the *elimination* process. Once that is done, we then *back substitute* the solutions into the previous equations until all unknowns are calculated.

Before trying to convert this process into a program, let's reformulate it in matrix form. The first step is to create an *augmented matrix,* consisting of the original coefficient matrix and the equation constants. For the preceding example, we have:

$$\left|\begin{array}{ccc|c} 2 & 3 & 4 & 29 \\ 1 & 1 & 1 & 9 \\ 2 & 1 & -1 & 3 \end{array}\right| \tag{16.18}$$

The last column in the augmented matrix consists of the equation constants, while the remainder of the matrix is formed from the coefficients. Now after we complete the elimination, the augmented matrix becomes:

$$\left|\begin{array}{ccc|c} 2 & 3 & 4 & 29 \\ 0 & -\frac{1}{2} & -1 & -\frac{11}{2} \\ 0 & 0 & -1 & -4 \end{array}\right| \tag{16.19}$$

The final step is to perform the back substitution. When completed, the matrix becomes:

$$\left|\begin{array}{ccc|c} 1 & 0 & 0 & 2 \\ 0 & 1 & 0 & 3 \\ 0 & 0 & 1 & 4 \end{array}\right| \tag{16.20}$$

In matrix notation the last column contains the values of the unknowns. For this reason, we want to use matrix notation to implement the Gaussian method.

There is one problem that we have not discussed—the difficulties that occur when one of the diagonal elements in the matrix is zero. For example, consider the following system of equations:

$$\begin{aligned} X_2 + X_3 &= 3 \\ X_1 + X_2 + X_3 &= 4 \\ X_1 + X_2 \quad &= 2 \end{aligned} \qquad (16.21)$$

This system has the solution ($X = 1, Y = 1, Z = 2$). However, if we attempt to apply gaussian elimination to this system, the fact that a_{11} equals zero will result in a division error. The fix is to rearrange the order of the equations so that the element involved in the division, called the pivot element, is not zero:

$$\begin{aligned} X_1 + X_2 + X_3 &= 4 \\ X_1 + X_2 \quad &= 2 \\ X_2 + X_3 &= 3 \end{aligned} \qquad (16.22)$$

This operation of reordering the equations so that the pivot element is nonzero is known as partial pivoting. In most algorithms for the gaussian elimination method the program will search the matrix to find the row with the largest value in the same column as the pivot element. The program then switches rows. For the previous example, before any elimination takes place, the program searches the first column to find the largest coefficient. It then switches the first and second rows.

Before presenting the program for implementing the gaussian elimination method, we should review the steps involved in the process:

1. Create an augmented matrix of size $n \times (n + 1)$ by placing the b_1 vector in the $n + 1$ column.
2. Convert the augmented matrix into an upper triangular matrix:
   First switch the pivot row and the row with the largest coefficient.
   After switching, proceed with elimination.
3. Back substitute, starting with the last equation.

The first subroutine that we present is the process for partial pivoting:

LS TO PERFORM PARTIAL PIVOTING

Algorithm:

1. Give the pivot element, locate the element in the pivot column below the pivot element that has the largest absolute value. Note the row number.
2. Swap the pivot element row with the row found in step 1.

PROGRAM SEGMENT:

```
      SUBROUTINE PARPVT(A,PVTROW,N)
      REAL A( 20, 21 ), PVTMAX
      INTEGER PVTROW,N,RMAX,I,J
      RMAX=PVTROW
      PVTMAX=ABS(A(PVTROW,PVTROW))
      DO 10 I=PVTROW+1,N
            IF (ABS(A(I,PVTROW)).GT.PVTMAX) THEN
                  PVTMAX=ABS(A(I,PVTROW))
                  RMAX=I
            ENDIF
10    CONTINUE
      DO 20 J=1,N+1
            TEMP=A(PVTROW,J)
            A(PVTROW,J)=A(RMAX,J)
            A(RMAX,J)=TEMP
20    CONTINUE
      RETURN
      END
```

We designed PARPVT to handle up to 20 equations. But because we pass N, the number of equations, we do not have to use necessarily all elements in the array. We have sized it instead for a worst case. PVTROW determines the active pivot row. Loop 10 and its body perform step 1 of the algorithm. Loop 20 performs the row swapping operation of step 2.

SUBROUTINE GAUSS performs the gaussian elimination and back substitution. We show this in the following LS:

LS FOR GAUSSIAN ELIMINATION

ALGORITHM:

1. Define augmented matrix, C, using A and B.
2. Set R (the pivot row) equals to 1.
3. Perform the partial pivoting operation using PARPVT.
4. For rows R + 1 through N, perform elimination.
5. Increment R by 1 and repeat until PVTROW equals N.
6. Perform back substitution.

PROGRAM SEGMENT:

```
      SUBROUTINE GAUSS(A,B,X,N)
      REAL A(N,N),B(N),X(N),C(20,21)
      INTEGER I,J,R,N
C----------STEP 1 OF ALGORITHM ----------------
      DO 10 I=1,N
            DO 5 J=1,N
                  C(I,J)=A(I,J)
```

```
5            CONTINUE
             C(I,N+1)=B(I)
10       CONTINUE
   C---------STEP 2 OF ALGORITHM ----------------
         DO 40 R=1,N-1
   C---------STEP 3 OF ALGORITHM ----------------
             CALL PARPVT(C,R,N)
   C---------STEPS 4 & 5 OF ALGORITHM ----------------
             DO 30 I=R+1,N
                 CONST=C(I,R)/C(R,R)
                 DO 20 J=R,N+1
                    C(I,J)=C(I,J)-CONST*C(R,J)
20               CONTINUE
30           CONTINUE
40       CONTINUE
   C---------STEP 6 OF ALGORITHM ----------------
         X(N)=C(N,N+1)/C(N,N)
         DO 60 I=N-1,1,-1
             SUM=0
             DO 50 J=I+1,N
                 SUM=SUM+C(I,J)*X(J)
50           CONTINUE
             X(I)=(C(I,N+1)-SUM)/C(I,I)
60       CONTINUE
         RETURN
         END
```

A MAIN program that uses the SUBROUTINE GAUSS can look like this:

```
C  MAIN PROGRAM DEMONSTRATING THE USE OF GAUSSIAN
C  ELIMINATION FOR SOLVING A LINEAR SYSTEM OF
C  EQUATION WITH 3 UNKNOWNS. FOR LARGER SYSTEMS,
C  CHANGE SIZE OF ARRAYS AND LIMITS ON LOOPS.
C ---------------------------------------------------
   REAL A(3,3),B(3),X(3)
   DO 10 I=1,3
         PRINT *,'Enter Coef. and Constant for eq.',I
         READ *,(A(I,J),J=1,3),B(I)
10 CONTINUE
   CALL GAUSS( A, B, X, 3)
   PRINT *,'The unknowns are ',X
   STOP
   END
```

We leave it to the reader to run the program and verify the results.

16.7 Validation and Error

The process of validating programs that return solutions to equations is straightforward. Simply plug the answers back into the equations. If they work, then the program is valid for that problem. A common method of checking the solution for a system of linear equations is to rewrite the matrix equation in the form:

$$\mathbf{AX} - \mathbf{B} = \mathbf{0} \tag{16.23}$$

In this equation 0 represents the zero vector. Multiplying **A** by the solution vector **X** should return the constant vector **B** if the solution is correct. Typically, the difference will not be zero due to round-off errors. However, the values will be small. For large systems of equation, checking the values within **A X** − **B** can be cumbersome. A simpler way to check this is to calculate the *magnitude* of the zero vector. We calculate the magnitude of a vector by taking the square root of the sum of the squares of the components of the vector:

$$|V| = \sum_{i=1}^{n} (V_i^2)^{1/2} \tag{16.24}$$

Note that the notation used to show the magnitude of a vector is similar to that used for the determinant. When the entity enclosed by the vertical bars is an array, it indicates the determinant. When they surround a vector, it indicates a magnitude. We can best implement the magnitude operation by a FUNCTION SUBPROGRAM, as shown in the following LS:

LS TO CALCULATE THE MAGNITUDE OF A VECTOR

ALGORITHM:

1. Sum the square of the elements.
2. Take the square root of the sum.

PROGRAM SEGMENT:

```
      REAL FUNCTION VMAG(VECTOR,N)
      REAL VECTOR(N)
      VMAG=0
      DO 10 I=1,N
         VMAG=VMAG+VECTOR(I)**2
   10 CONTINUE
      VMAG=SQRT(VMAG)
      RETURN
      END
```

We can rewrite the previous program for gaussian elimination now to include validation of the results. We do this by using the array multiplication LS from Chap. 9 with the VMAG FUNCTION presented here:

```
C
C     MAIN PROGRAM TO DEMONSTRATE HOW TO VALIDATE
C     RESULTS
C ----------------------------------------------------------
      REAL A(3,3),B(3),X(3),AB(3)
      DO 10 I=1,3
            PRINT *,'Enter Coef. and Constant for
     1                 eq.',I
            READ *,(A(I,J),J=1,3),B(I)
10    CONTINUE
      CALL GAUSS( A, B, X, 3 )
      PRINT *,'The unknowns are ',X
      CALL MULARR( A, X, 3, AB)
      DO 20 I=1,3
            AB(I)=AB(I)-B(I)
20    CONTINUE
      PRINT *, 'The |ERROR| =', VMAG( AB, 3 )
      STOP
      END
```

In the following examples we demonstrate how to use this approach to verify the calculations.

16.8 Applications

Many science and engineering problems require the ability to solve systems of linear equations. This section presents two problems: (1) solving an electric network consisting of voltage supplies and resistors and (2) determining the reactions of a simply supported two-dimensional structure.

ANALYSIS OF RESISTOR NETWORK

Figure 16.2 illustrates a schematic representation of a network of resistors and voltage supplies (batteries). If we wish to compute the electric currents $I_1, I_2, \ldots, I_6$ through the six resistances, we must apply the three basic principles of circuit analysis:

Ohm's law: Voltage drop across a resistor is proportional to the current ($V = IR$).

FIGURE 16.2

Analysis of Resistive Network

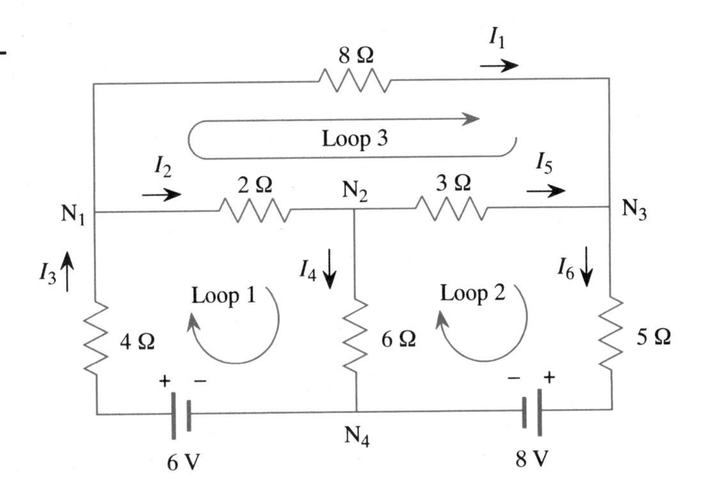

Kirchhoff's current law:	Current entering and exiting a node must sum to zero.
Kirchhoff's voltage law:	Voltage around any closed loop must be zero.

Current, designated by the symbol I, indicates the quantity of electrons flowing through the wire per second. We can make an analogy to water flowing through a pipe, where current is the flow rate in gallons per second. A *node* is a common connection point between two or more elements. We have labeled the nodes in the schematic as N1 through N4. The *voltage,* designated by V, is a measure of the energy stored per electron. Ohm's law states that the voltage drop across a resistor is proportional to the current flowing through it. The constant of proportionality is the resistance, R, in ohms. Kirchhoff's current law states that whatever current enters a node must also leave it. This means that electrons are not stored anywhere. Finally, Kirchhoff's voltage law states that whatever energy is added to the system must go into the components.

We can apply these three laws to the circuit to generate a system of six linear equations to solve for the six unknown currents, I_1 to I_6. First, we assign arbitrary directions for each of the currents associated with each of the resistors as shown in Fig. 16.2. Then we compute the sum of the voltages within a given loop. These must sum to zero. If a battery is within the loop, then going from − to + is a positive voltage change. Going through a resistor in the direction of its assigned current is a negative voltage change.

Now we are ready to proceed to generate the six equations. The first comes from loop 1 starting at the 6 V voltage supply:

$$6 - 4I_3 - 2I_2 - 6I_4 = 0$$

We use Ohm's law to get the terms corresponding to the voltage drops, such as $4I_3$ across the resistors. Next, we sum the voltages around loop 2, starting at the 8 V battery:

$$-8 + 6I_4 - 3I_5 - 5I_6 = 0$$

Note that the voltage across the battery is negative (+ to −) and the voltage across the 6 ohm resistor is positive. These come from going around the loop in the opposite direction of the assigned current.

In loop 3 there are no voltage sources, so we only need to compute the voltage drops across each resistor:

$$-8I_1 + 3I_5 + 2I_2 = 0$$

So far we were able to generate three equations using Kirchhoff's voltage law, but we still need three more; we will use Kirchhoff's voltage law. For example, at node N1, I_3 enters the junction, whereas I_2 and I_1 leave it. Since the sum of the currents through the node must equal zero, we have our fourth equation:

$$I_3 - I_2 - I_1 = 0$$

In a similar way, we can examine nodes N2 and N3 to derive the last two equations:

$$I_2 - I_4 - I_5 = 0$$
$$I_1 + I_5 - I_6 = 0$$

We now have six equations with six unknowns. We should rewrite them in the standard form to make it easier to enter into the program:

$$\begin{aligned}
-I_1 - I_2 + I_3 &= 0 \\
I_2 - I_4 - I_5 &= 0 \\
I_1 + I_5 - I_6 &= 0 \\
-2I_2 - 4I_3 - 6I_4 &= -6 \\
6I_4 - 3I_5 - 5I_6 &= 8 \\
-8I_1 + 2I_2 + 3I_5 &= 0
\end{aligned}$$

Although these equations appear to be simple to solve by hand, you will find it a formidable task. If you persevere, however, you will obtain the following results:

$$I_1 = -0.073 \qquad I_2 = +0.320 \qquad I_3 = -0.247$$
$$I_4 = +0.728 \qquad I_5 = -0.408 \qquad I_6 = -0.481$$

Now, let's run the program to see how our answer compares. Before sending this to SUBROUTINE GAUSS for solution, we must modify it for six variables. This is not difficult, and once done, the program execution would look like the following:

```
Enter Coef. and Constant for eq. 1
-1.0  -1.0   1.0   0.0   0.0   0.0   0.0
Enter Coef. and Constant for eq. 2
 0.0   1.0   0.0  -1.0  -1.0   0.0   0.0
Enter Coef. and Constant for eq. 3
 1.0   0.0   0.0   0.0   1.0  -1.0   0.0
Enter Coef. and Constant for eq. 4
 0.0  -2.0  -4.0  -6.0   0.0   0.0  -6.0
Enter Coef. and Constant for eq. 5
 0.0   0.0   0.0   6.0  -3.0  -5.0   8.0
Enter Coef. and Constant for eq. 6
-8.0   2.0   0.0   0.0   3.0   0.0   0.0
The unknowns are  -7.297298E-02  .320270  .247297  .728378  -.408108  -.481081
The |ERROR| = 2.215217E-07
```

We have used the method discussed in the previous section to verify our results. The reported error of approximately 10^{-7} is very small, and so we have confidence in the quality of the calculated currents. The values printed at the bottom of the output are the six current values, I_1 through I_6. Note that several of them are negative. The significance of the negative sign is that the calculations indicate that the current flows in the *opposite* direction from that assumed. For example, both I_5 and I_6 flow in the opposite direction from that shown in Fig. 16.2. Remember, when we constructed this figure, we made an assumption about the direction of flow. But, when we computed the results for the complete system, we found several of these directions to be in error.

SIMPLY SUPPORTED STRUCTURE

A simply supported two-dimensional structure is one which has a pin connection at one location and a roller connection at another location. In the illustration of Fig. 16.3 we show such a structure with the two support points in the same horizontal plane. The pinned end, illustrated by the triangular connection, exerts a force in both the vertical and horizontal directions. We designate

FIGURE 16.3

Forces Acting on a Simply Supported Frame

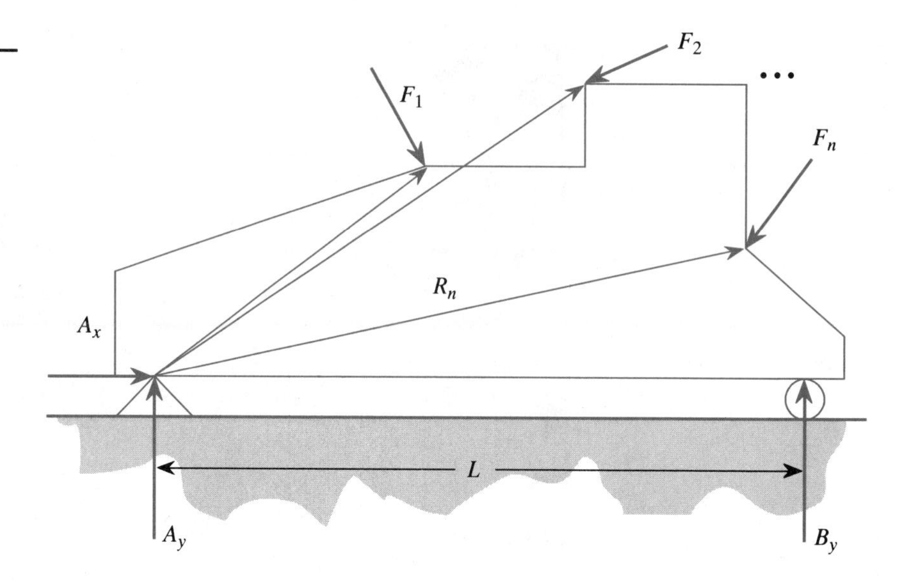

these reactions A_X and A_Y. The roller connection can only apply force in the vertical direction. We label that force B_Y. The applied loads, F_1, F_2, and F_3 along with the point at which they act are illustrated as vectors to highlight their directional nature. There can be a large number of such forces acting on the structure, although we show only three. In the program that we will develop, we may have as many as 20 forces acting simultaneously.

Our problem here is to solve for the reactions, A_X, A_Y, and B_Y for a given set of forces, $F_1, F_2, \ldots, F_n$. The process of solving for the unknown reaction involves equations of static equilibrium:

$$\sum F = 0 \qquad \text{(sum of the forces = 0)}$$

$$\sum M_0 = 0 \qquad \text{(sum of the moments = 0)} \tag{16.25}$$

The first of these equations states that all forces (which are vectors) must sum to zero for the body to be in equilibrium. The second equation states that all moments taken about any arbitrary point, O, must also sum to zero.

Because this is a two-dimensional problem, the force equation actually represents two separate scalar equations. One of these governs the sum of the forces in the X direction and the sum of forces in the Y direction. Each must individually and independently sum to zero. The moment equation, on the other hand, represents only a single equation that states that the moment about the Z axis must sum to zero. We need all three equations to solve for the three unknown reactions.

We start by summing all forces acting in the X direction:

$$A_X + \sum_{i=1}^{n} F_{X_i} = 0 \tag{16.26}$$

Next we sum all forces acting in the Y direction:

$$A_Y + B_Y + \sum_{i=1}^{n} F_{Yi} = 0 \tag{16.27}$$

Finally, if we take the moment about the origin, the only reaction that can contribute is B_Y. Writing the moment equation using vector notation, we produce the third equation:

$$L \otimes B + \sum_{i=1}^{n} R_i \otimes F_i = 0 \tag{16.28}$$

where the symbol $\otimes$ denotes the cross product of the two vectors. The reaction force B_Y is at a distance L from the hinge. Therefore, we use this value to calculate the moment. Similarly, the individual forces are at a distance, R_i, from the hinge, and so we must use a different moment arm for each.

Since this is a two-dimensional problem, the only term from the vector cross product that contributes to the moment will be the Z elements. Consequently, Eq. (16.28) represents only a single equation. The summation of the cross products between the position and force vectors are known values. Therefore, expanding the equation for the nonzero contributions will give us the final equation:

$$LB_Y + \sum (0, 0, 1) \cdot (R \times F) = 0 \tag{16.29}$$

The dot product $(\cdot)$ of $(0, 0, 1)$ with the cross product term returns only the Z element. To be able to generate these three equations, we must first establish a method for calculating the cross product. The most difficult term in this equation is the $(0, 0, 1) \cdot (\mathrm{R} \times F)$, commonly called the scaler triple product. Fortunately, we can calculate this product easily by using determinants. We use the following identity to do this:

$$n \cdot (r \times F) = \begin{vmatrix} n_X & n_Y & n_Z \\ r_X & r_Y & r_Z \\ F_X & F_Y & F_Z \end{vmatrix} \tag{16.30}$$

Now we only need to collect these three equations and rearrange them in our standard format:

$$\begin{aligned} A_X \qquad\qquad &= -\sum F_X \\ A_Y + B_Y &= -\sum F_Y \\ LB_Y &= -\sum (0, 0, 1) \cdot (R \times F) \end{aligned} \tag{16.31}$$

After we know the forces in the system, we can expand the right-hand side of each equation to complete the equations. The steps involved in setting up a program to solve this problem are:

1. Define the **A** matrix to be used for solving the system of equations.
2. Read in the number of externally applied forces.
3. Read in the positions and magnitudes for each applied force.
4. Define the constant reaction force vector **B** and perform the summation of forces and summation of triple products to form the right-hand side of each equation.
5. Solve the system of equations using gaussian elimination.
6. Report calculated reactions.

The best way to demonstrate how this works is to trace through an example problem by hand. For example, we assume the following data for a system with three forces and a distance of $L = 8.0$ separating the two support points:

FORCE 1: $F1(X,Y) = (-1,-1)$ acting at $R1(X,Y) = (2,1)$
FORCE 2: $F2(X,Y) = (0,-10)$ acting at $R2(X,Y) = (4,0)$
FORCE 3: $F3(X,Y) = (1,-1)$ acting at $R3(X,Y) = (6,1)$

The preceding data represents the X and Y components of each force and the distance in X and Y units from the origin. Now we expand the right-hand side of each equation given in Eq. (16.31):

$$-\sum F_X = -1(-1 + 0 + 1) = 0$$

$$-\sum F_Y = -1(-1 - 10 - 1) = 12$$

$$-\sum (0,0,1) \cdot (R \times F) = -1(-1 - 40 - 7) = 48$$

Then we substitute these quantities back into the system of equations:

$$\begin{aligned} A_X \qquad\qquad &= 0 \\ A_Y + B_Y &= 12 \\ 8B_Y &= 48 \end{aligned} \tag{16.32}$$

Notice that we have substituted a value of 8 for the variable, L. It is a simple matter to show then that this system has the following values:

$$A_X = 0 \qquad A_Y = 6 \qquad B_Y = 6$$

It will help you to understand the following program if you focus on the system of equations in matrix form:

$$\begin{bmatrix} 1 & 0 & 0 \\ 0 & 1 & 1 \\ 0 & 0 & L \end{bmatrix} \begin{bmatrix} A_X \\ A_Y \\ B_Y \end{bmatrix} = \begin{bmatrix} -\sum F_X \\ \\ -\sum F_Y \\ \\ -\sum (001) \cdot (R \times F) \end{bmatrix} \tag{16.33}$$

The square matrix on the left contains the coefficients of the unknowns A_X, A_Y, and B_Y. The primary job of our program is to calculate the constants in the vector on the right-hand side of the equal sign. These will change considerably from problem to problem and must be recalculated each time.

In the following program we give a MAIN program to perform the calculations summarized by Eq. (16.33). We use the gaussian elimination method since this is applicable to moderate size systems. The arrays in the programs have a limiting size corresponding to 20 different applied loads since larger systems should not be solved by this method. Finally, we introduce a FUNCTION SUBPROGRAM to calculate the triple product. You should use this FUNCTION only for the case of two-dimensional forces, and therefore is not for general use. So, be careful if you try to apply it to other problems. Here is the program for computing the reaction forces of a simply supported frame:

```
C     PROGRAM TO CALCULATE REACTION FORCES FOR A SIMPLY
C     SUPPORTED FRAME SUBJECT TO MULTIPLE LOADING
C ----------------------------------------------------------
      REAL R(20, 2), F(20, 2), A(3, 3), B(3), X(3)
      DATA ((A(I,J),J=1,3),I=1,3)/1,0,0,0,1,1,0,0,1/
      DATA (B(I), I=1,3)/0,0,0/
      PRINT *,'Enter distance between supports?'
      READ *,A(3,3)
      PRINT *,'How many applied forces?'
      READ *,N
      DO 10 I=1,N
            PRINT *,'For FORCE',I
            PRINT *,'Enter Rx,Ry for position and
     1                Fx,Fy '
            READ *,(R(I,J),J=1,2),(F(I,J),J=1,2)
 10   CONTINUE
      DO 20 I=1,N
            B(1)=B(1)-F(I,1)
            B(2)=B(2)-F(I,2)
            B(3)=B(3)-TRIPRO(R,F,I)
 20   CONTINUE
      CALL GAUSS( A, B, X, 3)
      PRINT *, 'Ax= ', X(1),' Ay= ', X(2),' By= ', X(3)
      STOP
      END
C ----------------------------------------------------------
C     FUNCTION SUBPROGRAM TO COMPUTE TRIPLE PRODUCT
C ----------------------------------------------------------
      REAL FUNCTION TRIPRO(R,F,I)
      REAL R(20,2),F(20,2)
      TRIPRO=R(I,1)*F(I,2)-R(I,2)*F(I,1)
      RETURN
      END
```

In the preceding program we used two-dimensional arrays to store the positions and magnitudes of the applied forces. This makes transfer of the data easier to handle. A typical session using the previous program would look like this:

BOLD = COMPUTER RESPONSE *ITALIC = USER RESPONSE*

```
Enter distance between supports?
     8
How many applied forces?
     3
For FORCE 1 Enter Rx,Ry for position and Fx,Fy
     2     1     -1     -1
For FORCE 2 Enter Rx,Ry for position and Fx,Fy
     4     0     0     -10
For FORCE 3 Enter Rx,Ry for position and Fx,Fy
     6     1     1     -1
AX= .000000     AY= 6.00000     BY= 6.00000
```

16.9 Very Large Systems of Equations

As stated earlier, due to round-off error, we can only use Cramer's rule on systems of equations of order 6 or less. For systems of intermediate size, up to 20 unknowns, we had to use gaussian elimination. There are applications, however, where the number of equations go into the thousands. Finite element methods, for example, require the solution of many equations. To solve such large systems, some special conditions need to be present.

Cramer's rule and gaussian elimination assume a full matrix. This means that nonzero values may be present at any location of the **A** matrix. Some problems, however, result in many elements being zero, typically leaving a band of nonzero elements down the diagonal. A full and banded matrix follow:

```
*  *  *  *  *      *  *  0  0  0
*  *  *  *  *      0  *  *  0  0
*  *  *  *  *      0  0  *  *  0
*  *  *  *  *      0  0  0  *  *
*  *  *  *  *      0  0  0  0  *
```

For banded matrices, special algorithms (some based on gaussian elimination), have been developed. Because so many terms are zero, the actual number of terms needed in the calculations are significantly reduced. This greatly reduces the number of equations to solve before round-off errors become a problem. We will not discuss these methods here, as they are beyond the scope of this text. But you should be aware that there are many such tools available, if you need to solve such complex problems.

Exercises

All exercises in this chapter require validation and appropriate documentation.

16.1 A statically determinant truss is typically constructed so that three links join at a pin connection as shown here:

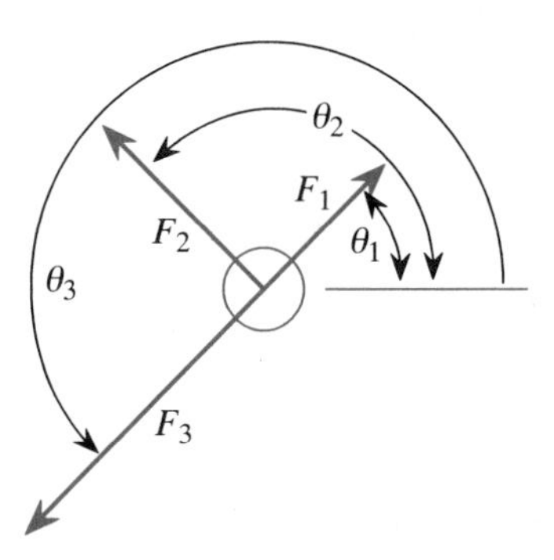

Let F1 be a force of known magnitude at a known orientation, and F2 and F3 be forces of unknown magnitude, but at known orientations. The orientation of the forces is described by $\theta 1$, $\theta 2$, and $\theta 3$ for the forces F1, F2, and F3, respectively. Using the sum of forces equals zero on the pin, write a program to solve for the unknown magnitudes of F2 and F3. Use Cramer's rule to solve the system of equations.

16.2 It was shown that the Z component of the cross product, $R \times F$ can be calculated using the following determinant:

$$(0,0,1) \cdot (R \times F) = \begin{vmatrix} 0 & 0 & 1 \\ R_X & R_Y & R_Z \\ F_X & F_Y & F_Z \end{vmatrix}$$

In a similar way, we can formulate the X component as $(1,0,0) \cdot (R \times F)$ and the Y component as $(0,1,0) \cdot (R \times F)$. Write a subroutine, **`TRIPLE (N, R, F, RXF)`** that returns the triple product for the desired component. Use the notation that N = the desired component, R = the vector of radius components, F = the vector of force components and RXF is the scalar value corresponding to the triple product.

16.3 It is possible to determine the coefficients of a polynomial of order N that passes through $N + 1$ points exactly. For example, to pass a second-order polynomial through the following three points:

X:	0	1	2
Y:	1	2	5

the following three equations can be generated using $Y = a + bX + cX^2$:

$$1 = a + b(0) + c(0)^2$$
$$2 = a + b(1) + c(1)^2$$
$$5 = a + b(2) + c(2)^2$$

where we have substituted the appropriate values of X and Y into the polynomial equation. This system of linear equations can be solved using any of the methods discussed in this chapter to determine the coefficients a, b, and c.

Write a program that can read in three arbitrary pairs of (X,Y) data points and returns the coefficients of a second-order polynomial. Use Cramer's rule to solve the system of equations. How can you validate the program?

16.4 Modify the gaussian elimination subroutine to be able to accept up to ten unknowns without changing the MAIN program. Using this version of the gaussian elimination method, rework the previous exercise for any number of data points up to 11.

16.5 In the chapter on curve fitting the topic of "best fit" was introduced. A straight line was fitted to a set of data points such that the error between the line and the data was minimized. This process can be extended to a second-order polynomial by solving the following system of equations:

$$\begin{vmatrix} n & \sum X_i & \sum X_i^2 \\ \sum X_i & \sum X_i^2 & \sum X_i^3 \\ \sum X_i^2 & \sum X_i^3 & \sum X_i^4 \end{vmatrix} \begin{vmatrix} c_0 \\ c_1 \\ c_2 \end{vmatrix} = \begin{vmatrix} \sum Y_i \\ \sum X_i Y_i \\ \sum X_i Y_i^2 \end{vmatrix}$$

In this equation X_i, Y_i are the individual data points, ΣX_i indicates a summation of all X values, and so forth. The quantities for which you are to solve are the coefficients, c_0, c_1, and c_2, the values for the "best fit polynomial":

$$y(X) = c_0 + c_1X + c_2X^2$$

Using this method, determine the second-order polynomial which provides the best fit to the following data points:

X:	0	0.05	0.1	0.15	0.2	0.25
Y:	(corresponding values of sin(X))					

Compare the best fit values to the actual values of the sine function. Is a second-order polynomial a good substitute for the sine function over this range?

16.6 In some instances a curve must be fit through a number of data points with a prescribed slope at the points. Consider the problem of fitting a third-order polynomial through two points using the equation:

$$Y = c_0 + c_1X + c_2X^2 + c_3X^3$$

Not only must the curve fit this polynomial, but the slope of the curve must simultaneously fit the derivative:

$$Y = c_1 + 2c_2X + 3c_3X^2$$

Using the following data points:

X:	0	5
Y:	0	0
Y:	1	−1

we can develop the following system of equations:

$$\begin{aligned} c_0 \quad\quad\quad\quad\quad\quad &= 0 \\ c_1 \quad\quad\quad\quad &= 1 \\ c_0 + 5c_1 + 5^2c_2 + 5^3c_3 &= 0 \\ c_1 + 2c_25 + 3c_35^2 &= -1 \end{aligned}$$

These four equations will provide the desired coefficients of the fitting polynomial with the desired properties. Write a program that can read in two data points along with the slopes at that point and return the coefficients.

16.7 Solve the currents, i_n, in the following resistive circuit. Use gaussian elimination. Assume that the currents flow in the directions indicated:

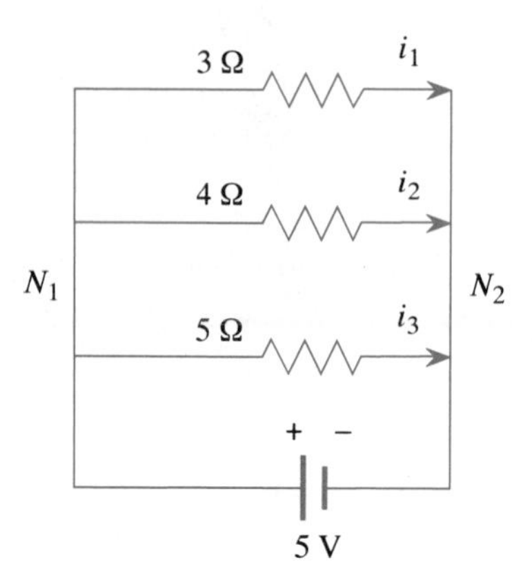

16.8 Determine the magnitude of the current flow through each of the resistors in the following circuit. Use gaussian elimination to solve the system:

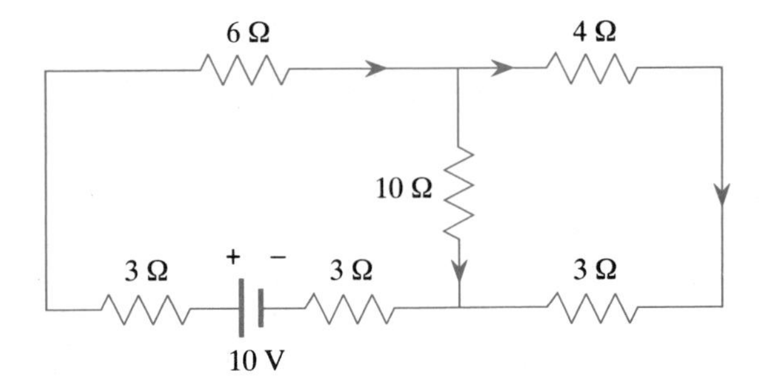

16.9 A panel is attached to a wall with two hinges as shown in the following diagram:

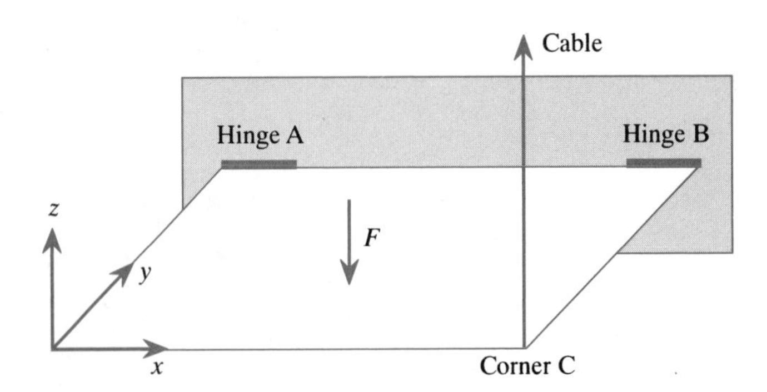

The dimensions of the panel are 20 in wide by 40 in long. Hinge A can generate forces in the X, Y, and Z directions. Hinge B can only generate forces in the Y and Z directions. Also, the panel is supported by a single cable attached to the corner "C." A vertical force, F, may be applied at any location (X,Y) on the panel. Using the sum of the forces and the sum of the moments equal to zero, generate a system of six equations to solve for the reactions at the hinges and the cable for any position and magnitude of the force, F. Run your program with $F = -100$ (down) at a location of $X = 10$, $Y = 10$.

16.10 The following figure illustrates two bars connected to each other and the ground by five springs having different spring constants k_1 through k_5:

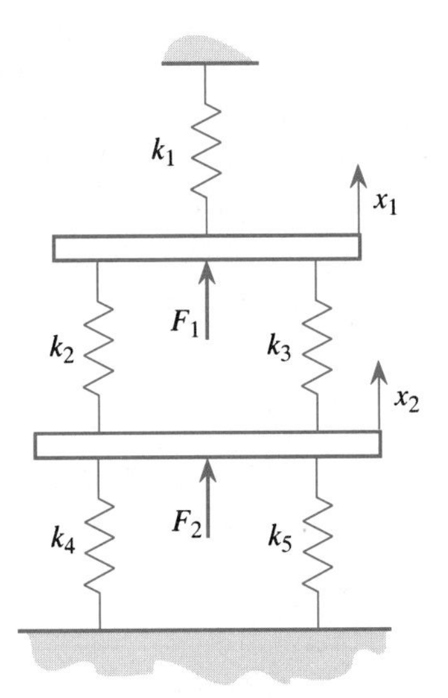

The free body diagram for each of the bars is shown here:

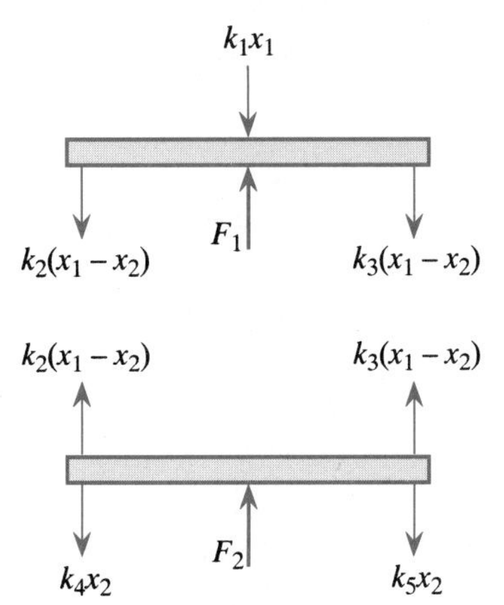

Use the fact that the sum of the forces equals zero for each bar, construct a system of equations to solve for the displacements X_1 and X_2, given the magnitudes of the applied forces and the spring constants.

Solve this problem for k_1 through k_5, all having the same spring constant of 100 lb/in and with applied forces of $F_1 = 25$ lb and $F_2 = 50$ lb. Use any technique for solving the system that seems appropriate.

16.11 The key to gaussian elimination is being able to subtract a multiple of one row of a matrix from another. Write a subroutine, ROWSUB(A, I, J, F), which subtracts F times row I of the matrix A from row J. The result is then stored in the Jth row of A. Let A be a 10 × 10 matrix.

Now, using the subroutine that you just wrote, write another subroutine, REDUCE(A, I), which subtracts a multiple of row I from all subsequent rows in matrix A, such that the resulting elements A(I + 1, I), A(I + 2, I), . . . , A(10, I) are all zero.

If you write the subroutines correctly, the following calling program will perform gaussian elimination on a matrix read from the data file named GAUSS.DAT:

```
      REAL A( 10, 10 )
      OPEN ( UNIT=2, FILE='GAUSS.DAT',
     1        STATUS='OLD' )
      READ(2,*) (( A( I, J), J=1,10), I=1,10 )
      DO 10 I = 1, 9
              CALL REDUCE ( A, I )
   10 CONTINUE
      PRINT *, (( A( I, J), J=1,10), I=1,10 )
      STOP
      END
```

CHAPTER

17

NUMERICAL DIFFERENTIATION AND INTEGRATION

17.1 Introduction

Differentiation of a function, using calculus, is the process of determining the function known as the derivative. An easy way to think of a derivative is the slope of a line tangent to the curve at a given point. To illustrate, we have plotted the function, $f(X) = X^2 + 2X + 3$ and its derivative over the range of $X = 1$ to $X = 5$ in Fig. 17.1. Notice that the value of the derivative changes for each point on the function. For example, at $X = 1$ the derivative has a value of 4, whereas at $X = 5$ the derivative has a value of 12. This change in the value of the derivative shows that the function has an increasing slope between these points, as you can verify qualitatively by examining the function.

In many areas of engineering and science the derivative has an important physical meaning. For example, acceleration is the derivative of speed. Speed, in turn, is the derivative of distance. Sometimes we can derive an analytical function for the derivative. For example, for the curve drawn in Fig. 17.1, we can explicitly write the formula for the derivative, $F'(X)$, as $F'(X) = 2X + 2$. Of course, to do this, we need to know something about differential calculus. But there may be times when it is easier to calculate the derivative numerically. This is especially true when we work with experimental data, where no explicit function is available to differentiate. Here we have no choice. We must obtain the derivative numerically.

The process of *integration* involves the determination of the area under a curve defined by a function. Figure 17.2 illustrates the *definite integral* for the function $f(X) = X^2 + 2X + 3$ for the range of $X = 2$ to $X = 4$. A simple definition of a definite integral is the area under the curve over a specific range of values.

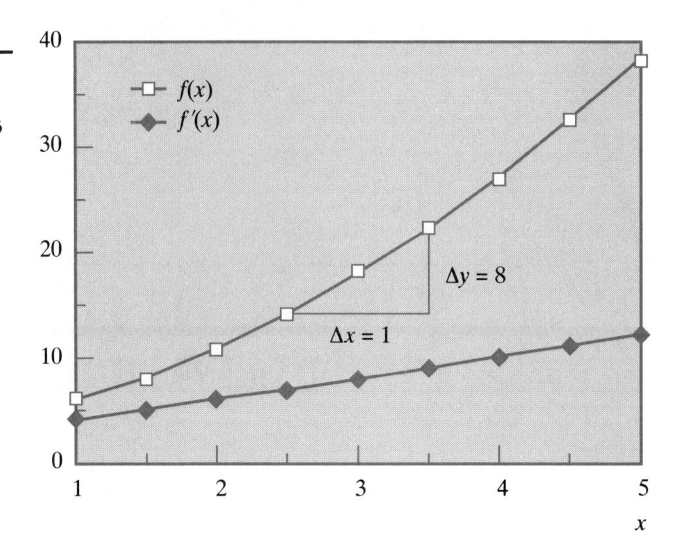

FIGURE 17.1

The Function $f(x) = x^2 + 2x + 3$ and Its Derivative over the Range $x = 1$ to $x = 5$. The Slope at $x = 3$ Is Illustrated

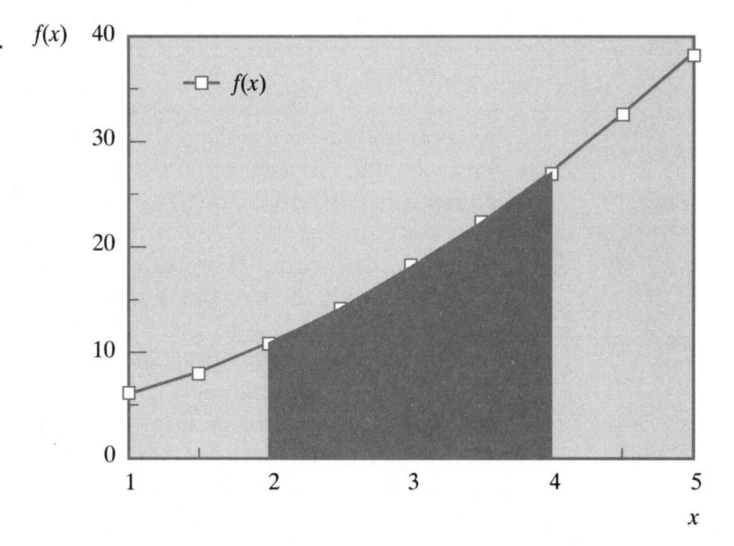

FIGURE 17.2

Integral of the Function $f(x) = x^2 + 2x + 3$ over the Range $x = 2$ to $x = 4$. The Shaded Area Equals the Value of the Integral

The integral also has physical significance in engineering and science. For example, the integral of acceleration is speed and the integral of speed is distance. Integration and differentiation can be thought of as ***inverse*** mathematical functions. In fact, we sometimes call integration "inverse differentiation" to highlight this relationship. Unlike differentiation, though, integration is sometimes difficult to do, even if an analytical equation is available to describe the data. Therefore, you are much more likely to need numerical methods for integration than for differentiation.

17.2 Approximating the First Derivative

There are many methods for obtaining a numerical approximation for the derivative of a function. The approach that we use in this section focuses on a graphical construction. We present three principal ways to construct a tangent line, which will then define the derivative. There are errors associated with such a construction, however, since it is only an approximation. Therefore, we need to examine the topic of truncation errors. Once we know how to approximate the derivative of a function, we can then apply the same method to calculate second and third derivatives. At the end of this section we will summarize the equations that you will need for these calculations.

THE FORWARD DIFFERENCE METHOD

We can define the derivative by the equation:

$$f'(X) = \lim_{\Delta X \to 0} (\Delta Y/\Delta X) = \lim_{\Delta X \to 0} ((f(X + \Delta X) - f(X))/\Delta X) \qquad (17.1)$$

For a straight line, the slope of a curve is $m = \Delta Y/\Delta X$. The value, ΔX, is the difference between any two values of X $(X_2 - X_1)$, and ΔY is the difference between the two values of Y $(Y_2 - Y_1)$ at the corresponding X values. A straight line has the same slope everywhere, so it does not matter how far apart the values of X are. If a function is not a straight line, though, we can still use this idea to estimate the slope. But now the calculation does depend on how far apart the two X values are. Therefore, in Eq. (17.1), ΔX must approach zero. The derivative of the function is $\Delta Y/\Delta X$, but ΔX must be very close to zero.

We illustrate this idea with the following calculations for the function $f(X) = X^3 + 2X^2 + 3X + 1$. If, for example, we wish to compute the derivative at $X = 3$, we choose values on either side and then compute $\Delta Y/\Delta X$. The true value for the derivative of this function at $X = 3$ is 42, but the accuracy of our approximation, based on $\Delta Y/\Delta X$, strongly depends on how small a value we use for ΔX:

ΔX	X_1	X_2	Y_1	Y_2	$\Delta X/\Delta Y$	Error
1.00	2.500	3.500	36.6350	75.3750	38.750000	3.250000
0.50	2.750	3.250	45.1719	66.2031	42.062500	−0.062500
0.10	2.950	3.050	52.9274	57.1276	42.002500	−0.002500
0.05	2.975	3.025	53.9569	56.0569	42.000625	−0.000625
0.01	2.995	3.005	54.7903	55.2103	42.000025	−0.000025

There are three ways that we can construct the estimate to the tangent line:

1. Connect the two points on either side of the point of interest—*the central difference method.*
2. Connect the point of interest to the next point to the right—*the forward difference method.*
3. Connect the point of interest to the next point to the left—*the backward difference method.*

We illustrate these methods in Fig. 17.3 for the approximation to the derivative at a general point X_I.

In many situations it does not matter which method you use to approximate the derivative. In these cases the following is true:

$$f'(X) = \lim_{\Delta X \to 0^+} \frac{f(X + \Delta X) - f(X)}{\Delta X} = \lim_{\Delta X \to 0^-} \frac{f(X - \Delta X) - f(X)}{\Delta X} \tag{17.2}$$

The basic approach is to use a finite, but small, value for ΔX. The forward difference method approximates the derivative by stepping forward by a quantity, ΔX, from the point of interest, X_i. Notice that the quantity $\Delta X = X_{i+1} - X_i$ is a positive quantity. Therefore, in Eq. (17.2), we designate this very small quantity as 0^+. The line drawn between these two points is an approximation to the derivative, if ΔX is very small as illustrated in Fig. 17.4. Based on this construction, it is a simple matter to show that the approximate value of the derivative is:

$$f'(X_i) \approx \frac{f(X_i + \Delta X) - f(X_i)}{\Delta X} \tag{17.3}$$

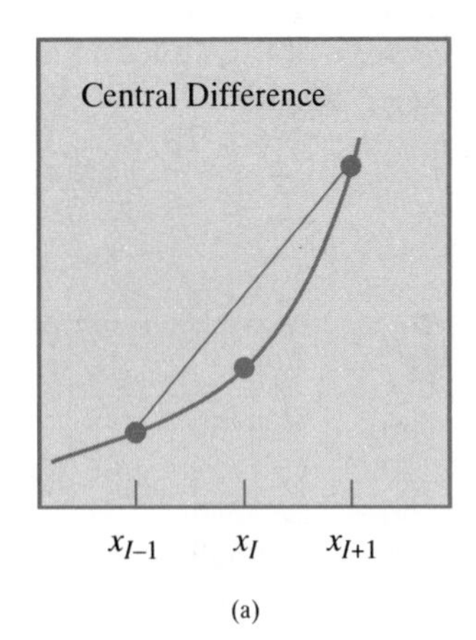

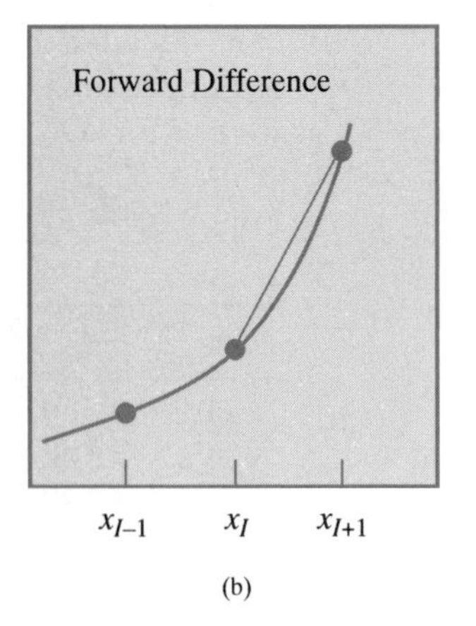

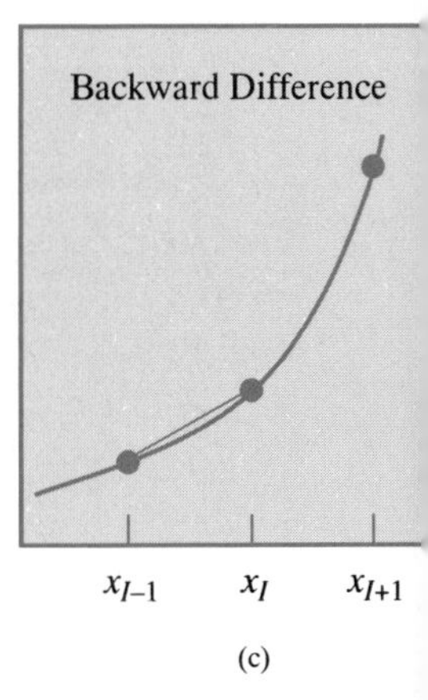

FIGURE 17.3

Illustration of Three Methods of Approximating the Derivative at a Point, x_i

FIGURE 17.4

Forward Difference Approximation to the First Derivative

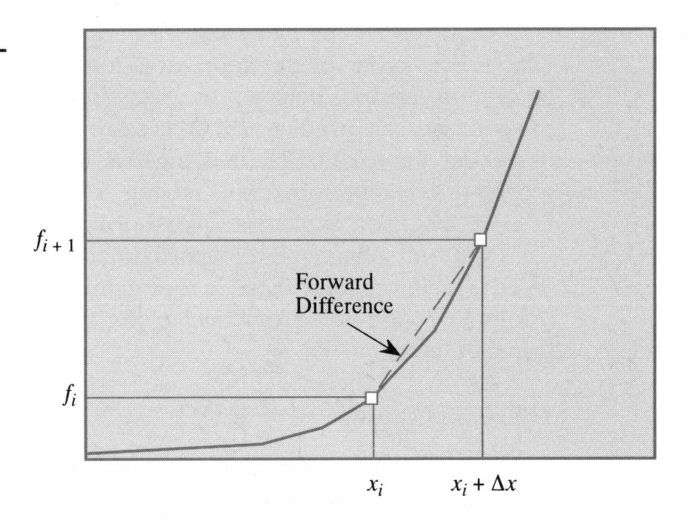

This equation represents the first term in Eq. (17.2). We rewrite this equation by using the notation:

$$f' = \left(\frac{df}{dX}\right)_i$$

$$f_i = f(X_i)$$

$$f_{i+1} = f(X_i + \Delta X)$$

Combining this notation with Eq. (17.3), we have the usual form of the forward difference approximation:

$$\left(\frac{df}{dX}\right)_i \approx \frac{f_{i+1} - f_i}{\Delta X}$$

To illustrate how this approximation works, we approximate the first derivative of $\sin(X)$ at $X_0 = 0$, using $\Delta X = 0.001$. First, we evaluate the function at the two points, $X = X_0$ and $X = X_0 + \Delta X$:

$$f_i = \sin(X_0) \qquad = \sin(0) \qquad = 0.0$$

$$f_{i+1} = \sin(X_0 + \Delta X) = \sin(0 + 0.001) = 0.0009999$$

Now we substitute these values into the approximation for the derivative:

$$f'(0) = \frac{f_{i+1} - f_i}{\Delta X} = \frac{0.0009999 - 0.0}{0.001}$$

$$f'(0) = 0.9999$$

This value is very close to the true value of the derivative, which is 1.000 at $X = 0$. We have specifically chosen an example that has a known derivative. This allows us to check the accuracy of the approximation before we try to apply it to other problems for which no analytical form of the derivative exists.

Equation (17.3) is a ***first-order approximation*** for the derivative. We can use the order of the approximation to show how the errors, called the truncation errors, behave. By knowing the order of the approximation method, we can predict how decreasing ΔX will affect the accuracy. For a first-order approximation, halving ΔX will reduce the error by half. For a second-order approximation, halving ΔX will reduce the error to $(1/2)^2$ or 1/4. We will soon explore a second-order method.

You will obtain better approximations by decreasing ΔX. We will see shortly, however, that there is a practical limit to this. Making ΔX too small results in round-off errors due to the fixed number of digits the computer stores in its memory.

THE BACKWARD DIFFERENCE METHOD

By using the same approach that we used to derive the forward difference approximation, we can develop the backward approximation. We start with the second term in Eq. (17.2) and substitute $f(X_i - \Delta X)$ and $(-\Delta X)$. The result is:

$$f'(X_i) \approx \frac{f(X_i - \Delta X) - f(X_i)}{-\Delta X} \tag{17.4}$$

As you can see from the diagram in Fig. 17.5, $f(X_i - \Delta X) - f(X_i)$ is equal to ΔY.

It is more convenient if we rearrange Eq. (17.4) to remove the negative sign from the denominator:

$$f'(X) \approx \frac{f(X_i) - f(X_i - \Delta X)}{\Delta X} \tag{17.5}$$

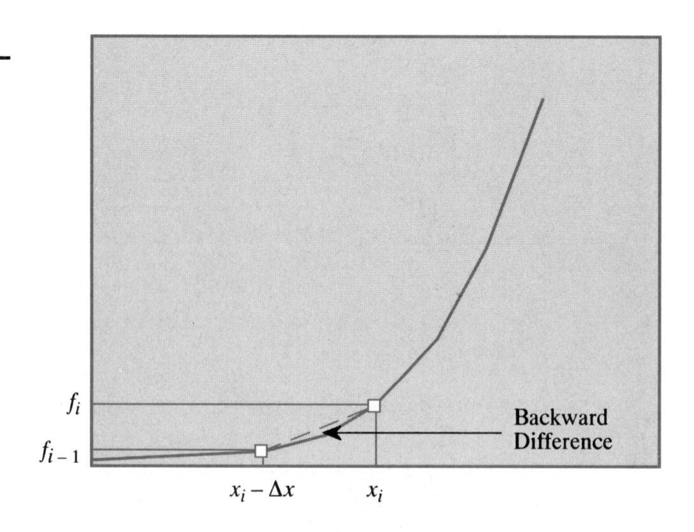

FIGURE 17.5

Backward Difference Approximation to the First Derivative

or in the notation of the previous example:

$$\left(\frac{df}{dX}\right)_i \approx \frac{f_i - f_{i-1}}{\Delta X}$$

We call the equation the *first-order backward difference approximation* for the first derivative. Like the forward approximation, this also is of first order.

Let's now apply this approximation to our previous example of the derivative of the sine function at $X = 0$ with $\Delta X = 0.001$. As before, we have $\Delta X = 0.001$. Thus, we need to evaluate the function at $X_0 = 0$ and $X = X_0 - 0.001 = -0.001$:

$$f_i = \sin(X_0) \qquad = \sin(0) \qquad = 0.0$$

$$f_{i-1} = \sin(X_0 - \Delta X) = \sin(0 - 0.001) = -0.0009999$$

Now we substitute these values into the approximation for the derivative:

$$f'(0) = \frac{f_i - f_{i-1}}{\Delta X} = \frac{0.0 - (-0.0009999)}{0.001}$$

$$f'(0) = 0.9999$$

Notice that we obtained the same result as we did with the forward difference method. This is what we should expect since both methods are attempting to obtain the same quantity. Therefore, the choice of which method to choose is usually a matter of convenience.

THE CENTRAL DIFFERENCE METHOD

By combining both methods (forward difference and backward difference), we can develop a second-order method. Recall that a second-order method has the advantage that the errors decrease more quickly as we reduce the step size. Therefore, it is worthwhile to develop. One way to do this is by averaging the estimates provided by the forward difference and the backward difference methods:

$$f'(X) \approx \frac{\text{forward difference} + \text{backward difference}}{2}$$

$$\approx \left(\frac{f_{i+1} - f_i}{\Delta X} + \frac{f_i - f_{i-1}}{\Delta X}\right) \Big/ 2$$

$$\approx \frac{f_{i+1} - f_{i-1}}{2\,\Delta X} \tag{17.6}$$

Of course, we could have derived this equation directly from Fig. 17.3, but we wanted to show you the relationship among the three approximations summarized in Fig. 17.6.

The central difference method is a second-order approximation. Thus, reducing the size of ΔX by 1/2 will decrease the truncation error by $(1/2)^2$ or 1/4 for the general case. As a result, the central difference method is pre-

FIGURE 17.6

Relationship between the Forward, Backward, and Central Difference Approximations to the Derivative

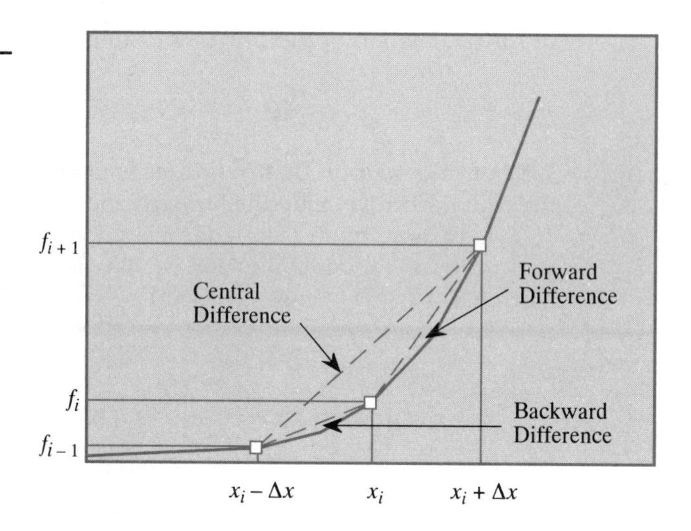

ferred for determining a derivative. The only times that you will want to use one of the other two methods is when f_{i-1} or f_{i+1} do not exist. This may happen at the beginning or end of a data set. In these cases you have no choice. You must use the appropriate alternate method.

There are several special cases that we must consider. All three methods will give an exact value for the first derivative if the function is a straight line. In addition, the value will be independent of ΔX, unless ΔX is so small, that computer round-off errors become significant. A less obvious special case is present for the central difference method. This method will provide exact values for the first derivative for polynomials of second order or less, which we demonstrate numerically for the function $f(X) = X^2$ at $X = 2$. The derivative of this function is $f'(X) = 2X$, which gives a value of $f'(2) = 4$. Now let's compare this with the value from the central difference approximation. To accentuate any potential errors, we choose $\Delta X = 1$, which usually would be an unacceptably large step size:

$$f(X + \Delta X) = f(2 + 1) = 3^2 = 9 \qquad f(X - \Delta X) = f(2 - 1) = 1^2 = 1$$

$$f'(X) \approx \frac{(f_{i+1} - f_{i-1})}{2\Delta X} = \frac{9 - 1}{2(1)} = 4$$

Although we chose a very large ΔX, we obtained the exact answer. We will see in a following section that special cases like the one just mentioned can play an important role in program validation.

17.3 Higher-Order Derivatives

We can extend the graphical method that we just presented for the first derivative to estimate higher-order derivatives. The exact derivations are not difficult, but they can be lengthy. Therefore, we do not present the derivations here. Instead, we simply list the resulting formulas in Table 17.1.

TABLE 17.1

Finite Difference Formulas for Higher-Order Derivatives

Forward Difference Approximations:

$$f'_i = \frac{f_{i+1} - f_i}{\Delta X}$$

$$f''_i = \frac{f_{i+2} - 2f_{i+1} + f_i}{\Delta X^2}$$

$$f'''_i = \frac{f_{i+3} - 3f_{i+2} + 3f_{i+1} - f_i}{\Delta X^3}$$

Backward Difference Approximations:

$$f'_i = \frac{f_i - f_{i-1}}{\Delta X}$$

$$f''_i = \frac{f_i - 2f_{i-1} + f_{i-2}}{\Delta X^2}$$

$$f'''_i = \frac{f_i - 3f_{i-1} + 3f_{i-2} - f_{i-3}}{\Delta X^3}$$

Central Difference Approximations:

$$f'_i = \frac{f_{i+1} - f_{i-1}}{2\Delta X}$$

$$f''_i = \frac{f_{i+1} - 2f_i + f_{i-1}}{\Delta X^2}$$

$$f'''_i = \frac{f_{i+2} - 2f_{i+1} + 2f_{i-1} - f_{i-2}}{2\,\Delta X^3}$$

In the equations in Table 17.1 the number of primes shows the order of the derivative. Thus, f''' means a third derivative of the function f. So, for example, if we wished to compute the third derivative of $f(X) = X^3 + 2X^2 + 3X + 1$ at $X = 3$, we can use the central difference formula:

$$f'''_i = \frac{f_{i+2} - 2f_{i+1} + 2f_{i-1} - f_{i-2}}{2\,\Delta X^3}$$

If we use $\Delta X = 0.01$, then we can easily calculate the required terms in this equation:

$$\begin{aligned}
f_{i+2} &= f(X_0 + 2\,\Delta X) = f(3.02) = 55.844408 \\
f_{i+1} &= f(X_0 + \Delta X) = f(3.01) = 55.421101 \\
f_{i-1} &= f(X_0 - \Delta X) = f(2.99) = 54.581099 \\
f_{i-2} &= f(X_0 - 2\,\Delta X) = f(2.98) = 54.164392
\end{aligned}$$

from which we obtain:

$$f'''(X = 3.0) = (55.844408 - 2(55.421101) + 2(54.581099) - \frac{54.164392}{2(0.001)^3}$$

$$\approx 6.0$$

This is identical to the true value for the third derivative, which is 6.0 for all values of X. Before proceeding, you should make sure that you understand how to apply these equations. Also, you must understand which formulation (forward, back, or central difference) to use.

17.4 Implementing Numerical Differentiation

The process of setting up a subprogram (preferably a function) to do numerical differentiation should be straightforward. We need only to evaluate the function at a number of points and then to perform a calculation to obtain the approximation of the derivative. Only one problem exists with our current repertoire of FORTRAN commands. For each function we wish to differentiate, we need to develop its own derivative function. Fortunately, a feature is present in FORTRAN that allows subprogram names to be passed as arguments.

The **EXTERNAL** statement is a FORTRAN statement that we can use to show that we want to pass a subprogram name to another subprogram. We illustrate this in the following program that uses the EXTERNAL statement with a forward difference approximation for the derivative in the form of a FUNCTION SUBPROGRAM:

```
C     MAIN PROGRAM TO ILLUSTRATE NUMERICAL
C     DIFFERENTIATION
C ------------------------------------------------------------
      EXTERNAL F
      EPS = 0.001
      PRINT *, ' ENTER THE VALUE FOR X '
      READ *, X
      PRINT *, 'THE DERIVATIVE OF F(X) AT X = ', X ,
     1           ' IS'
      PRINT *, DERV1( F, X, EPS)
      STOP
      END
C ------------------------------------------------------------
C     FUNCTION SUBPROGRAM TO CALCULATE THE FIRST
C     DERIVATIVE USING THE CENTRAL DIFFERENCE
C     APPROXIMATION
C ------------------------------------------------------------
      REAL FUNCTION DERV1( FUNC, X, DELTAX)
      DERV1 = ( FUNC( X+DELTAX ) - FUNC( X ))/ DELTAX
      RETURN
      END
C ------------------------------------------------------------
C     FUNCTION TO DEFINE THE FUNCTION FOR
C     DIFFERENTIATION
C ------------------------------------------------------------
      REAL FUNCTION F( X )
      F = X**2 + 2*X + 3
      RETURN
      END
```

Because the program declares the function named F to be external, it may be passed as an argument to other subprograms. The EXTERNAl statement must appear in the program that is passing the subprogram name to another subprogram. Thus, the statement, EXTERNAL F, appears only in the MAIN program. If the function that you wish to pass is a BUILT-IN FUNCTION, then you should use the INTRINSIC statement instead:

```
C    MAIN PROGRAM TO ILLUSTRATE THE USE OF THE
C    INTRINSIC STATEMENT
C -----------------------------------------------------------
     INTRINSIC SIN
     EPS = 0.001
     PRINT *, ' ENTER THE VALUE FOR X '
     READ *, X
     PRINT *, ' THE DERIVATIVE OF SIN(X) AT X = ', X,
 1             ' IS'
     PRINT *, DERV1( SIN, X, EPS )
     PRINT *, COS( X )
     STOP
     END
C -----------------------------------------------------------
C    FUNCTION TO PERFORM NUMERICAL DIFFERENTIATION
C    USING THE CENTRAL DIFFERENCE METHOD
C -----------------------------------------------------------
     REAL FUNCTION DERV1( FUNC, X, DELTAX)
     DERV1 =( FUNC(X + DELTAX) - FUNC(X - DELTAX))
 1          /(2 * DELTAX)
     RETURN
     END
```

In this second example we are sending the COSINE intrinsic function to the subprogram, DERIV1, for differentiation. Notice that we did not need to define any function here. We only had to list it in the argument list of the CALLing statement. We can do this since we have previously told the compiler our intention in the declaration statement at the beginning of the program.

In the first example we used the forward difference method, whereas in the second example we used the central difference method. As suggested before, the choice of method is up to you. But we recommend that you use the central difference method, which will give you more accurate results.

17.5 Numerical Integration

As we described in Sec. 17.1, the integral of a function is the area under the curve defined by that function. The easiest way to find this area is by graphical methods as we will show you in this section. We present three methods—the rectangular rule, the trapezoidal rule, and Simpson's one-third rule.

All three methods fall into the class of schemes known as the Newton-Cotes formulas. These formulas approximate a complicated function $f(X)$ by a polynomial that is easy to integrate. This is done by dividing the area into several smaller areas with an area that is easy to calculate. By adding together all the smaller areas, you can obtain an approximation for the total area. You have already done something like this before when you plotted a function on ruled graph paper and counted the number of blocks to estimate the area.

THE RECTANGULAR RULE

The rectangular rule is the simplest of the Newton-Cotes formulas. With this formula, we approximate the area under the curve by rectangles as Fig. 17.7 shows. To do this, we break the area into several rectangles of equal width, indicated in the figure as ΔX.

The height of the ith rectangle is $f(X_i)$ and the area is $f(X_i)\,\Delta X$. If we draw a rectangle so that one corner touches the curve, we have two possible ways of drawing it. If we draw it so that the left-hand corner defines the height of the rectangle, then the height becomes $f(X_i)$. But, if we choose the right-hand corner to define the height, we must use $f(X_{i+1})$ instead. One method will underestimate the area whereas the other method overestimates the area. Thus, the accuracy is not so good as other methods that we will discuss. When we calculate the total area under the curve with the rectangular rule, we obtain:

$$\text{Area} = \int_a^b f(X)\,dX \approx \sum_{i=0}^{n-1} f_i\,\Delta X \qquad \text{(using the left-hand limit)} \qquad (17.7)$$

or

$$\text{Area} = \int_a^b f(X)\,dX \approx \sum_{i=0}^{n-1} f_{i+1}\,\Delta X \qquad \text{(using the right-hand limit)} \qquad (17.8)$$

FIGURE 17.7

Approximating an Integral by the Rectangular Rule Method

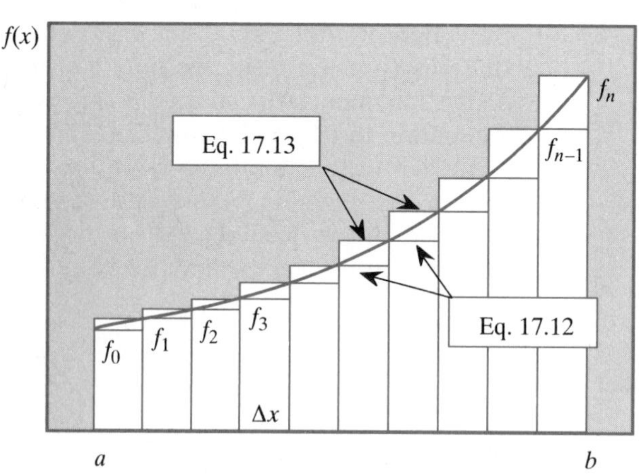

The mathematical symbol, $\int_a^b$, indicates the integral of the function over the range of $X = a$ to $X = b$. The integral is approximated by the summation of the individual areas of each rectangle. In Fig. 17.7 we have shown the rectangles drawn according to these formulas. As we suggested earlier, one formula overestimates and the other one underestimates.

As an example, we consider the integral of the function X^2 over the range of $X = 0$ to $X = 5$:

$$\text{Area} = \int_0^5 X^2\,dX$$

For this example, we break the area into five separate regions as Fig. 17.8 shows. First, we calculate the value of ΔX that will be the width of the rectangles. Since we have decided to break the region of $X = 0$ to $X = 5$ into five regions, ΔX is easy to determine:

$$\Delta X = \frac{b - a}{n} = \frac{5 - 0}{5} = 1$$

Next, we find the height of each rectangle. To do this, we must evaluate the function for each point, X_i for six different i values ($i = 0$ to 5):

i	X_i	$f(X_i) = X_i^2$
0	0	0
1	1	1
2	2	4
3	3	9
4	4	16
5	5	25

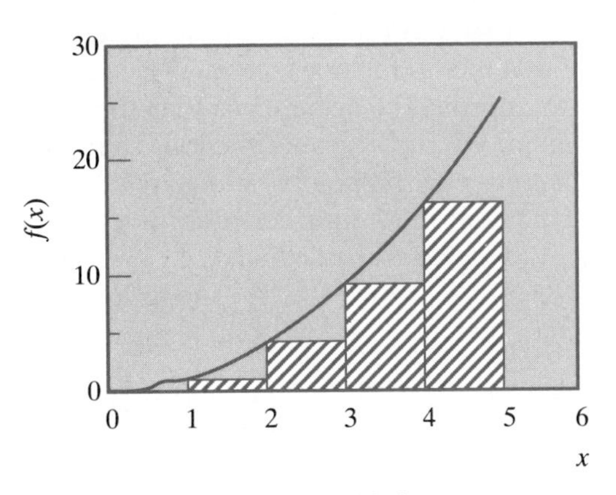

(a) Left-Hand Limit

(b) Right-Hand Limit

FIGURE 17.8

Integration of x^2 from $x = 0$ to $x = 5$

You should note that $X_i = (a + i\,\Delta X)$. This will come in handy later when we try to generalize the formulas. Now we are ready to add up the areas of each rectangle:

$$\text{Area} \approx \sum_{i=0}^{n-1} f_i \Delta X = (0 + 1 + 4 + 9 + 16)(1)$$
$$= 30 \qquad \text{(using left-hand limit)}$$
$$\text{Area} \approx \sum_{i=0}^{n-1} f_{i+1} \Delta X = (1 + 4 + 9 + 16 + 25)(1)$$
$$= 55 \qquad \text{(using right-hand limit)}$$

Because X^2 is an increasing function, the left-hand limit results in an answer lower than the exact value of 41.6666, and the right-hand limit results in a higher estimate. One way that we might lower the error is to take the average of the two results, which gives us a much better agreement of 42.5 versus 41.667.

We can obtain much better results by breaking the area under the curve into many more regions. For example, if we break it into ten regions, we get values of:

Area (left-hand limit) = 35.625

Area (right-hand limit) = 48.125

Area (average) = 41.875 (true value = 41.667)

If we break the area into even smaller rectangles, we obtain an even better estimate. However, the method is a first-order method. Thus, reducing the size of ΔX by 1/2 will reduce the error by only 1/2.

One way to think of the rectangular rule of the previous section is that we are approximating the function by a constant over a small region. From this we can construct a simple rectangle whose area is simple to calculate. Still, the accuracy of this method is not high, and we need something better.

One way to improve the accuracy is to assume that the function is linear over a small region. We replace the function over a small interval, X_i to X_{i+1}, by a straight line between the points (X_i, f_i) and (X_{i+1}, f_{i+1}). The area bounded by this straight line and the X axis defines a trapezoid as shown in Fig. 17.9.

The area, A_i, for one trapezoid is:

$$A_i = \Delta X \frac{f_{i-1} + f_i}{2} \tag{17.9}$$

As we did in the previous example, the total area is the sum of the areas of all trapezoids.

$$\text{Area} = \int_a^b f(X)\,dX \approx \sum_{i=1}^{n-1} \Delta X \frac{f_{i-1} + f_i}{2}$$
$$= \left(\frac{\Delta X}{2}\right)(f_0 + 2f_1 + \cdots + 2f_{n-1} + f_n) \tag{17.10}$$

FIGURE 17.9

Approximating an Integral by the Trapezoidal Method

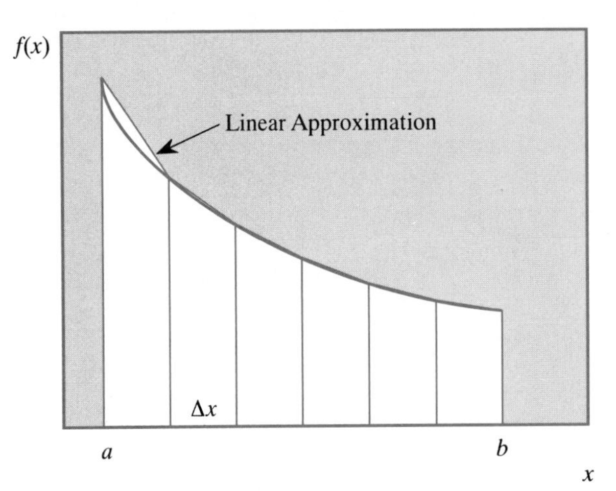

Now let's repeat the previous example for the integration of X^2 over the range of $X = 0$ to $X = 5$ to see if the trapezoidal rule provides a more accurate estimate. As before, we divide the range of integration into five sections as in Fig. 17.10.

The first step is to calculate the size, ΔX, of the interval:

$$\Delta X = \frac{b - a}{n} = \frac{5 - 0}{5} = 1$$

FIGURE 17.10

Use of Trapezoidal Rule to Integrate x^2 from $x = 0$ to $x = 5$

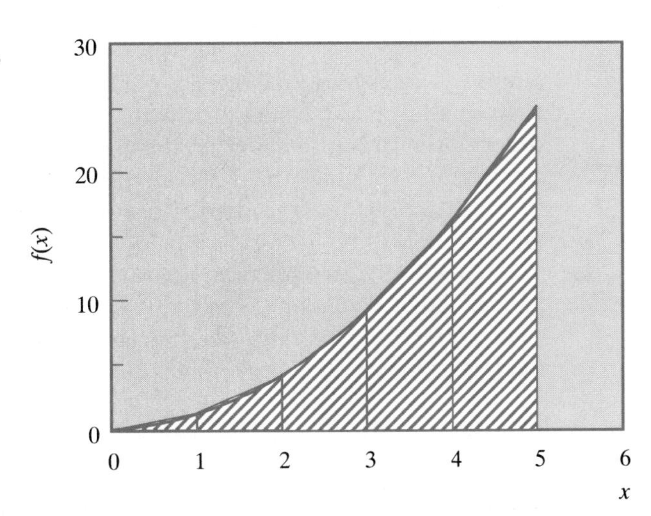

Now we evaluate the function at each point, X_i:

i	X_i	$f(X_i) = X_i^2$
0	0	0
1	1	1
2	2	4
3	3	9
4	4	16
5	5	25

Then we substitute these values into the summation of the individual areas:

$$\text{Area} \approx \sum_{i=1}^{n} \frac{\Delta X}{2}(f_{i-1} + f_i) = \frac{\Delta X}{2}(f_0 + 2f_1 + 2f_2 + 2f_3 + 2f_4 + f_5)$$

$$\approx (1/2)\,(0 + 2(1) + 2(4) + 2(9) + 2(16) + 25) = 42.5$$

The value of 42.5 represents a 2 percent error from the exact 41.667 value. Note that this is the same value that we obtained with the rectangular rule for the average of the left- and right-hand limits. This agreement is not an accident. If you expand the summations for the left- and right-hand rules, and then take the average, the trapezoidal rule will emerge. We leave this as an exercise.

The trapezoidal rule is a second-order method. Since the method fits straight-line segments to each interval of ΔX, the trapezoidal rule provides exact answers for linear functions just like the rectangular method.

SIMPSON'S ONE-THIRD RULE

Simpson's one-third rule uses a parabola to approximate the function over each small interval. Notice that with each new method, we are increasing the order of the fitting function. In the rectangular method we fit the function by a constant, or a polynomial of order zero. Then in the trapezoidal method we fit the function to a polynomial of order 1 or a straight line. In the current method we will be fitting the function to a polynomial of order 2, a parabola.

As with the two previous methods, we will divide the area into many segments. This calculation gives the width of the segment. But to complete the calculation of the area, we need three points to define the parabola. Consider the following second-order polynomial:

$$P(X) = UX^2 + VX + W \tag{17.11}$$

The three points that we use to define the constants U, V, and W will be (X_{i-1}, f_{i-1}), (X_i, f_i), and (X_{i+1}, f_{i+1}). For convenience and without any loss of generality, X_i will be taken as the origin ($X = 0$) for the polynomial. We now figure out U, V, and W as follows:

$$f_{i-1} = U(-\Delta X^2) + V(-\Delta X) + W$$
$$f_i = W$$
$$f_{i+1} = U(+\Delta X^2) + V(+\Delta X) + W \tag{17.12}$$

By rearranging these equations, we can solve for the values, U, V, and W:

$$U = \frac{f_{i-1} - 2f_i + f_{i+1}}{2(\Delta X)^2}$$
$$V = \frac{f_{i+1} - f_{i-1}}{2\,\Delta X}.$$
$$W = f_i \tag{17.13}$$

Once we have the values for U, V, and W, we can define the profile of the small area, A_p. This area is given by:

$$A_p = (\tfrac{2}{3})U(\Delta X^3) + 2W(\Delta X)$$

Now when we substitute back in the equations for U and W, we get an equation containing only f_i and ΔX terms:

$$A_p = (f_{i-1} + 4f_i + f_{i+1})\frac{\Delta X}{3} \tag{17.14}$$

We call this equation Simpson's one-third rule. The one-third terminology is because ΔX is divided by 3. This formula gives the area under two segments (width $= 2\,\Delta X$). To calculate the area over the range $a \le X \le b$, we sum over groups of two segments. Thus:

$$\text{Area} = \int_a^b f(X)\,dX \approx \sum_{i=1}^{n/2} A_i \tag{17.15}$$

In the preceding equation A_i is the area given by Eq. (17.14) for the jth group of two segments. Note that n must be an even integer to use this formula. Expanding the sigma notation yields:

$$\text{Area} = \frac{\Delta X}{3}\,[(f_0 + 4f_1 + f_2) + (f_2 + 4f_3 + f_4) + \cdots + (f_{n-2} + 4f_{n-1} + f_n)]$$

For programming purposes, it is more convenient for us to rewrite this equation as:

$$\text{Area} = \frac{\Delta X}{3}\left[f_0 + 4\left(\sum_{i=\text{odd}}^{n-1} f_i\right) + 2\left(\sum_{i=\text{even}}^{n-2} f_i\right) + fn\right] \tag{17.16}$$

The reason for rearranging the equation this way becomes apparent in the following section when we present the program.

To carry out this equation we only require ΔX and the value of the function at the different X_i values. We repeat the process for obtaining the integral of X^2 over the range from 0 to 5 with ten segments. Note that we had to divide the range into an even number of segments. Thus, we chose ten so that we can compare it to our previous results. As in the previous examples, we start by computing the size of each segment:

$$\Delta X = \frac{b - a}{n} = \frac{5 - 0}{10} = 0.5$$

Now we evaluate the function at each X_i:

i	X_i	$f(X_i) = X_i^2$
0	0.0	0.00
1	0.5	0.25
2	1.0	1.00
3	1.5	2.25
4	2.0	4.00
5	2.5	6.25
6	3.0	9.00
7	3.5	12.25
8	4.0	16.00
9	4.5	20.25
10	5.0	25.00

Then we substitute these values into Eq. (17.16):

$$\text{Area} = \frac{\Delta X}{3}\left[f_0 + 4\left(\sum_{i=\text{odd}}^{9} f_i\right) + 2\left(\sum_{i=\text{even}}^{8} f_i\right) + fn\right]$$

$$\text{Area} \approx \frac{0.5}{3}(0 + 4(0.25 + 2.25 + 6.25 + 12.25 + 20.25)$$

$$+ 2(0 + 1 + 4 + 9 + 16) + 25)$$

$$\approx 41.667$$

For this example, Simpson's rule returns an exact value. Since the method fits a second-order polynomial to the function, Simpson's one-third rule will return exact solutions for functions like our examples, which are second order.

The truncation error for Simpson's method is fourth order. Consequently, we can usually use larger step sizes with Simpson's rule, which makes it much faster than either of the two previous methods.

17.6 Implementing Numerical Integration

As with numerical differentiation, numerical integration can be conveniently implemented using subprograms. By using EXTERNAL and INTRINSIC statements, we can perform integration of functions with a single subprogram. We discuss this process in this section. We also discuss how to reduce the number of calculations required. This has two purposes: First, if we can reduce the number of mathematical operations, we can speed up the overall integration. More importantly, reducing the number of mathematical operations will improve accuracy since round-off errors will be minimized.

Recall from Chap. 12—"Why Numerical Methods?"—that we gave an example for numerical integration. That example used the rectangular method. We reproduce the program of that example here:

```
C     PROGRAM TO INTEGRATE THE FUNCTION SIN(X)/X
C     USING THE RECTANGULAR METHOD
C -----------------------------------------------------------------
      AREA = 0
      XMIN = 1
      XMAX = 10
      N = 9
      DELTAX = ( XMAX - XMIN ) / N
      DO 10 X = XMIN, XMAX - DELTAX, DELTAX
            Y = SIN( X )/ X
            AREA = AREA + Y * DELTAX
   10 CONTINUE
      PRINT *, ' AREA = ', AREA
      STOP
      END
```

This program, although it is not in the desired subprogram form, implements Eq. (17.7):

$$\text{Area} \approx \sum_{i=0}^{n-1} f_i \Delta X$$

Whenever an equation contains the sigma notation (Σ), a DO LOOP containing a summation process is indicated. In the sample program the DO 10 LOOP increments over the values of X_0 to X_{n-1}. The value of f_i in the summation is the term, $\sin(X)/X$. This loop can be very calculation-intensive since the value of n may need to be a large number to ensure high accuracy. Therefore, we should investigate ways to reduce the number of calculations. One method of doing this is to factor out constant terms. For example, we can factor ΔX out from inside the summation:

$$\sum_{i=0}^{n-1} f_i \Delta X = f_0 \Delta X + f_1 \Delta X + \cdots + f_{n-1} \Delta X = \Delta X(f_0 + f_1 + \cdots + f_{n-1})$$

$$= \Delta X \sum_{i=0}^{n-1} f_i$$

All that we have done is to move the constant, ΔX, outside the summation. This simple maneuver can reduce the number of multiplications from $(n - 1)$ to only one. This results in a simple change to the program:

```
C     PROGRAM FOR RECTANGULAR INTEGRATION WITH REDUCED
C     NUMBER OF MULTIPLICATION
C ----------------------------------------------------------
      AREA = 0
      XMIN = 1
      XMAX = 10
      N = 9
      DELTAX = ( XMAX - XMIN ) / N
      DO 10 X = XMIN, XMAX - DELTAX, DELTAX
         Y = SIN( X ) / X
         AREA = AREA + Y
   10 CONTINUE
      AREA = AREA * DELTAX
      PRINT *, ' AREA = ', AREA
      STOP
      END
```

In this modified program the only multiplication of ΔX takes place outside the loop. This is the only change to the program.

Now we will try to rework Simpson's rule to minimize the number of calculations. We start by examining Eq. (17.15):

$$\text{Area} = \int_a^b f(X)\, dX \approx \sum_{i=1}^{n/2} A_i$$

After fitting the function to a parabola, we collect terms into the somewhat unusual format:

$$\text{Area} = \frac{\Delta X}{3}\left[f_0 + 4\left(\sum_{i=\text{odd}}^{n-1} f_i\right) + 2\left(\sum_{i=\text{even}}^{n-2} f_i\right) + fn\right]$$

If Simpson's rule were to be programmed directly without this regrouping, the constant term $\Delta X/3$ would be evaluated and multiplied for each term in the series. In addition, the points at the boundary of adjacent segments would be evaluated twice. But the preceding equation avoids these pitfalls.

This will greatly improve the efficiency of the program. Here is how we implement this version of Simpson's one-third rule:

```
C     FUNCTION TO INTEGRATE WITH SIMPSON'S ONE-THIRD
C     RULE. TERMS HAVE BEEN GROUPED TO MAXIMIZE
C     COMPUTATIONAL EFFICIENCY
C ---------------------------------------------------
      REAL FUNCTION SIMP( F, A, B, N )
      REAL ODD
      IF ( MOD( N, 2 ) .EQ. 1) STOP 'SIMP ERROR:N was
  1       not even '
      DELTAX = ( B - A ) / N
C ---------------------------------------------------
C     ...COMPUTE END TERMS: F(X0) AND F(XN)
C ---------------------------------------------------
      END = ( F( A ) + F( B ))
C ---------------------------------------------------
C     ...COMPUTE ODD TERMS: F(X1) THROUGH F(XN-1)
C ---------------------------------------------------
      ODD = 0
      DO 10 X = A + DELTAX, B - DELTAX, 2 * DELTAX
            ODD = ODD + F( X )
 10   CONTINUE
C ---------------------------------------------------
C     ...COMPUTE EVEN TERMS: F(X2) THROUGH F(XN-2)
C ---------------------------------------------------
      EVEN = 0
      DO 20 X=A+2*DELTAX,B-2*DELTAX,2*DELTAX
            EVEN = EVEN + F( X )
 20   CONTINUE
C ---------------------------------------------------
C     ...CALCULATE AREA USING ONE THIRD FORMULA
C ---------------------------------------------------
      SIMP = DELTAX / 3 * ( END + 4 * ODD + 2 * EVEN)
      RETURN
      END
```

To use the SIMP FUNCTION, the EXTERNAL or IMPLICIT statement must be used since we are passing the function name as one of the parameters in the argument list. We have included an error trap in the FUNCTION to detect if N is an odd value. Recall that Simpson's rule must divide the range of integration into an even number of intervals. Line 4 of the SIMP FUNCTION does this check. The character string after the STOP statements is an option of the STOP command. If that STOP executes, the string is printed at the terminal, showing why the stop has taken place. This is a convenient option when trapping errors.

17.7 Validation and Total Error

We cannot overemphasize the importance of validation. In the preceding sections we have presented several methods to calculate derivatives and integrals. But before using any of these methods, we want to develop a means to see if they work correctly and to give accurate answers.

A few special cases exist for each method. These special cases will return exact answers, which give us an automatic check for the validity of the program. Table 17.2 summarizes the special cases for each method presented.

For example, if we want to validate our program for Simpson's rule, we can use a polynomial of second degree or less. Our previous example of the integration of X^2 would fall into the category of a suitable special case. Recall that when we evaluated this function by hand, we commented that Simpson's rule is precise for this function. So, if we run the program for this function between the values 0 and 3, we will get an answer of 9.0. Here is a MAIN program to test Simpson's rule:

```
C     MAIN PROGRAM TO TEST SPECIAL CASE OF SIMPSON'S
C     RULE FOR X**2
C ---------------------------------------------------
      EXTERNAL POLY2
      AREA = SIMP( POLY2, 0.0, 5.0, 10)
      PRINT *, ' EXACT AREA IS 9 '
      PRINT *, ' SIMP RETURNED A VALUE OF ', AREA
      STOP
      END
C ---------------------------------------------------
C     FUNCTION TO DEFINE THE FUNCTION
C ---------------------------------------------------
      REAL FUNCTION POLY2(X)
      POLY2 = X * X
      RETURN
      END
```

TABLE 17.2

Special Cases for Various Integration Methods

Function	Method	Exact Solution Available For
First derivative	Forward difference	Polynomial of order 1 or less
	Backward difference	Polynomial of order 1 or less
	Central difference	Polynomial of order 2 or less
Integration	Rectangular rule	Constant
	Trapezoidal rule	Polynomial of order 1 or less
	Simpson's rule	Polynomial of order 2 or less

Once we have validated the program, the next step is to determine an appropriate value for any numerical parameters used in the function evaluation. For example, the parameter, ΔX, controls the accuracy of the answer, but it has nothing to do with the function being integrated. We use the parameter ΔX only to do the integration. However, this parameter affects the accuracy of the method, so you must try to find the value of the parameter that gives the most accurate answer.

This leads us to the natural question, how can one find an appropriate value for a numerical parameter? To answer this question, we must examine the types of error that can occur. The accuracy of any numerical method depends on two kinds of errors—the truncation error and the round-off error. We refer to the combination of these two as the total error. The typical behavior of truncation, round-off, and total error is illustrated in Fig. 17.11.

Typically, as you reduce the step size, ΔX, the truncation error decreases, but the round-off error increases. We have already shown you with an example how the truncation decreases with smaller ΔX. The round-off error is due to the fixed number of digits the computer can store and use in calculations. For example, consider the integration process that adds the area of each strip to calculate the total area under the curve. If we use a very small value of ΔX, then at some point the area being added (a single strip), may be significantly smaller then the total area. Thus, the following may happen:

$$\text{Total area} = \text{previous area} + \text{additional increment} = 1234567.0 + 0.1$$

If the computer can store only seven significant digits, the addition of the 0.1 term will be lost. This is the source of the round-off error. This also occurs during numerical differentiation where ΔX may be so small that the difference between two function evaluations is negligible. This will result in a faulty estimate of the derivative.

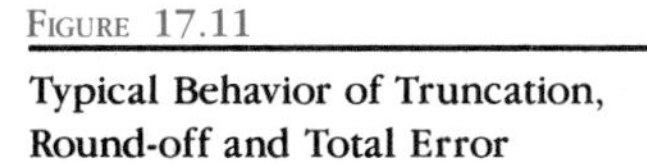

FIGURE 17.11

Typical Behavior of Truncation, Round-off and Total Error

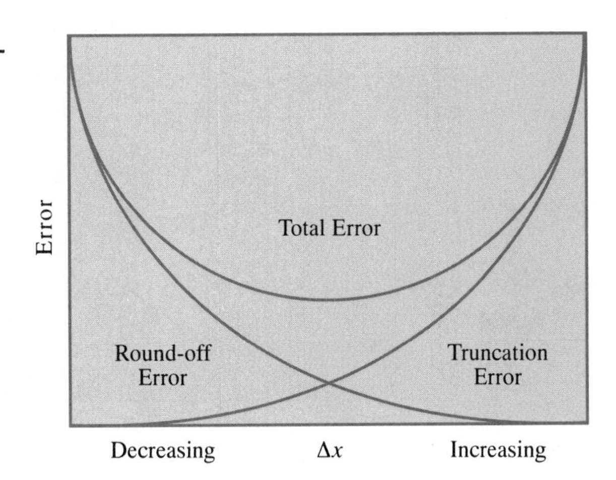

At some particular point an optimum ΔX occurs. The way you usually settle on a "reasonable" value for ΔX is to run the program several times with different values of ΔX. At some value of ΔX smaller values will not change the result significantly. How much is "much" depends on the necessary degree of accuracy that you need for your solution. To illustrate, we examine the problems of numerically determining the derivative for $f(X) = X^2$ at $X = 5$ with the forward difference approximation. Since this problem has a known analytical solution of 10, we can determine the total error. Figure 17.12 contains this plot. The program appears here:

```
C     MAIN PROGRAM TO EVALUATE THE DERIVATIVE OF
C     Y=X**2 AT X=5 FOR DIFFERENT STEP SIZE.
C ------------------------------------------------------------
      EXTERNAL F
      OPEN ( UNIT=1, FILE='X.DAT', STATUS='UNKNOWN')
      OPEN ( UNIT=2, FILE='Y.DAT', STATUS='UNKNOWN')
      READ *, X
      DO 10 POWER = -2, -8, -0.5
            EPS = 10 ** ( POWER )
            ERROR = ABS( 10 - DERV1( F, 5.0, EPS))
            WRITE(1,*) POWER
            WRITE(2,*) ERROR
            PRINT *, EPS, DERV1( F, 5.0, EPS),
     1                 ERROR, F( 5.0 )
   10 CONTINUE
      CLOSE( UNIT=1 )
      CLOSE( UNIT=2 )
      STOP
      END
```

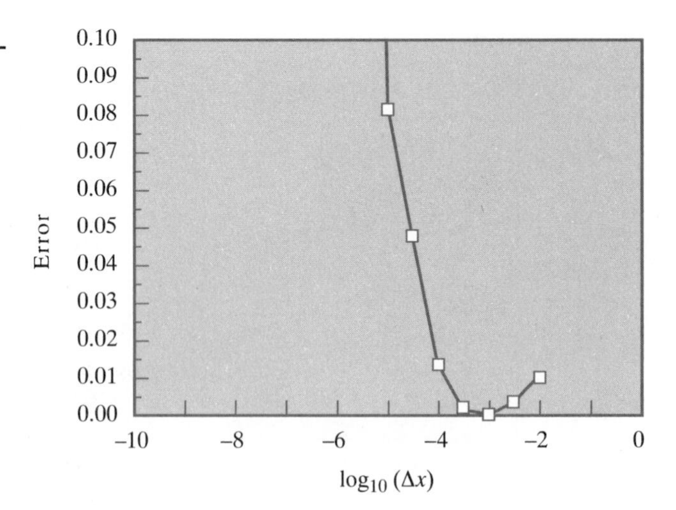

FIGURE 17.12

Plot of Total Error versus log (Δx) for Integration of the Function $f(x) = x^2$ at $x = 5$ Using the Forward Difference Method

```
C ------------------------------------------------------------------
C     FUNCTION TO COMPUTE THE DERIVATIVE BY FORWARD
C     DIFFERENCE
C ------------------------------------------------------------------
      REAL FUNCTION DERV1( FUNC, X, DELTAX)
      DERV1 = (FUNC(X + DELTAX) - FUNC( X )) / DELTAX
      RETURN
      END
C ------------------------------------------------------------------
C     FUNCTION TO BE DIFFERENTIATED
C ------------------------------------------------------------------
      REAL FUNCTION F( X )
      F = X**2
      RETURN
      END
```

We have plotted the results from this program in Fig. 17.12. From these results the optimum ΔX occurs approximately at 0.001 or 10^{-3}, which results in a total error of 2.29×10^{-4}. Extremely small ΔX values produced unacceptably large total errors due to round-off. Of course, in this example we knew the actual value of the derivative so that we could calculate the total error. Usually, however, we do not know the true answer, and so we must use other methods to select the optimum ΔX. One way commonly used is to find the region of ΔX where the solution is most stable. These methods are beyond the scope of this book, and so we will not discuss them further.

These results should convince you to choose your interval size with some care. If possible, run your program with many different values of ΔX, and choose a reasonable step size. An incorrect choice can contribute significant errors otherwise.

17.8 Summary

In this chapter we presented the topics of numerical differentiation and integration. We presented three methods for estimating the first derivative. These were the forward difference method, the backward difference method, and the central difference method. For each method, we described the order of the truncation error and the significance this plays in controlling the accuracy of the solution.

We also showed how to set up a differentiation function using the EXTERNAL and INTRINSIC statements. These statements allow you to pass subprogram names to other subprograms in the CALLing statement.

We presented three methods to perform numerical integration. These were the rectangular rule, the trapezoidal rule, and Simpson's one-third rule. To minimize computational time and round-off errors, we developed two methods to reduce the number of repeated calculations.

Program validation was emphasized in this chapter. The easiest way to validate a program is to use special cases that yield exact answers. Also, we discussed the topic of accuracy. The total error in a numerical approximation is made up of two parts—the truncation error and the round-off error. The truncation error has to do with the method of calculation. The round-off error, on the other hand, is due to the limited number of digits the computer can hold. There is some point where an optimum value exists for the numerical parameter. This value minimizes the total error. Whenever you use a numerical method, you should search for a reasonable value for the numerical parameter.

Exercises

17.1 Write a FUNCTION SUBPROGRAM named SIGMA to perform the sigma notation summing process on a function.

$$\text{Sum} = \sum_{i=1}^{n} F(X_i)$$

The arguments that should be passed to this function include the index lower limit, the index upper limit, array storing X values needed, and the function's name. Create an appropriate MAIN program to validate this function. (*Note:* You will need to use an EXTERNAL or INTERNAL statement.)

17.2 Write a FUNCTION SUBPROGRAM to perform differentiation using the central difference method. Validate your program by using one of the special cases outlined in Table 1.2. Finally, generate a total error versus a LOG10(ΔX) plot for the function exp(X) about the point $X = 0$. The actual value of the derivative is 1.0. Graph the curve using the plotting routines outlined in Chap. 10. At what value of ΔX is the best estimate achieved?

17.3 Implement the program given for the error as a function of ΔX in the text for the derivative of the function $Y(X) = X^2$. However, use DOUBLE PRECISION variables in place of REAL single precision variables. Plot the total error curve using the routine provided in Chap. 10. How does the optimum ΔX value and total error compare with the value reported in this chapter?

17.4 Derive the expression for the right-hand version of the rectangular rule truncation error per step. Then derive the total error expression.

17.5 Write a FUNCTION SUBPROGRAM that integrates a set of points stored in two arrays, X(I) and F(I), using Simpson's method. The spacing of the X values may not be uniform. How will you validate this function? Once you have constructed this function, determine the total distance traveled by a moving body from the time/velocity information tabulated in the following table:

Time (s)	0	1	2	4	7	10	14
Velocity (ft/s)	10	11	12	13	13	12	10

17.6 Verify that Simpson's rule returns exact solutions for polynomials of order 2. What is the optimum value for ΔX?

17.7 Find the value of $0 < X < 1.0$ at which the function $f(X) = \sin(X)\cos(X)$ is changing most rapidly.

17.8 In the time interval, $0 < t < 10$, a large vat is being filled at the unsteady rate of $q(t) = 10 \sin(0.1\pi t)^3$ kg/min. Simultaneously, fluid is being drained off at the steady rate of 1 kg/min. If the vat at time $t = 0$ held 100 kg, how many kilograms are in the vat after 10 s?

17.9 How accurate can finite difference methods be for computing the derivative if we consider smaller and smaller intervals? Consider again the ill-mannered function $f(X) = \sin(1/X)$ and find the derivative at points near 0 and at some point near 1.0 using successively smaller values of ΔX. Compare your results with the actual derivative $f'(X) = -(X^2)\cos(1/X)$. Does the accuracy depend also on where you take the derivative?

17.10 The voltage drop across an inductor is known to be proportional to the rate of change of current (Faraday's law):

$$V = L\frac{dI}{dt}$$

To verify this law the following data was taken:

V	1.0	2.0	3.0	4.0	5.0
I	0.1	0.4	0.9	1.6	2.5

Using this data, write a program to verify Faraday's law.

17.11 Occasionally in engineering we need to perform integration over an infinite interval. In theory this does not present a problem as long as the function we are integrating goes to zero at a sufficiently fast rate. Computationally, however, the problem is somewhat different from that of integrating over a finite interval. Write a program that integrates the function $f(X) = \exp(-X)$ from 0 to infinity. Use the trapezoidal rule and continue to add terms to the series until the final term you add is "negligible." You must decide upon a criterion for what constitutes a negligible contribution.

17.12 In the previous problem we saw that we could perform indefinite integration by establishing a termination condition that depended on the size of the individual terms of the series. Sometimes, however, we must have an even more sophisticated termination condition. Consider, for example, the integral of the function $f(X) = \exp(-X)\sin(X)$. This curve does not decrease monotonically nor is it always positive. Therefore, if we were to apply the trapezoidal rule, we might find that the sequential terms in the series are not necessarily decreasing! Nevertheless, the integral of this curve from infinity is finite. Write a program that integrates this function from 0 to infinity using a termination condition that will not terminate the series prematurely.

17.13 Area under a surface: The graphical integration techniques presented are also applicable to finding the area under a surface. Develop a program analogous to the rectangular method to compute the area under the surface $f(X,Y) = \sin^2(X)\cos^2(Y)$ for $0 < X < \pi$ and $0 < Y < \pi$.

17.14 The numerical approximation of the derivative is strongly dependent on the accuracy of the data. Any errors in the data generally lead to even larger errors in the estimate of the derivative. One way to diminish the effects of errors in the data is to apply curve smoothing before you compute the derivative. Write a program that does this.

APPENDIX

CHARACTER SETS

There are two character sets in common usage today. The first is the ASCII code used by most microcomputers. The other code is the EBCDIC, which is used primarily on larger computers such as mainframes and supercomputers. These character sets are used to answer such questions as—"Is Q less than q?". Computers answer these questions by converting the letter (or character) into the number corresponding to its position on either the ASCII or EBCDIC list. The question then will be answered by the position on the appropriate list. Therefore, the answer will be that it depends on which code your machine uses. Here is the complete listing of all 256 codes. Not all of them are used, and some indicate a control character, such as line feed or carriage return.

Decimal	ASCII	EBCDIC
000		
001		
002		
003	♥	
004	♦	
005	♣	
006	♠	
007	BEEP	
008		
009	TAB	
010	LINE FEED	
011	HOME	
012	FORM FEED	
013	RETURN	
014		
015		
016		
017		
018		
019		
020	¶	
021	§	
022		
023		
024	↑	
025	↓	
026	→	
027	←	
028		
029	↔	
030		
031		

Decimal	ASCII	EBCDIC
032	SPACE	
033	!	
034	"	
035	#	
036	$	
037	%	
038	&	
039	'	
040	(	
041	)	
042	*	
043	+	
044	'	
045	–	
046	.	
047	/	
048	0	
049	1	
050	2	
051	3	
052	4	
053	5	
054	6	
055	7	
056	8	
057	9	
058	:	
059	;	
060	<	
061	=	
062	>	
063	?	
064	@	BLANK
065	A	
066	B	
067	C	
068	D	
069	E	
070	F	
071	G	
072	H	
073	I	
074	J	¢
075	K	.
076	L	<
077	M	(
078	N	+

Decimal	ASCII	EBCDIC
079	O	
080	P	&
081	Q	
082	R	
083	S	
084	T	
085	U	
086	V	
087	W	
088	X	
089	Y	
090	Z	!
091	[	$
092	\	*
093	]	)
094	^	;
095	_	←
096	'	–
097	a	/
098	b	
099	c	
100	d	
101	e	
102	f	
103	g	
104	h	
105	i	
106	j	
107	k	'
108	l	%
109	m	_
110	n	>
111	o	?
112	p	
113	q	
114	r	
115	s	
116	t	
117	u	
118	v	
119	w	
120	x	
121	y	
122	z	:
123	{	#
124	|	@
125	}	'

Decimal	ASCII	EBCDIC
126	~	=
127		"
128	Ç	
129	ü	a
130	é	b
131	â	c
132	ä	d
133	à	e
134	å	f
135	ç	g
136	ê	h
137	ë	i
138	è	
139	ï	
140	î	
141	ì	
142	Ä	
143	Å	
144	É	
145	æ	j
146	Æ	k
147	ô	l
148	ö	m
149	ò	n
150	û	o
151	ù	p
152		q
153	Ö	r
154	Ü	
155	¢	
156	£	
157	¥	
158		
159	*f*	
160	á	
161	í	
162	ó	s
163	ú	t
164	ñ	u
165	Ñ	v
166	ª	w
167	º	x
168	¿	y
169		z
170	¬	
171		
172		
173	¡	
174	"	
175	”	
176		
177		
178		
179		
180		
181		
182		
183		
184		
185		
186		
187		
188		
189		
190		
191		
192		}
193		A
194		B
195		C
196		D
197		E
198		F
199		G
200		H
201		I
202		
203		
204		
205		
206		
207		
208	%	J
209		K
210		L
211		M
212		N
213		O
214		P
215		Q
216		R
217		
218		
219		

Decimal	ASCII	EBCDIC
220		
221		
222		
223		
224	α	\
225	β	
226	Γ	S
227	π	T
228	Σ	U
229	σ	V
230	μ	W
231	τ	X
232		Y
233	θ	Z
234	Ω	
235	δ	
236	∞	
237		

Decimal	ASCII	EBCDIC
238	ε	
239	$\cap$	
240	$\equiv$	0
241	$\pm$	1
242	$\geq$	2
243		3
244	⌠	4
245	⌡	5
246	$\div$	6
247	$\approx$	7
248	$\circ$	8
249		9
250		
251		
252		
253		
254		
255	FORCE FEED	

APPENDIX B

CHARACTER DATA MANIPULATION

B.1 Introduction

Sometimes programming involves the use of non-numerical data (e.g., arranging a list of names in alphabetical order). For such cases, FORTRAN has the capacity to input, manipulate, and print such data, commonly called character data, non-numeric data, or alphanumeric data.

A group of alphanumeric characters is known as a CHARACTER string. The computer processes character data similar to numerical data (i.e., REAL and INTEGER variables) in the sense of storage in labeled memory locations or variables names. All variable names used for string data must be identified using a DECLARATION statement at the beginning of the program. For example, the variable TEXT can be declared a string variable using the following DECLARATION statement:

```
CHARACTER TEXT
```

B.2 Declaration

The general form of the DECLARATION statement is:

```
CHARACTER name1*S1, name2*S2,...
```

In this statement, *name1* and *name2* indicates variable names. S1 indicates the maximum length string that can be stored in variable *name1*. The length of a string is the number of characters present. If a common length is to be assigned to a number of variables, the following option can be used:

```
CHARACTER*SO name1, name2, name3*S3,...
```

By indicating a length after the CHARACTER statement, the default length for name1 and name2 is taken to be S0. It is still permissible to declare a specific length to a variable as shown for name3.

Note. If no length is specified after CHARACTER or the variable name, then the default length used is one character.

B.3 Assignment Statements

To assign a value to a CHARACTER variable, the string must be enclosed between single quotes. The following program assigns string values to a number of variables:

```
CHARACTER NAME*10,SHORT
NAME='Bill'
SHORT='ABC'
PRINT *,NAME,SHORT
STOP
END
```

The variable NAME has been defined to hold a maximum of ten characters. You can think of a CHARACTER variable as a one-dimensional array having elements for each of its characters. Since NAME has been declared to have length 10, it would have ten such elements. When the string is loaded into memory, the first character goes into the first element, the second character into the second element, and so on. Unused elements are filled with blank spaces. Note that a blank space is a character. If the string is too long, then the excess characters are truncated and lost. The output from the previous program would look like this:

```
Bill      A
```

Since NAME has length 10, ten characters are printed ('Bill' with six trailing blanks). Since SHORT has length 1 (the default value) the string 'ABC' gets truncated to one character.

B.4 List-Directed Input and Output

The process of printing or writing string variables is not unlike other variable types. Free formatting (list-directed formatting) can be used, as was the case with the previous example. When reading values using free formatting, the string must be enclosed between quotes. The following program will read in a string and print it using list-directed formatting:

```
CHARACTER NAME*20
PRINT *,'ENTER YOUR NAME BETWEEN QUOTATION MARKS'
READ *,NAME
PRINT *,'Hello ',NAME
STOP
END
```

An execution of this program may appear this way:

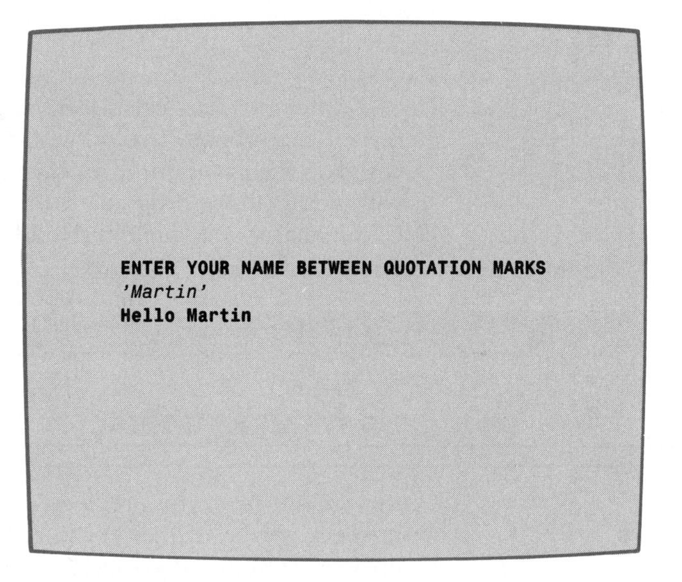

Note that in the preceding example the quotation marks are not printed during output.

B.5 Formatted Input and Output

For most cases list-directed output will be adequate. However, formatting can be used to make the input more user-friendly. By using formatted input, we remove the need for quotation marks. The specifier used for character data has the form *An*. *A* indicates alphanumeric, and *n* is used to specify the variable string length. Repeat factors can also be used. The following example cal culates the area of a circle and then prompts to see if you wish to exit the program. Using character data, "Y" or "N" can be entered:

```
      REAL PI,R,AREA
      CHARACTER YESNO
      PI=3.1416
   10 PRINT *,'Enter the Radius for the circle:'
      READ *,R
      AREA=PI*R**2
      PRINT *,'AREA is ',AREA
      PRINT *,'Do you wish to exit <y/N>'
      READ 20,YESNO
   20 FORMAT(A1)
      IF (YESNO.EQ.'Y'.OR.YESNO.EQ.'y') STOP
      GO TO 10
      END
```

After the area calculation is performed and the answer printed to the screen, the prompt 'Do you wish to exit ⟨y/N⟩' will appear. Entering 'y' or 'Y' will stop the program. Any other character (or simply hitting the return key) will reexecute the program. Note that you can compare character data using a conditional statement. You should be aware that the characters 'y' and 'Y' are different and must be tested.

The *A* specifier can also be used for character output, but for most cases it is rarely merited.

B.6 Comparing Character Data

As was shown in the previous example, conditional statements can be used to check character data. For the equal and not equal operators, the solution is obvious. For operations such as greater than, less than, and so on, the answer depends on your system. To answer such questions, you should obtain a copy of your computer collating sequence. This list indicates the "alphabetical" order for all characters on your system. To see an example of a collating sequence, refer to Appendix A. Most systems will, however, hold to the convention of characters 0 through 9 and A through Z, which are from the lowest to highest order:

'A' < 'B'	TRUE
'Z' > 'E'	TRUE
'H' < 'C'	FALSE

When comparing strings with multiple characters, the method is similar to alphabetizing. First, check the first characters. If they are the same, then check the next characters, and so on:

'AB' < 'AA'	FALSE
'AA' < 'AAA'	TRUE

Because of this convention, a subroutine used to sort a list of numbers in ascending order can be used to sort a list of names in alphabetical order. The only change necessary is to declare the array used to store string variables. An example of defining a one-dimensional character array is as follows:

```
CHARACTER NAMES(100)*10
```

This statement declared the array NAMES to have 100 elements, of which each element can store ten characters.

B.7 Character Data Manipulation

Two operations are possible with character variables: concatenation and substrings. Concatenation is the simplest operation. The operator for concatenation is //. It allows you to append two strings together:

```
CHARACTER*5 FIRST,LAST,FULL*10
FIRST='JOHN'
LAST='DOE'
FULL=FIRST//LAST
PRINT *,FULL
FULL=LAST//','//FIRST
PRINT *,FULL
STOP
END
```

Thus, in the above program, the fourth line will cause the character string contained in LAST to be added to the character string contained in FIRST. The sixth line in the program will perform a similar function. Note though that this time, we concatenate 3 character strings—LAST and FIRST with a comma between them.

The output of this program would be:

Since both string variables have length 5, a trailing space is present after JOHN for the first print statement. Note in the second statement using concatenation that constant strings can also be used, and the operator can be repeated.

B.8 Substrings

The concept of a character variable having many elements is much like a one-dimensional array. Each character resides in its own element. Substrings allow you to extract or replace a list of elements stored in a string variable. The general form of a substring is:

A(I:J)

where 1 <= I <= J <= length of string

If I is omitted, it is assumed to have a value of 1. If J is omitted, it is assumed to have a value equal to the length of the string.

Consider the following example program:

```
CHARACTER ALPHA*26,ONE*3,TWO*5
ALPHA='ABCDEFGHIJKLMNOPQRSTUVWXYZ'
ONE=ALPHA(:3)
TWO=ALPHA(2:6)
PRINT *,ONE
PRINT *,TWO
PRINT *,ALPHA(25:)
ONE(1:1)='E'
TWO(2:3)='2'
PRINT *, ONE
PRINT *,TWO
STOP
END
```

The output from this program would be:

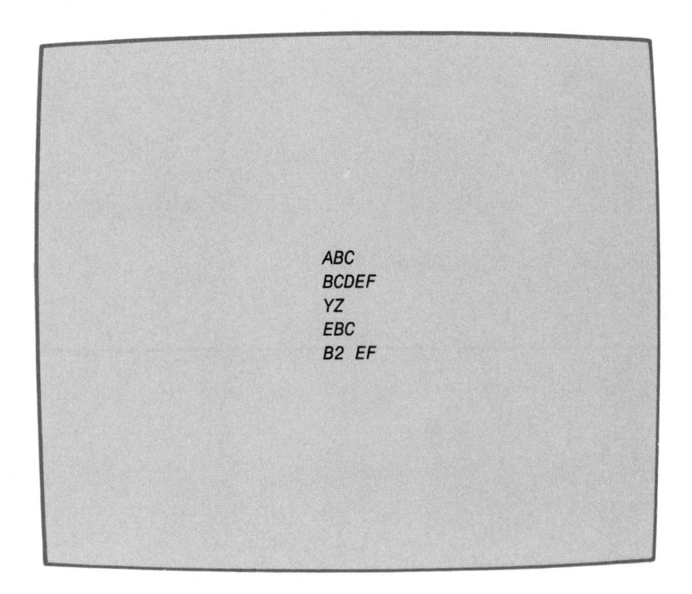

The first statement assigns the characters 1 through 3 of ALPHA to the variable ONE. The second assignment statement assigns characters 2 through 6 of

ALPHA to TWO. The print statement with ALPHA(25:) will print characters 25 through the end of ALPHA. The last two assignment statements illustrate the use of substrings as the target variable. In the third assignment statement character 1 of ONE is changed to character 'E'. In the last assignment statement characters 2 through 3 of TWO have the value '2' assigned to it. Since the string being assigned is smaller than the receiving substring, a blank is added at the end.

B.9 Summary

Most programming encountered for physical science and engineering problems usually do not require character data manipulation. In some cases, however, character data can be used to make your program more user-friendly. For example, a request for a run identification can be entered at the beginning of a program execution and printed in the output file. To illustrate this, an example is given in which a Y/N request is made, rather than something like "0 for no or 1 for yes." Such features make programs more user-friendly, and easier to use.

INDEX

INDEX

INPUT/OUTPUT

GENERAL FORM	EXAMPLE	REFERENCE (Page)
PRINT format,output-list	PRINT *,'A EQUALS ',A PRINT 20, X, Y PRINT '(1X,F6.2)', X	93 , 116
statement-label FORMAT(specifiers)	20 FORMAT(1X,'X=',F8.3)	116
READ(device#,format [,END=label] [,ERR=label])input-list	READ (1,10) YESNO READ (*,*) A READ (1,*) X, Y READ (5,*, END=20) N READ (1,'(A1)') ANSWER	115
READ format,input-list	READ *, X READ 20, YESNO	93 , 115
WRITE (device#,format) output-list	WRITE (1,*) X,Y WRITE (1,20) A, B, C WRITE (*,*) ANSWER	133